The Politics of Middle Eastern Oil

J.E. PETERSON

Middle East Institute

Washington, D.C. 1983

Library of Congress Catalog Card Number 83-050826
ISBN 0-916808-22-X Hardback
ISBN 0-916808-24-6 Paperback

Designed by Maria Josephy Schoolman
Manufactured in the United States of America

Contents

Acknowledgements

The idea for this book originated in the seminars which I have given at Bowdoin College and the College of William and Mary on the subject of "The Politics of Middle Eastern Oil," where it became quite apparent that there was considerable need for a collection of readings on the topic, conveniently gathered in one place and supplemented by comprehensive bibliographic references and handy facts and figures on the Middle East and its oil. I am grateful for the many helpful comments and suggestions offered by the students in these seminars.

I also wish to thank the Department of Government at the College of William and Mary for the assistance provided me during the preparation of this manuscript, my research assistants, Charles Andrew Waters and William Allred, for their help in the preparation of the appendices, and the publications staff of the Middle East Institute, for its hard work and patience. Appreciation is also due to the various authors and publishers of the selections reprinted here for their cooperation, and especially to those authors who were kind enough to provide revisions and updating where necessary.

Contributors

Mohammad Abu al Khail is Saudi Arabia's Minister of Finance and National Economy.

Saeed Salman Abu Azreh is a government official in the United Arab Emirates.

M. A. Ajomo is Professor and Dean of the Faculty of Law at the University of Lagos, Nigeria.

Jahangir Amuzegar is Professorial Lecturer at the Johns Hopkins University School of Advanced International Studies, and Consultant to the Managing Director of the IMF. Among his many publications are *Iran: An Economic Profile* (Washington: Middle East Institute, 1977), and *Comparative Economics: National Priorities, Policies, and Performance* (Cambridge, Mass.: Winthrop, 1981).

Lisa Anderson is Assistant Professor of Government and Social Studies, and Director of Student Programs at the Center for International Affairs, Harvard University. She conducted research in Libya in 1979 and has published a number of scholarly articles on Libyan politics and history.

John Duke Anthony is President of the National Council on U.S.-Arab Relations, Washington, D.C., and consultant to various U.S. government agencies. He is the author of numerous articles and *The Arab States of the Lower Gulf* (Washington: Middle East Institute, 1975), and editor of *The Middle East: Oil, Politics and Development* (Washington: American Enterprise Institute, 1975).

Ali Attiga is Secretary-General of the Organization of Arab Petroleum Exporting Countries.

James A. Bill is Professor of Government at the University of Texas, Austin. He is the author or co-author of five books including *Politics in the Middle East* (Boston: Little, Brown, 1979; rev. ed., 1983), and *The Politics of Iran* (Columbus, Ohio: Charles E. Merrill, 1972).

J. S. Birks is an Honorary Research Fellow in Economics at the University of Durham. Together with C. A. Sinclair, he is the author of *Arab Manpower* (London: Croom Helm; New York: St. Martin's Press, 1980); *International Migration and Development in the Arab Region* (Geneva: International Migration and Development in the Arab Region (Geneva: International Labor Organization, 1980); and, with others, *Manpower and International Labor Migration in the Middle East and North Africa* (London: Oxford University Press, for the World Bank, 1983).

Harold Brown has been President of the California Institute of Technology, Secretary of the Air Force under President Johnson, and Secretary of Defense throughout the Carter administration. He is now Distinguished Visiting Professor of National Security Affairs at the Foreign Policy Institute of the Johns Hopkins University School of Advanced International Studies.

William R. Brown is the author of *The Last Crusade: A Negotiator's Middle East Handbook* (Nelson Hall, 1980), and is Dean of Arts and Sciences at Central Connecticut State University.

David Curry is Deputy Director of the Office of Regional and Resources Policy, U.S. Department of the Treasury.

Eric Davis is Associate Professor of Political Science at Rutgers University and author of *Challenging Colonialism: Bank Misr and Egyptian Industrialization, 1920–1941* (Princeton: Princeton University Press, 1983). He is presently coordinating a project for the Social Science Research Council on the impact of oil wealth on social change in the Middle East.

Adeed Dawisha is the Deputy Director of Studies at the Royal Institute of International Affairs in London. He is author of *Egypt in the Arab World: The Elements of Foreign Policy* (The Halsted Press, 1976); *Saudi Arabia: Search for Security* (Adelphi Paper 158, International Institute for Strategic Studies, 1979); *Syria and the Lebanese Crisis* (St. Martin's Press, 1980); and co-editor of *The Soviet Union in the Middle East: Policies and Perspectives* (Holmes and Meier, 1982).

Dale F. Eickelman is Associate Professor of Anthropology at New York University. Among his books are *The Middle East: An Anthropological Approach* (Englewood Cliffs, N.J.: Prentice-Hall, 1981).

Hermann Frederick Eilts is University Professor of International Relations, Director of the Center for International Relations, and Chairman of the

Department of Political Science at Boston University. He was a career foreign service officer until 1979, serving as Ambassador to Saudi Arabia (1965–1970) and Egypt (1973–1979).

Paul Hallwood is a member of the Department of Political Economy, Kings College, University of Aberdeen, Scotland, and co-author with Stuart Sinclair of *Oil, Debt and Development: OPEC in the Third World* (London: George Allen & Unwin, 1981).

Mary Elaine Hegland is a candidate for a Ph.D. in social-cultural anthropology from the State University of New York at Binghamton, and is author of " 'Traditional' Iranian Women: How They Cope," *Middle East Journal* (Autumn 1982).

Jacqueline S. Ismael is an Associate Professor in the Faculty of Social Welfare of the University of Calgary, and the author of *Kuwait: Social Change in Historical Perspective* (Syracuse: Syracuse University Press, 1982).

Abbas Kelidar is a Lecturer in Middle East Politics at the School of Oriental and African Studies, University of London, and editor of *The Integration of Modern Iraq* (London: Croom Helm; New York: St. Martin's Press, 1979).

Walter J. Levy is a consultant to industry and governments. A selection of his articles and memoranda has been published recently under the title, *Oil Strategy and Politics, 1941–1981* (Boulder, Colo.: Westview Press, 1982).

Phebe Marr is Associate Professor of History at the University of Tennessee, Knoxville. She is author of the forthcoming *Modern History of Iraq* (Boulder, Colo.: Westview Press), and has worked previously for the Arabian American Oil Company, Dhahran, Saudi Arabia, and for the Foreign Service Institute, Washington, D.C.

John R. Nellis is Professor of Public Administration and Development Planning at the Maxwell School, Syracuse University.

Peter R. Odell is Director of the Centre for International Energy Studies at Erasmus University, Rotterdam. He is author of *Oil and World Power* (Harmondsworth: Penguin, 1970; 7th ed., 1983); with L. Vallenilla, *The Pressures of Oil* (London: Harper & Row, 1978); and, with K. E. Rosing, *The Future of Oil* (London: Kogan Page, 1980; 2nd ed., 1983).

Edith Penrose is Professor of Political Economy at the Institut Européen d'Administration des Affaires, Fontainebleau, France; and Emeritus Professor of Economics at the School of Oriental and African Studies, University of London.

J. E. Peterson has been a member of the Departments of Government at Bowdoin College and the College of William and Mary, and during 1983–1984 a Fellow in International Security Affairs at the Foreign Policy Research Institute, Philadelphia. Among his publications are *Oman in the Twentieth Century* (1978), *Conflict in the Yemens and Superpower Involvement* (1981), and *Yemen: The Search for a Modern State* (1982).

Eric Rouleau is the chief Middle East correspondent and editorialist for *Le Monde,* Paris.

C. A. Sinclair is an economist in the World Bank's EMENA Education and Manpower Development Division, and an Honorary Research Fellow in the Department of Economics at the University of Durham. Together with J. S. Birks, he is the author of a number of articles and books on labor migration in the Middle East.

Stuart Sinclair is a consultant and contributor to various publications of the United Nations and the Economist Intelligence Unit. He is the author of *Urbanization and Labour Markets in Developing Countries* (London: Croom Helm: New York: St. Martin's Press, 1978); and, with Paul Hallwood, *Oil, Debt and Development: OPEC in the Third World* (London: George Allen & Unwin, 1981).

John Townsend is with Business International, Geneva, Switzerland, and the author of *Oman: The Making of a Modern State* (London: Croom Helm, 1977).

Christopher Van Hollen was a Senior Associate at the Carnegie Endowment for International Peace when this article was written in 1981. He is currently Executive Director of Islam Centennial Fourteen. A former career foreign service officer, he was Deputy Assistant Secretary of State for Near Eastern and South Asian Affairs from 1969 to 1972, and Ambassador to Sri Lanka from 1972 to 1976.

Valerie Yorke is the author of *The Gulf in the 1980's* (London: Royal Institute of International Affairs, Chatham House Papers, 1980), and is writing a study on the domestic politics of Syria, Jordan and Israel and regional security implications for the International Institute for Strategic Studies, where she is a Research Associate.

Introduction

In 1983, the oil price revolution was a decade old. Ten years earlier, the fourth Arab-Israeli war of October 1973 had served as a catalyst, allowing the members of the Organization of Petroleum Exporting Countries (OPEC) to quadruple the posted price of oil in the space of a few months from $3.01 per barrel (October 1973) to $11.65 a barrel (January 1974). This action caused consternation in many countries around the world, which were heavily dependent on oil imports from OPEC states. Immediate impact was felt in severe economic dislocations, increased inflation, deepening recession, additional pressure on the shaky international financial system and an increased burden on the developing countries.

Gradually the world learned to adjust to permanently higher energy costs and terms such as "petrodollars" and "recycling" were added to vocabularies. Because most OPEC states faced limited absorptive capacities, immense current account surpluses were deposited and invested in the West. Initial fears that this would lead to foreign ownership of many Western commercial and industrial concerns were soon proven unfounded as the oil states found themselves reeling from a flood of imports and spiraling inflation. Gradually they began to reduce their surpluses as new development programs took hold and as the OPEC share of the world oil market diminished.

A second shock to the international economic system occurred with the new price spiral in 1979. In the five years since the 1973–1974 jump, the real price of oil had declined, the absorptive capacity of the OPEC countries had increased and world dependence on OPEC supplies had remained just as heavy. In the space of a year, the marker price of OPEC oil shot up from $13 a barrel to $28 per barrel before worldwide recession and the emergence of new sources of oil around the globe combined to reduce demand and increase supply. By the beginning of 1983, OPEC was experiencing great difficulty in maintaining its official price of $34 a barrel in the face of rapidly declining demand and thus OPEC production. For the first time, an effort was made to introduce production quotas for OPEC members and nearly all producers—with the notable exception of Saudi Arabia—were resorting to unofficial discounting.

After considerable discussion, OPEC members finally reached agreement in March on a reduced market price of $29 a barrel and a total production level of 17½ million barrels per day (down from a high of 30 million in 1979). It was widely speculated, however, that this agreement would not prove durable, since non-OPEC producers quickly moved to undercut the OPEC price. Furthermore, the new production ceiling did not accurately reflect actual OPEC production, which had fallen to approximately 14 million barrels a day.

The decline in OPEC's share of the world oil market and its growing inability to control the price of oil provides only one indication of the limited extent of its power. The thirteen members of OPEC are all developing countries, some of them among the world's least-developed countries. In common with many other Third World states, they are heavily dependent on the export of a single commodity as the basis of their economies.

The producers of coffee, bauxite, copper and other raw materials have also sought to arrest their deteriorating terms of trade and acquire the necessary foreign exchange to finance their development programs by drastically increasing the prices of their commodities. The difference between these producers and the oil producers lies in both the heavy concentration of vast oil reserves in only thirteen countries and in the uniquely vital importance of oil in the world economy. This combination of circumstances allowed OPEC countries to increase the price of their single major export tenfold in less than a decade. But this new cost of oil has spurred exploration for and production of other supplies of oil, as well as the development of alternative energy sources. Because of this competition and the nature of oil as a nonrenewable resource, the OPEC countries have only a few short decades in which to use their oil effectively in providing the basic means for development.

OPEC's success was viewed as a triumph by the whole of the Third World, as it constituted the South's first major victory in redressing its inferior economic and political status *vis-à-vis* the industrialized countries of the North. A handful of hitherto-disregarded countries had seemingly struck a shattering blow in the name of the Third World. Even as this psychological boost to the developing countries has remained important, however, the bill for vastly higher oil prices has hit them proportionately much more severely than the northern countries. The setback to economic growth and development in the South has only been partially offset by enlarged OPEC aid programs, especially as the developed countries have continued to decrease their contributions to foreign aid. Attempts by OPEC to use oil as the opening salvo in the South's construction of a new international economic order have had little success; the experience of oil has provided little spillover to other areas, another indication of the severe limitations of OPEC's new power.

These limitations hold even more true for the OPEC states' direct relations with the West. Rather than a case of OPEC having a stranglehold on Europe, Japan and the United States, the oil producers have become even more dependent on the West for purchase of their oil, as havens for their investments, and particularly as sources of essential goods and services necessary to their development. Because of the oil producers' new position as enhanced participants in the world economy, they have just as much a stake in its well-being as do oil importers and consequently they face just as heavy constraints on their economic and political activities. Thus, in many ways, the costs in unleashing the vaunted "oil weapon" may far outweigh any benefits. Despite the 1973–1974 oil embargo and threats to reinstitute it, the Arab oil producers have generally failed to force a reorientation of American policy in the Arab-Israeli conflict. Rather than having the upper hand in their political relationships with East and West, the Gulf oil producers remain vulnerable and unsteady pawns in an East-West competition for their assets.

Almost overnight, the Persian (or Arabian) Gulf was transformed from an unknown backwater into a key concern of American and Western strategic policy. But attempts to protect the Gulf from the clutches of the Soviet Union and perhaps even from itself have undeniably been flawed. Uncritical U.S. support for the Shah backfired when revolution consumed Iran and resentment of American interference in Iranian affairs resulted in the taking of the hostages. Fears that revolutionary upheaval would spill over to the Arab littoral of the Gulf led to pronouncements of American determination to prevent either outside or internal forces to gain control of these states. But the U.S. is faced with a fundamental dilemma in the Gulf: while it needs to defend its vital interests there by whatever means necessary, its enhanced presence may provoke circumstances which directly reduce its ability to defend those interests.

Most readers are already acquainted with the tremendous continuing effects of the oil price revolution. Much less familiar are the motivations and effects of this revolution on the OPEC countries, not to mention basic awareness of OPEC membership. In particular, most Europeans and Americans know remarkably little about the Middle Eastern oil producers, their transformation and problems as a result of the sudden infusion of oil wealth, and the consequences (both positive and negative) that these changes have had for the rest of the world. Only gradually has the world come to accept the more forward and substantial role that the Middle

Eastern oil producers have come to play in regional and global political and economic arenas.

In addition to providing an overview of the global effects of the oil price revolution, an important objective of this book of readings is to present an analysis of the various impacts of oil wealth on the Middle Eastern oil producers and to outline the resultant changes in these producers' relations with a variety of external actors, ranging from other Middle Eastern states to the superpowers. Along the way, I hope some contribution will have been made in eliminating some of the more egregious stereotypes surrounding this area. It is long past time to bury the clichés of the "greedy oil sheik" and "bloodthirsty Moslems." There is no justification for columnists to write glibly of "desert despots," "cruel and sybaritic Saudis" and of a "turbulent Islamic society pregnant with nasty surprises."

The substance and organization of this book reflect the belief that an adequate understanding of the politics of Middle Eastern oil goes far beyond the immediate issue of oil-pricing politics but must include the domestic changes transforming the oil producers and their new status and power in the world. With this in mind, Part I is devoted to the role of Middle Eastern oil in the international economy. The first readings in this section outline the evolution of the Middle Eastern oil industry from its beginnings in Iran in 1908, and the banding together of the oil-producing states into OPEC—and the Arab producers into the Organization of Arab Petroleum Exporting Countries (OAPEC). This section also includes an overview of the oil price revolution, its causes, its immediate impact on the global economy and financial system, the longer term effects of necessary adjustment to permanently higher energy costs, and the heightened interdependence and need for cooperation between producers and consumers.

While the oil price revolution has obviously resulted in substantial change in the West, the impact on the Middle Eastern oil producers themselves has been extraordinary. Part II examines how the vastly increased oil revenues completely transformed the economies of these states. The effect on their societies, even if less directly obvious, has been just as dramatic and undoubtedly more profound. This, of course, has direct implications for political evolution—or revolution, as the Iranian case has demonstrated. Since the first discovery of oil in the Middle East at the turn of the century, the number of Middle Eastern oil producers has steadily increased. Ten countries are of direct concern here: Algeria, Bahrain, Iran, Iraq, Kuwait, Libya, Oman, Qatar, Saudi Arabia, and the United Arab Emirates (UAE). Seven of these are members of OPEC; and all but Iran and Oman belong to OAPEC. Several other Middle Eastern states, including Egypt, Israel, Syria, Tunisia and Turkey, also produce and export oil, some at even higher levels than the smaller OPEC states. They are excluded from primary consideration here, however, since their oil revenues do not provide the major portion of their GNP, unlike the case in the ten states mentioned earlier.

While Part I looks at the international economic aspects of Middle Eastern oil and Part II is concerned with the domestic changes occurring in the oil-producing states, the focus of Part III is on the new political relationships emerging from the oil price revolution. Thus, the readings in the final section not only analyze the changing balance-of-power between the Middle Eastern oil producers and other actors in the Middle East, the Third World and the West, but also discuss the emerging concern of both superpowers over control of the Gulf with its immense concentration of oil reserves. Fittingly, an examination of the problem of Gulf security and the search for a viable American response to real or perceived threats to the Gulf complete the readings.

Together, the selections presented in these pages offer a comprehensive insight to the Middle Eastern oil producers and their place in the world. As a matter of deliberate design, the readings have been selected for their diversity in terms of authors' backgrounds, topics covered, approaches displayed and range of viewpoints advanced. The writers represent nearly a dozen Middle Eastern and Western countries and are drawn from the ranks of both scholars and "practitioners." Among the former are to be found economists, anthropologists, political scientists, sociologists and historians. The "practitioners" include government officials, former ambassadors, cabinet ministers, oil and business consultants, and journalists. With one or two exceptions, all the material presented here has been previously published in scholarly journals, current affairs periodicals or full-length books.

Because of the wide variety of original forms of publication and authors' purposes in writing, these readings unavoidably display a vast range of styles, tense usages, spellings of Middle Eastern names and terms, and dates of authorship. Care has been taken, both in the choice of contributions and in the chapter introductions, to minimize the inevitable effects of aging as much as possible. Updating has been carried out through the insertion of editorial notes in the text and through the glossary, chronologies and tables located in the appendices.

A number of appendices and maps have been included at the end of the book for ease of reference. Basic data on the Middle Eastern oil producers, their oil production and foreign aid will be found there, as well as a glossary and dictionary. Other appendices provide essential background on the U.S. Rapid Deployment Force and the Iran-Iraq war; chronologies have also been included on the war, the Iranian revolution and international oil developments over the last decade. Finally, effort has been made to compile a bibliography which is both comprehensive and readily usable by the reader interested in pursuing a particular topic further. Consequently, bibliographic entries have been arranged in strict alphabetical order and numbered; those entries relevant to the topic of a particular chapter have been numerically listed at the end of the chapter introduction. Since this book touches on a wide range of topics and thus there is a vast pool of potential entries, it has been necessary to keep the bibliography down to a manageable size through rather arbitrary methods. Emphasis has been placed on publications of a general, rather than specialized, nature which have appeared in the last five years, excepting a few sources of particular value falling outside these guidelines. The appendices are current through the end of 1982.

PART ONE

OIL, THE MIDDLE EAST, AND THE INTERNATIONAL ECONOMY

The Middle Eastern Oil Industry

Oil has been produced in and exported from the Middle East since the turn of the century and the region's position of dominance in the world oil industry began to appear following World War II. As Appendix B shows, nearly a third of all world production today takes place in the Middle East and the region contains two-thirds of all proven reserves. Consequently, it is not surprising that most people in Western countries instantly associate the Middle East with oil. There is very little corresponding awareness, however, of the origins and development of the oil industry in that part of the world and the conditions allowing it to grow to its present position of importance. The following articles outline the beginnings and emergence of the Middle Eastern oilfields as an exclusive Anglo-American reserve dominated by the "Seven Sisters"—the giant international oil companies. As Professors Odell and Penrose show, only gradually were the "independents"—a number of other oil companies, large and small—able to penetrate the region.

At the same time, there has been a steady expansion of the host countries' ability to control their own oil resources, whether in the terms of the concessions granted, the rate of production or the price charged for oil. The picture that emerges is one of a steady shift in control over this vital energy source, from the consuming nations and their companies to the producing states. At the same time, the relative importance of Middle Eastern oil in the world economy was dramatically demonstrated in the decade of the 1970s.

The first selection below provides a concise overview of the role of the West in gaining control of Middle Eastern oil, from the acquisition of the first gigantic concessions by the "Seven Sisters" to the emergence of the Anglo-American struggle for dominance in the Gulf. But Professor Odell also outlines the ensuing stage of increasing producer insistence on their control over their own oil, beginning with the appearance of the posted-price system and the formation of both the national oil companies and OPEC. In the following piece, Edith Penrose takes up the theme of the steadily advancing transfer of control over oil production and ownership from the international companies to the host governments. Here are clearly shown the various stages in this process: the posted-price system, institution of profit-sharing and imposition of income taxes on the producing companies, the first demands for nationalization and then acceptance of the principle of participation, by which ownership was gradually assumed by the host governments.

Much has happened of course since these articles were written (Odell's soon after the 1967 Arab-Israeli war and Penrose's shortly following the October 1973 war). Participation has been nearly completed and the international oil companies reduced to the status of buyers, operators and consultants in their old concessions. The national oil companies, set up by the host governments to operate existing oilfields and exploit new ones, began to expand their activities from exploration and production to such "downstream" operations as petrochemical industries, tanker fleets and marketing. The oil price revolution, which opened with that quadrupling of prices by early 1974, continued to march on. After a quiescent phase in the mid-1970s, another sharp price spiral in 1978 and 1979 boosted prices to more than $30 a barrel.

By this time, the international oil industry had become much more diversified. The "Seven Sisters" found themselves in intense competition with a multitude of American, European, Japanese and even Third World independents. The dramatic rise in prices meant that older oilfields around the world, once too expensive to operate, were put back into production. The search for new reserves was intensified and by the end of the decade significant discoveries and subsequent production in the North Sea, Alaska and Mexico helped to ease the shortfall in supply caused by seemingly insatiable world demand. Thus, by the beginning of the 1980s, a combination of factors, including new sources of production, the worldwide recession, and use of alternative sources of energy, produced a global oil glut. At the end of 1982, OPEC's posted price of $34 per barrel was being undercut by a number of exporting countries and OPEC members, confronted with falling demand for their production. While no one expected the glut to last forever, it seemed to mark a watershed in the oil price revolution and provide breathing space for consumers and producers alike.

Bibliographic Sources for this Chapter:
15, 23A, 54, 65, 72J, 96, 128, 184, 189B, 206, 219A, 220M, 231, 232, 290, 299, 309K, 367, 374J.

The Significance of Oil

PETER R. ODELL

The Middle East without oil would be a very different region. In addition to changing the face of large parts of the area, oil has helped shape the policies and alignments of all the countries of the Middle East—not only with each other but also with the world's great powers. These outside powers, in their turn, have found their oil interests and their oil ambitions in the Middle East spilling over into their more general relationships with each other, thus helping a process which has made the region a centre of international tension over long periods of time.

Early British oil interests in the area merely reflected late nineteenth-century British interests of a more general nature. These arose out of imperial connections with India and the Far East which demanded the establishment of coaling stations and of territorial enclaves to protect both them and routeways such as the Suez Canal. As a result of these imperial developments, Britain then took an earlier interest than any other outside power in the possibilities of oil from the Middle East. Even so, the region was well behind other parts of the world in the development of oil resources. By the time the Anglo-Persian company finally discovered oil in Persia in 1908, and started to work the discoveries under the stimulus of participation by the British government, which sought British owned and controlled oil for its navy, the oil industry elsewhere in the world had almost half a century of history behind it. Britain's strategic and military domination of all areas of the Middle East outside Turkish control acted as a restraint on oil activities, other than those blessed officially by Britain, and thus from 1908 to 1919, apart from the steady development of resources in Persia, little in the oil sector took place to disturb the traditional economies of the Middle East.

Reprinted by permission of the publisher, SAGE Publications, from *Journal of Contemporary History,* Vol. 3, No. 3, July 1968, pp. 93-110.

After the first World War, the division of the former Turkish empire into British and French spheres of control led to an understanding between the two countries to divide the spoils as far as oil was concerned. Thus, although Persia remained wholly British in its oil development, neighbouring Iraq, where the possibilities of oil appeared good, lay on the border of British and French influence and hence necessitated the participation of both British and French capital in a consortium created to initiate oil exploitation. Those parts of the Middle East wholly within the French sphere turned out to offer little by way of oil possibilities—given the technology of the period—and hence French oil interests were limited to a share of Iraqi operations and, at a later period, to participation in the crude oil pipelines that were built from Iraq to the Mediterranean. On the other hand, Britain gradually extended its oil interests to other parts of the Persian Gulf where its earlier informal political authority was formally confirmed in 1919. In the inter-war period British firms discovered oil resources along this coast, particularly in Kuwait. The demand for oil at that time, however, was not great enough to give a strong incentive to develop these facilities, for Iran and Iraq were quite capable of producing whatever oil could be sold out of the Middle East.

The world oil-pricing system, dominated by the United States, and the existence, informally before 1928 and formally after that, of an international oil cartel, effectively inhibited the potential producing areas of the Middle East. The USA, Mexico, and to a lesser degree Venezuela and the Dutch East Indies, maintained their dominant role in world oil markets and expansion of known Middle East oil resources was thus restrained. But attempts to make Middle East oil more important in an international context also turned on the political and commercial efforts of Washington and of American companies to win access to the oil wealth of the region. Eventually, as a result of both political and commercial pressure, the US secured an entry to Middle East producing areas, and although the British government managed to keep American interests clear of its most important producing and refining facilities in Persia, it was forced to allow them into Iraq and the Persian Gulf states. In Iraq the consortium, in which British and French companies retained an interest, came to be dominated by a group of US oil companies. In Kuwait, the concessionary company became 50/50 British and American. Thus by the outbreak of war in 1939 the United States had won for itself a position of virtual parity with British interests, and one potentially stronger than that of the longer-standing French oil influence in the region.

In this period up to 1939 one can view the structure of the Middle East oil industry as lying wholly within the framework of a classical colonial situation. The countries there were not colonies like British or French possessions elsewhere in the world, but British and French political control over virtually all the territories concerned was practically complete. In addition, oil wealth provided the basis for economic colonialism, while the power of the United States in general and of its oil industry in particular brought in a third outside power as an economic force of significance. As shown already, however, one should not overstress the significance of oil in this period. Middle East oil remained relatively unimportant in world markets, and even by 1939 it was contributing only 5 per cent of total world oil production and its exports were limited to countries within the immediate region and, via the Suez Canal, in western Europe. But even in western Europe oil from the Middle East accounted only for under 20 per cent of total imports, which remained dominated by those from the United States and Mexico. The real significance of pre-1939 developments in the Middle East, however, is that they established the framework for the post-1945 expansion of oil activities in the region. With the cessation of hostilities, British, French, and United States oil companies rushed in to take advantage of the concession areas which they had tied up for themselves before 1939.

All companies concerned participated in the rush to expand Middle East oil production and exports. The post-war world was short of energy; US oil was in demand at home; Mexican oil was required to service the increasing demands of Mexico itself under the process of industrialization; Venezuelan production had expanded rapidly towards the end of the war but it was physically unable to meet the demands of both the Americas and western Europe; whilst in the Far East, in the formerly important oil-exporting areas of the Dutch East Indies and Burma, the industry had been destroyed in the war. War-time oil development in the Middle East had been limited, but there had been sufficient work to demonstrate that the area was potentially prolific, and there was little hesitation in investing capital in Iran, Iraq, Kuwait, and Saudi Arabia, because of the virtual certainty of discovering new oil resources and finding an outlet in world markets. The rush to invest was not limited to the development of producing facilities but was accompanied by expansion of the area's refining facilities and, more significantly, by arrangements for more effective transport facilities for crude oil to the markets of western Europe. Investment here included expenditure on the Suez Canal to ensure its usefulness for larger tankers which were now being built as a result of war-time design developments, and also on the construction of pipelines from the producing areas of Iraq and the Gulf to Mediterranean ports in the Lebanon and Palestine—pipelines which in the period before the advent of large tankers reduced the cost of getting the oil to western Europe and hence improved the profitability of the companies concerned. Thus, looking at the Middle East as a whole, one notes that production increased from under 50 million tons in 1945 to almost 100 million tons by 1950 and 250 million tons ten years later. At the same time the declared reserves discovered by the drill rose from the estimated 3000 million tons in 1945 to nearly ten times this level by 1960, and it is well known that these declarations of reserves by the companies concerned represent only a percentage of the total, for some of them have been and, in fact, still are under no obligation to reveal the extent of the resources they have discovered. From providing only about 10 per cent of the world's total oil production in 1945, the Middle East's importance increased steadily throughout the 1940s and 1950s; by 1960 it was responsible for no less than 25 per cent of the world's total. This breathtaking expansion of the physical facilities to produce and transport Middle East oil fell almost entirely within the framework of the operations of the seven major international oil companies which before 1939 had constituted an international cartel and which in the post-war period 'understood' each other so well that they could be relied upon not to take decisions which might operate against the best interests of their fellow oil companies.

The Middle East thus achieved a significant status in the rapidly growing international oil industry. The economies of western European countries and of many countries in other parts of the world became dependent upon Middle East oil resources. The investment of the companies grew by hundreds of millions of dollars every year, but this increasing involvement in the Middle East, on the part both of consuming nations and of international companies, occurred at a time when changing circumstances were bringing to an end the political stability of the inter-war period. The defeat of the French in 1940 lost them their influence and prestige, and Syria and Lebanon quickly claimed their independence in the immediate post-war period. In the rest of the region there was significant Anglo-American pressure to keep the area under political control. As oil became more important, so ran the British and American argument, so greater control became more necessary. But control could not always be maintained. Certain territories declared notional independence to be effective independence. In Iran this declaration took the form of the nationalization of the Anglo-Iranian Oil Company's facilities. The international oil industry—wholeheartedly backed by Washington and London—retaliated by isolating Iranian oil from world markets and for three years little crude oil from the producing fields and few products from the massive refinery at Abadan moved on to world markets. The cessation of Iran's exports did not cause the world supply difficulties that had been anticipated, for they were quickly made up by additional production and exports from neighbouring countries whose oil development thus received a considerable stimulus. Iran was thus obliged to accept a compromise whereby the Anglo-Iranian Oil Company, later renamed British Petroleum, along with American and other interests, were allowed back to work the prolific Iranian fields on behalf of the National Iranian Oil Company. The formula appeared to concede the national ownership of resources that Iran claimed, but in reality the work of the Consortium at first differed little in essentials from the earlier situation.

Anglo-American efforts to secure a degree of direct or indirect political control over Middle East countries were only partly motivated by their desire to ensure the "rights" of their companies to explore for and develop oil. In part they were dictated by the constant fear that the Middle East might become increasingly susceptible to external political intervention by other nations. It was thought that the USSR had not only a political but also an economic interest in securing control over the Middle East oil producing countries. Observers noted the rapidly increasing demand for energy within the Soviet Union and assumed that this would require more energy than the Soviet Union could produce within its own boundaries. From this assumption observers predicted that the USSR would seek access to Middle East oil not on commercial terms through purchases from the producing companies, but within the framework of a political attempt to capture the growing nationalist movement in these countries. Given the international situation at that time, the coupling of Soviet interest in Middle East oil with a more general view of Moscow's expansionist tendencies led to a real fear of the Soviet Union's intentions in the region. Contemplating this 'threat', the US and UK governments forgot their earlier differences on the sharing of the Middle East's oil wealth and, emphasizing the need to present a united front in the area to meet the external enemy, attempted to sell to the Middle East countries the idea of the peripheral anti-Soviet alliance as developed in western Europe through NATO. It was from these efforts that CENTO emerged, in which one of the most important oil producing countries of the Middle East—Iran—decided to participate. Behind this screen provided by Turkey, Iran, and Pakistan, Anglo-American influence in the Middle East was felt to be more secure, and both diplomats and oil investors breathed a little more freely once the treaty had been signed.

In retrospect, one can see that this concern about Soviet intentions in the Middle East was exaggerated, at least on the oil side. The USSR proved to have more than enough oil and gas within its own borders to satisfy its own requirements. The rapid development of the Soviet oil industry enabled the country not only to change from a coal-based to an oil- and gas-based economy, but also, by the mid-1950s, to become a potential rival to the Middle East in terms of oil exports. Middle East oil now faced a new competitor, particularly in western Europe, and the Soviet oil export agency, in order to break into the markets controlled in the main by the companies producing oil in the Middle East, cut prices where necessary. The USSR also sought oil outlets in the developing countries such as Brazil, Guinea, India, Ceylon, etc. The steady growth of both Soviet production and oil export potential quickly dispelled the idea that the USSR needed the Middle East to provide fuel and power for its economic development. Its recent decision to purchase large quantities of natural gas from Iran confirms this view, for the gas can be used in the Soviet Union to replace domestic oil, which will then be available for export, thus giving the USSR a net gain in hard currencies.

Defence of Anglo-American oil interests in the Middle East, however, was not restricted to warding off the 'threat' from the USSR. Elsewhere in the world, countries and companies looked enviously at the oil wealth of the region and viewed with disfavour the control exercised by a limited number of Anglo-American companies which, although producing oil cheaply, were able to sell it expensively because of their control over transport, refining, and marketing facilities. As a reaction against this, French, Italian, Japanese and other outside interests attempted to secure influence in the Middle East. Companies from these countries, both state and private, started to tender for the right to explore for oil in areas offered for exploration by the Middle East governments. The degree of profit-sharing, state participation, and guarantees of various other kinds they were prepared to concede, was greater than that made by the international companies, which had not bargained for being outbid by these outsiders. The defence against them rested not only on the technical and commercial expertise of the Anglo-American companies; it was backed by diplomatic and political pressure by the American and British governments. In certain instances the pressure was successful, but the growing hostility of some of the nationalist groups in the Middle East (encouraged by diplomatic counter-offensives by the French, the Japanese, and the Italians), coupled with the greater commercial and economic attractiveness of the newcomers' proposals for oil acreage, eventually secured rights in the area for oil interests other than those of the United States and the United Kingdom.

The first move in this direction emerged out of Iran's expropriation of the Anglo-Iranian Oil Company's interests in 1951. The 1954 agreement negotiated with the Consortium limited the latter's right to produce and explore for oil to those areas in which some work had already been done. The rest of the country's oil areas was vested in the National Iranian Oil Company, which since 1957 has signed contracts with several other foreign companies and other (sometimes state) entities for exploration and development work. In particular, the Italian state oil organisation, ENI, has secured acreage in areas formerly conceded to Anglo-Iranian. More recently, Kuwait has insisted that the Kuwait Oil Company (owned 50 per cent by BP and 50 per cent by Gulf Oil) give up its "rights" in almost half the country. The acreage relinquished is now being allocated to other companies. Most recently, Iraq has taken the most extreme action in this respect by unilaterally declaring that the rights of the Iraq Petroleum Company—which had the whole of the country under concession—extend only to the areas it actually worked. As these amount to only about one per cent of the country's area, the remaining 99 per cent of Iraq is now "on offer" to other parties. And in spite of the IPC's claim that the Iraqi government's action is illegal and that no other party has any right to work oil in the country, Iraq has not been unsuccessful in interesting other potential oil seekers. Though it is reported that strong diplomatic pressure from Britain and the US (whose oil companies are dominant in IPC) secured the withdrawal of an offer from ENI to develop acreage in Iraq, this has not prevented a French state oil company from concluding an arrangement to develop some of Iraq's oil resources hitherto controlled by IPC. Most recently, even Saudi Arabia has ended the monopoly rights of the original concession holder—ARAMCO—and has accepted tenders for oil exploration and development from other companies. Thus by 1968—in spite of the considerable commercial power of the traditional companies and the diplomatic pressure of their governments—the former simple pattern of oil concession arrangements in the Middle East, essentially countrywide arrangements within the hands of eight companies working either alone or, more frequently, in alliance with each other, had been replaced by a complex jig-saw pattern in which over 50 companies and institutions are involved.

These new concessions are important not only because they have brought in companies and interests other than British and American (even India now has an interest in Iran), but also because the Middle East governments have taken the opportunity, when agreeing to new concession arrangements, to insist on some degree of national control over the eventual exploitation of any resources discovered. Typically, the new-style agreements involve initial obligations by the companies concerned to undertake a rapid exploration of their areas, and to move equally rapidly into developing any fields discovered. The state then claims a share in the development on payment only of an appropriate share of the development costs. Governments thus secure the right not only to share in the profits but also to participate in deciding how the field shall be operated: not an unimportant right in respect of decisions on the speed at which production shall be built up and off-take increased—for an international company can restrict development in order to protect its investments elsewhere. Through this approach to direct participation in oil ventures, state investment becomes virtually riskless, for it makes no contribution to the risky exploration expenditure. The new-style agreements often also insist on the companies agreeing to build a local refinery to process part of the crude oil produced and also on their own tanker fleets (another recent development by some of the producing countries) being given first option on the transportation of some of the oil. All this, of course, adds up to much more than merely sharing the profits of an operation run entirely by the oil company; it represents the achievement of significantly enhanced economic status by the countries concerned, not only *vis-à-vis* the companies but also in relation to the outside powers to which the companies owe their allegiance.

To date, however, the increasing number of concession arrangements and the establishment of new terms for newly negotiated agreements have been less significant than the changes which the Middle East governments have been able to secure in the terms of the old concessions, from which some 75 per cent of Middle East oil is still produced. The pre-1950 concession arrangements merely reflected the semi-colonial status of the area.

Supported by their governments, the companies sought concessions, offering a royalty of modest size on any oil that might eventually be produced, and receiving in return the exclusive right to search for oil—usually over an area covering the whole of the national territory concerned.

By 1950 the international oil companies elsewhere in the world were already having to accept a changed status as a result of nationalist pressure. In 1938, Shell and Esso interests in Mexico had been nationalized because the companies had been unwilling to renegotiate their concession agreements on terms acceptable to the government. In 1943 the Venezuelan government passed new oil legislation that marked a turning point in government-company relationships. It introduced the concept of profit-sharing between government and companies, and of company participation in the domestic economy over and above that demanded by the mere production of petroleum. Given political change and awareness in the Middle East in the immediate post-war period, it was only a matter of time before similar nationalist reaction to the companies began to make itself felt. Nationalization in Iran in 1951 was the most obvious manifestation of this, but the crunch effectively came a year earlier, in 1950, in Saudi Arabia, which by then had become a very successful producing area with an even greater potential remaining to be developed.

Given the low-cost nature of Saudi Arabia operations—due not only to favourable geological conditions but also to the proximity of the fields to the coast and hence to ocean transportation facilities—and the value of this crude to the American companies concerned in serving their rapidly growing markets of the eastern hemisphere, the joint concessionary company, ARAMCO, was not unwilling to accede to the government's demand that the recently introduced "posted price" for Saudi Arabian oil should be used for calculating total profits (= posted price per barrel × number of barrels produced minus the costs of producing the oil and getting it to the export terminal). Total profits thus determined would then be divided equally between company and government. In this way Saudi Arabia achieved a profit-sharing status similar to that won by Venezuela some years earlier, and in addition established the principle of their 50/50 split between government and company. Once ARAMCO had accepted this new form of relationship with Saudi Arabia, there was nothing that oil companies elsewhere in the Middle East could do to prevent their own concession terms being renegotiated. They accepted the inevitable and sought only delay. Delays proved to be very short, for by 1954 all major producing countries in the area were on a 50/50 profit sharing basis.

Although profits to the companies remained high under the new formula—by virtue of low production costs and high posted prices—the effect of the agreements was to reduce significantly the gross returns made by the oil companies in their producing operations. This fundamental change in government/company relationships might not have been achieved so easily had it not been for decisions by the United States—and later Britain—to allow their oil companies to offset all tax payments to the governments of the producing countries against their tax obligations to the US and British treasuries. Thus the net additional tax burden on the oil companies was greatly reduced and it was, in effect, the taxpayers of the US and Britain who suffered the main consequences of the new financial arrangements between producing countries and companies.

Since the Middle East countries took this action to secure a greater and more direct interest in oil operations in their territories, government/company relationships have become a more or less continuing dialogue in which first one government and then another has sought to improve the terms of the oil concession arrangements. The manoeuvres have been mainly concerned with ways in which pricing and profit-sharing should be carried out.

Profit-sharing based on posted prices had guaranteed Middle East countries rising revenues from oil whilst the demand for oil continued to rise and whilst posted prices remained high. After 1957, however, the world energy situation changed, bringing increasing competition between energy sources. In their search for markets, oil companies started to cut prices. At first price cutting was achieved by shaving profit margins, but soon companies producing oil in the Middle East started to offer crude to independent and state-owned refineries at less than the officially posted prices. As this practice grew, it produced an incentive for the companies to bring down the posted prices in order to reduce their tax obligations to host governments. The companies had certainly expected full freedom to change prices, as they had done for a quarter of a century in the US and the Caribbean, and indeed in the Middle East from time to time in the 1950s; then, however, the price changes had been essentially upwards in sympathy with changes in the western hemisphere. Small reductions in posted prices in 1958 and 1959 produced some indications of disapproval from certain Middle East governments, but it was not until major cuts—of the order of 10 to 15 per cent—were announced in 1960 that a storm broke over the heads of the companies whose decisions would reduce the oil revenues of the countries concerned by 5 to 7½ per cent. The companies at first insisted on their right to take such action when necessitated by commercial considerations, but they were eventually obliged to reappraise their position under the combined onslaught of all the oil producing countries, where the strident nationalist reaction threatened not only the interests of the companies but also the standing of Britain and the US in the difficult post-Suez period. The posted price issue was, indeed, a sufficient spur to the producing countries for them to act collectively and form the Organization of Petroleum Exporting Countries.

International Oil Companies and Governments in the Middle East

EDITH PENROSE

INTRODUCTION

One cannot help feeling that in the future when better perspectives on our age are achieved historians will look back and see in the development of the present economic "crisis" over energy one of the most striking ironies of history. It has been increasingly clear over at least the last twenty years that an indispensable proportion of oil, on the lavish use of which the continued growth and prosperity of the industrialized world has depended, would soon, and for a considerable period continue to come from a small number of countries in the Arab Middle East and North Africa.

Over the same period, the control which the great international oil companies had exercised for half a century over output and offtake prices of Middle Eastern oil was rapidly passing into the hands of the governments in that area and out of the hands of the heretofore dominant Western companies. Also during this period, but as a result of entirely unconnected political developments, the governments of the industrial countries actively supported, or in practice condoned or acquiesced in, policies which were inexorably to lead to open conflict with those governments on whose good will oil supplies depended.

Portents and signs of what was to come with respect both to the shift of power in the oil industry and to the conflict over the Israeli occupation of Arab territory became increasingly abundant as the 1960s gave way to the 1970s but were resolutely and complacently ignored. The October War took all by surprise, as strangely enough, did the decision of Arab governments to use their control over oil as a diplomatic supplement to their military effort to obtain the implementation of UN Resolution 242, the withdrawal

Reprinted by permission of the publisher, American Enterprise Institute for Public Policy Research, from John Duke Anthony, *The Middle East: Oil, Politics, and Development.*

of Israel from Arab lands, and some sort of justice for the Palestinian people displaced by the Israelis who took over their lands and farms.

Today these three developments—the increasing dependence of the world on Middle Eastern, and specifically Arab, oil; the assumption of control over their industry by the oil-exporting countries; and the establishment and expansion of Israel—have reached a critical point in their historical evolution at roughly the same time, partly independently and partly because of interaction one with the other. In particular, the determination of the Arabs to force a satisfactory resolution of the role of Israel in the Middle East has drastically accelerated the rate of change in the position of the companies. In this paper I shall examine the changing role of the international oil companies in the Middle East.

COMPANY GOALS AND POLICIES: THE TRADITIONAL APPROACH

Historically, two objectives have been of equal importance in the policies of the international oil companies: (1) to find and develop oil resources adequate for their needs and (2) to control existing or potential supply in order to ensure "orderly marketing" and prevent "excessive" price competition. This in practice required a high degree of cooperative action. The history of efforts to find and develop oil reserves is a colorful and romantic record of entrepreneurship and ingenuity, courage and determination, remarkable financial success and disastrous failure, impressive technological accomplishments, and great organizational skill. The history of company efforts to gain and maintain control over oil reserves is somewhat more obscure but is becoming clearer as the relevant records become available. It appears to be a record of equal entrepreneurship and ingenuity, considerable negotiating skill and/or skullduggery, and effective exercise of political influence.

Although it is the history, or at least the conjectured history, of this second type of activity which has given rise to so much suspicion of the international companies, one must always remember that to a very large extent their activities were supported, or at least condoned, by their governments. The United States, for example, has often permitted its own multinational corporations to act abroad in ways which would not be permitted at home on the ground that many of their activities, otherwise illegal under the antitrust laws, were in the "national interest" or that action against them would prejudice "national security." There has been rivalry among governments as well as among companies for control over oil supplies, a control often considered necessary in order to ensure access to such supplies—itself a somewhat unusual situation in international economic relations in recent times. Moreover, the ability of the international companies to develop and sell oil and its products freely in world markets, in competition with each other and outsiders, has been substantially restricted by governments of importing countries intent on discriminating among sources of supply or on protecting their own indigenous oil or coal industries. Finally, the tax policies of the home governments of the multinational companies have themselves been a powerful influence on the operational and financial policies of the international companies, providing incentives which largely determined the size and forms of payment to the producing countries and even the price structure of the industry, which in the end eased the transition from control by the companies to control by host governments.

In other words, many aspects of the policies of the multinational corporations that have shaped the development of the Middle Eastern oil industry, and have become the subject of much criticism, were not independent of the policies adopted by their parent governments, of which those of the U.S. government have been by far and away the most important. It is true that the policies of governments are themselves often influenced by the pressure, open or discreet, of companies, sometimes corruptly using venal officials to subvert government processes, but this neither relieves government of its responsibility nor absolves it from blame.

It is thus misleading to use history for the narrow purpose of bringing charges and condemning the oil companies. Fundamentally, the problem involves the operation of a particular economic and political system of which the multinational companies are only a part and which is changing rapidly. The oil industry is only one of a number of industries in which the role of the multinational corporation is changing in a similar direction and for similar reasons. It happens to be concerned with a particularly vital natural resource which, partly for geographical reasons and partly because of the policies pursued by the oil companies, is peculiarly exposed to effective monopolistic control by the governments of the major producing countries. It holds the center of the world stage at the moment, but times will change.

THE MARKET STRUCTURE: POSTED PRICES AND PROFIT SHARING

The oil industry of the world outside North America and the Soviet Union has long been dominated by seven large multinational corporations—five American, one British/Dutch, and one British.[1] By 1935 the

1. These companies, ranked in order of size, are now Exxon, Royal Dutch/Shell, British Petroleum, Gulf Oil, Texaco, Standard Oil of California, and Mobil Oil. Royal Dutch/Shell is 60 percent Dutch and 40 percent British. These companies owned nearly 90 percent of the world's oil outside North America and the Communist countries in the early 1950s, and by 1972 this percentage had fallen only to 72 percent. For reference purposes it might be helpful to indicate the important changes of name for the major companies: Exxon, formerly Standard

concessions attained by these companies covered such a large percentage of the territory of the area that any serious entry of newcomers was practically precluded. The financial terms were extremely favorable for the companies, and markets were well in hand. Nevertheless, as the profitability of the industry became increasingly apparent, and also, therefore, the real value of the concessions held by the companies, the pressures from rulers and host governments for greater shares in the revenues could not be easily resisted. It was the Americans who hit upon what seemed to be an ideal solution to the problem, one which would give the governments very large increases in revenues at very little cost to the companies—or at least to the American companies. The solution was the abandonment of royalties as the principal method of paying host governments and the substitution of profit sharing on the basis of published, or "posted," prices for crude oil. In the end, this was to provide the governments with an extremely effective tool for exercising control over the companies, but the "end" would not come for another twenty years and was naturally not foreseen at the time.

Before World War II, very little crude oil moved in international trade. Most of it was refined in or near the areas where it was produced, and the refined products were shipped to markets. After the war, for a variety of reasons refineries were increasingly established in or near the chief centers of consumption and refining gradually shifted away from the producing areas.[2] Thus, as the great upsurge of Middle Eastern oil began to flow from the newly developed fields of Kuwait and Saudi Arabia, most of it was shipped as crude to refineries in Europe and the United States. However, since these refineries were owned by the companies producing the crude oil, there was no publicly known international price for crude.

In other words, crude oil was not generally bought and sold in free markets. The actual price of the crude as it was transferred from the producing affiliates in the Middle East to the parent companies depended on particular arrangements made for each affiliate. In Iraq, for example, Iraq Petroleum Company (IPC) was set up as a nonprofit-making company, and crude was transferred at cost plus one shilling a ton to the parents. Aramco, on the other hand, was set up as a profit-making affiliate, and crude was transferred at various prices depending on the circumstances. Sales to the U.S. Navy, accounting for 90 percent of Aramco's sales to outsiders, were made at relatively high prices, and arrangements with par-

---

Oil (New Jersey); British Petroleum, first Anglo-Persian Oil Company and then Anglo-Iranian Oil Company; Texaco, formerly Texas Oil Company; Mobil Oil, formerly Socony Vacuum and then Socony Mobil.

2. As product markets expanded and the optimum size of refineries increased, it became economical to transport the crude oil and refine it close to the product markets since crude was cheaper to transport than products. In addition, many consuming countries considered that the security of their product supplies was greater when refineries were located outside the main areas of crude oil production. In the event of interruptions from any one area, other sources of crude could be substituted.

ent companies apparently varied. In Kuwait still different arrangements prevailed, particularly in relation to the long-term contracts with other major companies. Thus, it was necessary to find an acceptable and accepted way of pricing oil if revenues to the host governments were to be calculated with reference to profits made on crude oil sales.

Because parent companies do not make profits on interaffiliate sales, their consolidated income statements can only show profits when sales are made outside their own group. The transfer prices which govern interaffiliate sales in any multinational company are likely to be related to the impact of taxation and are necessarily arbitrary unless there is a reasonably free competitive market against which to appraise them. This did not exist for crude oil. Hence an agreed price had to be established.

Before the war, the three largest international companies—Anglo-Persian, Jersey Standard, and Shell—had joined in a cartel, the famous "As Is" Agreement of 1928,[3] by which prices of products were to be determined with reference to delivered prices from U.S. ports along the Gulf of Mexico. In other words, products in the Middle East were supposed to be priced as if they had been produced in and shipped from the United States. This naturally gave large profits to sellers in other lower cost areas of the world as well but at the same time prevented them from expanding markets by reducing prices.

During the war, the British auditor general objected to the British Navy's having to take bunker fuel from the Persian Gulf at prices which reflected shipping costs from the United States. The companies therefore reduced prices of bunker fuel bought in the Persian Gulf to the level prevailing in the Gulf of Mexico, which of course still reflected higher U.S. production costs. After the war, this time under U.S. government pressure, crude oil prices in the Persian Gulf were still further reduced, but they remained connected to the movement of U.S. prices, a link which was quickly broken when circumstances changed.

Thus, posted prices adopted after the war for the purposes of calculating profits to be attributed to crude oil for tax purposes were based on prices which had emerged in response to U.S. government pressure on the companies. These prices perhaps also reflected the companies' willingness to encourage the use of Middle East oil by reducing delivered prices from the Middle East relative to those from the United States. The posted prices thereby established were still very high in relation to costs of production in

3. The "As Is," or "Achnacarry," Agreement of 1928 was an agreement between the Big Three companies of the time—Shell, Standard of New Jersey, and Anglo-Persian—designed to check the growing competition in world markets and to strengthen weakening prices. It was a fairly comprehensive cartel agreement for market sharing and price fixing and provided the basis for a number of agreements in local markets, but in general it was not very effective. It was an attempt to ensure that the market position of the several companies would be maintained "as is" throughout the world.

the Middle East and were intended to set recognized norms which would be observed by all companies and accepted by governments as the "proper" price of crude. It was assumed that oil would be both sold in the market and transferred to affiliates at these prices, which the companies could change as market conditions dictated.

The large difference between posted prices and production costs implied large profits on crude oil from the Middle East. Most of these "profits" were imputed profits since little crude was sold to third parties and the arrangement made between the American company in Saudi Arabia and the government permitted the Saudis to levy a 50 percent income tax on these profits. Aramco is subject to American income taxes, but under U.S. law, taxes imposed by foreign governments could be deducted from a company's U.S. tax liability on foreign income. Hence, the effect of the so-called 50/50 arrangement in the Middle East was simply to transfer Aramco's tax payments from the U.S. Treasury to Saudi Arabia, so long as the payments could be defined as income taxes. Moreover, the U.S. tax laws provide a subsidy for crude oil production in the form of a depletion allowance against taxable income. It therefore paid integrated U.S. oil companies to put as much of their integrated profit as possible at the crude oil level (or at least, in the case of foreign production, up to the limit of the tax credits obtainable).

It was in 1950 that Saudi Arabia "imposed" the first income tax on the oil companies in the Middle East. The law was designed to capture 50 percent of the profits attributed to crude oil (*including* royalty payments, thus imposing no extra fiscal burden on the company). Mobil Oil Company announced the first posted prices in the area in the same year. Since royalties were included in the assessment of the income tax in contrast to the normal accounting practice which treats royalties as an expense deductible from pre-tax profits, the entire payment could be offset against the company's U.S. income tax liability, leaving it free of tax in the United States. This method of paying Saudi Arabia spread rapidly to all other Middle Eastern governments, although not without some difficulty for British companies, whose government did not at that time give such generous tax deductions for foreign income taxes.

As a result of the new arrangements, the revenues of the oil-producing countries soared, and their governments seemed reasonably satisfied except for the conflict between the Persian government and the Anglo-Iranian Oil Company (AIOC) over the terms on which the company should operate. This conflict culminated in the nationalization of the AIOC in 1951 and in the creation in 1954 of a consortium which included all of the major companies plus a few independents. Iranian oil was virtually off the market for four years.

In spite of this, however, output from the Middle East as a whole rose rapidly, and toward the end of the decade, oil surpluses began to press on prices. Even the major companies began to discount the effective delivered price of their crude oil. As competition intensified, prices in product markets fell.

Falling product prices with unchanged posted prices for crude oil resulted in an increasing proportion of the companies' integrated profits going to the governments in taxes. In consequence, the companies reduced the posted prices in 1959 and again in 1960. Although these cuts were partially restored, it was too late to avoid a hostile response; the formation of the Organization of Petroleum Exporting Countries was the immediate reaction of the exporting countries.

THE PRODUCING COMPANIES ORGANIZE TO COOPERATE

The formation of OPEC in 1960 marked the beginning of the gradual assumption of control by the producing countries of their own industry, a process which was to be completed in 1974. The first success of OPEC was to prevent the companies from making any further cuts in posted prices, even though market prices continued to fall. OPEC succeeded in obtaining the acceptance by the companies of royalty expensing (which increased revenues of the governments by roughly half the amount of the royalties), the reduction and final elimination of marketing allowances for tax purposes, and the relinquishment of unexploited territory (which made possible the entry of new companies and the receipt of higher revenues and improved concession terms from them). Most of these gains were obtained without much difficulty, for the governments had a strong case which the companies recognized. But perhaps even more important were the stirrings of uneasy recognition by the companies that the governments of the OPEC countries would become increasingly powerful and pressing and that some flexibility in dealing with them might therefore be wise. Even so, there was considerable disagreement among the major companies over the most appropriate course of action.

OPEC continued to develop a common oil policy which would assist cooperation in collective action, and at its sixteenth conference in June 1968 the organization adopted a general "Declaration of Petroleum Policy." This document proclaimed the "permanent and inalienable sovereignty" of the governments over their natural resources and asserted their right to obtain the greatest possible benefit from them. It also proclaimed the "doctrine of changing circumstances" which justified unilaterally enforced changes in contractual relations with foreign concessionaires when it was in the country's interest to make such changes. The producing countries' objectives were clearly stated: direct exploitation by the state through contracts which ensured maximum control for the governments, "reasonable" equity for the state in oil operations, progressive and accelerated relinquishment by the companies of concession territory not being exploited, and determination by the governments of tax reference prices.

Although some progress was being made toward a number of these goals in some countries, the companies were not particularly worried by the declaration. They could see little indication that any individual country was in a position to make sweeping gains of this sort without paying a cost which would in the end prove unacceptable. Nor did they think that the OPEC countries as a group were likely soon to achieve a degree of unity sufficient to enforce such demands. On the other hand, they had the example of the determination and insistence of Iraq throughout the decade and were soon to face even more effective action in Libya.

CONFLICT IN IRAQ

During most of the late 1950s the Iraqi government had been pressing for modifications of its agreement with the IPC group. As early as 1956 the minister of national economy had objected to unilateral cuts in posted prices by the companies for crude oil shipped from Basrah and in particular to some of IPC's accounting practices and to the slow rise in Iraqi output after 1955. The government continued to press the companies for improvements in the terms on which they operated, and a few days before the overthrow of the monarchy in July 1958 the companies finally agreed to a substantial increase in output and to relinquishment of some of the concession territory, but by then it was too late.

One of the first acts of General 'Abd al-Karim Qasim's new republican regime was to confront the companies not only with long-standing demands relating to various accounting matters but with new and more far-reaching demands as well. As time went on, the negotiations were accompanied by increasing publicity and popular agitation in which the cry of "nationalization" was frequently expressed. Even while the negotiations were in process, the companies, with extraordinary insensitivity, unilaterally reduced posted prices and expressed their disapproval of the imposition by the government of cargo dues (essentially an export tax) on oil at Basrah by reducing shipments. These acts further embittered relations.

Iraqi demands included an improved share of profits, equity participation in the company, relinquishment of concessionary rights in nonproducing areas, and acceptance of government rights to natural gas. The government insisted on a "package" settlement, and it seems clear that one of their objectives was to bring about substantial changes in the pattern of Middle Eastern oil concessions. In October 1961, the negotiations broke down, and in December the government expropriated over 99 percent of IPC's concession area, leaving the company with only those areas then in production.

The negotiations were doomed to failure, not only because the government was striking at the very heart of the arrangements governing the established operations of the major companies, but also because it was acting alone in so doing and the companies could for the moment risk the loss of Iraqi oil. Nevertheless, within the next few years provisions for equity participation, an increased share of profits, and automatic relinquishment of unexploited concession areas were to become almost standard in new concession agreements and were even incorporated in modifications of existing agreements—not only in Iraq but elsewhere in the Middle East. The companies had been willing to go a considerable distance to meet Iraqi demands but not on the issues which the regime in Baghdad considered the most essential. With the unilateral expropriation of IPC concession rights by Law 80 of 1961, the companies felt compelled to stand on their legal and moral position and refused to "accept" the law, demanding arbitration instead (as was their legal right).

By 1965, however, events throughout the Middle East had gone far in changing the older pattern of concession agreements (as well as in changing the governments of Iraq), and it may have seemed to the companies that more might be gained by forgetting grievances and accepting unpalatable terms than by further resistance to expropriation. Furthermore, they very much wanted to get back the North Rumaila field in southern Iraq. In any event, early in 1965 an agreement was arrived at between the companies and the government under which the companies to a large extent accepted the terms of Law 80 but were to regain North Rumaila. With the exception of Jersey Standard, they also agreed to take up new concessions in the expropriated areas in partnership with the government. This was a very great departure for the major companies. It might well have led to far-reaching changes in concession terms throughout the Middle East much earlier than in fact occurred since most concession agreements contain a "most favored nation" clause under which better terms offered by the companies in one country can be claimed by others.

At the time of the Arab-Israeli war in 1967, however, the signing and publication of the agreement had not yet taken place because of difficulties more closely related to Middle Eastern and Iraqi politics than to the companies specifically. After the war the government abandoned all attempts to come to an agreement with the IPC and passed a further law which reserved all the unexploited areas of Iraq for exploitation by the state-owned Iraq National Oil Company (INOC). The state company was empowered to enter into arrangements with foreign companies for the exploitation of reserves.

To develop these areas the Iraqis turned to the French, whose attitudes during the June 1967 war had been more sympathetic to the Arabs than those of the British and American governments, and to the Russians. North Rumaila, however, was reserved exclusively for INOC. By the end of 1967 the Soviet Union had agreed to cooperate in an extensive development program for Iraqi oil, and by the summer of 1969 two agreements had been negotiated. One provided for a large Soviet loan, which was to be repaid in crude oil beginning in 1973; another provided that the Soviet Machino-Export Organization would supply machinery, equipment, and technical assistance to Iraq on a barter basis. Over the next four years Iraqi oil was to be developed under government control with East European assistance and financed largely by claims on Iraq's future crude oil output.

In spite of a variety of attempts, Iraq and the IPC were unable to resolve their differences. As a result, in June 1972 the government nationalized the IPC proper in the north but left the company's affiliate in Basrah untouched. Though negotiations over compensation were prolonged, in March 1973 an overall settlement was reached which ended nearly twelve years of bitter conflict between the companies and the government.

LIBYA TAKES THE LEAD

In the meantime very great changes had taken place in the relationship between the companies and the other exporting countries in the Middle East. By the time of the Iraqi nationalization, it had become clear that other governments could also control or nationalize the companies operating in their countries if they so desired and that they could now do so without the prolonged cost which Iraq had had to bear. In particular, this had been demonstrated by developments in Libya, where in less than twelve months the government had destroyed the principle of equality in profit sharing, even in concessions with the major companies, and had effectively asserted its control over posted prices by forcing the companies to accept substantial increases against their will.

The development of Libyan oil had followed close on that of Algeria, where exports on a significant scale began in 1960. But whereas Algerian oil had been largely in the hands of French companies and was destined primarily for the French market, Libyan oil was in the hands of international companies, among which American independents were predominant. Thus, the major companies were not in a position to control the development of oil in Libya as they had elsewhere in the Middle East.

Exports began in 1961 and production rose steadily, exceeding that of Kuwait by 1968 and reaching a peak in 1970, after which it declined as a result of government restrictions. These were adopted partly to put pressure on the companies during the 1970 negotiations and partly because conservation measures were, with reason, thought to be desirable for some of the fields.

The first Libyan oil law was drawn up in 1955 in collaboration with the interested companies, and its terms were extremely favorable to them. In particular, taxes were based on prices actually realized from a company's sales instead of on the higher posted prices as elsewhere in the Middle East. It may be that such attractive terms were necessary at first to encourage rapid development, but as the extent and nature of the country's oil resources became clearer, the government imposed increasingly stringent terms, which, by and large, were readily accepted by the companies until 1965. The Libyan government then decided to impose taxes on the basis of posted prices in accordance with OPEC principles.

The American independents—notably Continental Oil and Marathon, which with Amerada formed the Oasis group and accounted for about half of Libya's 1964 output—had been making inroads into European oil markets largely by price competition. Higher taxes per barrel, therefore, would be particularly damaging for them, and they resisted strongly. The major companies, however, quickly accepted (and probably even encouraged) the proposal, in spite of the fact that it was a violation of the concession terms, because it could be expected to help contain the price competition of the independents by making price reductions more costly for them. The independents were defeated in the end.

Nevertheless, Libya's oil was still very attractive to oil companies, and in 1966 another U.S. independent, Occidental Oil, obtained a concession. Occidental was very successful in discoveries and by 1969 was the largest single producer in Libya, accounting for one-fifth of Libya's exports that year. For Occidental the retention of Libyan oil was of great importance, leaving her extremely vulnerable to government pressure.

Very soon after forcing the companies to post their prices for tax purposes, the government concluded that the postings were far too low in comparison with those of other crude oils in the Middle East and North Africa. The closure of the Suez Canal after the war with Israel in June 1967, and the sharp increase in freight rates from the Persian Gulf countries that followed, intensified Libya's determination to raise posted prices and, because her proximity to Europe gave her an increased advantage over the sellers from the Gulf, strengthened her bargaining position. In addition, the government became concerned about the rate at which some companies, notably Occidental, were depleting oil reserves, and in 1968 it imposed curbs on output in one of the fields. Also in 1968, a national oil company was established, and the government announced that no more "old-style" concessions would be granted. Henceforth foreign companies would be required to enter into joint ventures with the national oil company, although existing concessions could continue.

In September 1969 came the military coup. The new government, headed by a Revolutionary Command Council (RCC), presided over by Colonel Muammar Qadhdhafi, stated that it would make no spectacular changes in the existing arrangements with the companies but would impose stricter controls on the industry as a whole. The RCC continued to press the demands of the previous regime, especially those regarding the increase in posted prices. It was soon clear that both external and internal developments had given this relative newcomer to the international oil industry both the opportunity and the will to impose severe terms on its concessionaires.

At the beginning of 1970 the government approached the companies with demands to raise both tax rates and posted prices. The companies refused to accept the government's case for substantial changes but did offer a small increase in posted prices. The major companies, with extensive oil production elsewhere in the Middle East, were very much concerned to keep the settlement with Libya as moderate as possible, for they knew that any gains for Libya would have to be offered almost immediately to the other countries. The independents, on the other hand, were not burdened by such considerations, but because they had little oil elsewhere, they were extremely vulnerable to Libyan pressure.

The negotiations lasted for nine months, with the government insisting on its very stiff demands, threatening nationalization and forcing reductions in output as a means of exerting pressure on the companies. Close cooperation was maintained with Algeria, which not only unilaterally raised its own tax (posted) prices but also took over some of the foreign companies operating in the country. Algeria's action and her support of Libya gave added credence to the latter's threats. By September, Occidental had decided that the time had come to capitulate. It accepted an increase in the tax price of oil of 30 cents a barrel, with annual escalations for five years thereafter and an increase in the tax rate from 50 percent to 58 percent. Other companies had little choice but to follow. Settlements with the different companies differed in important ways, but all had to accept increased tax rates and tax prices.

Libya had decisively won the day. It is not yet clear why the companies backed away from an outright confrontation. Libya had the encouragement of OPEC countries, but it was impossible to know what steps they would in fact have taken in her support. There were supply shortages in the Mediterranean, and prices were rising, so that further reductions in deliveries would have been uncomfortable, especially for Europe, and a complete stoppage of Libyan oil could have had serious consequences. It has been stated that the United States feared that Libya would nationalize the American companies if they resisted too long and that European companies would then take over to ensure their own supplies.[4] One oil company executive has testified that the interest of the United States in gas imports from Algeria inhibited it from antagonizing either country.[5] It is possible that some U.S. authorities believed that higher oil prices were in the country's interest in view of its declining production capacity and its need for greater investment in exploration and development. Finally, there is the very simple explanation that Libya held all the trump cards: her financial position would have enabled her to do without further revenues for a period long enough to disrupt Europe seriously without herself suffering the eco-

4. James E. Akins, "The Oil Crisis: This Time the Wolf Is Here," *Foreign Affairs,* April 1973.

5. See G. H. M. Schuler, *Petroleum Intelligence Weekly,* 22 April 1974, Suppl., p. 4.

nomic or political consequences that had befallen Mussadiq in Iran in the early 1950s, whereas the companies were divided and vulnerable.

It is difficult to know what weight to give these various explanations, but it does seem likely that by this time the United States was beginning to be aware of the extent of its dependence on Middle East and North African oil and of the importance of establishing firmer and more satisfactory relations with these countries.

As expected, both companies and governments immediately moved to generalize the terms. Algeria, which had been advising Libya, claimed that parity required a substantial increase in the posted prices of her crudes. The companies offered 55 percent tax rates to the gulf states and to Nigeria and raised the posted prices for Iraqi crude at Mediterranean terminals and also the posted prices for Iranian and Kuwaiti heavy crudes. Similar offers were made to Saudi Arabia and Abu Dhabi.

These countries were not, however, satisfied with these offers and demanded greater increases, but before the issues could be resolved the Twenty-first OPEC Conference was held in Caracas. At this conference the OPEC countries laid down the terms that they would accept: a minimum 55 percent tax rate, uniform increases in tax prices in accordance with market conditions, and the elimination of disparities in posted prices. A rigid and speedy timetable was drawn up to govern negotiations, and it was backed by threats of common action to ensure compliance of the companies.

The companies, which formed their own common negotiating group, tried unsuccessfully to get OPEC to negotiate a settlement for *all* of the OPEC member countries together, for they were afraid of "leapfrogging"—if a settlement were made for individual countries, or a group of countries, they feared that it would encourage further demands from others, setting in motion a process whereby countries continually jumped over each other's backs in a series of escalating demands. The producing countries were adamant in their demands and very militant, threatening to deny oil to companies which refused to accept their terms. In the end the governments obtained nearly all they asked for, but the companies did succeed in getting a five-year agreement (reluctantly agreed to even by the Libyans), which they had very much wanted in the hope that they could look forward to some stability in prices.

PARTICIPATION AND AFTER, 1972–73

Any expectation that the Tehran and Tripoli agreements would bring some respite from government pressure on the companies was soon dispelled. OPEC took up at once, and in earnest, the old issue of equity participation for the governments in the oil-producing affiliates of the major companies operating in the Middle East. As noted above, the OPEC Declaration on Petroleum Policy in 1968 had clearly proclaimed this as an important objective. In June 1968 the oil minister of Saudi Arabia underlined the demand in a major speech at a Beirut seminar. In September 1971, with the price problem temporarily settled, talks led by Saudi Arabia on behalf of five Gulf states were opened with the companies on the question of participation, and by the following March the companies had accepted the objective in principle. A draft General Agreement on Participation was worked out by October 1972, which provided for an immediate 25 percent share for the governments, rising to 51 percent by 1983, for tax prices rising as envisaged by the Tehran agreement, or perhaps slightly faster, and for arrangements for the pricing of the "participation crude" to which the governments would be entitled by virtue of their ownership shares but which they would sell back to the companies for some time.

The General Agreement on Participation provided the framework within which individual countries were expected to negotiate detailed agreements with their own concessionaires. Saudi Arabia and Abu Dhabi signed it in December with only minor modification, but Iraq would have nothing to do with it and it was rejected by Kuwait's legislature. Nevertheless, this agreement, together with the Tehran and Tripoli price agreements, might have provided the foundation for future peaceful collaboration between the companies and governments for several years to come, had it not been for the high rates of world inflation, the instability of monetary parities, and the very large increases in the market prices of crude oil, primarily resulting from the high bids of American independent companies. In addition, more and more responsible people in the Arab oil-exporting countries were beginning to realize that the changed circumstances increased the likelihood that they would be able to use their growing control over oil to induce consuming countries, and the United States in particular, to bring pressure on Israel to accede to the terms of UN Resolution 242 and to give up the territories she conquered in 1967.

Special modifications were made in 1972 and early 1973 in the Tehran agreement to take account of the devaluation of the U.S. dollar, but these did not deal with the continued rise in the market price of oil, which was rapidly undermining and destroying the primary object of the agreement. The purpose of the Tehran and Tripoli price agreements had been to raise the governments' share of oil revenues through an increase in tax rates and tax prices and to establish a reasonably stable basis for sharing profits. For this purpose it was necessary that tax prices should remain above market prices. Between the spring of 1970 and the summer of 1973, the market price of crude oil measured in dollars more than doubled, but the fiscal income of the host governments accounted for only about half of the increase. The rest was due to market pressures, as U.S. companies, freed from import quotas and controls, joined other users of crude oil in the rush to a market made even tighter by the reluctance of some major companies, uncertain of their own supplies, to sell their own crude to third parties.

As market prices rose to exceed posted prices, the share of the governments in total revenues did not reflect the rise in market prices. It was estimated that the "national profit split on realized prices between governments and companies may have changed from 80–20 in the governments' favor at the time of the Tehran agreement to around 64–36" by the summer of 1973.[6] This outcome had not been anticipated by either companies or governments, and it was clear that as prices continued to rise the agreements would work increasingly to the disadvantage of the exporting countries. There were several attempts in OPEC conferences to abrogate the agreements, but the Iranian and Saudi Arabian governments resisted. These governments insisted that they did not want unilaterally to go back on what they had agreed. As time went on, however, the assumptions about prices and monetary parities which *both* they and the companies had held when the agreements were made became increasingly unrealistic. The competitive advantage of the major companies over the independent companies grew, since the former had access to crude oil on Tehran terms.[7] At the same time the opportunities open to the governments to take advantage of market conditions were restricted by their price agreements.[8]

Similarly, the "buy-back" provisions of the participation agreements, which enabled the companies to buy back a large proportion of the oil to which governments were entitled by virtue of their equity holdings, came under strain as the governments received offers to sell their oil at prices above the buy-back prices agreed upon. Kuwait, whose national assembly had never ratified the participation agreement approved by the government, demanded a greater equity share; Libya forced the independent companies to give up a 51 percent share to the government on very onerous terms and nationalized 51 percent of the interests of the major companies. Saudi Arabia began discussions with Aramco to revise buy-back prices and to increase the amount of crude available to the government. Negotiations over the revision of both the price and the participation agreements were in process when the Syrians and Egyptians attacked the Israeli forces occupying their lands on 6 October 1973.

PRICES AND THE OCTOBER WAR

Arab spokesmen, and particularly King Faisal, had increasingly warned the United States during the spring and summer of 1973 that if the United

6. *Middle East Economic Survey,* 21 September 1973, p. 2.

7. *Petroleum Intelligence Weekly (PIW)* calculated in August 1973 that the cost advantage of the majors over the independents with Saudi Arabian crude oil was some 93 cents a barrel (*PIW*, 20 August 1973).

8. *PIW* also calculated that the real cost of importing oil for several major European countries and Japan was still in August 1973 below 1957 levels. For countries whose currencies had depreciated, however, this was not true (*PIW*, 24 September 1973).

States did not change its position toward the Arab-Israeli conflict Saudi Arabia would be forced to adopt a harder position. On the outbreak of war, in response to a cable from the U.S. secretary of state, Dr. Henry Kissinger, asking him to use his good offices to bring an end to the fighting, King Faisal replied that the United States should move to contain Israeli aggression. As the fighting intensified during the first week, he sent a message warning President Nixon of the gravity of the situation, but there was no satisfactory response, and on 17 October the Arab countries instigated their oil restrictions. This naturally very much intensified the shortage in the market and increased the pressures on market prices, especially for oil which was not subject to destination restrictions.

So far as posted prices were concerned, the governments abandoned negotiations and simply announced their new prices. These were about the same as the ones presented to the companies during the negotiations which had broken down on the 16th of October just before the imposition of the embargo. From this time on the governments would determine both prices and output; the transition from company to government control was complete.

Nevertheless, the October War not only accelerated the transition but produced conditions which brought about such a very large and abrupt increase in prices that a "crisis" was created for the importing countries. In January 1974, posted prices were again sharply increased by OPEC to over $11 a barrel, largely at the instigation of the Shah of Iran, who was impressed by the very high prices the American companies were prepared to bid to obtain oil. There is little reason to think that prices would have risen to this extent until the end of the decade, if then, had the war, with its accompanying embargo and restrictions, not taken place. There is no underlying shortage of oil resources in the world, although there is at present a near monopoly of sellers in the world market. As early as the middle of 1974, the underlying surplus was beginning to make itself felt, partly as a result of increased production capacity and partly as a result of reduced demand in response to the high prices.

As a result, the pressure on market prices disappeared, and the governments began to find difficulty in selling their oil in world markets at the prices they had decided upon. The major companies, with their cheaper buy-back, or equity, oil could undercut the prices that the governments tried to maintain in world markets. To remedy this situation, the governments in the last half of 1974 moved first to raise the cost of oil to the companies and introduce a unified price system and then to take them over entirely. The companies will soon be simply operators in the Middle East and refiners and marketers elsewhere. Their unrivalled expertise will be used to find and produce oil for government account; in the Middle East the government companies will soon be the only sellers of their own crude.

CONCLUSION

In this paper I have tried to lay out an important part of the framework against which one must appraise the development of the oil industry in the Middle East and the role of the international companies in it. The history of the oil industry is primarily the history of a very few international capitalist firms (each, of course, with its own nationality) possessing the finance and expertise required to develop one of the world's most essential industries on a global scale. These few companies sometimes operated cooperatively, sometimes collusively, and often as rivals; but collectively they long successfully resisted the incursions of newcomers and obtained very extensive control over the industry and over the oil resources of the nonindustrialized countries. In the early days, the host countries received a very small share of the rewards, but their share grew rapidly as they became increasingly able to assert their views and as the economic value of their resources underlined the importance to the companies of keeping them reasonably content with their foreign concessionaires. As the exporting countries' share of the industry's financial prosperity rose, they increasingly used their revenues to promote their own economic and political advancement, with the result that within a few decades after the development of their resources they were able to take control of the industry from the companies.

For a while before World War II, the great companies joined in a straightforward cartel aimed at controlling output, maintaining prices, and sharing world markets among themselves. After the war they joined in consortia for the same purposes. These arrangements to a considerable extent made possible the maintenance of a post war order similar to that before the war. But the effectiveness of the companies' control did not last long in the face of their own rivalry, the entry of newcomers, and the pressures of their host governments. When conditions made it possible for the governments of the oil-producing countries to assert their power, at first tentatively and then in full vigor, a structure of prices and marketing arrangements ideally suited for their purposes had already been created by the companies. Moreover, the governments could now call on the experience and expertise of the international companies to administer the policies which the governments determined.

While the financial resources made available by the very success of the oil companies were bringing about the development of the economic capacities and political strength of the oil-exporting countries, the counterpart of these resources—the oil supplied to the industrialized countries—was bringing about the development of vast and intricate economic structures and consumer habits which depended almost entirely on oil, and increasingly on oil from these very countries. The extent of its dependence on Middle East oil was realized by the United States only in the early 1970s, at almost exactly the same time as the exporting countries had been successfully testing their new economic power and finding it good. But even then there might have been a reasonable basis for cooperation had it not been for a crisis in the long course of events surrounding the establishment and expansion of an Israel in Palestine against bitter Arab opposition. A serious oil "crisis" was the immediate consequence of the subsequent conflict, as was also an acceleration of the change, inevitable in any case, of the position of the oil companies in the Middle East.

Producer Cooperation and Organization

The most important reason for the shift in the control of the Middle East's oil from the Western countries and companies to the producing countries undoubtedly was the banding together of the major producers in 1960 to form the Organization of Petroleum Exporting Countries (OPEC). This group's rise to center stage in world economic and political issues and the emergence of the oil price revolution form the material of the next chapter. But despite near universal recognition of OPEC's name, there is remarkably little understanding of what it comprises and who constitutes its membership. All too often, the prevalent image has been that of a sinister conspiracy by greedy "oil sheiks" to take over the world. This is the stuff of fantasies but reflects popular oversimplification and distortion of complex issues.

OPEC's thirteen members comprise a widely divergent set of nations which essentially have only two things in common: exports of oil and membership in the group of developing nations. This makes it difficult to regard OPEC as a single, cohesive unit. Its organization may be divided into four main features:

1. OPEC is an intergovernmental organization restricted exclusively to countries heavily dependent on oil exports for foreign exchange and economic development;
2. membership is in practice limited to developing countries;
3. decisions are reached by a ministerial conference, require a unanimous vote, and are subject to the approval of all member governments;
4. political matters are at least formally outside the scope and purview of the organization.[1]

1. Zuhayr Mikdashi, "The OPEC Process," in Raymond Vernon, ed., *The Oil Crisis* (New York: W. W. Norton, 1976), p. 206.

OPEC's one significant area of coordination has been in setting posted prices for its crude oil, but even here deep differences continue to exist. There are no legal obligations for members to abide by mutually agreed prices: a two-tiered pricing system resulted in 1976 from a basic deadlock between opposing OPEC states and unofficial discounting was prevalent in the early 1980s. (Appendix C contains a chronology of OPEC price decisions and related matters.) The frequent characterization of OPEC as a cartel ignores the fact that its members only attempt to reach agreement on prices and have not found success in regulating production or assigning individual markets. As an organization, OPEC is no more an authoritative body for its member states than the International Energy Agency is for the consuming countries.

While it is often difficult to isolate the economic aspects of international oil decisions, OPEC has consciously attempted to keep away from political matters. It was not responsible for the cutbacks in oil supply and the embargo of several countries in 1973–1974; these were decisions made by individual members. In large part, this reluctance to delve into politics is only pragmatic: there is little common basis for agreement among the thirteen members. Rather, the predominant impression is one of differences. Most obviously, the OPEC countries are scattered around the world. Despite popular misconceptions, only eight of the thirteen are located in the Middle East and only seven are Arab. The Arab members are Algeria, Iraq, Kuwait, Libya, Qatar, Saudi Arabia and the United Arab Emirates; Iran is the other regional member. Elsewhere, Asia is represented by Indonesia, Africa by Nigeria and Gabon, and South America by Venezuela and Ecuador.

The differences are not just geographical, or even cultural. There are both republics and hereditary monarchies. Politics range from the conservative to the radical; economic organization from *laissez-faire* to socialist. While all the members claim non-aligned status and are aggressive champions of the Third World's cause, some exhibit close ties to the United States and poor or non-existent relations with the Soviet Union; others display the opposite imbalance.

Even though all are developing countries, development requirements differ greatly from country to country. There are wide variances in sizes of countries and their populations, extent of non-oil resources and, most importantly, levels of development. This results in different absorptive capacities: some countries receive more income from their oil than they can effectively use at present; others do not have enough current income and must borrow funds to carry out development plans. Consequently, OPEC meetings have often generated two opposing camps: members with balance-of-payments deficits form the "hawks" and press for higher prices to pay their bills, while capital-surplus countries are generally among the "doves," both in terms of their stance on oil prices and levels of production.

These kinds of differences between OPEC states in general are also to be found among the Middle Eastern members, as will be shown in the chapters of Part II. Disagreement on political, as well as economic, matters all too often seems to be the norm, even when the only non-Arab country, Iran, is excluded. Under the Shah, despite its status as a capital-surplus producer, Iran was often a hawk on pricing matters, in opposition to its Arab neighbors; sharp differences have continued into the era of the Islamic Republic of Iran.

The appearance of the oil glut at the beginning of the 1980s only compounded OPEC's problems in maintaining cohesiveness. Even as the established price of $34 a barrel was routinely ignored by some members, OPEC's share of world production plummeted from 51% in 1977 to less than a third in early 1983. (Libyan production had been particularly hard hit: from 1.7 million barrels per day [mbd] in late 1980 to 660,000 barrels per day in early 1982.) As a result of sharply falling income, most OPEC states eventually found it necessary to resort to unofficial discounting in order to compete with cheaper non-OPEC suppliers. For the first time, unsuccessful attempts were made within OPEC to establish production quotas to put an end to fratricidal price-gouging. Iran, trapped in a long and expensive war with Iraq, was only the most flagrant of a long list of quota violators. Saudi Arabia, which had usually been able to play a pivotal role in mediating between OPEC factions because of its large amount of surplus production, found its output falling from the 11 mbd or more of the late 1970s to 5.5 mbd in mid-1982 and even 4 mbd in early 1983. Strong political differences between the two countries worsened their rift within OPEC, as the Iranians demanded that the Saudis lower production even more and the Saudis responded with a threat to lower their prices and flood the market—an extreme action that could spell the end of OPEC. The failure of the December 1982 OPEC meeting to resolve the dilemma and the obvious fragility of the March 1983 price-cut agreement seemed to mean that the best hope for a solution lay with an anticipated upswing in worldwide demand sometime in 1983.

The two selections in this chapter chronicle the efforts of oil-exporting states to formalize cooperation among producers by establishing permanent organizations. M. A. Ajomo examines the background to OPEC and discusses its structure and institutions in some detail. He also includes a study of the organization's international legal status and attempts to evaluate its achievements. Ali Attiga's contribution is somewhat more narrowly focused on the Arab producers and the formation of the Organization of Arab Petroleum Exporting Countries (OAPEC), whose ranks include Bahrain, Egypt (expelled in 1979) and Syria, in addition to the Arab members of OPEC. To some extent, OAPEC's role has been taken up by the Gulf Cooperation Council (GCC), founded in 1981. However, while the GCC does work to increase economic cooperation and regional development planning among its six members—Kuwait, Saudi Arabia, Bahrain, Qatar, the United Arab Emirates and Oman—its purpose goes far beyond that of OAPEC in attempting to coordinate political policies and security goals. (A more detailed discussion of the GCC is contained in Valerie Yorke's contribution to Chapter 9.)

Bibliographic Sources for this Chapter:
6, 7, 9, 15, 19, 29, 57, 90, 116, 138, 145, 189H, 221, 238, 251, 272, 309D, 347, 371, 374L, 416.

An Appraisal of the Organization of the Petroleum Exporting Countries (OPEC)

M. A. AJOMO

INTRODUCTION

The oil industry has developed from humble beginnings. From the 1859 modest discovery at Drake's well in Titusville,[1] it has grown to become the main industry of the twentieth century. Its fortune has fluctuated over the decades. It has brought many nations, great and small, from obscurity to sudden affluence. Although it has caused revolutions and wars, it has resulted in compromise of principles and ideologies. Oil is a raw material subject to individual as well as state ownership, yet it has international repercussions. Its use has such versatility that all efforts to find a single substitute for it have failed. It is no surprise, therefore, that its utilization and control should be the subject of the formation of an intergovernmental organization.

The Organization of the Petroleum Exporting Countries (OPEC) did not originate the first concerted action to control oil supplies entering into world trade. Some degree of marketing cooperation between major companies, American and British, existed up to the end of World War II, though it was not entirely effective. Public outcry against such restrictive agreements was a major factor contributing to the abandonment of such efforts.

It is significant to note that Britain and the United States, during World War II, went as far as signing an agreement on oil which other states would have been invited to adhere.[2] This, it was felt by some oilmen, would have perpetuated and brought under the control of governments both the "Red Line" arrangements about joint concession seeking in the Middle East and the "As Is" agreements to moderate commercial competition in the inter-

1. J. E. Hartshorn, *Oil Companies and Governments* 299 (1962).
2. Tugendhat, *Oil, The Biggest Business* 9 (1968).

Reprinted by permission of *Texas International Law Journal,* Vol. 13, No. 11, 1977, pp. 11-31.

national sphere between the major companies.[3] That also did not materialize. Nevertheless, a search for some workable arrangement continued. In August 1944 Britain and the United States eventually reached agreement to promote an international pact that, among other things, would have provided for a regime to which both producers and consumers could be a party and which was to ensure a just and equitable exploitation and distribution of oil.[4] Oil producers within the United States were very suspicious of this move; it failed to get the support of the United States Senate and so the move again died naturally.[5]

In the following decade, there were other proposals put forward by Western countries for the "international control of oil resources" or for the "regional development of the Middle East." The Italian proposals for a financial pool for the economic development of the area and President Eisenhower's proposal for an Arab Development Institution are examples of such efforts. The latter was even discussed in the United Nations, but nothing came of it.[6] The Arab countries were not enamoured by the idea of an "oil institution" sponsored by the West. By 1948 the Arabs had come to realize the potential of oil as a weapon of sanction when the Arab League, during the Palestine War, prevailed upon the government of Iraq to force the suspension of shipments through the pipeline to the Haifa refinery at the time, and later, over several years, put pressure on oil companies marketing in Israel to withdraw.[6a] The first real attempt at the formation of a joint body to control the supply of oil came in 1951 from the Arabs. In that year the League of Arab States established an Oil Experts Committee. In 1954 the League set up a permanent Petroleum Department within the League's Secretariat.[7] All these moves were aimed at forging a unified Arab oil policy to emphasize that Arab oil belonged to the Arab nation as a whole.[8] The Arabs were aware that such an association could only be effective if it included Arab as well as non-Arab large exporters of petroleum, notably Iran and Venezuela. The Petroleum Department of the League of Arab States organized the first Arab Petroleum Congress in Cairo in April 1959, and invited the representatives of Iran and Venezuela as observers.[9] Thus until the late 1950's, oil activities outside the technologically developed countries gained prominence only in the Middle

3. The Red Line Agreement was the most famous example of an arrangement to curtail competition ever made in the international oil industry. It was the basis of an undertaking by shareholders in the Iraq Petroleum Company not to compete with each other for concessions in the area of the late Ottoman (Turkish) Empire.

4. J. E. Hartshorn, *supra* note 1, at 301.

5. *Id.* at 300.

6. *Id.* at 301.

6(a). See Shihata, *The Case for the Arab Oil Embargo,* Beirut, 1975. Also Shihata, "Destination Embargo of Arab Oil: Its Legality under International Law" 68 AJIL 591 (1974).

7. J. E. Hartshorn, *supra* note 1, at 301; *see also* F. Rouhani, *A History of OPEC* 75 (1971).

8. *See* D. Hirst, *Oil and Public Opinion in the Middle East* 106 (1966).

9. *See* F. Rouhani, *supra* note 7, at 76.

Eastern countries of Iran, Iraq, Kuwait and Saudi Arabia, and in the South American country of Venezuela.

The international oil companies dominating the industry at that time were referred to as the "seven sisters."[10] These companies were organized into several consortia in the different countries in which they operated. All the producing countries which initiated the idea of OPEC depended very heavily on revenues from petroleum. In some cases, as much as 90% or more of governmental revenue was derived from crude oil export.[10a] Therefore, a country's economic well-being depended almost entirely on the whims and caprices of whichever of the members of the strong international oil cartel operated within its territory. The situation was much more serious than can be imagined because these producing countries had no influence whatsoever on the conduct of operations or in the determination of the prices paid for the oil. Consequently, governmental revenues were determined more or less unilaterally by the oil companies; any attempt by any producing country to interfere was considered an affront and was ruthlessly dismissed. In the absence of a unified resistance by the producing countries, the oil companies had no trouble manipulating production exports and governmental revenue from oil or applying other diabolical devices to undermine the sovereignty of the producing countries.

Although an appreciation of this fact had for some time, even as far back as 1949, induced contact for solidarity among some oil producing countries, these contacts did not assume any appreciable importance until 1959. It was in that year that the foundations of the economy of the major oil producing countries were shaken by the unilateral action taken by the oil companies to reduce "posted prices."[11] This was done successively in 1959, and again in 1960.[12] This was the last straw, and like the wage earners of old, the exporting countries realized that they had to hang together or be hanged separately.[13] On September 14, 1960, in response to the unilateral reduction of the posted price of crude oil, five of the major world oil exporting countries, Venezuela, Iran, Iraq, Kuwait and Saudi Arabia, met in Baghdad and formed OPEC, which has since evolved into a permanent, but loosely organized, producers' intergovernmental organization which coor-

10. The seven sisters were Esso, Gulf, Texaco, Mobil, Socal, British Petroleum and Royal Dutch Shell. These companies, five American, one British and one Anglo-Dutch, have continued to dominate the oil world. *See generally* A. Sampson, *The Seven Sisters* (1975).

10(a). In the case of most Middle Eastern countries and Venezuela the figure ranges from 91% in the case of Algeria to 99% in the case of Saudi Arabia (See *OPEC, Facts and Figures,* Vienna, 1983, p. 16).

11. *See* Shihata, *Arab Oil Policies and the New International Economic Order,* 16 Va. J. Int'l L. 261, 269 (1976).

12. This action cost the producer countries a $93 million loss for the year 1960 alone. Compare this with $59.3 billion surplus in 1974. 7 OPEC Weekly Bull. No. 4, at 1 (Feb. 2, 1976).

13. *See* Frankel, *The Current State of World Oil,* Middle East Economic Survey Supp. 7 (Sept. 6, 1968).

dinates members' oil policy.[14] The primary motive that led to the creation of OPEC was the desire to restore crude oil prices to their pre-1959 level.[15] OPEC was the first institution created on the international level aimed at safeguarding and defending the common petroleum interests of underdeveloped producer countries. Saudi Arabia and Iraq were the two Arab countries which played the most prominent role in developing an awareness of oil affairs which led to the formation of OPEC. Unlike most political/economic transformations, OPEC began almost unnoticed in the world at large. Since then a new era of constructive cooperation has prevailed among the oil-producing countries.

A development concurrent with the establishment of OPEC was the encouragement given by the United Nations to the newly developed nations through several resolutions and declarations regarding states' sovereignty over natural resources.[16] Encouraged by the support for anti-colonialism in the councils of the United Nations,[17] oil producing countries saw an opportunity to divest international or multinational corporations operating in host countries of the Third World of some of their wide powers. These companies usually had more assets than the states in which they operated and, by their sheer size and power, had been able on some occasions to influence political events in many host countries. The host countries rightly viewed the presence of the oil companies as vestiges of colonialism and any opportunity to reduce their influence was readily seized.

In the formative years of OPEC, the five founding members of the Organization accounted for well over 70% of all the crude oil moving into the international market and well over 65% of the world's oil production.[18] In addition, the members had certain common characteristics. They were, for the most part, developing countries who were large exporters of petroleum, dependent upon oil revenues to finance their economic development projects and to balance their budgets, unable to exploit their oil resources without the assistance of foreign technology, weak when compared to the complex industrial societies of the parent countries of the oil companies, sovereign governments and members of the United Nations.[19]

14. *See* Agreement Concerning the Creation of the Organization of the Petroleum Exporting Countries (OPEC), *done* Sept. 14, 1960, 43 U.N.T.S. 247.

15. *Id.* OPEC Res. I.1, para. 2.

16. *See* G.A. Res. 1314, 12 U.N. GAOR, Supp. (No. 18) 27, U.N. Doc. A/4090 (1958), which created a Commission on Permanent Sovereignty over Natural Resources with responsibility for conducting a survey of the status of the permanent sovereignty of peoples and nations over their natural wealth and resources; G.A. Res. 1515, 14 U.N. GAOR, Supp. (No. 16) 9, U.N. Doc. A/4684 (1960), G.A. Res. 1803, 16 U.N. GAOR, Supp. (No. 17) 15, U.N. Doc. A/5217 (1962).

17. *See, e.g.,* G.A. Res. 1514, 14 U.N. GAOR, Supp. (No. 16) 66, U.N. Doc. A/4684 (1960), granting independence to the colonial countries and peoples.

18. *See OPEC and the Oil in the Middle East* (1962). According to data compiled in 1973, OPEC accounted for 56% of the world's oil production and 87% of the world's oil exports. Statistics Unit—OPEC (Jan. 29, 1976); See also *OPEC, Facts and Figures* (Vienna, 1983).

19. *See* F. Rouhani, *supra* note 7, at 82.

The position has not changed materially, even as of today. Since the establishment of OPEC in 1960 the Organization has been strengthened by the admission of eight other countries, of which Nigeria is one. The Organization now has a total membership of thirteen. Arranged in order of 1974 oil earnings, they are the following: Saudi Arabia, Iran, Venezuela, Nigeria, Libya, Kuwait, Iraq, United Arab Emirates,[20] Algeria, Indonesia, Qatar, Gabon, and Ecuador. OPEC thus has all the world's significant crude oil exporters in its fold, except for the US, China, the Soviet Union, Canada, Great Britain and Mexico.

The Eighth Conference (Extraordinary) of OPEC, held in Geneva in 1965, adopted the Statute of the Organization.[21] As an intergovernmental organization OPEC is by definition limited to consultative and advisory functions. This view is reinforced by article 2(A) of the OPEC Statute which states that the principal aim of the Organization shall be the coordination and unification of the petroleum policies of member countries and the determination of the best means for safeguarding their interests.

The original and present objectives of OPEC are the stabilization of crude oil prices, the pursuance of the right to bilateral agreement on prices, the obtaining of equitable prices, and the availability of a forum for regular consultation among its members for coordinating and unifying their policies, and for taking concerted action.

Membership Categories

The OPEC Statute provides for three categories of members: founder members, full members and associate members.[22]

Founder members are those countries which were represented at the First Conference held in Baghdad from September 10–14, 1960, and which signed the original agreement which brought the Organization into being.[23] These countries are Iran, Iraq, Kuwait, Saudi Arabia and Venezuela.

Full members are the founder members as well as those countries whose application for membership has been accepted by the Conference.[24]

Associate membership may be conferred on a net petroleum exporting country which does not qualify for full membership.[25] Gabon became an

20. OPEC approved the transfer of membership of Abu Dhabi to the United Arab Emirates at the Thirty Seventh (Extraordinary) Conference held in Geneva in January, 1974. *See* OPEC Res. XXXVII.167 (Jan. 9, 1974).

21. OPEC Res. VIII.56 (April 1965), *reprinted in* 4 Int'l Legal Materials 1175 (1965) [hereinafter cited as OPEC Statute]. The original text of the Statute was approved by the Conference in Caracas in January, 1961. OPEC Res. II.6 (Jan. 1961). The OPEC instrument was deposited with the Secretariat of the United Nations in November, 1962 in accordance with U.N. Charter art. 102. The 3603-word statute, which contains forty-six articles, came into effect on May 1, 1965. OPEC Statute, art. 46.

22. OPEC Statute, art. 7.

23. *Id.* art. 7(A).

24. *Id.* art. 7(B).

25. *Id.* art. 7(D).

associate member in November 1973 and a full member on June 11, 1975.[26] Associate members may be invited by the Conference to attend meetings of some or all organs of the Organization and to participate in their deliberations, but without the right to vote.[27] However, associate members, like full members, are entitled to all the general facilities of the Secretariat, including its publications and the right to the library in Vienna. With the admission of Gabon as a full member by the Conference at Libreville on June 11, 1975, the Organization now has thirteen full members.[28] There is at present no associate member on the roll of OPEC.

Conditions for Admission as a Full Member

It appears that the two most important conditions which a state must fulfill before it can be considered for admission are that it must be a "substantial" net exporter of crude oil and that it must share fundamental interests with the other member countries of OPEC.[29] If the rule of literal interpretation is applied to the word "substantial" in article 7(C) of the Statute of the Organization, great oil producers such as the United States, the Soviet Union, Canada, Mexico and Argentina would qualify for membership in the Organization. However, the practice of OPEC regarding admission shows that the two requirements must be read conjunctively. Two fundamental interests which the present OPEC members share with each other are that both the financing of their development projects and the balancing of their budgets depend on revenues from oil, and that they were and are still unable to exploit their oil resources without the assistance of foreigners, especially major oil companies.[30] While all OPEC member countries share these interests, the same could not be said of great producers such as the United States, the Soviet Union, Canada and now Great Britain. These are technologically developed countries in their own right and do not require outside help, as a matter of necessity, to exploit their natural resources.

Another condition for admission is that a prospective applicant must be accepted by a majority of three-fourths of full members, including the concurring vote of all the founder members.[31] The implication here is that the five founder members of OPEC enjoy some sort of hegemony within the Organization similar to that enjoyed by the five permanent members in the Security Council of the United Nations. The provision is in conflict with

26. *See* OPEC Press Release No. 4, at 1 (1975).
27. OPEC Statute, art. 7(E).
28. These countries and the dates of their admission are as follows: Iran, Iraq, Kuwait, Saudi Arabia and Venezuela, September 1960; Qatar, January 1961; Indonesia and Libya, June 1962; Abu Dhabi, November 1967; Algeria, Nigeria, and Ecuador, July 1969; Gabon, June 1975. The United Arab Emirates replaced Abu Dhabi in January, 1974.
29. OPEC Statute, arts. 7(C) & 7(D).
30. *See* F. Rouhani, *supra* note 7, at 81.
31. OPEC Statute, art. 7(C).

the principle of equality of states embodied in article 3 of the Statute. It is also in conflict with the principle established for budget appropriations under the financial provisions of the Organization which require apportionment on an equal basis among the members, irrespective of the rank of production.[32] At a time when there is a general outcry against the special powers of veto enjoyed by the five permanent members in the Security Council of the United Nations, this provision is rather difficult to justify. Moreover, while such ranking of states is detested in a political organization like the United Nations, where budget appropriations are based on capacity, it is more difficult to justify in an organization such as OPEC where ranking is not reflected in budget appropriations.

Termination of Membership

It would appear from the wording of article 8 of the Statute that no member may be expelled. There is a provision for withdrawal by a member provided the member gives the required one year's notice and has fulfilled all the financial obligations resulting from its membership.[33] A member who withdraws under article 8 may apply for readmission under the provisions of article 7(C). Since the unanimity rule (of all full members) applies to all decisions of the Conference other than on procedural matters, it would appear that the conditions for admission stipulated in article 7(C) are special, nonprocedural conditions limited specifically to admission only.

Internal Disputes

The absence of a provision for settlement of disputes by, for example, the traditional methods of arbitration, conciliation and judicial settlement, is either deliberate or accidental. Most intergovernmental agencies of this nature make contingency provisions in case of disagreement, and it cannot be said that one cannot envisage a situation which may lead to a disagreement within OPEC. Possibly the framers of the Statute have been overoptimistic about the chances of the Organization. If so, the operation of the Statute and the workings of the Organization have not yet borne out this optimism.

The recent meeting in Bali, Indonesia, which ended without reaching a definite decision on price increase for crude oil, clearly shows that members can and do disagree.[34] Similarly, the events in Qatar have shown that there is a level beyond which non-economic matters should not be allowed to influence oil price determination. Saudi Arabia's attempt to impose its

32. *Id.* art. 38(C).
33. *Id.* art. 8.
34. *See* 7 OPEC Weekly Bull. No. 22, at 2 (June 8, 1976); *see also* Z. Mikdashi, *The Community of Oil Exporting Countries* 34 (1972). For other examples of discord among the OPEC members, *see* Note, *From Concession to Participation, Restricting the Middle East Oil Industry,* 48 N.Y.U. L. Rev. 774, 791–93 (1973).

price decision on the Organization led to a split in OPEC's pricing policy and the emergence of a two-tier pricing system. Under a situation which would have justified a rise of 20%, Saudi Arabia and the United Arab Emirates requested an oil price increase of 5%; the other eleven members increased their oil prices by 10% effective January 1, 1977.[35] In fact, following the recent recession in the West which has affected oil consumption patterns, OPEC, after near collapse, eventually agreed to price cuts and production quota allocations which it appears will hold till 1985. The path to the eventual agreement was fraught with many obstacles which nearly led to the disintegration of the Organization. The lingering question of differential[36] has also led to accusations between members concerning unilateral setting of prices for crude oil.

INSTITUTIONS OF THE ORGANIZATION

The Conference

The Conference is the supreme authority of the Organization,[37] and consists of delegations representing the member countries. Each member country is represented by a delegation of one or more members. If the delegation consists of more than one person then the nominating country appoints one of them as the head of the delegation. The head of the delegation is usually the minister of petroleum of the country. The practice of OPEC has been to limit the number in a member's delegation to a reasonable level.[38] The Conference holds two ordinary meetings in a year.[39] A nonmember country may be invited to send a representative to the Conference as an observer, either at its own request or at the instance of the Organization, if the Board of Governors considers this desirable. Under this rule, Algeria and Nigeria attended Conferences as observers well before their formal admission into OPEC. Similarly, Trinidad and Tobago have on some occasions been granted observer status. A quorum of three-quarters of member countries is necessary for holding the Conference.[40] The President normally holds office for the duration of the meeting. Decisions of the Conference, other than on procedural matters, require the unanimity of all full members.[41] The unanimity requirement enables a

35. *See* Business Times, Lagos, Jan. 4, 1977, at 6.
36. "Differential" is a term used to describe the price attached to different types of oil based on sulphur content, quality, etc. Although the Saudi Arabian light (34° API) crude oil is used as a standard for OPEC crude oils, the Algerian and Nigerian oil are reputed to be the lowest in sulphur content.
37. OPEC Statute, art. 10.
38. In the United Nations General Assembly, representation of each member country is limited to five. U.N. Charter art. 9, para. 2.
39. OPEC Statute, art. 12.
40. *Id.* art. 11(B).
41. *Id.* art. 11(C).

single member to veto any Conference resolution. The disadvantage of this is obvious; such power conflicts with the principle of democracy in an international organization and renders the working of the Organization tedious.

When a resolution has been approved unanimously, it is legally binding and becomes effective thirty days after the conclusion of the meeting or any other period specified by the Conference, unless within the period any member notifies the Secretariat otherwise. Absence is not regarded as a negative vote. This accords with the practice which has become crystallized in the Security Council of the United Nations. There is, however, nothing to indicate what would happen if a member were present but abstained. One would not regard this as a negative vote either, following the practice in the Security Council. It is more logical to regard abstention of a member present at deliberations as a tacit consent than to do so in respect of an absent member. An interesting provision in the voting procedure is one which enables an absent member to veto a resolution of the Conference merely by communicating this to the Secretariat before the resolution becomes effective.[42] This is definitely a drawback to a decisionmaking body such as the Conference. A revision of this procedure is advocated.

The main functions of the Conference are to formulate the general policy of the Organization, to determine the manner of its implementation, to decide upon any application for membership in the Organization, to confirm the appointment of members of the Board of Governors, to appoint the principal officers of the Organization, to direct the Board of Governors to submit reports or make recommendations on any matters of interest to the Organization, to consider or decide upon the reports and recommendations submitted by the Board of Governors on the affairs of the Organization, to consider and decide upon the budget of the Organization, and to approve any amendments to the Statute of the Organization.[43] All matters not expressly assigned to other organs of the Organization are reserved to the Conference.[44] In other words the Conference has residuary powers while other organs have their powers specifically defined.

The Board of Governors

This body is composed of Governors nominated by the member countries and confirmed by the Conference.[45] The Board's main duties are to direct the management of the affairs of the Organization, to implement the deci-

42. *Id.*
43. *Id.* art. 15.
44. *Id.* art. 16.
45. Before the revision of the Statute in 1965, each original member was entitled to nominate a Governor. Nonfounder members were only entitled to nominate one Governor collectively. OPEC Res. II.5 (Jan. 1961), *reprinted in* 1 Resolutions Adopted at the Conferences of the Organization of the Petroleum Exporting Countries 7 [hereinafter cited as OPEC Resolutions]. This disability was removed in 1965. *See* OPEC Statute, art. 17(B).

sions of the Conference, to consider and decide upon any reports submitted by the Secretary-General, to submit reports and make recommendations to the Conference on the affairs of the Organization, to draft the budget of the Organization for each calendar year and submit it to the Conference for approval, to approve the appointments of the Chiefs of Departments of the Organization, to nominate the Auditor, to consider the Auditor's report, and to prepare the agenda for the Conference.[46]

The Chairman of the Board of Governors is appointed by the Conference for a period of one year, in accordance with the principle of alphabetical rotation. Precedence is given to the date of membership in the Organization over the principle of alphabetical rotation whenever it becomes necessary. The Board meets at least twice a year in regular sessions. A quorum of two-thirds of membership of the Board is necessary for a meeting, and decisions are approved by a simple majority vote,[47] unlike the Conference where unanimity is the rule.

The Board may decide, by two-thirds majority vote, that a Governor's continued membership on the Board is detrimental to the interests of the Organization and may remove him. The Governor is immediately replaced by a substitute nominated by his home country.[48] This provision has not yet been utilized against any Governor, and the chance of it being invoked is very remote in view of the sensitivity of oil politics in general. Even if it were to be utilized, the provision seems too subjective. There is nothing to indicate that objective criteria will guide the Board in the exercise of its power in the absence of categorized conditions such as inefficiency, old age, and disloyalty. In the case of incapacitation of a Governor article 24 provides that the corresponding member country shall nominate a replacement.

The Secretariat

The Secretariat consists of the Secretary-General, the Deputy Secretary-General, and other staff as may be required.[49] Until 1969 the Statute did not lay down any conditions concerning the qualifications of the Secretary-General although his tenure of office was limited to one year. Resolutions have not only filled the qualifications gap and significantly improved the position of the Secretary-General but have increased the tenure of his office from one year to three years,[50] thus allowing for continuity of management.[51] This is very salutary. Among the qualifications desired of the

46. OPEC Statute, art. 20.
47. *Id.* art. 17(D).
48. *Id.* art. 23.
49. *Id.* art. 28.
50. OPEC Res. XX.117 & XII.129.
51. These new stipulations were added to the Statute by the Conference in Algiers in June 1971. Since appointment to the post of Secretary-General is on a rotational basis, convention prevents the rejection of a candidate nominated by a member-country due to lack of the

Secretary-General are that he be at least thirty-five years of age, hold a degree in law, economics, science, engineering, or business from a recognized university, and that he have at least fifteen years of experience with ten years in the oil industry. The remuneration is 65,000 Austrian shillings, approximately £21,000 sterling.

The Secretary-General is the Chief Executive of the Secretariat. In that capacity, the executive responsibilities assigned to the different departments of the Organization are exercised on his behalf and under his authority in accordance with the Statute and under the direction of the Board of Governors. The fact that the Secretary-General is appointed suggests that his post was conceived primarily as administrative. The Secretary-General organizes and administers the work of the Organization, coordinates and ensures the execution of the functions of the different departments, prepares reports for the Board of Governors, and acts as a liaison between the Chairman and other members of the Board on the activities of the Organization, particularly the implementation of the resolutions and recommendations of the Conference.[52] He is the legal representative of the Organization, and also deals with matters of protocol.[53]

Officers of the Secretariat are appointed by the Secretary-General after nomination by their respective governments.[54] The paramount consideration in the appointment, transfer or promotion is the necessity to secure the highest standard of efficiency, competence, and integrity. At the initial appointment, however, due consideration is given to achieve an equitable nationality distribution among the member countries. Member countries propose candidates for appointments to the posts of Chiefs of Departments and their deputies. The international character of the staff is guaranteed by article 32 of the statute in a manner similar to that of the staff of the United Nations Secretariat.[55] They are considered to be international civil servants. In the performance of their duties they have no allegiance to their home governments that had nominated them for service with OPEC.

The discipline of members of the staff is the responsibility of the Secretary-General who is assisted by an Administrative Committee comprised of three Chiefs of Departments designated by the Secretary-General for each case as it arises. The procedure laid down for the committee closely resembles that of the United Nations Administrative Tribunal.

The Secretary-General is assisted by the Deputy Secretary-General, a Division of Research, a Personnel and Administration Department, a Pub-

---

requisite qualifications. In the Lagos Conference of December 1972, Algeria put forward a doctor of medicine who was approved in preference to candidates with better petroleum qualifications and experience nominated by Iran, Iraq and Venezuela.

52. OPEC Statute, art. 29.
53. *Id.* art. 27.
54. *Id.* art. 31(B).
55. U.N. Charter art. 100, para. 2.

lic Relations Department, an Economics Department, a Legal Department, an Information Department, a Technical Department, and a Statistical Unit.[56] The Conference is free to create more departments as the need arises. The office of the Secretary-General assists him in carrying out contacts with governments, inter-governmental organizations, and delegations, and in the enforcement of the resolutions and recommendations of the Conference.[57] Notwithstanding the provisions of the Statute, the Board of Governors may authorize the Secretary-General to allocate functions to and make minor adjustments between departments. For example, in June 1968, at the Twenty-Second meeting, the Board of Governors transferred the Statistical Unit from the Information Department to the office of the Secretary-General. The Unit's duty is to collect and collate statistical information from member countries, the oil industry and other sources.

Article 39(A) vests the Secretary-General with power to commission consultants and experts outside the Secretariat to advise and conduct studies on specific matters when necessary. The Secretary-General may also engage experts and specialists, regardless of nationality, when occasion demands subject to the approval of the Board of Governors and subject to provisions being available within the budget.

The Administration Department

The Administration Department is responsible for the Organization methods, general and conference services, personnel matters, budgets, accounting and internal control. It studies and reviews general administrative policies and industrial relations methods used in the oil industry in member and other countries, and advises member countries of any possible improvements. It keeps abreast of the current administrative policies and policy changes occurring in the international petroleum industry which might affect the Organization or be of interest to it.[58]

The Economics Department

This department reviews the economic aspects of the world petroleum industry and carries out a continuous program of economic studies, for the purpose of achieving a speedy and successful realization of the aims of the Organization. It undertakes special economic studies of particular aspects of the international petroleum industry with a view to ascertaining where the best interests of the member countries lie.[59] Its studies form the very basis of the essential decisions of OPEC.

56. For the definitions of the duties of these departments, *see generally* OPEC Statute, arts. 34, 35, 36, 37 & 38.

57. *See* art. 33(2) of OPEC Res. XII.76 (Dec. 1966), *reprinted in* 6 Int'l Legal Materials 358 (1967) [hereinafter cited as Amended OPEC Statute], amending arts. 2(B), 33, 34, 36 & 37 of OPEC Statute.

58. Amended OPEC Statute, art. 34.

59. *Id.* art. 35.

The Legal Department

This department reviews the legal aspects of the world petroleum industry and carries out a continuous program of legal studies for the purpose of achieving a speedy and successful realization of the aims of the Organization. It undertakes special legal studies of particular aspects of the international petroleum industry with a view to ascertaining where the best interests of the member countries lie.[60] The responsibility for monitoring the action taken by member countries in the execution of resolutions of the Organization was taken away from the legal department and vested in the Secretary-General following the amendment to article 33 of the Statute in 1966.[61]

The Information Department

This department maintains an Information Center, including a comprehensive library and a documentation center. It establishes the proper procedure for its most efficient utilization as a central pool of information for the member countries in particular and the public in general. It maintains a program of general and technical publications and periodicals and seeks to expand appropriate relations with other oil industry institutions with a view to the further enrichment of the Organization as a center for information about the industry.[62] In June 1968 at its Twenty-Second meeting, the Board of Governors transferred the press and public relations functions from the office of the Secretary-General to the Information Department.[63]

The Technical Department

This department reviews the technical aspects of the world petroleum industry and carries out a continuous program of technical studies for the purpose of achieving a speedy and successful realization of the aims of the Organization. It undertakes special technical studies of particular aspects of the international petroleum industry with a view to ascertaining where the best interests of the member countries lie.

The rest of the Statute deals with consultative meetings and specialized organs, financial provisions, and other miscellaneous matters. One or two of these call for particular mention. There is, for example, an interesting provision concerning the apportionment of budget appropriation on the basis of equality irrespective of production capacity. It is difficult to justify this method of appropriation when one recalls that equality of membership is not reflected in article 7, which deals with admission of members. There the founder members enjoy a special privilege in the Conference in that their concurrence is necessary before a decision obtained by three-fourths

60. *Id.* art. 36.
61. *See* OPEC Res. XII.76 (Dec. 1966), *supra* note 57.
62. Amended OPEC Statute, art. 37.
63. *Id.* art. 33(3).

majority of members becomes valid. It is a great contrast to the practice in the United Nations, where apportionment is based on budget capacity at the national level.[64] It is easier to appreciate in the United Nations the disproportionate influence the Great Powers wield in the Security Council since they contribute the bulk of the budgetary appropriations of the organization.

Article 44 deals with the procedure for amendments to the Statute of OPEC. Because the Organization is a child of circumstances, numerous amendments have been proposed to meet the changing circumstances in the last decade or so, such that the Statute has been almost rewritten.

Article 45 provides that all resolutions contrary to the context of the Statute shall be abrogated. What is the implication of this provision? Does an inconsistent resolution stand abrogated automatically, or will a special procedure have to be put into motion? Who decides the consistency of a resolution? The Statute provides no direct answers to these questions. Since thirty days must elapse before a resolution becomes effective, one can only assume that within this period a member would have noticed the inconsistency and notify the Secretariat accordingly.

One other institution which deserves mention is the Economic Commission. The Conference is authorized to set up specified organs and committees when the need arises. These committees function within the Secretariat. The Seventh Conference of the Organization held at Djakarta in November 1964 established the OPEC Economic Commission as a specialized body to examine the position of petroleum prices on a continuing basis.[65] The Commission collects the necessary price information through contacts with both public and private institutions in the oil industry, and presents this information to the member countries in the form of recommendations and findings on a monthly basis.[66] The Economic Commission is composed of a Commission Board, a Commission Staff, and national representatives who act as the liaison officers with the Commission Staff.[67] The national representatives usually go before the meetings of the Board of Governors as a committee of experts. The scope of the Commission's functions has been recently extended to cover other areas apart from the issue of prices.

Location of Headquarters

In accordance with article 5 of the Statute, the Second Conference chose Geneva as the Organization's headquarters in 1961[68] but moved after four

64. *See* G.A. Rules of Procedure, Rules 160-1.

65. OPEC Res. VII.50 (Nov. 1964), *reprinted in* 1 OPEC Resolutions, *supra* note 45, at 69.

66. *See* M. al-Otaiba, *OPEC and the Petroleum Industry* 75 (1975); M. Mughraby, *Permanent Sovereignty Over Oil Resources* 127 (1966).

67. *See* art. 6 of OPEC Res. VIII.50 (Aug. 1965), *reprinted in* 5 Int'l Legal Materials 157 (1966).

68. OPEC Res. II.10 (Jan. 21, 1961).

years of residence in Swiss territory to Vienna in 1965.[69] The choice of a European city, and a nonmember of the Organization, for the headquarters is without precedent. So also is the choice of English as the official language of the Organization when the trend in international organizations is towards bilingualism or even trilingualism. The choice of Geneva or even Vienna as the Organization's headquarters was probably due to a desire to offer the possibility of more direct contact with oil importing governments.[70]

INTERNATIONAL LEGAL STATUS OF OPEC

The resolution that brought OPEC into existence makes no reference to the international legal status of the Organization. Nor do any of the subsequent resolutions of OPEC clarify either its international juridical personality or the privileges and immunities that should be extended to its organs and officials.[71] This is hardly surprising, for only rarely will the constitution of an international organization provide an express answer to the question of whether the organization is endowed with international personality. To ascertain whether OPEC has international juridical personality, it becomes necessary to examine what constitutes personality itself.

In the municipal jurisdiction of each country the notion of juridical personality has a precise definition, but in international law such precision in definition is lacking. The notion and content of "personality" are susceptible to different interpretations. The state is the typical legal person under international law. Other entities may be considered to possess "personality" depending upon such circumstances as capacity to perform certain legal acts, insofar as these entities possess the power of independent action on the international plane.[72] When states therefore agree to establish an international organization, such an organization may be endowed with international personality in order to enable it to perform its allocated functions with greater efficiency. This trend of according personality to inter-governmental entities has become particularly discernible since the establishment of the United Nations in 1945 and the subsequent proliferation of international organizations. Normally, therefore, whether an institution possesses international personality or the quantum of personality that has been granted in each case can only be measured by reference to the type of express or implied functions, powers, privileges and immunities, the indicia of personality, defined in the constitution of the organization and developed in its practice.

69. OPEC Res. IX.60 (July 13, 1965).

70. *See* F. Rouhani, *supra* note 7, at 140.

71. *See* F. Rouhani, *supra* note 7, at 129; on the 1946 Convention on the Privileges and Immunities of the United Nations, *see* 1 U.N.T.S. 15.

72. *See* 1 D. O'Connell, *International Law* 91 (1970); M. Sorensen, *Manual of Public International Law* 595 (1968).

As has been noted above, with the development of international institutions in the last sixty to seventy years there has been a discernible trend in international law towards the attribution of some degree of international personality to them. A crucial decision concerning the juridical personality of international institutions is the Advisory Opinion of the International Court of Justice in the *Reparation for Injuries Case.*[73] Compensation was sought for the damages suffered by a United Nation's official, Count Bernadotte, while in the service of the United Nations in the Middle East. He had been assassinated by a gang of terrorists in Israel while serving as a mediator between the Jews and the Arabs in Palestine. The Court, in discussing the issue of international personality of international organizations, stated that the United Nations is a corporate body within international law entitled to enjoy rights and privileges in accordance with the wishes of members of the organization and has the capacity to maintain its rights by bringing international claims.[74] The Court added that is not the same thing as saying that it [*i.e.,* the UN] is a state, which it certainly is not, or that its legal personality and rights and duties are the same as those of a state. Still less is it the same thing as saying that it is "a Super State." Whereas a state possesses the totality of international rights and duties recognized by international law, the rights and duties of an entity such as the United Nations must depend upon its purposes and functions as specified or implied in its constituent documents and as developed in practice under international law. The Organization must be deemed to have those powers which, though not expressly provided in its Charter, are nevertheless conferred upon it by necessary implication as being essential to the performance of its duties.[75]

The inference which one can draw from that opinion is that several states can by a treaty create an entity with a view toward coordinating their common interests by such an entity. Such an entity, although not a state, may nevertheless possess international rights and duties. However for its public international status to be recognized the entity must fulfill three requirements, viz:

(i) it must be established by an international agreement, i.e. one between States;
(ii) it must have organs and
(iii) it must be established under international law.

73. Advisory Opinion on Reparation for Injuries Suffered in the Service of the United Nations, [1949] I.C.J. 174 [hereinafter cited as Reparation for Injuries Case].

74. *Id.* at 179.

75. *Id.* at 179–82. On the international personality of international organizations, *see generally* Seyersted, *International Personality of Intergovernmental Organizations,* 4 Indian J. Int'l L. 1 (1964); Rama-Montaldo, *International Legal Personality and Implied Power of International Organizations,* 4 Brit. Y.B. Int'l L. 111 (1971), M. el-Sayed, *L'Organisation des Pays Exportateurs de Pétrole* 145 (1967); Scelle, *Some Reflections on Juridical Personality in International Law,* in *Law and Politics in the World Community* 49 (1953).

Particularly in the case of OPEC there is no doubt that the members regard the Organization as possessing the quality of a juridical person. No member country would ever deny to OPEC the right to conclude agreements, to acquire property, or to sue or be sued on its territory.[76] But as regards non-member countries, is the position the same? Can it be said that OPEC possesses objective international personality *erga omnes* in the sense that this quality would generally be valid *vis-à-vis* third countries?

In the *Reparation for Injuries Case,* the Court stated that the international community has the power in conformity with international law to bring into being an entity possessing an objective international personality, and not merely a personality recognized by them alone, together with the capacity to bring international claims.[77] While the attribution of objective personality may be applicable to the United Nations because of its near universality, it is doubtful whether this observation of the Court could be applicable automatically to all organizations, whatever their nature. Moreover there is no conventional or customary rule of international law to the effect that an international organization acquires *ipso jure* by its creation an objective international personality. The extent of each organization's personality is determined by taking into consideration the terms of the treaty which created it.

Everything depends upon the willingness and courtesy of third states in any given situation. Third countries may recognize the international personality of an international institution or acquiesce to the exercise of international functions by such an entity in relation to themselves. Each state is free to make such a judgment by reference to the permanence of the institution in the form of its structure and special organs, the framework of its functions and powers, the degree of its autonomy and independent will, and the extent to which it can contribute to the formation of rules of international law.

For example, Switzerland expressly recognized the international personality of the League of Nations although it was not a member.[78] Austria expressly recognized the international legal status of OPEC when it affirmed the juridical personality of OPEC and, in particular, its capacity to contract, to acquire and dispose of movable and immovable property, and to institute legal proceedings.[79] Although OPEC never entered into any specific agreement with Switzerland concerning the recognition of international personality, the permitted existence of OPEC activities in Swiss

76. *See* F. Rouhani, *supra* note 7, at 129.

77. Reparation for Injuries Case, *supra* note 73, at 185.

78. Communications from the Swiss Federal Council Concerning the Diplomatic Immunities to Be Accorded to the Staff of the League of Nations and of the International Labour Office, 7 League of Nations O.J. 1422 (1926).

79. Agreement Regarding the Headquarters of the Organization of the Petroleum Exporting Countries, *signed* June 24, 1965, art. 4, Austria-OPEC, 589 U.N.T.S. 135.

territory from 1960 to the middle of 1965 may be regarded as tacit recognition by Switzerland of the international personality of OPEC.

In terms of its Statute, its common practice, and its similarities to and specific agreements with other international organizations, there can be no doubt about the powers of OPEC on the international plane, powers distinct from those of its individual members. By the exercise of its powers, the Organization attains its goals of coordinating and unifying the oil policies of its member states and determining the appropriate means of safeguarding the common interests of its member countries.[80] OPEC fulfills most if not all the criteria developed by international jurisprudence for attributing personality to an intergovernmental agency. One can therefore safely conclude that OPEC possesses international juridical personality, although of a limited character.

ACHIEVEMENTS OF OPEC

The stature of OPEC on the international oil scene and its increasing influence in international oil affairs cannot be challenged or easily defied. Member countries contribute overwhelmingly to world trade in petroleum. For example, some 76% of the world oil export in 1971 was from OPEC countries (excluding Nigeria, which was then not a member), while in 1970, crude oil produced by OPEC countries was 49% of total world production.[81]

Since its formation, OPEC's objectives have been expanded as a result of the confidence generated by its earlier success. This success has been guided by a high regard for fair play. One of the principles of fair play which OPEC acknowledges is that the "status quo" must change with "changing circumstances."[82] Guided by this principle, OPEC has forced the reexamination and renegotiation of existing contracts concerning royalty payments and taxation, and has firmly won the acceptance by the international oil industry of the principle of the inalienable rights of countries to participate in the exploitation of their natural resources even where such rights had not

80. *See* OPEC Statute, art. 21; *see also* preamble to Res. 1.1 of September 14, 1960, *supra* note 14; M. el-Sayed, *supra* note 75, at 149.

81. At the end of 1973 the figure was 55.8% at thirty-one million barrels per day. Total world production was fifty-five and one half million barrels per day. At the end of 1974, total proved reserves of crude in OPEC member countries was estimated at 65% of world total. *See* Memorandum submitted by Algeria to the Conference of Sovereigns and Heads of State of OPEC Member Countries, Table VIII(a), at 255 (Mar. 1975).

82. This principle has been reaffirmed in several United Nations resolutions and those of the Organization of African Unity (O.A.U.) and OPEC. *See, e.g.,* G.A. Res. 1803, 17 U.N. GAOR, Supp. (No. 17) 15, U.N. Doc. A/5217 (1963), G.A. Res. 2692, 25 U.N. GAOR, Supp. (No. 28) 63, U.N. Doc. A/8028 (1971), G.A. Res. 3171, 28 U.N. GAOR, Supp. (No. 30) 52, U.N. Doc. A/9030 (1974); S.C. Res. 330, 28 U.N. SCOR, Supp. (1973) 3, U.N. Doc. S/INF/29 (1974). O.A.U. Resolution on the Permanent Sovereignty of African Countries over Their Natural Resources, in 1 *Bull. on Foreign Affairs* (Nigeria) 45 (July 1, 1971); OPEC Res. XVI.90 (June 1968), *reprinted in* 7 Int'l Legal Materials 1183 (1968).

been provided for in existing contracts. OPEC has influenced the formation of national oil companies, encouraged active training programs of indigenous personnel, and has, through specialized departments of its Secretariat in Vienna and its Economic Commission, carried out studies and issued useful information and data on international oil activities for the benefit of its members.

OPEC's achievements cannot be measured only in terms of material gains to its member countries. Apart from the achievement of greater income for its member countries, other goals have been pursued. Its biggest achievement lies in the fact that it has helped to generate so much confidence among member countries that for many "sovereignty" now assumes new meanings and dimensions. It has transformed the role of the major international oil companies from that of the "bully-boys" of the 1940's and 1950's to that of partners in progress. This transformation has met with continuous resistance from the oil companies.

The Organization has played an important role in stabilizing "posted" or "tax-reference" oil prices and checked further unilateral price reductions by the oil companies. For example, on December 9, 1970, the members of OPEC demonstrated a remarkable cooperation when they passed a resolution at the Twenty-First Conference at Caracas which has constituted one of the most comprehensive as well as effective measures taken by the Organization for attaining its fundamental objective of upward revision of posted or tax-reference prices in member countries.[83] The result of this cooperation has been a shift from what economists would term a "buyers market" to a "sellers market,"[84] thus assuring the stabilization of high levels of crude oil prices through collective intergovernmental cooperation. This goal has not been achieved without struggle. The oil companies had been reluctant to reorganize in a way that would give the producing countries more power, more involvement and more income.

OPEC has recently brought about agreements with the oil companies providing for higher crude oil prices[85] and increased tax payments by oil companies to the producer governments, thus contributing to its members' gross national products, balances of payments and governmental income. More recently OPEC has secured from the oil companies compensation

83. *See* OPEC Res. XXI.120 (Dec. 1970), *reprinted in* 10 Int'l Legal Materials 240 (1971); *see also* Shihata, *supra* note 11, at 268.

84. For terminology, *see* L. Tarshis, *Modern Economics* 77–97, 131–33 (1976). A buyers market is one in which the oil companies who represent the consumers in the producing countries dictate the price they are willing to pay per barrel of oil. Before 1970, this usually was determined unilaterally by one company, and the other companies quickly fell in line by adjusting their prices for purposes of uniformity. A sellers market is one in which the producers dictate the price by simply informing the companies to comply. The producers have accomplished this by the market economics of supply and demand.

85. For example, OPEC countries' earnings per barrel of oil rose from $1.15 in 1958 to $10.45 in the first quarter of 1975. Memo to Summit of Heads of State of OPEC from Algeria, Table V(b), at 196 (1975).

with respect to the eroded parity of the dollar; in fact, the most recent discussion is that the U.S. dollar may no longer be a unit for tax computation. OPEC has brought about agreements with the oil companies treating oil royalties as an expense item rather than a credit against income tax liability. Member countries now have greater participation in the industry's gross profit through direct taxation.[86] OPEC has brought about the complete elimination of the "marketing allowance" or sales discounts for income tax purposes, which the oil companies deducted as a cost at the production stage.[87] OPEC has drawn up and urged the adoption of identical principles in member countries' hydrocarbon laws; in particular, OPEC has followed Libya's lead in pushing for a short duration of new concessions.[88] It has helped its members in assessing and analyzing conditions in the international petroleum field and trained nationals in the technical and economic aspects of the industry in order to ensure that the controlling power in the hydrocarbon industry is firmly in the hands of the oil producing nations.

OPEC has participated actively in international and regional congresses dealing with the oil industry and has adopted a declaratory resolution expressing the determination of member countries to follow uniform lines of petroleum policy.[89] A basic policy principle embodied in that declaratory statement relates to greater participation by producing countries in the ownership of operating companies, especially participation in major concessions before expiration dates. The members of OPEC have agreed on a pro-forma Regulation for the Conservation of Petroleum Resources.[90] The general pattern has already been adopted by some member countries. OPEC also assisted in coordinating various national oil companies of member countries.

Thus in under five years OPEC has succeeded in doing what it could not do in the past 23 years of its existence; it has fundamentally altered the structure of world economic power. Since its creation in 1960 the Organization has had two main goals: to raise the taxes and royalties earned by member governments from crude oil production and to assume control over production and exploration from the major international oil companies. With host government revenue at a peak of $40 per barrel in 1979, a

86. Between 1960 and 1975, the royalty has risen from 12½% to 20% and taxes from 50% to 85% in most cases. *See also* OPEC Res. IV.33 (June 1962), *reprinted in* 1 OPEC Resolutions, *supra* note 45, at 47.

87. *See* OPEC Res. XXI.120, *supra* note 83.

88. *See, e.g.*, the Libyan Petroleum Law No. 25 of 1955 and Nigerian Petroleum Decree No. 51 of 1969.

89. Declaratory Statement of Petroleum Policy in Member Countries, OPEC Res. XVI.90 (June 1968), *reprinted in* 7 Int'l Legal Materials 1183 (1968). The oil companies viewed this as nationalization by installments. *See also* OPEC Res. XXIV.135 (July 1971), *reprinted in* 10 Int'l Legal Materials 1082 (1971), which called on member nations to initiate talks to implement the principle of participation advocated in Resolution XVI.90.

90. See note 105 *infra*.

twenty-fold increase over the price at inception of OPEC in 1960,[91] and with Libya, Saudi Arabia and Kuwait moving to complete or having achieved 100 per cent acquisition of the companies producing properties, these goals are close to achievement.

It is significant to note that OPEC member countries have as far back as 1975 been awarded 15 per cent voting rights in the World Bank and 10% in the International Monetary Fund, compared with their previous 5% voting rights in each body.[92] This is in itself a testimony to the growth in lending to the two world bodies and to special bilateral facilities through which the OPEC countries have launched substantial aid programs for certain of the less developed countries of the world. Indeed this is a test of OPEC's recognition in the international money market. With the constant meetings and seminars, the empirical research programs, the training facilities for members of the Secretariat, the studies, publications, the Organization fosters economic cooperation among member states and also offers an opportunity for individual member countries to exchange ideas, information and experiences in a forum of long and detailed dialogue which is essential in maintaining international comity. In fact, nothing can foster international understanding more than the prompt dissemination of ideas, views and shared experiences. This has now been made possible by the establishment, within the Secretariat, of the OPEC News Agency (OPECNA) which began operating from the Secretariat in Vienna on November 3, 1980 with a daily news service, which is transmitted to 350 destinations in 70 countries, covering five continents. It is OPECNA's objective, first and foremost, to make OPEC's stand-point, its activities, policies and decisions clear to everyone, in an effort to dissipate the misunderstandings, and misconceptions which have accompanied the actions of the Organization since its birth.[92a]

Perhaps the most encouraging development in the history of the Organization was the creation of the OPEC Special Fund which has established multilateral financial facilities exclusively designed to assist nonmember Third World countries,[93] with the donors standing to gain no economic return. Originally envisaged as a temporary aid instrument, the fund was made permanent in January 1980, when OPEC Finance Ministers decided to convert it into "an international agency for technical assistance to other developing countries" and to endow it with an international personality. It is now formally known as "The OPEC Fund for International Develop-

91. Western consumers have attributed world inflation to this increase, but OPEC has effectively refuted this. *See* Organization of Petroleum Exporting Countries: Declaration Concerning the International Economic Crisis, *reprinted in* 14 Int'l Legal Materials 566 (1975); *see also* 7 OPEC Weekly Bull. Nos. 42 & 43, at 1, 2 (Oct. 25 & Nov. 2, 1976).

92. 6 OPEC Weekly Bull. No. 40, at 8 (Oct. 20, 1975).

92(a). N. Nguema in OPEC (Bull.) Oct. 1981, p. 24.

93. *See also* Shihata, *supra* note 11, at 275; Kaplan, *International Economic Organizations: Oil and Money,* 17 Harv. Int'l L. J. 203, 212 (1976).

ment" with its headquarters in Vienna. The scope of the Fund's operations covers the financing of balance of payment deficits, the financing of a specific development program or project, and lending to international development agencies such as the International Monetary Fund (IMF) and the projected International Fund for Agricultural Development (IFAD), the beneficiaries of which are developing countries. The Fund had the sum of US $800 million at its disposal for initial disbursements. The Ministerial Committee on Financing and Monetary Matters of OPEC which met in Vienna on January 16, 1980 approved the second replenishment of the resources of the Fund by a sum of $800 million. It is interesting to note that, net aid disbursements by OPEC countries in 1978 stood at 2.8 per cent of their GNP. That of OPEC's Development Assistance Committee was only 0.4 per cent of their GNP, which was much less than the one per cent of GNP aid agreement within the United Nations frame-work. What is more, a typical OPEC donor country does not provide its aid out of a renewable annual income or as an attempt to promote its exports, as is usually the case with industrialized country donors. OPEC grants aid without strings.

There have been many reasons for the achievements of the OPEC countries. Among them is their near monopoly control of world oil supplies,[94] and the threat, often articulated, to act in unison in withholding these supplies if their demands are not met.[95] Moreover, despite the efforts of the major consuming countries to develop other sources of energy, some of the new substitute sources of energy have not been able to play any significant role in satisfying world energy requirements and are not likely to be able to do so until at least the end of this century.[96] The production costs of the most promising of the nonconventional energy sources are indeed presently estimated at about twice the actual price of crude oil.[97] Oil remains the most convenient and relatively the cheapest source of energy in the world today. The bargaining power of each member has been increased. Through OPEC the exporting countries can for the first time confront the international

94. 1974 Figures supplied by the Bank of England indicated that OPEC countries account for some 90% of the world's petroleum exports. Economic Progress Report, H.M.S.O. No. 64, at 1 (July 1975); *see also* 7 OPEC Weekly Bull. No. 4, at 50 (Feb. 2, 1976).

95. For an interesting debate concerning the legality of such action, see Paust & Blaustein, *The Arab Oil Weapons—A Threat to International Peace,* 68 Am. J. Int'l L. 410 (1974); Shihata, *Destination Embargo of Arab Oil: Its Legality Under International Law,* 68 Am. J. Int'l L. 591 (1974); Paust & Blaustein, *The Arab Oil Weapon: A Reply and Reaffirmation of Illegality*, 15 Colum. J. Transnat'l L. 57 (1976); *cf.* Boorman, *Economic Coercion in International Law: The Arab Oil Weapon and the Ensuing Juridical Issues*, 9 J. Int'l L. & Econ. 205 (1974); Dempsey, *Economic Aggression and Self Defense in International Law: The Arab Oil Weapon and Alternative American Responses Thereto,* 9 Case W. Res. J. Int'l L. 253 (1977).

96. *See* Note, *Cartel Pricing in the International Energy Market: OPEC in Perspective,* 54 Ore. L. Rev. 643, 657 (1975), for a discussion of alternative energy sources.

97. *See* Readings in Petroleum Literature, 7 OPEC Weekly Bull. No. 44, at 2 (Nov. 8, 1976); *see also A Digest of Energy Reports,* 1 OPEC Rev. 7–10 (October 1976).

major oil companies on equal footing, and through collective bargaining achieve their objective of promoting a more just world economic system.[98]

Yet OPEC must proceed with caution. Its policies in the future must call for continued restraint in pricing, in development and in production. Such is the summary of the Solemn Declaration of Sovereigns and Heads of State of OPEC countries which met in Algiers from March 4–6, 1975.[99] While the declaration visualizes the wider role which OPEC will be called upon to play on the international stage, it at the same time envisages a new economic order in which developed and developing countries will participate equally in decision-making and in the world's natural resources, taking into account the needs and legitimate interests of all peoples.

98. Through OPEC, the exporting countries could for the first time face the international oil companies which were in a position, through joint ownership of concessionaire companies in the Middle East, to practice collective bargaining in their dealings with the exporting countries. Mikdashi, *supra* note 34, at 69. *See also* Amuzegar, *OPEC in the Context of the Global Power Equation,* 4 Den. J. Int'l L. & Pol. 221, 223 (1974), for a discussion of other factors contributing to OPEC's success.

99. *See* note 91 *supra.*

Oil and Regional Cooperation Among the Arab Countries

ALI ATTIGA

While there is a wide scope for co-operation within OPEC, its chief limitation is the diversity of its members in terms of geographic location and socio-economic conditions and political affiliations. But within OPEC there are seven Arab countries whose conditions are theoretically more favourable to regional co-operation. From only three Arab members of OPEC that established OAPEC in January 1968, the organization has grown to ten Arab countries. As a regional institution it includes countries with contiguous borders (except where broken up by the Israeli occupation of Palestine) and stretches from Algeria in the West to Iraq in the East. It has more than 70% of the population of the Arab world and generates more than 80% of its GNP. OAPEC members constitute a region within the Arab world which shares common values and social institutions with other Arab countries that are not members. The Arab countries as a group have many historical, cultural, religious and linguistic ties, besides their complementary natural and human resources.

OAPEC deals only with hydrocarbons and its main objectives are as follows:

1. The promotion of the co-ordination of the petroleum economic policies of its members and their co-operation in working out solutions to problems facing them in the petroleum industry.
2. Exchange of information, harmonization of their relevant legal systems in the oil sector, and the promotion of training and exchange of expertise and employment opportunities in the petroleum industries of its member countries.

Extracted by permission of the publisher, Oxford University Press, from Robert Mabro, ed., *World Energy: Issues and Policies,* Oxford Energy Seminar, St. Catherine's College, 1980.

3. The use of the resources and common potentialities of its member countries in establishing joint projects in various phases of the oil industry in which all or some of OAPEC members may participate. Such joint projects take the form of commercial companies aimed at creating among other things, a common Arab capability in both upstream and downstream activities of the oil industry.

OAPEC has created four joint companies, one training institute and it is working towards the formal establishment of a Judicial Board. It is also undertaking studies for the establishment of an engineering consultant firm and a number of projects in the petrochemical industry such as lube oils, detergents, synthetic rubber and carbon black for domestic consumption. OAPEC is active in promoting regional co-operation in the field of energy for domestic consumption in the Arab countries. It is also engaged in promoting better international relations between oil exporters and importers through the dissemination of information and the organization of direct contacts concerning the development needs of less developed countries, including the oil exporters.

The main objectives of the OAPEC companies are to enable their member countries to acquire common capabilities and experience in the management of their oil resources. Two of the four companies are concerned with oil transportation and ship-repair facilities and skills. The third company deals with the promotion of investment of petroleum industries within the OAPEC member countries and the Arab world as a whole. Its main objective is to promote the development of hydrocarbon resources, especially downstream activities on a regional basis. For this purpose, it is required to give priority to equity participation in joint oil and gas projects. It can also provide such projects with direct loans and help to mobilize additional financial resources for their initial establishment or expansion.

The fourth OAPEC-sponsored company operates in the general field of petroleum services associated with exploration and production. It is a holding company with plans to establish a number of subsidiary companies either wholly owned or in partnership with foreign companies possessing well-established technology. It has started with the establishment of a joint drilling company, which was expected to be in operation before the end of 1979.

OAPEC-sponsored companies are legally incorporated as independent companies operating under the supervision and direction of their shareholders and boards of directors. OAPEC as such has no direct or indirect control or supervision over the policy directions or administrative performance of these companies. Theoretically, they are expected to follow the overall objectives of OAPEC as specified in their basic statutes. In practice, however, it is rather difficult to ensure this kind of policy direction without regular consultative meetings and policy guidelines from the

OAPEC Council of Ministers. So far these arrangements have not been institutionalized, except for an annual routine meeting between the OAPEC Secretariat and the company managers.

While these companies taken together with their sponsoring organization provide an adequate institutional framework for regional co-operation in the development of hydrocarbon resources, their actual performances are greatly limited by the tendency of their member countries to plan and manage their oil industry individually and, at times, even in competition with each other. This tendency has resulted in the neglect of joint projects and common policy formulation in the petroleum sector. In fact, it is common for a member country to promote its own separate projects and facilities, while not doing the same for the joint ventures. Yet we have already noted that real and lasting success in the management and development of oil resources requires regional co-operation among the producers for exports and for domestic markets.

The Arab Petroleum Training Institute was formally established in 1979. Its main objective is the training of prospective instructors for teaching in the training institutes of their respective countries. It is also expected to promote the production of teaching materials and various aids, as well as the use of Arabic as the language of instruction. The Institute is expected to become a regional centre for technical meetings, seminars and special short courses designed to promote professional contacts and efficiency among the member countries. It is open to the nationals of all Arab countries.

Although the non-oil sectors are outside the jurisdiction of OAPEC, oil revenues play a decisive role in promoting co-operation in these sectors. Prior to 1973, oil wealth tended to have a divisive impact on relations between oil and non-oil Arab countries. Beginning with the colonial history of oil concessions and the world power struggle, associated with that history, the Arab world became divided into separate political and legal entities with strong local and foreign vested interests protecting the new political and social situation. At that time oil revenues were just enough to create divisive trends, but not enough to provide for significant aid and assistance designed to initiate trends favouring regional co-operation within the Arab world. But after the 1973–74 oil price adjustment it became possible for the oil states to provide bilateral and multilateral assistance to other Arab countries. Many national and regional development funds, banks and companies have been established during the last five years. Billions of dollars have been invested in these institutions and in other national and bilateral development projects. However, the regional aspect of intra-Arab aid has so far been rather limited.

Regional co-operation is highly essential for the viable development of the oil-exporting countries. It is also essential for the most obvious needs of the other Arab countries. Even the proper development and management of the hydrocarbon resources are dependent upon adequate regional co-operation and policy co-ordination among oil-exporting countries. For the region as a whole there are sufficient complementary resources for sustained economic and social development. For individual Arab countries working alone there are serious limitations on their balanced economic and social development regardless of how much capital or manpower or other natural resources they may possess.

The International Oil Factor

In late 1973, the long lines of cars at gas stations everywhere seemed to provide dramatic proof that the "energy crisis" had definitely arrived. While the immediate spark for this crisis was the fourth Arab-Israeli war of October 1973, continued uncertainty over oil supplies and escalating prices point to other serious causes. In fact, the October war was simply the catalyst for a number of inter-related developments which resulted in emphasis on the "oil factor" during the last decade. Ian Smart has delineated three separate processes in this turn of events.[1] First, the accelerating depletion of world oil supplies and continued growth in demand had raised fears of an energy crisis well before October 1973. Secondly, the potential importance of the Middle East because of the concentration of much of the world's oil there was graphically demonstrated in late 1973 by the oil embargo and production cutbacks imposed by various Arab states as a result of American and other Western support for Israel. The third process has been the series of enormous price increases irregularly instituted since the early 1970s by the thirteen members of OPEC and matched by other non-OPEC producers.

The result has been a transfer of wealth from the industrialized West to the OPEC states on a scale unprecedented in history. Early fears that this small group of underdeveloped countries would use their tremendous reserves of "petrodollars" to take over the world were soon proven baseless. Despite widespread hardships—especially in the Third World—the world learned to adjust. Workable patterns of "recycling" petrodollars evolved, and OPEC financial managers proved to be particularly adept at productive investment of the funds at their disposal. At the same time, rampant inflation in the OPEC countries combined with vastly increased imports of Western goods and services to erode their income available for investment.

1. "Oil, the Super-powers and the Middle East." *International Affairs* (London), Vol. 53, No. 1 (January 1977), pp. 18–19.

Adjustment was not easy, however, nor complete. The energy crisis remained a continuing "crisis" for a number of reasons. Demand for oil continued to grow despite the calls for conservation. In 1972, the United States imported 28% of its oil; the figure shot up to 37% in 1975 and 42% in 1976. Higher prices brought more oil into production but most of these sources were marginal: there was general agreement that it was only a matter of years before world demand would outstrip total supply. While a number of alternative energy sources were proposed and some implemented, they either had minor effect on the overall picture, or remained technically and/or economically unfeasible, or politically unacceptable. Because of the relative inelasticity of demand, successive OPEC price hikes forced more and more resources to be devoted to payment for growing imports of oil.

Besides the purely economic aspects of the "oil factor," there continued to be widespread apprehension over the potential disruption of existing supply patterns. Increased attention was focused on the Middle East. Despite the Camp David process and an Egyptian-Israeli peace treaty, the Arab-Israeli conflict was still explosive and carried with it the real possibility of another embargo. In 1979, revolution in Iran triggered another crisis as production in that country was drastically curtailed. Gulf oil exports were further reduced in 1980 when Iran and Iraq went to war.

The shortages caused by the Iranian revolution, whether real or only anticipated, and the resultant crisis served well to point out the fragile nature of the changed international oil system. It also demonstrated the continued deep dependence of the West on imported oil, especially from the Middle East. Even though increased supply and lessened demand led to an "oil glut" at the beginning of the 1980s and a consequent decline in OPEC production levels, the long-term picture seemed to remain one of Middle Eastern predominance in a tight world oil market.

The three selections in this chapter approach the international oil factor and its economic effects from very different directions. In the first, Mohammed Abu al Khail, the Saudi Minister of Finance and National Economy, lays the background for the 1973–1974 spiral and assesses its impact on both oil importers and exporters. His words are of particular interest insofar as they may be assumed to represent an "OPEC point of view." His characterization of the 1974–1979 period as a "wasted respite" helps to explain OPEC justifications for the 1978–1979 price spiral. Next, David Curry focuses on the shocks to the international financial system brought about by the oil price explosion and examines the complex patterns of investment brought about as the result of large current account surpluses held by some OPEC members. As he notes in his conclusion, the size of these surpluses had peaked by 1978 as OPEC's share of the world oil market decreased and its members' absorptive capacities grew larger. In the third selection, Walter J. Levy views the situation at the beginning of the

1980s and assesses the long-range impact of the oil price revolution on Western economies. He concludes by outlining some necessary policies for importing countries to follow in the future.

Bibliographic Sources for this Chapter:
5, 6, 9, 15, 19, 23B, 23D, 27, 29, 43, 46, 48, 55, 57, 58, 67, 74, 85B, 93, 106, 111, 116, 136, 156, 178, 179, 180, 189, 201, 208, 222, 232, 233, 234, 235, 249, 294, 296B, 296M, 301, 325, 329, 330, 332, 347, 348, 351, 352, 353E, 357, 358, 374, 391, 414.

The Oil Price in Perspective*

MOHAMMED ABU AL KHAIL

Much has been written about the rise in oil prices since 1973, the effects of this on economic activity and inflation in oil importing countries, and the implications of continuing oil price increases for the future health of the world economy. Not all of this discussion has, however, taken proper account of the longer-run availability of the oil resource and the need for a proper balance of the interests of both oil producers and oil users and indeed of the interests of the world as a whole. The purpose of this article is to review the questions posed above from these viewpoints, together with the policies followed by the oil exporting countries in facilitating international adjustment to higher levels of oil prices.

THE SITUATION BEFORE 1973–74

The world oil situation in the years preceding the recent increases in oil prices was marked by a number of features. The oil industry had, of course, been long dominated by major international companies that have derived their power from their vertically integrated operations and from joint ventures and interlocking directorates in one or more of the integrated stages of the industry. Moreover, they had the backing, directly or indirectly, of the political and economic power of their respective governments. They controlled exploration and production of crude oil and, through their control over transportation and refining, the distribution and marketing of the products. The major companies co-operated closely in the making of decisions concerning price and supply; until the growth of the independents and

*Most of the statistical information has been derived from IMF sources, such as *International Financial Statistics*, *Annual Reports* and *Direction of Trade*.

Reprinted by permission of the publisher, The Royal Institute of International Affairs, from *International Affairs*, Vol. 55, No. 4, October 1979.

the wider prevalence of arm's-length sales, the ability of the major companies to price their products within the integrated system was unquestioned.

Throughout the 1950s and 1960s the oil price remained at what should now be generally viewed as low levels. This was the period of an extraordinary pace of exploration and development of oil, particularly in the Middle East and North Africa. At the same time demand was buoyant. Postwar reconstruction and economic development efforts created rapidly growing markets for energy in many areas. The low price enabled oil to maintain a competitive edge over substitute energy sources and to make rapid gains at the expense of coal in these markets. The major companies considered it to be in their interest, particularly after the dispute over the control of the Suez Canal in 1956–57, to expand supply potential, as this lessened their dependence on any one particular area and strengthened their bargaining position *vis-à-vis* the governments of the producing countries. Moreover, a concessionaire company is not expected to take a long view; to it the present value of any oil that might be left in the ground in the distant future is necessarily very low. Because of the growing energy demand, maintenance of low prices for oil greatly expanded the volume of sales, increasing total profits.

Production and demand reinforced each other; the easy availability of oil was no doubt an important factor in making it possible for the European countries and Japan to stage a dramatic recovery and expansion in their economies within a short period after the Second World War. And there evolved a life-style in the industrial societies that was based substantially on the abundance and cheapness of oil.

Towards the end of the 1950s several new developments occurred that upset the calculations and the pricing policy of the major companies. The more important among these were the rising number and growing strength of independent companies in the world oil market, mandatory import controls in the United States virtually closing that market to increased supplies of foreign oil, and the impact of oil exports from the Soviet bloc, and particularly the Soviet Union, to Western Europe. Oil was being widely sold at substantial discounts below posted prices, particularly by independents who found their American market outlets closed off. The governments of the importing countries saw little justification for the majors charging their affiliates the posted prices; for the companies themselves a reduction was needed to maintain their share of the market and bolster profits. On the other hand, the interest of the oil exporting countries was to freeze posted prices, because on these were based the taxes and royalties owing to them. In February 1959 the companies chose to cut the price of Middle Eastern oil without consulting with the governments concerned; and despite the uproar that this caused, they made a further unilateral reduction in the price in August 1960.

This prompted the formation of OPEC in the following month. While these price cuts dramatised the need for a concerted stand against the decline of prices and hence of government revenues, there were other factors which had by that time brought the exporting countries closer together. Most of these countries were dissatisfied with their concession agreements—which they considered to be inequitable and to have been granted in many cases under duress, and they sought ways of amending them. At the same time political and other leaders in these countries were becoming increasingly conscious of the importance and irreplaceable character of their hydrocarbon resources, and the crucial role that oil revenues must play in the development of their economies and the improvement of the living standards of their people.

OPEC's achievements in the next few years were rather modest. It prevented a further decline in the posted prices; for the next ten years these prices remained at the August 1960 level, or some 12 per cent below the 1958 level. The rise in export prices of industrial countries between 1958 and 1970 was over 15 per cent, so that the real value of crude oil and of per barrel government revenues declined during this period by some 25 per cent or more. The purchasing power of oil revenues in terms of development equipment fell even more sharply, for the prices of imported capital goods rose a good deal faster than general import prices in this period.

Under conditions of excess productive capacity the buyers' market continued well into the late 1960s, and efforts by OPEC to secure increases in the posted prices of crude were of no avail. The OPEC countries had not at that time acquired sufficient political authority and cohesion among themselves to push through their claims. Market prices began to firm up after June 1967 when several factors—including the transformation of the Suez Canal into a military frontier—began to pose difficulties for oil purchasers. However, despite supply problems and an upsurge of demand for oil products that was beyond expectations, posted prices remained unchanged until 1970–71 when Libya's successful negotiations with the oil companies set in motion negotiations by other oil exporting countries. These and the following years saw a gathering strength among these countries as the withdrawal of foreign power and foreign military presence was completed, and as their divisions began to be narrowed. The long era of cheap oil and a buyers' market came to an end. At long last the producing countries began to exercise legitimate control over the use of their own resource. The price of oil, as is well known, rose in several stages over 1971–73 and quite sharply at the beginning of 1974.

THE WASTED RESPITE 1974–79

The period before 1974 was one of unbridled optimism concerning oil. For the consuming countries there was no thought of an energy crisis in even the distant future to be brought about by the diminishing supply of oil.

The near-term situation of excess supply did not encourage a stocktaking for later years. As the growth of demand had for years been more than compensated by the growth of supply, the possibility of an energy gap was, if anything, an academic curiosity. It must also be added that the producing countries for their part cannot be totally absolved of their share of the responsibility. Even allowing for their diverse circumstances and interests, and their political situation and weak bargaining power, they lost valuable time in controlling the operations of the oil companies in their countries and asserting their interests in the determination of the price.

The succession of events leading to the 1974 price increase brought about a much-needed, although belated, awakening to the facts of life of world oil. There have been numerous assessments since then of the demand and supply arithmetic of oil. These assessments are necessarily tentative, depending on the many assumptions that must be made for the future years. On the demand side, there are the questions of the rate of growth of the economies of the oil importing countries, of the energy coefficient of growth—which would depend on conservation measures as well as the development of technology—and of progress in bringing to fruition plans to develop substitute energy sources. All of these raise complex issues not only with respect to new investment and scientific innovations, but also with respect to ecological and safety considerations.

On the supply side there are the questions of a correct estimation of existing oil reserves and, on the basis of geological assessments and probability calculations, of the rate of discovery of new proven reserves to stretch the life-span of world oil. There is, further, an important consideration, very often ignored as a factor of any consequence in the deliberations of Western analysts, concerning how rapidly the oil producing countries themselves should allow the depletion of their oil resource. While each new discovery of oil potential, whether in Mexico or in China, creates a new burst of optimism and complacency, more realistic estimates inevitably point to the existence of limits in ultimate reserves which assures a transition period of only a short number of years before a global energy balance depending substantially less on oil must be attained. Many of the functions that are now performed by oil, functions that are critical for the economic machinery that the world has built over centuries—for its industrial and agricultural growth and for modern societies, however organised—will have to be performed in the not too distant future by other and newer energy sources.

Looked at from this perspective, reflecting the hard facts of world demand and supply, what should be the appropriate price for oil? The economic range within which the price should be set is a very wide one, with the lower limit defined by the cost of production and the upper limit by the cost of producing alternatives to oil. It is this wide range that has created confusion in the minds of people who believe that the price of oil in recent years has not followed the hallowed rules of demand and supply. Until 1970 the price was closer to the lower limit although well above that level, but considering the particular characteristics of oil, it should have been at or near the upper limit. The 1973–74 price corrections therefore had to be of a somewhat drastic character, being the first decisive attempt by the oil exporting countries to set things in the right direction. It was an attempt to begin the process of removing the longstanding distortion of the oil price *vis-à-vis* prices of other commodities, to signal the need for conservation and the efficient use of oil resources, and to pave the way for a concerted drive to close the gap between longer-term energy needs and their supply through the development of other energy sources.

In the four years following 1974, the OPEC countries exercised considerable restraint in their price policy, even at the cost of their own interests, in order to afford the consuming countries the time necessary for the required adjustments. Between 1973 and the end of 1977, the price of oil barely kept up with other prices, the real value of oil remaining steady or falling somewhat. In 1978 its real value fell sharply as the dollar price of oil remained relatively stable in the face of world inflation and the falling value of the American currency. There have been increases in the price in the current year and, as a consequence of the adjustment in June 1979, for the first time since 1974 the oil price has risen relative to the increase in other prices over the same period.

Unfortunately, the breathing spell offered by the four years since 1974 of a fall in the real value of oil did not foster a sense of urgency in the large oil using countries to take the necessary steps to curb their voracious appetite for oil. Instead, in many countries, it revived to some degree the easy complacency of the previous era—and valuable time has thus been lost in setting firmly the fundamentals of a viable energy policy. Rather, the governments of importing countries have sought temporary palliatives by bringing pressures on the oil exporters to raise existing capacity to meet their consumers' relatively unrestrained demand. Immediate exigencies have continued to hold sway over longer-run interests. Nevertheless, perhaps the events of 1979 will create an impetus towards achieving a satisfactory international energy policy.

THE ECONOMIC EFFECTS OF THE 1974 OIL PRICE INCREASE

The 1973–74 increases in the price of oil understandably caused a tremor throughout the oil importing countries. In the confusion of those days, there was endless talk of the approaching doom for Western civilisation as the industrial world would have to make a rapid transfer of a large part of its wealth into the hands of a few, largely small and undeveloped, countries that would not know how to use responsibly their newly acquired affluence and the considerable power that it would give. Five years later, one hears little of these doomsday fantasies.

This is not to say that the rise in the oil price did not have its effects or that the adjustments in the importing countries to new, considerably higher, oil prices have been completed. Take, for instance, the question whether the oil price rise of 1974 made it substantially harder to control inflation in the oil importing countries.

It has no doubt affected the structure of prices—indeed this was a desirable change insofar as it removed previous distortions and brought the new set of prices into closer conformity with the real value of petroleum products. This effect is not very different from that of an upward adjustment from an artificially low level in the price of capital to reflect its scarcity in an economy, or in the price of an important consumer good such as bread from a very low subsidised level. What gives an oil price increase particular significance is that it affects a large number of countries at the same time and that the use of oil is pervasive in the processes of production and consumption. On the other hand, from the point of view of an efficient use of resources, it was necessary for the oil price to be increased relative to other prices.

It is nevertheless true that the impact of the 1974 oil price increase on the general rate of inflation ran its course within a year or two in most countries. The prices of goods and services that used petroleum as a source of energy or as a raw material were marked up to reflect most, if not all, of the increased costs. The indirect effects, namely pressure for higher wages in response to higher consumer prices resulting from the oil price increase, are more difficult to trace. Consumer prices have been affected by various other factors in the periods both prior to, and subsequent to, the oil price increase. These factors have had considerably more pronounced effects on consumer prices than the oil price increase, and the main motivation for wage push—where this was a strong element in the spiralling of inflation—would appear to lie in the general factors sustaining inflation rather than in the specific matter of the price of oil.

Inflationary forces were already very strong before 1974, both in the industrial and in the developing countries. The upsurge in global demand in 1972–73 came on top of an inflationary momentum which had been in existence for a number of years and which had its primary source in American fiscal policy in the late 1960s. Public attitudes and institutional practices became geared to an assured expectation of continuing advances in costs, prices, and rates of remuneration. The policies followed by governments to mitigate inflation were inadequate to cope with these forces, and the rate of inflation reached a high level during 1973. It rose further in 1974 and 1975, and the oil price increase was an element—but not a major one—in this rise. In the three following years 1975–78 the average inflation rate of about 7 per cent in the industrial countries remained below that of

1973, although it was still unacceptably high. The relatively slow rise, and later the virtual freezing of oil prices in this period contributed, if anything, to moderating the forces of inflation, which were fuelled by a complex of factors including institutional rigidities.

Has the higher oil price contributed to the slow pace of economic recovery in the industrial world, or hindered the economic development efforts of the non-oil developing countries, or disrupted the functioning of the monetary system? Various incomplete arguments have been advanced tending to answer these questions in the affirmative. We shall look at them in turn.

One argument is that the oil price increase has exerted a deflationary force on aggregate demand. At first sight this appears to be a plausible analysis. This argument runs as follows: after the increase consumers in the importing nations pay higher prices for energy and, given that demand for petroleum products does not respond significantly, if at all, to price changes, they therefore purchase less of other goods. These cutbacks in purchases spread through the economy, having their effects on sales, output and jobs, all of which are reduced in the consumer goods industries. These initial reductions in turn cause further reductions in demand, output, and employment through the familiar multiplier process.

There were also other arguments pointing to further worldwide disturbing effects of the oil price increase. The aggregate current account surplus of the oil exporting countries implied an equivalent deficit for the importers. However, individual countries might not wish or might find it difficult to maintain these deficits—indeed some countries might regard as unwise the incurrence of substantial debt to finance the deficit. Further, the recycling of funds from OPEC countries is unlikely to match up with the distribution of deficits among the non-OPEC countries—so that not all countries are able to borrow, or if able, to borrow on reasonable terms. Attempts by such countries to reduce their deficits by reducing non-oil imports or expanding exports to other oil consuming countries might improve their trade balance, but only by worsening the trade balance of others. Such attempts do not help international adjustment and, if followed generally, they would lead the world down the path of a progressive reduction of effective demand and of world trade.

The fear has also been expressed that the surplus of the OPEC countries might pose a threat to the stability of the international monetary order. Over a short period, it was forecast in 1974, a substantial proportion of the liquid reserves of the world would be owned and controlled by a few countries. They would thus be in a position to shift the funds around between money centres in ways that might cause serious disruption in the world monetary system.

So much for the arguments as to the disturbing effects of the oil price increase. The fact is that events since 1974 have lent little support to these arguments and have belied the gloomy prognostications. A crucial assumption behind them was that the oil exporting countries, being undeveloped, would have little productive use for their newly acquired wealth. Consequently they would continue to accumulate their growing surplus revenue—which then would represent a threat to world demand and to the adjustment of international payments. In mid-1974 a World Bank estimate placed OPEC's investable surplus in 1980 at some $650 billion. Several other estimates at that time also pointed to very large cumulative building of OPEC assets. In fact, however, it has turned out that, if the current account positions of the major oil exporters are aggregated, the cumulative surplus is of the order of $185 billion over the period 1974–78 (see table below). Of particular interest is the fact that the aggregate surplus on the current account of this group of countries fell from some $68 billion in 1974 to as low as $7 billion in 1978.

If we look at the development of official reserves, the total of such reserves held by major oil exporting countries amounted to about $13 billion at the end of 1973; this figure rose to some $56 billion at the end of 1978, an increase of $43 billion over a period of five years. The share of these countries' reserves in the total of world reserves rose from about 7 per cent at end-1973 to 16.5 per cent at end-1978—certainly not a disproportionate increase when account is taken of the relatively rapid growth for the economies of these countries. The share of the total imports of these countries in the world total has, for example, more than doubled in these years, from 3.7 per cent in 1973 to 7.8 per cent in 1977.

Payments Balances on Current Account, 1973–78

(In billions of US dollars)

	1973	1974	1975	1976	1977	1978
Major oil exporting countries	7	68	35	40	33	7
Other developing countries	−11	−30	−38	−26	−21	−32
Industrial countries	19	−4	23	5	−1	29

More will be said about these developments in the oil exporting countries in the following section. The point that needs to be emphasized here is that, as the increased earnings of the oil exporters were used to purchase a larger volume of goods and services from other countries, the initial deflationary effect on demand as a result of the rise in oil prices in these countries was offset by the subsequent growth of demand for their products. Similarly, as higher outpayments on account of oil brought in their wake higher receipts from exports of goods and services, the problem of payments adjustment did not arise.

Moreover, by comparing the current account surplus and the reserves increase for the OPEC countries it can be seen that as much as four-fifths of the current account surplus was offset by an outflow through the capital account of the balance of payments. These were capital exports from the oil producing countries to the rest of the world, whether through purchases of longer-term assets or through loans. These capital exports not only helped the receiving countries to finance their current account deficits, but those governments and individuals who came to be in possession of the funds increased their own demand for goods and services—thus neutralising the initial dampening effect on demand induced by the oil price rise.

The development of the current account balances of various groups of countries presented in nominal terms, as in the preceding table, does not give an adequate view of the real change that has taken place in the structure of international payments between the pre-1973 period and the present. In the 1977 annual report of the International Monetary Fund, the annual averages of current account balances for the major groups for the years 1967–72 (a period described in the report as having little bias in cyclical conditions) are scaled up to 1977 levels of world trade prices and real economic activity in the respective groups. These adjustments yield the following results:

Comparison of Adjusted Current Account Balances

(In billions of US dollars)

	1967–72 Actual Average	1967–72 Average rescaled to 1977 levels	1977 Actual	1978 Actual
Major oil exporting countries	+0.7	+3	33	7
Industrial countries	+10.2	+31	−1	29
Non-oil developing countries	−8.1	−28	−21	−32

With the indicated adjustments, the current account deficit of the non-oil developing countries in 1977 was thus lower than what it was in the 1967–72 period, while the pre-1973 surplus of the industrial countries was replaced by a surplus for oil exporting countries. Extending this analysis to 1978 by rescaling the 1967–72 averages to 1978 levels of trade prices and output, it can be seen that the 1967–72 averages correspond fairly closely with the actual structure of current account balances in 1978.

This line of analysis yields a very important conclusion. By 1978 the adjustment of world payments balances to the 1973–74 oil price increases had been largely completed. The industrial countries did not resume their traditional role of providing real resources or financing to the rest of the world until 1977. However, the surplus funds accruing to the oil exporting countries were placed in the financial markets of industrial countries, and from these markets they were available to the non-oil non-industrial countries to finance their deficits. But in 1978 the industrial countries returned to their traditional surplus, and the current account position of the oil exporting countries in real terms also approached that of the period before 1974.

The non-oil developing countries have clearly shown much courage and determination in adjusting to higher oil prices and maintaining the momentum of their economic growth. The growth rate over the past five years in these countries has averaged about 5 per cent. This is somewhat lower than the rate attained in the years before 1973, and it certainly cannot be considered adequate from the point of view of the needed improvement in the standard of living—particularly in countries with high rates of population growth. But when account is taken of the sluggish demand for their products in the industrial world as a result of slow economic recovery, of the obstacles placed by many wealthy nations in the way of their exports, and of the lack of interest of the developed countries in the expansion of direct transfers of resources, the maintenance of this rate of growth in the non-oil developing world must be regarded as a commendable performance. In contrast to the generally unfavourable trends and attitudes encountered in the industrial world, these countries have benefited substantially from large-scale grants and other development assistance from the OPEC countries and the greatly enlarged flow of remittances from their nationals employed in the OPEC states.

THE ABSORPTION OF THE OPEC SURPLUS

It is useful to look more closely at the developments that have led to the very rapid shrinkage of the surplus that accrued to the major oil exporting countries in 1974. The greatly enlarged earnings from oil export had the initial effect of virtually lifting the financial constraints on their plans to restructure their economies. There was a sharp acceleration in expenditures on economic development, in particular on building the domestic infrastructure. At the same time major programmes for social improvements were undertaken at government initiative. A large part of this expansion was met by rapid increases in imports, which were facilitated by the maintenance of a liberal import policy. The rapid progress of imports is reflected in the following figures:

Imports of Major Oil Exporting Countries

	1973	1974	1975	1976	1977	1978
Value (US$ billion)	20	36	56	68	84	98
Percentage increase in value		77	57	21	24	17
Percentage increase in volume		38	42	19	14	4

The value of imports registered a very steep rise in 1974 and 1975; in 1975 it was nearly three times the level of 1973. After 1975 the growth in import values slowed, but it was still substantial. The value figures, of course, show the effect of continued inflation in import prices. But even in volume terms the increase in imports remains striking. The increase in real imports has, moreover, been strongly reinforced by an increase in net payments by the oil countries for external services—over the 1974–78 period this increase was as much as $22 billion. There has thus been a remarkably rapid absorption by these countries of increased amounts of both goods and services, which has been the principal reason for the shrinkage in their surplus.

Despite the increase in imports, the oil exporting countries have had to contend with strong inflationary pressures, particularly in the early part of this period. Domestic spending by governments raised demand well above what could be met from the increase in domestic output and in imports. Moreover, the rising prices of imports themselves added fuel to the inflationary fires. In this situation, most of the oil exporting countries in 1975 and 1976 gave priority to the control of inflation by restraining domestic spending. In some countries, as in Saudi Arabia, much weight was given to the improvement in the standard of living of all citizens, and hence to the provision of educational, medical, housing, and other social services. In these countries measures were taken to reduce the impact of inflation by expanding free health and welfare programmes and increasing subsidies on essential items of consumption and housing. Along with these policies, the governments took steps to relieve the bottlenecks in transportation and distribution, particularly those that impeded the flow of imports. They gave special emphasis to the expansion of port capacity by constructing new facilities and by taking measures to speed up the handling of goods.

Greater attention to the control of inflation has not prevented the oil exporting countries from sustaining relatively high levels of economic growth. The non-oil sector in these countries has grown in the 1974–78 period by over 10 per cent on the average. The rate was higher in the early part of the period but tapered in later years to some 8 per cent in 1978. Some countries were no longer in surplus and were therefore not in a position to maintain the pace of growth; a part of the decline in 1978 also reflected political developments in Iran which impaired economic activity. Contrary to the general trend, the growth rate of the non-oil sector in Saudi Arabia has risen from year to year, averaging as much as 18 per cent in the five years ending 1977–78.

In implementing ambitious development plans, a critical problem—particularly in countries with small populations—has been the shortage of manpower, not only of professionals and skilled workers but also of unskilled labour. The oil exporting countries in the Middle East and North Africa have followed a liberal immigration policy, attracting workers from other Arab countries and also from several non-Arab countries, particularly in South and East Asia. The demand for expatriate labour in the oil exporting countries has thus led to the growth of a viable international labour market for many developing countries. In some of these countries foreign exchange earnings from this labour market have financed a sizable proportion of their imports; in a few cases these earnings have been substantially higher than their export receipts. The total amount of remittances has increased over time; on the basis of estimates in a recent World Bank staff study the volume of remittances received by 12 major beneficiary countries in 1977 was at least $7 billion, and was probably much higher.[1]

How has the surplus that has accrued to the oil-exporting countries been disposed of? At first the net cash surplus was mainly invested in liquid assets in the major industrial countries—bank deposits and short-term government securities. But over time there has been an increasing switch in favour of longer-term and less liquid forms of investment. These latter investments have formed the bulk of the net placements of funds in the last two years. There is no evidence that the governments or the monetary authorities in the oil exporting countries have altered the currency or asset composition of their holdings in response to short-term fluctuations in exchange markets; or that their actions otherwise have caused an aggravation of exchange rate instability. But there has, of course, been a tendency gradually to diversify holdings.

A sizable proportion of the surplus has consisted of official grants and loans to developing countries. Precise information on the amount of these flows is hard to find, but rough estimates of bilateral aid, contributions and capital subscriptions to regional and international development agencies, investments in the 1974–75 oil facility of the International Monetary Fund, and direct purchases of World Bank bonds would bring the total to an annual average of approximately US$7 billion in the 1974–78 period. As a proportion of the gross domestic product of the oil producing countries this amount is several times the United Nations target for official development assistance—which has so far remained well beyond the reach of the industrial countries. If the annual value of remittances is taken in conjunction with this assistance, there has been a very substantial flow of financial resources from oil exporting countries to other developing countries in the years since 1973.

To channel this development assistance effectively the Arab countries have set up a number of institutions either as agencies of their own governments or as regional intergovernmental agencies.[2] The first national institution in the Arab region for providing financial and technical assistance—the Kuwait Fund for Arab Economic Development—was set up as early as 1961; its operations have greatly expanded since 1974. This was followed in

1. Some of these remittances were from workers employed in Europe.
2. OPEC also has established a Special Fund for making loans to developing countries on easy terms.

the 1970s by the Abu Dhabi Fund (1971) established to provide loans and guarantees or acquire equity participation in projects within Arab countries, and by the Saudi Fund (1974) for the purpose of financing projects in developing countries throughout the world. Among the regional institutions are the Arab Fund for Social and Economic Development (1971), established by members of the League of Arab States to assist the Arab countries, the Arab Bank for Economic Development in Africa (1974), also established by members of the League to finance development projects and provide technical assistance to member countries of the Organization of African Unity, and the Islamic Development Bank (1975), sponsored by members of the Islamic Conference to foster the economic development and social progress of its member countries and of Muslim communities. The lending by these institutions—both national and regional—is made available on highly concessionary terms and serves to finance mainly economic infrastructure projects. Total loan commitments in 1977 amounted to about US$2.0 billion, and the commitment rate is higher today.

Looking to the future, an important question is whether the oil exporting countries could or should maintain the rapid pace of economic growth attained in the recent years. These countries need to take stock of their recent development and analyze the contribution that this is likely to make in providing a secure future basis for growing income and employment. Rapid growth inevitably brings in its train problems that have important social and political repercussions. Continuing high inflation causes distortions and inequities. Migration into the towns, the relative decline of agriculture and traditional pursuits, and the need for a continuing increase in the employment of foreign labour—all of these have important effects on social equilibrium and cohesion and the maintenance of traditional values. It would seem prudent at the present juncture for many of the oil exporting countries to temper the feverish pace of activity and concentrate on the consolidation and rationalisation of what has been achieved.

PROSPECTS FOR THE FUTURE

For the present and the coming years, it is important that a common worldwide understanding is reached concerning the use and valuation of oil. Such an understanding can be based on the simple fact that use of oil is the use of a fixed stock of capital. For the consuming as well as the producing countries, it is borrowing from future generations. This intertemporal consideration should remain paramount in our decisions concerning the use of oil. It is to the common interest of all countries that the period of transition to non-oil energy sources be stretched as far as possible, and that this period is used effectively to develop these sources. For the producing countries, whose main source of subsistence for the foreseeable future is oil, the oil revenues must be invested to generate productive capital either domestically or abroad—investments that yield a rate of return commensurate with the real value of their resource.

The current disarray in the markets for oil—for which the oil companies and the importing countries must assume a large share of the responsibility—must be ended. Beyond that, it should be realised that dependence on increased production of oil to meet an uncontrolled expansion of demand will be a highly unrealistic policy. A major reason for the limitation on the level of output of oil stems from technical factors—there is a capacity constraint on the efficient optimal rate of production. But even apart from this there is, as we have seen, an optimal rate at which the economies of the oil producing countries can be further expanded. This rate should be guided not only by economic considerations of efficiency and of the availability of the complementary factors needed for production growth, but also by considerations of social equity and social balance. Nor are there the appropriate opportunities abroad at any particular time that would justify the rapidly growing investment of oil revenues. Thus a continued fast increase in oil production would be imprudent and would place an unwarranted burden on future generations.

The lesson from all this is self-evident. If resolute action is not taken on the side of demand, the price of oil will continue to rise. The increase in the price this year is a clear reflection of this. Much attention has recently been given to the upsurge of prices in the spot market, but it should be realised that the volume of transactions in the spot market is limited and that the major part of the oil continues to be sold under longer-term contracts at the official selling prices of the exporting governments. Contrary to the exaggerated reactions that the recent higher prices in 1977–78 have evoked, the direct effect on growth and inflation rates of the increase that has taken place has been quite small—well below one per cent in respect of growth and somewhat more than that in respect of inflation. Although the indirect effects may be considerable, the recent oil price increase should not pose an insurmountable obstacle to the control of inflation and the maintenance of steady growth of the world economy. It is, however, a reminder of the fundamental fragility of the present situation and of the urgent need in the consuming countries to conserve energy and develop alternative sources. To hope that the oil exporting countries will expand production to stabilise prices is merely to postpone the basic correction.

As the supplier of a vital resource, the oil exporting countries recognise their responsibility of ensuring that it is made available in sufficient quantity so that there is no serious setback to the continued growth of the world economy. But as countries struggling to develop their economies, they expect to benefit from a transfer of technology and experience from the industrial nations, from free access of their newer exports to the markets of these nations, and the maintenance of stable prices for equipment and other goods that they need to import. In this, their interests do not differ from those of other primary-producing developing countries. These latter not only need Western 'know-how' and appropriate conditions of trade, but because of the balance of payments constraints that they face in their development effort, they also need large-scale transfer of financial resources on appropriate terms.

OPEC has consistently sought Western co-operation since 1973 on a number of issues—the transfer of technology, the protection of the real value of overseas assets, and effective and sustained aid to the non-oil developing world. What is needed is a major new thrust toward a co-ordinated approach to world development. This is the rationale of the effort to establish a new international economic order—a project which has generated endless discussion but has made little tangible progress. Neither the North-South dialogue nor the discussions between industrial and developing countries in UNCTAD and at other international forums have succeeded in bringing about a significant change of attitudes and policies in the industrial world. The oil issue is one of a complex of issues that now divide the industrial and developing countries. Its solution, no less than that of the others, requires a common and balanced approach that rises above selfish national interests.

Petrodollars and the International Payments System

DAVID CURRY

This short paper discusses the financial impact of the fourfold increase of oil prices in late 1973–January 1974. These price increases represented a large and sudden shift in the international terms of trade in favor of a handful of oil exporting countries which were not in a position to spend immediately all of their export receipts on imports of goods and services. More broadly, they set in motion profound changes in the world economic and financial system. Adjustment to the shock of significantly higher petroleum prices is, in fact, still incomplete.

SHIFT IN CURRENT ACCOUNT PATTERNS

The vulnerability of the oil consuming countries to the unilateral petroleum price increases imposed by members of the Organization of Petroleum Exporting Countries (OPEC) in 1973–74 varied sharply. Some countries, like the United States, had indigenous energy resources which filled at least part of their domestic needs. Others, like Japan, had to rely on imports of energy for almost all of their consumption. Nonetheless, the oil importing nations, individually and as a group, faced the prospect of changes in their international payments position on a scale without historical precedent in a peacetime environment.

Partially as a result of the oil price increases and resulting policies adopted by governments of oil importing nations the structure of world payments balances on current account shifted markedly, particularly between 1973 and 1974. Only four OECD nations—Belgium/Luxembourg, Germany, the Netherlands and Switzerland—experienced surpluses on current account (totaling about $9.9 billion of which $9.7 billion accrued to Germany) in 1974, as opposed to twelve nations in surplus (a collective total of about

Originally published in: U.S. Congress, Joint Economic Committee, *The Political Economy of the Middle East: 1973-78; A Compendium of Papers* (Washington: USGPO, 1980).

$8.2 billion) on current account in 1973. As a group, OECD countries moved from approximate balance on current account in 1973 ($3 billion surplus) to an aggregate deficit of $33 billion in 1974.

Between the early and middle years of the seventies, the shift in the payments positions of various country groups also was quite dramatic. In the period 1971–73, the average levels of surpluses and deficits were relatively small as shown in Table 1. After the oil price increases, international payments disequilibrium, as reflected by current account balances, was quite pronounced.

OPEC countries as a bloc—after years of approximate balance with other countries—accrued surpluses averaging nearly $50 billion a year over the period 1974–76. OECD countries moved from a surplus position on current account averaging in the neighborhood of $6 billion a year from 1971–73 to a deficit position averaging around $22 billion during the period 1974–76.

Table 1. World Payments Patterns, Balance on Goods, Services, and Private and Governmental Transfers

(In billions of dollars)

	Average 1971–73	Average 1974–76
OPEC	+1.4	+47.0
OECD	+5.6	−22.0
LDC	−4.1	−23.0
Other[1]	−2.7	−12.3
Residual[2]	−0.2	+10.3

1. Israel, South Africa, and non-market economies of Eastern Europe, the USSR, and the People's Republic of China.
2. Over the past several years the "residual" item has grown significantly. Recorded exports exceed imports by a sizable sum for any calendar year due to goods in transit and inflation. The OECD staff has estimated that $12–14 billion is involved in these factors. These factors would by themselves make the residual negative rather than positive. Both the IMF and OECD attribute the positive residual to asymmetries in the reporting of service transactions; that is an over-reporting of service imports, or under-reporting of service exports.

Source: National Advisory Council on International Monetary and Financial Policies, Special Report to the President and to the Congress on U.S. Participation in the Supplementary Financing Facility of the International Monetary Fund.

The average deficit on current account incurred by non-OPEC developing countries increased more than five-fold, jumping from about $4 billion over 1971–73 to $23 billion per year over 1974–76. Other countries, primarily non-market economies, experienced an increase in current account deficits from about $3 billion a year in 1971–73 to a more than $12 billion average in 1974–76. Changes in the current account position of countries within each of these groups were uneven. The bulk of the increase in the OPEC surplus accrued to the Middle East producers, while the changes in the deficits were most pronounced, for example, in the U.K., France, Italy and Brazil.

FINANCING THE DEFICITS

In the aggregate, it was anticipated that oil deficits—per se—would be financed by the OPEC countries since the value of deliveries of OPEC petroleum to oil importing countries in excess of OPEC imports from these countries would, of necessity, take the form of oil importing country liabilities to OPEC members. However, OPEC capital flows to other areas were distributed in a markedly different fashion from the distribution of current account deficits that arose in the wake of the oil price increases and resulting policies adopted by governments of oil importing nations (Table 2).

Table 2. Distribution of Net Payments Deficits, by Area, on Goods, Services, and Private and Governmental Transfers—Disposition of OPEC Surpluses

(In percent)

	1971–73	1974–76
OECD	38	75
LDC	40	15
Other	22	10

The bulk of the funds accruing to OPEC countries were placed in a handful of major industrial nations and in the Euro-banking market. This disparity gave rise to a dramatic increase in the intermediation of capital on a global basis with banks and other private market participants providing 75 percent of the financing of current account deficits over the 1974–76 period (Table 3).

Table 3. Estimated Sources of Funds for Countries in Current Account Deficit—1974–76

(In billions of dollars)

	Estimates	Percent
International Monetary Fund	$15	7
Other official sources	40	18
Private sources	170	75
Banks	(106)	(47)
Other	(64)	(28)
Total	225	100

Private financial institutions were in a position to expand the level of their international activity. Many institutions were competing eagerly for new customers as they sought to establish themselves in new activities and new geographic areas, and endeavored to broaden their scope of operations to spread risks and diversify portfolios at a time when domestic loan demand had slackened relative to immediately preceding years. Also, many large banks served as residual repositories of surplus OPEC funds due, in part, to a general lack in OPEC countries of personnel with financial expertise to effectively and fully utilize the very large amount of funds accruing to these nations.

More than half of the estimated $105 billion of OPEC surpluses flowing to industrial countries was placed with banks (see Table 4). However, while

Table 4. Estimated Disposition of OPEC Investable Surplus

(In billions of dollars)

	Revised				First half
	1974	1975	1976	1977	1978
United States	12.50	9.50	12	9.25	—
of which					
Treasury securities:					
Bills	5.3	0.4	(−1.0)	(−0.9)	(−0.8)
Bonds and notes	0.2	2.0	4.2	4.3	−.8
Other marketable U.S. bonds	.9	1.6	1.2	1.7	.6
U.S. stocks	.4	1.6	1.8	1.4	.6
Commercial bank liabilities	4.1	.6	1.6	.4	−.1
Subtotal (banking and portfolio placements)	10.8	6.3	7.8	6.9	−.5
Other (including direct investment, prepayments on U.S. exports, debt amortization, etc.)	1.7	3.2	4.2	2.3	(−.5)
Euro-banking market	22.50	8	11	12	1.25
United Kingdom	7.50	.25	−1	.75	−.25
Other developed countries	6	7.75	8	8	2.50
Less developed countries[1]	6	7.25	7.50	8.50	2.25
Non-market countries	.50	2	1.25	1.50	.25
International financial institutions (including IMF oil facility)	3.75	4.25	1.75	.50	—
Total allocated	58.75	39	40.50	40.50	6
Estimated current account surplus	70.50	34.25	40	34	4.75
Adjustment for lag in receipt of oil revenues	−11.25	+1	−4.50	+3	3.50
Estimated gross borrowings	.50	4	8	12	2.75
Cash surplus plus borrowings	59.75	39.25	43.50	49	11
Discrepancy in estimates	1	.25	3	8.50	5

1. Includes grants.

OPEC deposits of funds directly with financial institutions may indicate the extent of direct exposure of these institutions to OPEC nations, the financial intermediation of OPEC surpluses accomplished by these institutions in fact exceeded OPEC deposits. In a broad sense, where oil importing nations (on current account) accrued surpluses or incurred deficits smaller than their net imports from OPEC and received OPEC investments in bank and nonbank instruments in excess of their current account deficits with OPEC members, the money and capital markets—including banks—in the capital importing countries clearly intermediated part of the excess OPEC investments.

The bulk of all OPEC financing of the worldwide oil payments deficits has been supplied by the Mideast OPEC members as a group (Table 5). In 1974, non-Mideast OPEC countries accrued current surpluses of about $14 billion, or 20 percent of the total, but since then virtually all of the surpluses have accrued to the Mideast oil exporting countries.

Table 5. OPEC Payments Balances on Goods, Services, and Private Transfers

(In billions of dollars)

	1974	1975	1976	1977	Estimate 1978
Total OPEC surplus	70.6	34.3	40.0	34.0	13.4
Members outside Mideast	14.4	−0.9	0.7	−1.4	−9.6
Surpluses of Mideast OPEC members	56.2	35.2	39.3	35.4	23.0
Official Development Assistance	3.0	5.0	4.5	3.6	—
Financing at market terms	53.2	30.2	34.8	31.8	—
of which to					
Official international financial institutions	3.4	3.6	1.6	.5	—
United States	7.1	10.6	12.6	9.6	—
Bank and money market balances and portfolio securities	5.5	7.6	8.9	7.2	—
Other placements in the U.S.[1]	1.6	3.0	3.7	2.4	—
Non-oil developing countries	3.0	2.2	3.0	4.9	—
Other Areas	39.7	13.8	17.6	16.8	—
Bank deposits	22.0	9.0	8.7	8.0	—

1. Largely prepayments on U.S. exports and amortization of outstanding loans.

Source: Treasury Staff Estimates.

INVESTMENT PATTERN OF MIDEAST OIL EXPORTING COUNTRIES

Due in part to the historical role of the U.S. dollar as the preeminent international medium of exchange and store of value, the bulk of Mideast

OPEC member investment has been in dollar-denominated instruments. The proportion of their external assets denominated in U.S. dollars is estimated to have fluctuated between 75 percent and 80 percent since 1974.

Based on U.S. statistics on international capital movements (the only publicly available and systematically collected statistics distinguishing between short- and long-term assets held by Mideast oil producers) the oil exporting countries in the Mideast have progressively increased the proportion of their investments placed in long-term assets.

With the initial, sudden surge of oil revenues accruing to these countries, 73 percent of their financial investments in the U.S. were placed in short-term assets (Table 6). Thereafter, the proportion of short-term assets acquired at the margin dropped precipitously to 24 percent in 1975 as their combined surplus declined from about $56 billion in 1974 to an estimated $35 billion due to extraordinary growth in their imports and recession in the rest of the world which dampened global demand for oil.

By 1977, the Mideast oil exporting countries were, on net, converting their short-term assets in the U.S. into long-term instruments. This lengthening of the maturity structure is consistent with development of more sophisticated investment planning and better definition of the domestic economic objectives of these countries which permits them to better appraise the amount of funds needed for economic development expenditures

Table 6. Mideast Oil Exporting Country Placements in Bank and Money Market Balances and Portfolio Securities Holdings in the United States[1]

(Dollars in billions)

	Commercial bank liabilities	Treasury securities	Other portfolio securities	Total	Of which short-term
1974	$1.9	$2.3	$1.2	$5.5	$4.0
1975	1.1	3.3	3.2	7.6	1.8
1976	1.8	4.1	3.0	8.9	1.6
1977	0.4	3.7	3.1	7.2	−0.4
1978 (first half)	−.5	−1.4	1.2	−0.7	NA
			(Percent of total)		
1974	35	42	22	100	73
1975	14	43	42	100	24
1976	20	46	34	100	18
1977	6	51	43	100	−6

1. Totals may not add due to rounding.
NA—Not available.

Source: Treasury Bulletin.

in the near-term and thus the amount of funds available for longer-term investments yielding higher returns. The development of greater expertise in the management of large sums of capital by the Mideast oil exporting countries is reflected in U.S. statistics which show (to some extent) that these countries have increasingly assumed the burden of assessing relative risks.

The data also suggest that the movement by the Mideast oil exporting countries toward long-term U.S. assets was accompanied by a decline in the proportion of their surplus funds invested in bank deposits and an increase in the proportion of funds placed in other money and capital market instruments. Allowing for seasonally large receipts of oil revenues in the colder months of each year and for cyclical fluctuations in the demand for oil, the data show that the mix of Mideast oil exporting country investments in bank deposits, money market instruments and portfolio securities began shifting toward private sector instruments as early as 1976. By the Spring of 1977, the news media were reporting increased flows of Mideast funds directly to U.S. corporations through their private placements of notes with Mideast investors (most notably with the Saudi Arabian Monetary Authority).

PRESENT AND PROSPECTIVE PATTERNS

By the first half of 1978, the downward trend in Mideast oil exporting country current account surpluses began to be reflected in the investment statistics. As the ability of most of the Mideast oil exporting countries to absorb imports increases and the oil importing nations continue to effect adjustment of their economies to higher energy prices, the strains on the international financial system arising from large international payments imbalances will probably diminish further and some of the Mideast OPEC members will begin (or continue) to draw upon their accumulated external assets to cover current expenditures. An outgrowth of continuing reductions of Mideast OPEC member current account surpluses, together with more direct management of their external assets, is likely to be a continuing decline in the relative (if not absolute) size of Mideast oil exporting country investments in the traditional instruments they have used since the oil price shocks of 1973–74.

Oil and the Decline of the West

WALTER J. LEVY

The year 1979 was one of grievous setbacks for the future security of the oil supply of the Western world, its economic and financial prospects, its strategic capabilities, and its political stability.

To begin with, we were completely unprepared for the collapse of the Shah's regime in Iran, even though there were many early warning signals. And our government and business community share a substantial measure of responsibility for shortsighted policies which contributed to the destabilization of the Shah's regime.

What followed in terms of oil alone should now be a familiar story. During the course of 1979, the decline in Iranian oil production was more than made up by other producing countries. And taking account of a virtually unchanged world oil demand, importing countries were able to increase their oil stocks during the year by well over one million barrels a day, more than has been achieved during recent periods. Nevertheless, a temporary decline in world oil production led to apprehensions by importing countries and their oil companies that they might be unable to cover their future needs. Accordingly, importers tried to obtain added supplies and to increase stocks at almost any cost. This, in turn, resulted in panic buying at largely uncontrolled and escalating spot oil prices.

Obviously, the Organization of Petroleum Exporting Countries (OPEC) would not maintain its official quotations at a lower level if an increasing proportion of its oil was purchased on the spot market by importing countries or companies at much higher prices. In due course, OPEC increased its official prices to more than double their previous levels. Also, OPEC countries would naturally try to sell more and more of their oil directly or through intermediaries at the higher spot prices. This would mean that the established customers for OPEC oil could only obtain smaller quantities of the oil they had previously been able to acquire directly from the producing countries. Accordingly, more and more companies and countries were unable to purchase oil from their former regular major oil company suppliers; instead, they were forced to turn to producing countries or trading companies to assure themselves of supplies, and they showed themselves willing to pay them practically any price that they asked for.

As early as February 1979 it was painfully clear that this state of affairs posed a massive danger for the world oil economy and that it required coordination and cooperation among importing countries and among their companies if this buying panic were to be stopped. As a minimum, the major importing countries would have had to establish a firm policy for themselves and their companies not to buy oil at above OPEC price levels; at the same time they had to be willing to establish an international and national allocation system that would assure all countries and companies an equitable share in the oil that was available at OPEC prices. Without such arrangements, higher spot prices would sooner or later be incorporated into higher official OPEC prices. This would be especially damaging because OPEC prices are not freely fluctuating market prices. Once raised, they are unlikely to come down again, because any future softness in prices would be countered by a cutback of OPEC production.

The governments of the major importing countries were, of course, aware of this problem. But either they did not really comprehend its importance or they lacked the will and leadership qualities to act. The relevant international institutions, such as the Organization for Economic Cooperation and Development (OECD) or the International Energy Agency (IEA) either felt that they did not have the constitutional power to deal with this issue, or that they would not have the support of their member governments. Industry generally opposed any course that might involve national or international official controls or interference.

While it is, of course, not certain whether any national or international arrangements, had they been made, would have succeeded, in actual fact nothing was even attempted to stem the tide of upward-spiralling prices. It would appear that most of the wounds of 1979 have been self-inflicted.

So we lived through 1979 with Iran in disarray, with the oil economy in turmoil (even though there was no overall sustained shortage), and with the Soviets invading Afghanistan. It was indeed a year illustrating the impotence of Western power and the failure of national and international leadership.

II

While 1979 has passed, the problems it created or aggravated are going to stay with us, probably for the indefinite future. Let us first review the medium term to 1985.

With a recession accompanied by a substantial retardation in world economic growth, and with continued conservation and sustained development of energy from all available non-OPEC sources including coal and nuclear energy, it now appears that world energy requirements *could* be met *if* OPEC production, after a decline during 1980 (as compared with 1979), would then increase again and reach about the 1979 level by 1985. This would require, in effect, that political and economic conditions in the major oil-producing areas, notably the Persian Gulf, be such as to lead to sustained and increased production to such levels. It would also require the adherence by the Western nations grouped in IEA to the oil import limitation targets for 1985 and intervening years, which in turn would require greatly increased conservation and further development of indigenous energy sources.

But this is only the statistical supply and demand part of the story, and even this much depends on the doubtful and difficult conditions just stated. Moreover, even under these conditions, the financial problems of recycling petrodollars and particularly meeting the needs of the less developed countries (LDCs) are sure to take on a much more serious dimension than in the 1974–78 period. This is a conclusion universally shared today by those in positions of public or private responsibility, including members of the banking community who assessed the situation in the earlier period and believed that the private banking community could to a very great extent handle the problem. Today, however, there is solid agreement that the omens are far darker.

Specifically, at current OPEC prices and assuming future increases at even somewhat *less* than the anticipated rate of inflation (the most favorable possible assumption), OPEC countries might accumulate a surplus of about $115 billion during 1980 and of some $350 to $450 billion between 1980 and 1985. This would imply that the importing countries would suffer a corresponding balance-of-payments deficit during this period, which would have to be financed somehow by the recycling of the petrodollar surplus.

For the year 1980, the deficit of developed countries has been estimated at about $50 billion and that of less developed countries at nearly $70 billion. The foreign debt of LDCs might reach about $440 billion in 1981, compared with about $385 billion in 1980 and $150 billion in 1975. And with petrodollar surpluses continuing to accumulate, the financial and balance-of-payments position of oil-importing countries would deteriorate at an accelerating rate. It is most unlikely that our national and international financial system can cope with this problem without risking sustained recessions, a slow rate—if any—of economic growth, high rates of inflation, widespread unemployment, industrial and national bankruptcies, and political upheaval.

The debt problem could, of course, be solved if the values of the currencies in which the debts are incurred decline in their purchasing power to such an extent that the repayment of debts, when due, can be done at a fraction of their original value. All of this did in fact occur between 1974 and 1978, a period which has been portrayed by some as the "golden age" of petrodollar recycling.

But it seems now that the jig is up. The recycling "success" in the past was predicated on some confidence in the value of the currencies involved, on the willingness of oil-producing countries to spend ever-increasing amounts of petrodollars for national development and military purchases, and on their readiness to suffer substantial declines in the real price of oil and to accept between 1974 and 1978 the virtual disappearance of their annual petrodollar surplus.

Now the producing countries are trying to establish a policy under which oil prices will not only go up regularly in step with the rate of inflation, but also will increase on a regular basis at rates exceeding the rate of inflation by an additional amount depending on the increase in the Gross National Product (GNP) of the OECD. In short, they are seeking a regular advance in the real price of their oil. Many of them have also cut back their earlier economic development programs so that the former wild increases in annual import expenditures have recently been coming down substantially.

Furthermore, the producers are now placing even greater emphasis on increasing the value of their oil exports by adding refining, petrochemical and transport operations to their national programs. This would of course further add to the foreign-exchange burden of oil purchased by the importing countries and would thus increase the petrodollar surplus even more.

More important, perhaps, the question arises whether the producing countries would really be prepared to continue to produce oil at a rate that would continuously generate surplus petrodollars for them. As they are well aware, they would in effect be exchanging oil in the ground, whose value will certainly advance at least with inflation, for monetary assets whose value would tend to decline with inflation.

If, alternatively, they were to reduce their production, they would not only stretch the life of their oil reserves, but as real prices in tight oil markets are likely to increase, their revenues would go up. It is true that the world economy would suffer very badly by any resulting oil shortage, but the producing countries might argue that, at least to some extent, their loss from the world's depressed economic activity would be less than the erosion in the value of their surplus financial assets that they would suffer through inflation.

The minimum protection the producing countries could be expected to demand, if they were to continue to produce oil at levels needed by the importing countries, would be an indexation of their petrodollar accumulations in line with inflation. To take care of this problem, I have recently suggested that this might be done by making available to them inflation-indexed energy bonds that would be issued by an international agency. As any additional cost that could result from such indexation would have to be underwritten by the oil-importing countries, the financial burden on the latter might, obviously, increase.

III

Let us now turn to some of the other major problems that affect the security and availability of oil supplies.

The producing countries, having taken full control over their national oil operations, in fact do not recognize as binding supply or price arrangements even if freely concluded by them. Recently they have gone so far as to change agreed-upon prices retroactively. This, they argue, they are entitled to do under the doctrine of sovereign control by producing countries over their natural resources. Because the contracting party in the producing country is also a government-owned oil company, they contend that any change the government requests in the terms and conditions affecting crude oil production and sales would in fact constitute a sovereign act, so that the government oil company that had entered into a contract would be entitled to claim *force majeure vis-à-vis* its customers.

But, undoubtedly, the increasing unilateral application of this doctrine is also related to the ease with which the host countries have been able to apply it, and to the lack of opposition by the affected private or public interested parties against unilateral expropriation or the cancellation of legal and contractual rights. Because of the fear of being arbitrarily cut off from supplies, Western nations and their companies now accept within a wide range practically any economic or political terms that a producing country may impose on them. This subservience, however, rather than safeguarding the remaining rights and position of the companies, in fact encourages the host countries to continue to proceed as they see fit. We have thus entered a period in international oil of near "lawlessness" in the relationship between producing countries, the oil companies and the importing countries.

The issues are not only supply and price stability. They also include exploration and development efforts that are now exclusively dependent on policies of producing countries, which obviously are not interested in creating any surplus of supplies that might endanger OPEC prices.

Moreover, especially since 1979, producing countries have cut back the oil they supply to the major international oil companies, frequently below the level of their direct requirements. The "Internationals," therefore, can no longer provide oil supplies to third parties as in the past. More and more of the oil is sold directly (or sometimes through trading companies) by producing to importing countries. As a matter of fact, the share of the

Internationals in world oil trade has declined from 78 percent in 1974 to about 44 percent in 1979, and is declining even further.

At the same time, the terms imposed by producing countries for oil supplies include more and more political and other extraneous conditions, related, for instance, to the interest of the producing countries in the Palestinian problem, or in their nuclear capabilities or in the political postures of their government customers.

Thus, oil companies no longer perform an effective independent role as a buffer in the relationship between oil-producing and oil-importing countries. They have practically no bargaining leverage with regard to any decisions affecting their operations in producing countries. Moreover, any action by OPEC on supplies or prices that would lead to a higher cost of its oil would, as OPEC is fully aware, also tend to benefit the oil companies, as their own non-OPEC production would also become more valuable.

This is at best a messy situation all around, where the producing countries can have it nearly all their way, because there are no countervailing powers they have to consider. The only question they face is whether, because of their own interest in the economic and political well-being of their customers, it would be prudent for them to exercise self-restraint.

This interdependence is indeed a very weak reed to lean on. It is unlikely to become a controlling factor in the decision-making process of OPEC countries as long as the various importing countries and their oil companies, in the spirit of *sauve qui peut,* are willing to go to practically any length in order to secure their individual oil supplies.

IV

Ideally, one might perhaps have hoped that the major importing countries would have formulated an effective energy policy, and also established a coordinated approach to OPEC with regard to supply and financial problems caused by its actions. Instead, what has dominated international oil relations has been the fear of importing countries and their companies that OPEC may cut off their supplies, or impose even stiffer conditions at the slightest sign of resistance or of a common policy approach by the importing countries. The latter have acted as if they were in such a weak position that in order to obtain continued supplies they must act separately and try to gain favors by ingratiating themselves with OPEC countries by any means feasible.

This policy has been a failure. And the experience of 1979, previously described, has demonstrated how ineffectiveness resulting from fear and lack of common purpose has badly hurt the West, and in fact accelerated the disintegration and Balkanization of the world oil economy. Now again individual importing countries move in various directions, each one suspicious of the other, while pretending on the surface some desire for a unified approach.

The International Energy Agency (IEA), which was specifically established to deal with emergency supply situations, operates under terms of reference which do not allow it to deal with the actual situations as posed now or for that matter in 1979. And when its terms of reference did in fact permit it to intervene, the Agency apparently tried its utmost to avoid any actions, perhaps out of fear that OPEC might not like it.

Without going into details, it would appear that in order to cope with current problems where oil importers might be confronted by OPEC countries with extraneous or detrimental terms on supply or on pricing, the importing countries should agree among themselves on a common policy, backed up by an allocation scheme—as we had earlier suggested as the appropriate response to the problems of 1979.[1] Interestingly enough, Saudi Oil Minister Zaki Yamani recently suggested that the developed and developing countries should jointly adopt a scheme for the equitable distribution of world energy consumption.

Under such a scheme, all the quantities available on regular terms should be shared equitably among the various countries. It would at least reduce the fears of some countries or companies of being left out of the supply stream unless they were prepared to make special deals on basically unreasonable terms—which, when once accepted, might set the pattern for all later OPEC transactions.

However, it is clear that as a result of recent developments affecting the international oil trade, it has now become much more difficult for importing countries and their oil companies to participate in an international oil allocation program. Any allocation scheme would require some redistribution of the flow of world oil; it is predicated on a substantial degree of flexibility in the world oil supply system that had previously been assured by the dominant role of the international oil companies in controlling and directing most of the movements of oil in world trade. As mentioned earlier, this role has now been greatly reduced.

Instead, restrictions on destinations in many recent OPEC export contracts and the proliferation of direct oil supply deals between the governments of importing countries and OPEC national oil companies might deprive the importers of the necessary flexibility for the diversions of oil shipments. Moreover, the importers might well fear that if they arrange any such diversion, the producing country might cancel their oil supply arrangements. And in those cases where the consuming country has obtained oil supplies only by granting the OPEC country special political or economic advantages, it would, in any case, most likely be reluctant to make such oil available for reallocation to other countries.

1. See Walter J. Levy, "A Warning to the Oil-Importing Nations," *Fortune,* May 21, 1979, p. 48.

There is thus a clear and present danger that recent developments in the structure of world oil trade may have undermined the basis for the emergency reallocation system of the IEA and for a self-defense program of the importing countries as suggested here.

This is a matter for urgent consideration by the governments of the importing countries. They are now confronted with a challenge to their ability to try to cope as a group with major oil supply shortages or with a continued deterioration of the terms under which OPEC oil is made available to them.

The importing countries might well be unable to agree on a joint policy. But if they should fail to take a common stand on protecting their freedom to reallocate purchased oil among themselves, they would in fact endanger their capacity for an effective self-defense against a loss of vital oil supplies and hence seriously jeopardize their future economic, political and strategic viability.

V

The medium-term prospects for the security and availability of oil and for the economic and political stability and strategic security of the Western world are indeed disturbing. If we now consider the longer term outlook, the picture looks equally forbidding. Not only are all the economic and political problems to which we have already referred likely to remain with us; but in addition, even by the year 2000, the massive challenge of moving from a mostly oil-based energy economy worldwide to one that can largely draw on other more amply available energy resources will probably still remain largely unresolved.

A comprehensive and authoritative projection of the overall energy picture up to the year 2000 that is currently available is the Exxon publication, *World Energy Outlook,* released in early 1980. This study postulates a very high rate of expansion of energy production from non-oil sources during this period. Specifically, it forecasts a nearly 120 percent increase over 1980 in the non-communist world's use of coal; a more than quintupling of nuclear power; a more than doubling of hydro-power including solar, geothermal, etc.; and a buildup of synthetic fuel production from coal, tar sands, and oil shale, from virtually zero to seven million barrels per day.

In addition, the Exxon projections reflect a highly successful effort in improving the efficiency of energy use. Between 1980 and 2000, the average annual growth of non-communist world Gross National Product, adjusted for inflation, is estimated at about 3.5 percent, while energy consumption is projected to grow at only 2.5 percent per annum. This reflects a substantially expanded conservation effort from the past. For example, according to Exxon statistics, between 1965 and 1973 energy use in the non-communist world grew more rapidly (5.5 percent per annum) than economic output (5.0 percent per annum).[2]

Even on such assumptions, however, the Exxon study concludes that there would remain in the year 2000 a demand for oil of 60 million barrels daily, which would be 15 percent above the 1978 demand level. Oil and gas combined would by the year 2000 contribute 86 million barrels daily oil equivalent, or 52 percent of total non-communist world energy demand as compared with 72 percent in 1978. Exxon also assumed that:

> The combined net oil exports from the Soviet Union, Eastern Europe, and the People's Republic of China to the rest of the world would remain roughly constant during the period, at 1.0 million barrels daily, with declines in availability from the Soviet Union offset by increasing amounts from China;
>
> total non-communist world oil production outside OPEC would increase by some 7 million barrels per day to 26 million barrels daily by the year 2000; and
>
> production (including natural gas liquids) still required from OPEC countries would, at 33 million barrels daily, be 10 percent above its 1978 level and amount to 55 percent of total non-communist world oil supply in the year 2000.

The average annual addition to oil reserves through new discoveries in the non-communist world, between 1980 and 2000, was projected by Exxon to range between 9 and 15 billion barrels. To assess the reserve-to-production ratio for the year 2000, let us use the figure of 12 billion barrels, the midpoint of this range. Accordingly, the reserve-to-production ratio would decline from 31 years in 1980 to roughly 20 years by the year 2000. For most of the producing countries, except for a few of the oil-rich countries, the reserve-to-production ratio by the year 2000 would be 10 years or less.

In the meantime, with a continued decline in the reserve-to-production ratio during the 20-year period, it is almost inevitable that real prices for oil

2. In the case of the United States, the change is even more dramatic. For the period between 1978 and 1990, the economic growth rate in the United States is projected at about 2.5 to 3 percent per year, while the growth in total energy demand is projected at a rate of only 0.8 percent per year. This of course reflects the high consumption of energy in the United States and the resulting possibilities for a greater degree of conservation than is possible in other industrialized countries. In the same period, for example, energy demand in Europe is projected to rise by 2 percent annually, and in Japan by 2.9 percent. For other nations than the United States, Canada, Europe and Japan, the projected growth in energy demand up to 1990 is 5.8 percent per year. The result, naturally, is a substantial progressive change in the percentages of total world energy demand by various countries and areas—with the United States dropping from 41 percent to 33 percent, and Europe from 27 percent to 26 percent, while the "other" countries increase their share from 19 percent to 28 percent. Exxon, *World Energy Outlook*, pp. 6–7, 26–31.

would take off. The ensuing financial and economic problems could far exceed those which would be incurred between 1980 and 1985 and would—within the framework of our society and institutions—appear to be beyond our current capacity to cope. Moreover, many of the producing countries would certainly, for reasons of economic or political policy or because of physical limits set by productive capacity, reduce their production when their reserve-to-production ratio declines below a certain minimum. There is thus a "self-destruct" element in the relationship between the projected size of oil production and the economic and political consequences that it could entail.

VI

The question thus arises: If indeed the problems can be kept manageable, what must the importing countries do to assure their future political, economic and strategic viability?

From all that has been said so far, it would appear that the oil importers would probably need 30 years or more to create an energy economy based overwhelmingly on sources other than oil—and even this goal might not be achievable if we wished to sustain a viable rate of economic progress. During the 30 years-plus when we must surely still very much rely on oil, a large part of that oil must be obtained from the oil-rich countries of the Persian Gulf. Even in the unlikely case that within the next 10 to 15 years massive new discoveries should be made, comparable to those of the Middle East, it would take perhaps some 10 additional years before they could be fully developed.

It is thus imperative—during the period when the importing countries must procure each year more than 25 million barrels a day from OPEC producers and some additional substantial quantities from non-OPEC sources such as Mexico—that the real price of oil be kept at a manageable level. This may well mean that for some time to come prices should not increase at a much higher rate than that of inflation. The underlying concept would have to be that oil must be conceived in terms of a "common heritage of mankind" that must serve both the welfare of the producing countries and that of the importing countries.

That might well be considered to be an unrealistic utopian fantasy, especially as it would have to be based on an accommodation between oil producers and oil importers where both parties take full account of each other's vital interests. Above all it presupposes that the United States, the Soviet Union and possibly China agree on common principles that would apply to their policies affecting the area, and thus would be willing to abstain from any actions that might interfere with the maintenance or establishment of peace and tranquillity in the Middle East.

Specifically, it would require the conclusion of effective arrangements between oil-importing and producing countries on the size of needed oil exports, on a sustainable level of prices, on the planning and implementation of development programs for producing countries and on many other areas of concern to all the parties involved.

It would also imply that the threat of instability within producing countries and that of intraregional quarrels in the Middle East would be contained and neutralized. Within the Middle East, the Arab-Israeli conflict must be concluded, which in fact means that the autonomy issue and the Palestinian problem must be equitably resolved. But there are many other potentially explosive quarrels affecting the region—such as those centering around Lebanon; or the relationship between Iran, Iraq and other Arab countries; that of South versus North Yemen; the isolation of Egypt within the Arab world; the problems that are posed by the Libyan attitude towards its neighbors; and so on.

Within the various Middle East countries, the modernization of their political systems and of their economies must proceed without upsetting their traditional values. The new Orthodox Muslim movement, as it might be spreading from Iran, might otherwise engulf all those forces in the area that are seeking to achieve reasonable political and economic progress.

If all this were possible, we could probably solve the major problems that are related to the world's long-term dependence on Middle East oil. However, this kind of scenario does not reflect the world as it is now and is likely to be in the future; and we have not much reason to hope that the various producing countries, the importing countries, and the superpowers will ever agree on such a rational course of conduct—especially as the immediate dependence of the Soviet Union on Middle East oil is less than that of the United States and of most other Western nations.

Instead, we will probably be confronted by a series of major oil crises which might take any or all of several forms: fighting for control over oil resources among importing countries or between the superpowers; an economic-financial crisis in importing countries; regional conflicts affecting the oil-producing area; or internal revolutions or other upheavals in the Middle East. At best, it would appear that a series of future emergencies centering around oil will set back world progress for many, many years. And the world, as we know it now, will probably not be able to maintain its cohesion, nor be able to provide for the continued economic progress of its people against the onslaught of future oil shocks—with all that this might imply for the political stability of the West, its free institutions, and its internal and external security.

* * *

In surveying the international oil situation as it appeared at the very beginning of the decade of the 1980s, Walter Levy takes a rather gloomy view of the oil price revolution's effects, OPEC's aggressiveness in pushing prices up again, and the West's lack of an adequate response. To some extent, Levy's pessimistic outlook appears to have been conditioned by the atmosphere prevailing at the time of writing, soon after the 1978–1979 price spiral. In the next few years, the general mood in the West became much more positive as "oil crunch" turned to "oil glut." Some observers even went so far as to predict the imminent death of OPEC as its production levels slipped from the 25 million barrels per day cited above by Levy to less than 18 million barrels per day in 1981 and 14–15 million barrels per day in 1982. In part, this turnabout was caused by the upsurge in worldwide production, including both the opening of new fields in the North Sea, Alaska, Mexico and elsewhere, and the renewed exploitation of older, more expensive and marginal fields. At the same time, a combination of global recession, improved conservation measures and utilization of alternative sources of energy worked to dampen demand. To most analysts, however, the glut was only a temporary phenomenon and sooner or later demand would again outstrip supply. As Levy points out elsewhere,

> Even in the short-to-medium term no firm reliance can safely be placed on the future availability of the required volumes of Middle East oil at manageable prices. If nothing else, the experience of the 1970s should have taught us this. In spite of the present world oil glut, the outlook for most of the 1980s still looks to be highly precarious. . . .[1]

1. "Oil: An Agenda for the 1980s," *Foreign Affairs,* Vol. 59, No. 5 (1981), p. 1085.

Economic Interdependence Between Producers and Consumers

The ultimate consequences of the immense shift in the control of international oil go far beyond the immediate dislocations in the global economy. The last decade has also seen the gradual development of a grudging spirit of cooperation between producers and consumers, of a growing awareness of the need for understanding the other side's point of view. This process has not been an easy one. For the industrialized countries, it has involved a painful recognition of the permanence of the changes in the international economic system, as well as a psychological adjustment to the emerging power of OPEC.

The gradual unfolding of the ramifications of the "oil crisis" forced several changes in the West's perception of what had happened.

> At first, in the two or three months after October, 1973, the crisis was generally seen as a threat to the security of the oil importers' supplies. Then, around the beginning of 1974, after the price of crude oil had moved violently upward, the focus of concern in most countries shifted to the question of price. By midyear of 1974, however, the concern over price had been converted into a concern over the international monetary mechanism as a whole; considerable doubt arose over whether institutions existed or could be created that would be capable of handling the massive shifts in financial resources that were developing.[1]

Some observers have maintained that the hostile reaction of many in the West to OPEC's actions was motivated by political objections as much as economic.

1. Raymond Vernon, "An Interpretation," in Raymond Vernon, ed., *The Oil Crisis* (New York: W. W. Norton, 1976), p. 1.

> The OPEC action was unusual and unprecedented because it was taken by a group of Third World, primary-product-exporting countries. For the first time in modern history, some nonindustrialized countries had succeeded in securing market power in world trade in their export commodity, and they had used this market power to impose a substantial improvement in their terms of trade with the industrialized West.[2]

Were claims that an unchecked oil revolution would destroy the international economic system, and with it the national economies of the industrialized countries, justifiable? Were fears that producing states would use their newly acquired "muscle" to exert political blackmail on the West realistic? Fortunately, the dire predictions of economic collapse and the thinly veiled threats of military action to secure continued access to oil were soon proven equally groundless. While the world will continue to be dependent on OPEC oil for the foreseeable future, the OPEC countries will be just as dependent on the rest of the world. This includes not only the great bulk of their income but also for investment opportunities and for the goods and services necessary for their development.

Consequently, both sides need to recognize clearly the extent of economic interdependence between producing and consuming nations. They must also work for a durable, constructive relationship, involving not only cooperation on oil matters but a considerable Western involvement in the development of the producing countries and recognition of the particular problems and vulnerabilities of both sides. The OPEC states, of course, must bear their fair share of the responsibility for harmonious interdependence. They must remain fully aware of the dangers of continued inflation and recession in importing countries caused by spiralling oil prices. There must also be recognition that the price rises have been felt most acutely in the Third World and in the countries least able to absorb sharply rising requirements for foreign exchange.

The key to the successful transformation of the global economy since 1974 seems to have been the emergence of a workable system for recycling petrodollars. In large part, this process has been automatic. Producing countries can do only two things with their oil revenues: (1) purchase goods and services for domestic use, or (2) invest their income abroad. The rampant inflation in OPEC countries caused by the mad rush for development and the massive imports of development and consumer items has resulted in far more expenditure than anticipated, spurring those OPEC states with capital account deficits to demand higher and higher prices for their oil even as the number of capital-surplus countries has been reduced.

While the total amount of OPEC financial reserves remains considerable, the pattern of their investment abroad has generally contributed to the maintenance of a stable economy. The early course of placing oil

2. Norman Girvan, "Economic Nationalism," in Vernon, ed., *The Oil Crisis,* p. 145.

payments in bank deposits and treasury bills in the West is still popular. But it has been supplemented by extensive portfolio investments, by direct investment in some Western corporations and by real estate acquisitions. Most of this equity investment has taken place in Europe rather than the U.S., which remains particularly sensitive to foreign ownership.

To assist countries in correcting balance of payments deficits due to increased oil bills, existing multilateral institutions have been mobilized, as in the IMF's Special Oil Facility, and new ones created, such as the Afro-Arab Oil Assistance Fund. The West's large international banks have emerged as an important means for recycling OPEC deposits into loans to both industrialized and Third World countries. OPEC's special responsibility to the Third World has been acknowledged with the provision of substantial amounts of development aid, as indicated in Appendix D. This aid has been dispersed in various ways: (1) through multilateral institutions, such as the Islamic Development Bank and the African Development Bank; (2) bilateral institutions, such as the Saudi, Kuwait and UAE Funds for Economic Development; and (3) through *ad hoc* loans and grants by OPEC states to Third World states. While the latter are obviously susceptible to political objectives, bilateral institutions have been noted for the rigid economic criteria established for loans and grants and the wide geographical dispersion of their recipients.

In the past few years, considerable attention has been given to the "oil weapon," the concern being prevalent that the Arab oil-producing states might reinstitute an oil embargo against the United States and perhaps other importing countries in the event of another Arab-Israeli war, as was done in 1973. Often overlooked is the fact that the 1973–1974 embargo was only decided upon with obvious reluctance and even then was not unanimous. The producing states are well aware of the considerable drawbacks, both economic and political, of such a renewed policy. The two contributions to the present chapter approach this dilemma from substantially different angles.

First, Eric Davis stresses how the interests of the producing states converge with Western interests much more than they diverge. In large part, economic interdependence stems from the producers' reliance on a single-commodity export economy: they depend on Western markets in which to sell their oil. At the same time, they must also spend or invest the bulk of their revenues in those same countries. Professor Davis points out that this necessity of interdependence affects the radical Arab producers just as much as it does the conservative monarchies of the Gulf. William Brown's article was written somewhat later than the preceding one, after the appearance of the oil glut and a major downswing in Western dependence on OPEC oil. This development added another facet to the assertion that there is no "oil weapon": not only had the 1973–1974 embargo been ineffective but the oil glut now signalled the death of OPEC. Professor Brown assesses this argument in light of the events of the early 1980s and then sketches the outline of what he sees might be a viable Arab oil strategy.

Bibliographic Sources for this Chapter:
48, 67, 75, 85, 136, 166, 178, 180, 189J, 190, 196, 204, 205, 233, 234G, 293, 296H, 296I, 296J, 296K, 297, 309F, 309G, 326, 354D, 374I, 374M, 406.

The Political Economy of the Arab Oil-producing Nations: Convergence with Western Interests

ERIC DAVIS

As demand for oil and oil prices continues to rise in the West, interest in the oil producers' cartel, the Organization of Petroleum Exporting Countries (OPEC), has similarly increased. Most Western analysts have been concerned about a possible recurrence of an energy crisis such as that which followed the October, 1973, Arab-Israeli war. The increasing dependence of the United States and other Western powers on oil produced in Arab states and the threat by Arab leaders to use the "oil weapon" for political ends have caused Western analysts to stress the potential for conflict between the Arab oil producers within OPEC and the Western industrialized nations. The notion of an adversary relationship between OPEC and the West continues to dominate Western thinking, in part because Arab oil producers have linked the supply and price of oil to continued Western support for Israel.

This study is concerned with a number of considerations which have been largely neglected by students of OPEC and the "oil crisis." In their preoccupation with the potential for conflict between Arab oil producers and Western consumers, Western writers have frequently failed to appreciate the extent to which the interests of the powerful core of OPEC—the Organization of Arab Petroleum Exporting Countries (OAPEC)[1]—converge

An earlier draft of this article was presented as a paper to the 1977 annual meeting of the American Political Science Association. I would like to thank Leonard Binder and the Center for Middle Eastern Studies of the University of Chicago for a grant which allowed me to conduct research in Saudi Arabia during 1974. I would also like to acknowledge the valuable comments of Robert Kaufman, Michael Levy, Walter Weiker, and Abraham Yeselson.

1. OAPEC, established in January of 1968, contains ten member states: Algeria, Saudi Arabia, Bahrayn, Egypt, Iraq, Kuwayt, Libya, Qatar, Syria, and the UAA. As of January 1, 1976, OAPEC possessed 49.5 percent of world oil reserves—clearly dominating the larger OPEC cartel.

with, rather than diverge from, those of the West. Analysts have not given due consideration to the hypothesis that increasing economic interdependence between the OAPEC states and the West will tend to mitigate the possibilities of conflict. A second hypothesis, which reinforces the central argument about economic interdependence, is that the cleavage within OAPEC between those states which are ideologically radical and hawkish on oil prices and those which are politically conservative and more moderate is more a consequence of internal problems than of conflict with the West. This cleavage is exacerbated by Soviet support for the radical bloc within OAPEC and the threat that the conservative OAPEC states feel such support presents to their own regimes.

Table 1. Crude Oil Production in Arab Oil-Producing Nations, 1975–1976 (in 10^3 b/d)

Country	Year 1975	Year 1976	Annual Change (%)
Saudi Arabia	7075.4	8575.1	+21.2
Iraq	2261.7	2279.7	+ 0.8
Kuwayt	2084.2	2143.6	+ 0.5
Libya	1479.8	1932.1	+30.6
UAA	1663.3	1927.8	+15.9
Algeria	1020.3	1076.2	+ 5.5
Qatar	437.6	487.1	+10.8
'Uman[1]	328.3	366.0	+ 7.6
Egypt	206.0	328.0	+37.2[2]
Syria	172.9	175.0	+ 1.2
Bahrayn	61.9	58.0	− 6.5
OAPEC Total	16791.4	19348.6	
Total OPEC	27192.2	30576.2	
% OPEC	61.8	63.3	
Total World		57430.7	
% Total World		33.7	

1. Although 'Uman is not officially a member of OPEC and OAPEC, it has followed the lead of the oil cartel in setting prices.
2. This increase is due to Israel's return of the Abu Rudayis oil fields in the Sinai Peninsula.

Source: *Petroleum Times,* April 1, 1977, pp. 15–16; *Middle East Economic Digest Annual Review,* December 31, 1976, p. 3.

To argue that increasing interdependence between the economies of the Arab oil producers and Western consumer nations will tend to decrease the potential for conflict between the two is not to argue that the Middle East is moving toward an era of political tranquility. As long as the Arab-Israeli dispute remains unresolved, a renewed conflagration between the Arab states and Israel is always a possibility. Another Arab-Israeli war would inevitably politicize the economic ties which exist between the Arab oil producers and Western consumers. Furthermore, the arguments proffered are not intended to imply a harmony of interests between the Arab oil-producing nations and the West. While conflict in the Middle East thus remains a distinct possibility, it is hypothesized that increased interdependence as well as intra-OAPEC conflict may influence the degree to which the Arab oil producers, particularly conservative states such as Saudi Arabia, will be willing to initiate hostile actions against the West similar to the 1973 oil embargo.

The thesis that interdependence reduces the potential for conflict between the Arab oil-producing nations and the West needs another qualification. Even if a comprehensive Arab-Israeli peace settlement were reached in the near future, peace in the Middle East would not be assured. It cannot be assumed that the only basis for conflict between the OAPEC states and the West is Western support for Israel. The massive arms race currently taking place in the Persian Gulf between Saudi Arabia and Iran also poses a threat to stability in the region. Since the United States has supplied arms to both Saudi Arabia and Iran, it may very well be drawn into any conflict that might develop between these two oil-producing states. The intention of the United States to supply F5E fighter aircraft to Egypt raises the possibility of the United States getting embroiled in a war by proxy with the Soviet Union—should Egypt and Libya, a Soviet client, engage in hostilities. The intensifying conflict between Algeria and Morocco over former Spanish Sahara poses problems for the United States. Algeria is a major producer of oil and natural gas and a recipient of Soviet arms, while the United States is a major arms supplier to Morocco. Conceivably, the United States could become involved in intra-regional conflicts over which it would have little direct control. These conflicts could lead individual oil producers to cut off oil supplies to the United States and to urge other OAPEC states to do the same. These factors must be kept in mind when one considers the hypotheses which are proposed.

INTERDEPENDENCE IN OIL MARKETS

The growing economic interdependence between the Arab oil-producing nations and the West is most apparent in the oil market. Were it not for the increasing demand for oil in the West, oil would not have attained such extraordinary significance as a political commodity. Until the late 1960s and early 1970s, when demand for oil began to outstrip supply, many oil-producing countries, including Saudi Arabia, experienced budgetary deficits.[2] A sharp increase in the demand for oil in the West in the early 1970s

2. In 1968 and 1969, for example, Saudi Arabia ran budget deficits of 549 million riyals ($2.03 billion) and 535 million riyals ($1.80 billion), respectively (*MEED, Special Report,* December 1976:24).

accounts for the present surpluses which have accrued to many OPEC countries, especially the OAPEC states. According to the American Petroleum Institute, while 67 percent of United States oil imports in 1973 came from OPEC nations, only 21.6 percent came from Arab states within OPEC. By the end of 1977, 84 percent of United States oil imports came from OPEC member nations with the OAPEC share having risen to 42 percent of total United States imports (*New York Times,* Dec. 29, 1977). Europe and Japan are likewise highly dependent upon Arab oil. Western Europe, for example, absorbed almost 44 percent of Saudi Arabia's oil exports in 1976 while Japan accounted for nearly 20 percent of such exports. Overall, Japan imports over 80 percent of its oil from the Middle East (*MEES,* Feb. 11, 1977; Hurewitz, 1976:138).

The increased dependence of the West upon Arab oil would seem to be a source of reassurance for the OAPEC states. Nevertheless, one of the greatest fears among OAPEC producers is the specter of a depression in the West which would lead to a precipitous drop in the price of oil.[3] Except for a recession in 1974 and 1975, producer fears have not been realized. However, demand for oil in the West has slackened, with United States consumption of OPEC oil expected to rise only 3.5 percent in 1978 as opposed to 6 percent in 1977. Furthermore, as a result of the decline of the value of the U.S. dollar, it has been estimated that OPEC members lost approximately $5 billion between September 1977 and March 1978. Any attempt to raise oil prices during periods of weak economic growth in the West would be self-defeating since it can only further erode the value of Western currencies. While emphasizing Western dependence upon Arab oil, it should also be kept in mind that Arab oil producers are as dependent upon Western markets as are other single commodity economies of the underdeveloped world (*New York Times,* Mar. 5, 17, 1978).

In the longer term, however, increased interdependence due to oil sales to the West would greatly enhance OAPEC power. Despite a sluggish world oil market during 1977 and 1978, long term projections indicate that the West will grow more, not less, dependent upon Arab oil. As the largest OPEC and OAPEC producer, Saudi Arabia is using the amount of oil it produces to wield subtle economic and political influence in the West. Although the Arabian American Oil Company (ARAMCO), which is still 40 percent U.S. owned, has developed a $22 billion plan to increase Saudi oil capacity to 16 million barrels per day during the early 1980s, the Saudi government has indicated that, in the future, it will be more responsive to

3. Indicative of this fear are remarks made by Kuwayti Minister of Finance, 'Abd al-Rahman al-'Atiqi, in a recent interview with the *Middle East Economic Survey* (MEES). Arguing that the oil producers and the West have to find a way to stabilize the "vicious inflationary process" in manufactured goods, money, and oil, al-'Atiqi warned, "if the industrialized countries do not step down from their pedestal and reach an understanding with us, we shall drown together for we are all in the same boat" (MEES, 1977:16).

world market conditions than to the specific needs of the United States. According to Saudi oil minister, Shaykh Ahmad Zaki Yamani in 1978, for the next two years, Saudi production was to exceed its present output of 8.5 million barrels per day only if world oil demand increased. Energy specialists of the U.S. General Accounting Office predicted a global energy shortage in 1983 or 1984, unless the Saudis substantially increased their oil production within the next few years (*MEED,* Jan. 20, 1978:29). Thus the Saudis, and their allies within OAPEC, have acquired an important bargaining chip to use in their economic and political dealings with the West, particularly the United States.[4] If Western demand for oil does show a marked increase within the next decade, the Saudis will probably be less likely to hold down oil prices as they did in 1977 and 1978.

One of the central reasons why the West remains so dependent upon OAPEC oil is its inability to develop either alternative energy resources or a comprehensive energy policy which stresses conservation. For the United States, which has been the least successful of the advanced industrialized nations in energy conservation, continued importation of large quantities of foreign oil has also meant the erosion of the U.S. dollar. The decline in the value of the dollar has angered many of the oil-producing countries. Within OAPEC, Saudi Arabia and Kuwayt have threatened to call upon OPEC to shift from the dollar to a basket of Western currencies as the basis for oil payments. Such statements have only further weakened the dollar and have led to a lowering of stock prices in the United States.[5] On other occasions, the Saudis and their allies have expressed their continued confidence in the dollar. The impact that statements by OAPEC officials can have upon the value of the dollar not only illustrates the increasing interdependence between oil producers and oil consumers but also demonstrates the increasing influence of OPEC and OAPEC in Western economic affairs.[6]

A second area of increased economic interdependence stems from the oil sales surpluses accumulated by certain OAPEC states. The United States Treasury estimated that OPEC surpluses from 1973 to 1976 were between $135 and $145 billion dollars. Of this total, $32.5 billion was invested in United States Treasury securities ($13.8 billion), federal and corporate bonds ($3.7 billion), corporate stocks ($4.1 billion), bank deposits ($8.8 billion) and direct investments in corporations, banks and real estate ($2.1

4. For example, in exchange for the sale of fifty F-15 fighter aircraft, Saudi Arabia was reported to have guaranteed oil shipments to the United States totalling 2.5 million barrels per day for the next three years.

5. For example, the Dow Jones industrial average fell 12.30 points on November 29, 1977, after it was reported that OPEC was considering raising the price of oil (*New York Times,* 1977).

6. It is interesting to note that King Khalid reportedly informed President Carter during his trip to Saudi Arabia on January 3, 1978, that he did not want to see any further erosion of the U.S. dollar. The following day the United States Treasury stepped in to support the dollar (*MEED,* 1978:4–5).

billion). The same Treasury report placed OPEC investments in European money markets at $22.5 billion in 1974, $8 billion in 1975 and $10.5 billion in 1976. Of total OPEC investment in the United States, Saudi Arabia alone accounts for about $20 billion of this amount. The fact that most of these surpluses or "petrodollars" are invested in short-term, highly liquid assets, such as bank deposits, Treasury notes and corporate securities, makes them particularly vulnerable to inflation and serves to check too rapid a rise in oil prices which would feed inflation (*MEED Annual Report,* 1976:14, 31; *MEED Special Report,* 1976:27; *MEES*, May 30, 1977:17; *New York Times,* July 3, 1977, Sept. 22, 1977). Thus, Arab oil producers have gained a *vested interest* in the continued economic health of the West.

While technically retaining control over their foreign investments, Arab states have ceded much of the funds' management to a relatively small number of Western financial institutions. Most of the almost $47 billion in foreign assets controlled by the Saudi Arabian Monetary Agency (SAMA) in 1976 was thought to be managed by five American banks: Chase Manhattan, Morgan Guaranty Trust, Bank of America, First National Bank of Boston and Citibank (*MEED Special Report,* 1976:27). Despite the opening of 28 new Arab banks in London and 18 in Paris since the 1973 oil price increases, Arab investments of petrodollars in Europe are largely managed by European or joint-venture banks. Clearly Western financial managers are still more highly trusted than their Arab counterparts (*New York Times,* Nov. 27, 1977). Neither the economies of the OAPEC countries nor those of the non-oil-rich Arab states such as Egypt and Syria are capable of absorbing the petrodollar surpluses (*MEES*, May 30, 1977:14). The need to use Western nations as a source of investment for petrodollars serves to check the resort to hostile actions on the part of the OAPEC states. This desire to avoid conflict with the West should intensify as the Arab states shift more of their Western investments into longer-term holdings such as banks, real estate, and public utilities.[7] Indeed, there is an added incentive to invest in Western countries in order to help maintain the stability of their currencies.

Concurrently, the investment of huge petrodollar surpluses in Western consumer nations has made these countries much more attuned to OPEC and OAPEC interests. A United States Senate Foreign Relations Committee report published in September, 1977, warned that the $50 billion in short-term assets which OPEC countries have invested in the West could be withdrawn in the event of a new Middle East crisis; this would cause considerable political and economic instability. Although the Arab oil-producing countries need the West as a source of investment, they can wield considerable influence by shifting funds from one oil-consuming country to

7. As part of the trend toward longer term investments, one can cite the recent acquisition of a 20 percent share of Financial General Bankshares, Inc., a Washington-based bank, by investors from Kuwayt, Saudi Arabia, and the UAA (*The New York Times,* 1978).

another. Perhaps the most obvious use of petrodollar surpluses was the Saudi demand for a seat on the executive board of the International Monetary Fund as a *quid pro quo* for becoming the second largest contributor to the IMF's Witteveen facility (*MEED,* Aug. 19, 1977:30; Sept. 2, 1977:24).

A third level of increasing economic interdependence rests on the fact that in each oil-producing country oil exists in finite quantities and will eventually be depleted. In Saudi Arabia, which has the largest proven reserves, oil is expected to last another 80 years, while in other countries, such as Algeria, oil reserves will be exhausted within the next 25 years.[8] The fact that most of the Arab oil-producing nations were, prior to the discovery of oil, primitive tribal societies with no significant economic base, with a small if not non-existent agricultural sector, and sparse populations, has not been forgotten by the hegemonic classes which presently dominate these states. Algeria and Iraq, which have relatively large populations and developed agrarian sectors, face more acute problems of development than do the tribal monarchies of the Arabian Peninsula and Libya. All OAPEC states have therefore embarked upon massive development programs to industrialize their societies and to make them as agriculturally self-sufficient as possible before their oil is exhausted.

The scramble to industrialize and develop modern agrarian sectors has meant that not only have the OAPEC states expanded their economic ties with the West through increased oil exports but they have also drastically increased their imports from the West. Saudi Arabia's imports from Common Market nations increased from $43.8 million per month in 1973 to $231.3 million in 1975. The corresponding figures for all OECD countries were $122.4 per month in 1973 and $604.9 million in 1975. By the end of 1976 the Arab world had replaced the United States as the EEC's main trading partner, accounting for 13 percent of EEC exports as against 11 percent for the United States. For its part, the United States was expected to sell over $14 billion in goods to 19 Middle Eastern and North African countries in 1977 which represented a 27 percent increase over 1976 (*MEED*, Feb. 4, 1977:13, Feb. 18, 1977:15). The vast majority of such Arab imports include capital goods and advanced technology, manufactured commodities, arms, food and luxury items.

THE IMPACT OF INDUSTRIALIZATION

The massive import of capital goods and Western technology is linked to the intense desire of OAPEC states to initiate the rapid industrialization of their countries. Since most of the OAPEC nations have very small populations, Arab planners have sought to stress industrial projects which are

8. Rustow and Mugno (1976:128, Table 1): It should be noted that other sources put Saudi Arabia's reserves at about 60 years while Blair feels that true reserves are much higher (*MEED,* 1976:19; Blair, 1976:18–19).

capital intensive wherever feasible. Given the sophisticated nature of these industries, most require a relatively long period of time to be implemented. One such project involves the Saudi coastal towns of Jubayl and Yanbu. The Saudi Arabian Bechtel Company, a subsidiary of the Bechtel Corporation, the giant American construction firm, received a twenty year contract in 1976 to transform the towns into industrial centers. The ultimate cost of the Jubayl site, which will include the world's largest desalinization plant, petrochemical installations, an aluminum smelter, an iron and steel complex, power generating plants, and a huge telecommunications center, is placed at $45 billion dollars. Literally hundreds of Western firms have sub-contracted to complete various aspects of the project. Some are extremely costly projects such as Jubayl; others are more modest ventures such as the Umm Sa'id fertilizer plant in Qatar or the huge drydock facilities in Dubayy and Bahrayn. But all such projects require a long time for completion, hence a considerable Western presence in the Arab oil-producing countries for some years to come.[9] In addition, the OAPEC states do not have the requisite managerial expertise to operate these sophisticated industrial projects once they are completed, which means that foreign managers must be employed. Since the United States and Great Britain have been the major suppliers of Western technology, capital goods, and expatriate managers to the OAPEC states in general, and to the Arabian Peninsula in particular, OAPEC members would be highly reluctant to jeopardize their industrialization programs by engaging in hostilities with the West. The United States has repeatedly warned that a renewed Arab oil embargo would be considered an act of aggression. Were such a boycott enacted, it seems certain that the forced exodus of American nationals from the participant countries, which would result, would have disastrous ramifications for these fledgling development programs.

In addition to the advanced technologies, capital goods, and manufactured products, which the OAPEC states import, significant quantities of their food also come from the West. Libya, for example, produces only 14 percent of its domestic needs while importing the remaining 86 percent from the EEC (*MEED* Special Report, 1977a:vii). States such as Saudi Arabia and Kuwayt have invested heavily in agricultural joint-ventures with Egypt and Sudan. Although oil producers such as Saudi Arabia and Libya hope to be self-sufficient in food production during the next decade, most experts feel that such goals are unrealistic. Since too-rapid development in agriculture (e.g., through indiscriminate irrigation) could lead to the salinization of topsoil and hence exhaustion of the land, anticipated self-sufficiency in food production probably will not occur until the next century. Meanwhile, growth in domestic populations and the increasing

9. A large proportion of the population of Jubayl will be comprised of expatriate managers and technicians (*New York Times,* 1978; *MEED,* 1976:20, 11; *MEED Special Report:* Sea Transport, 1977:vi–vii).

influx of expatriate labor employed in domestic industry and construction may increase, not decrease, OAPEC dependence on food from the West for at least the remainder of this century (*MEED Special Report,* 1976:22).

Increased economic interdependence is perhaps most dramatically evident in the area of arms sales. As the Arab oil-producing nations accumulate ever larger surpluses, and invest substantial amounts of capital in domestic development, fears of internal and external subversion greatly increase. This has been especially true in the Arabian Peninsula where Saudi Arabia, Kuwayt, the United Arab Amirates (UAA) and 'Uman have been particularly wary of the expansion of Soviet influence in the region—through Iraq, the People's Democratic Republic of Yaman (PDRY) and Ethiopia.

Table 2. Population and Per Capita Income of the Arab Oil-Producing Nations, 1974

	Population (in thousands)	Per Capita GNP (in dollars)
Algeria	15,175	650
Bahrayn	245	2,250
Egypt	36,350	280
Iraq	10,770	970
Kuwayt	991	11,640
Libya	2,240	3,360
Qatar	190	5,830
Saudi Arabia	8,008	2,080
Syria	7,127	490
UAA	198	13,500
'Uman	750	1,250

Source: *Middle East Economic Digest,* August 20, 1976, p. 28.

To a lesser extent, the Arab arms buildup has been a response to the massive defense program undertaken by Iran. What is most significant for our present concerns is that the vast majority of arms purchased by Arab oil-producing states, especially those in the Arabian Peninsula, has been from Western countries, primarily the United States, Britain, and France. By August, 1976, the United States had sold $6 billion worth of arms to Saudi Arabia. Under the most recent proposal of the United States government, Saudi Arabia will purchase 60 F-15 Eagle fighters (plus training and maintenance facilities) at an additional cost of $2.5 billion (*MEED, Special Report,* 1976:31; *New York Times,* Feb. 15, 1978). Arab awareness of dependence upon Western arms has prompted attempts to develop domestic arms industries such as the proposed $10 billion complex at Al-Kharj, Saudi Arabia, which is being planned by Edward Durell Stone Associates of New York. In 1978, France signed an arms accord with Egypt, Saudi

Arabia, Qatar and the UAA, which were part of a Cairo-based Arab arms consortium, to build French arms under license in Egypt. Rather than making Arab states independent of Western arms, such agreements only serve to create greater interdependence between Western and Arab armed forces (*New York Times,* Feb. 18, 1978, Mar. 15, 1978).

As with purchases of capital goods and technology, arms sales to Arab oil-producing countries have recycled a considerable amount of Arab petrodollars in the West. Such sales have created dependence among Arab armed forces on spare parts, technical advisors and maintenance personnel from the West. Indeed, this has been precisely one of the policy objectives of the United States government since 1973 (*New York Times,* Feb. 17, 1978). Of the $10 billion dollars that Saudi Arabia allocated to defense in the 1976–1977 budget, a considerable portion was spent on the construction of military installations and on army training and maintenance programs. While most construction in Saudi Arabia is being carried out by the U.S. Army Corps of Engineers, training programs are being conducted by a host of Western corporations such as Lockheed, Northrop, Litton Industries, Raytheon, Vinnel and British Aircraft Corporation (BAC) (*Meed Special Report,* 1976:31; *New York Times,* Feb. 24, 1975). These corporations have also provided substantial numbers of maintenance personnel. It is therefore possible that the armed forces, especially the air forces, of the Arabian Peninsula oil-producing states would be severely hindered, if not rendered ineffective, were Western military advisors and technicians to withdraw from these countries.

The nature of increased economic interdependence between the Arab oil-producing nations and the West becomes more apparent if we consider the problems of human resources facing most of the OAPEC states. Ambitious industrialization projects, a rapid expansion of the domestic infrastructure, and drastic increases in the size of Arab armed forces have necessitated a large-scale mobilization of educational resources. Large numbers of Arab students have been sent abroad to Western universities, while Western firms and educational institutions have been employed to establish massive vocational training progams as well as new programs of higher education in the OAPEC countries (*MEED,* Dec. 3, 1976:8–10; *MEED Special Report,* 1976:16; *MEED Special Report,* 1977a:ix). A shortage of technical expertise has meant that large numbers of Western managers and technocrats have found employment in the Arab oil-producing states. The British firm of Career Group of Companies stated in an April, 1977, report that 59 percent of all overseas technical appointments by British employment agencies were in the Middle East, primarily in Kuwayt, Saudi Arabia, Iran, Iraq and Qatar (*MEED Special Report,* 1977c:iii). Some countries, e.g., Libya, expected to reduce their dependence on foreign manpower by 1980. But while Libya's dependence on foreigners was to decrease from 60 percent to 40 percent of the workforce, Saudi Arabia's dependence was expected to increase from a level of 24.4 percent to 53.5 percent in 1980 (Wells, 1976:41).

Much of the technical-managerial manpower shortage in the conservative OAPEC states is exacerbated by the tendency of royal family members to avoid pursuing public and private sector careers which require technical training or managerial expertise. Saudi princes, for example, prefer positions in the armed forces and foreign ministry or provincial governorships to positions in the planning ministry or the national monetary agency. It is true that the Saudi royal family is limited in numbers and that the Saudi regime gives high priority to placing members of the royal family in key positions relating to the defense and national security of the state. Yet, it seems clear that members of the Saudi royal family enjoy their oil wealth far more than they like to work to create it.[10] Similar attitudes prevail in the other conservative monarchies. In Kuwayt, where Kuwaytis form only 40 to 45 percent of the population, members of the royal family and the native populace prefer plush public sector directorships, which require little work and which yield lucrative salaries, to positions in the private sector (*MERIP Reports,* 1975:6). The results of such attitudes, in Kuwayt and elsewhere, is dependence on expatriate expertise.

A final dimension of the increased economic interdependence between the Arab oil producers and the Western oil-consumer nations involves multinational capital. The interaction of Western and Arab capital is most evident in the oil industry. Other OAPEC states were induced to emulate Libyan nationalization of some foreign oil companies and demand for participation in others. In most instances, host governments did not opt for nationalization but rather demanded participation rights. Thus 60 (but not 100) percent of ARAMCO was taken over by the Saudi government. No less significant are joint-venture enterprises in banking, commerce, and industry. To date, joint-venture enterprise is most prominent in the banking sector. Examples include the London-based Saudi International Bank, which is controlled by SAMA and includes American, British, Japanese, Swiss and private Saudi capital, and the Paris-based Banque Arabe et Internationale d'Investissement which incorporates American, French, British, German, Qatari and Kuwayti capital as well as capital provided by the Abu Dhabi government (*MEED,* Sept. 24, 1976:38; *New York Times,* Nov. 28, 1977). The commercial and industrial spheres include enterprises such as the Anglo-Saudi firm of Laing-Wimpey Alireza—which includes the prominent 'Ali Riza family of Jidda—and Algosaibi Grandmet SDCC

10. Although no systematic data have yet been accumulated on political recruitment in Saudi Arabia, this is definitely the impression I received in interviews with officials at the Saudi Arabian Monetary Agency and with a member of the prominent 'Ali Riza family of Jidda in August 1974. Also note the composition of the present Saudi cabinet in *MEED Special Report:* Saudi Arabia (1976:1) and *Who's Who in Saudi Arabia* (1977:263 et passim). It should be noted that of the 38 princes listed, only seven are actively engaged in commercial affairs and only two of them have had formal training in business and finance.

Services, which was formed by the Algosaibi (al-Ghusaybi) family of Bahrayn with British and Danish capital (*MEED,* July 2, 1976:20–21; *MEED Annual Review,* 1976:28).

While need to industrialize and develop their armed forces has made the Arab oil-producing states highly dependent on the West, it has similarly created a dependence of the West upon Arab markets. The development of massive markets for manufactured commodities, capital goods, arms and manpower (especially during a period of sluggish economic growth) has created a bonanza from which none of the Western industrialized countries wishes to be excluded. As a result, OAPEC members, particularly more powerful producers such as Saudi Arabia, have been very effective in playing various Western firms off against one another. OAPEC has also used its highly valued markets to force Western firms to comply with the Arab boycott of Israel (*MEED,* Sept. 26, 1977:13). In addition, Arab oil producers have sought to use larger numbers of firms from friendly underdeveloped nations in order to gain better terms from Western firms. In 1977, the Saudi Ministry of Industry abruptly cancelled bids from Western firms for rural electrification projects claiming that the projected costs were too high. Instead, contracts were awarded to firms from Pakistan, India, Taiwan and South Korea (*MEED Annual Review,* 1977:78). Western firms were put on notice that they would lose additional contracts in Saudi Arabia and other OAPEC countries if they offered inflated bids. While the Arab oil producers need Western industrial goods and technical expertise, only through the continued purchase of large amounts of Western industrial commodities by the Arab oil-producing nations can the consumer nations hope to recycle their petrodollars—thereby limiting their trade deficits and maintaining healthy economies.

THE RADICAL BLOC WITHIN OAPEC

In discussing increased economic interdependence between the Arab oil-producing nations and the West, the focus has been primarily on the conservative pro-Western monarchies of the Arabian Peninsula. However, OAPEC also includes a so-called radical bloc comprised of Algeria, Iraq and Libya. In light of the professed hostility of these countries to Western interests, and their strong ties with the Soviet Union, these countries might provide an exception to the hypothesis that increased economic interdependence should lessen the likelihood of conflict with the West.

Superficially, it would appear that the larger populations and developed agrarian sectors of Algeria and Iraq, as well as the access to Soviet arms by all three members of the radical bloc, would make them immune to dependence on the West. Closer scrutiny of the economies of these countries shows that this is not the case. The members of the radical bloc within OAPEC, particularly Algeria and Iraq, feel the need to achieve rapid economic development more acutely than do the conservative OAPEC states. In Algeria and Iraq, large populations and relatively limited oil reserves necessitate higher oil prices to finance rapid economic development. Libya, which has been more reserved in its call for increased prices, seems to be influenced more in its relations with the West by the ideological concerns of President Mu'ammar al-Qadhdhafi and the ruling Revolution Command Council than by economic considerations. Nevertheless, Libya is highly dependent on Western manpower and technology.

The problems facing Algeria and Iraq stem in large measure from their increasing indebtedness resulting from the imports of large quantities of Western capital goods and technology. As with many other developing countries, this indebtedness has been exacerbated by the inflationary spiral in the West. Thus the Algerian government has given high priority to heavy industry, such as a proposed multi-billion dollar complex which will produce machinery for the steel, chemical, petrochemical, electric and mining industries. Dubbed "Little Pittsburgh," this project will be constructed by a foreign consortium and is expected eventually to employ 25,000 workers (*MEED,* July 9, 1976:18).

While its development plans are impressive on paper, Algeria is finding it increasingly difficult to finance such projects. During 1976, the United States Import Export Bank restricted any further loans to Algeria, except those on projects bringing a quick return on investment. The bank claimed that Algeria was too heavily in debt and could not continue to support its debt service on large loans when its oil revenues were lower than had been forecast. Oil output increased only 3.1 percent between 1975 and 1976. The delay in implementing many development projects was also cited by the bank as a reason for Algeria's poor credit rating (*MEED,* Sept. 10, 1976:12).

Not only has Algeria been adversely affected by insufficient oil revenues but also by the refusal of the Federal Power Commission to ratify the shipment of large quantities of liquefied natural gas to the United States. Algerian proven natural gas reserves, which comprise 2,800 billion cubic meters, are the most extensive in the world. Under the Second Development Plan (1974–1977), the Algerian government allocated 40.6 percent of investment capital to the development of the hydrocarbon industry. The inability of Algeria to ship liquefied natural gas to the United States in any significant amount has meant the loss of its largest potential market. Most observers feel that American reticence is based upon national security questions concerning Algeria's socialist orientation (*MEED,* April 1, 1977:18).

Like Algeria, Iraq is engaged in an ambitious development program. In 1977, 38 industrial projects were expected to be completed at a cost of $1.04 billion by the state supervisory body, the General Organization for Planning and Industrial Construction (*MEED,* Nov. 23, 1977:6). Iraq has also experienced problems resulting from a sluggish oil market and a rising import bill. Thus Iraqi oil production in 1976 increased only 0.8 percent over that of 1975 (see Table 1). Further indications of economic difficulties are apparent: a budget deficit of $475 million in 1975, the small 4.2 percent increase in state spending in 1977 as compared to 1976, and a public debt of over $6.5 billion. Government officials are also concerned about productivity in Iraqi industry. Minister of Industry and Minerals, Fulayyih Hasan al-Jasim, informed a Baghdad symposium in September, 1976, that while investment in industry during the past five years had been three and one-half times as great as during the previous twenty years, returns on investment in Iraq were among the lowest of all developing states. According to al-Jasim, Iraqi industry was running a deficit of $410 million and operating at only 30 percent of its productive capacity. The Iraqi government's concern with this situation was underlined by Vice President Saddam Hussayn who asserted that there would be no wage increases for either workers or management until productivity in industry improved (*MEED,* Sept. 17, 1976:13).

The third of the radical states within OAPEC is Libya. Having ousted King Idris and the al-Sanussi family in 1969, the Libyan regime of Mu'ammar al-Qadhdhafi is an anathema to the royal families of the Arabian Peninsula who fear a fate similar to that of the Libyan monarchy. While wealthier in per capita terms than either Algeria or Iraq, Libya has experienced a lag in export earnings since 1970 due to a steady decline in oil production—which only began to reverse itself in 1976 (*MEED Special Report,* 1975a:v). Libya shares with Algeria and Iraq the problem of acquiring sufficient revenues to meet the rapid escalation in the price of imported capital and manufactured goods necessitated by a comprehensive development program. Unlike Algeria and Iraq, Libya does not have adequate manpower to meet its development needs. Large numbers of Egyptians and Turks are already employed in Libya, and the Libyan government is seeking to import still more workers, especially Turks (*MEED,* Mar. 25, 1977:23; *MEED Special Report,* 1977c:xiv). Many Western analysts feel that Libya's ability to implement its development program will be determined by whether or not it can acquire the requisite manpower resources.

The desire of this radical bloc within OAPEC to raise oil prices has led to a struggle with the conservative oil producers. The tension finally led to open confrontation during the January, 1977, OPEC meeting in Qatar and the emergence of the so-called two-tier price system. The moderates led by Saudi Arabia and its client, the UAA, agreed to only a 5 percent increase in the price of oil during 1977 while Algeria, Libya, and Iraq, together with Iran and Kuwayt, increased their prices 10 percent in January and agreed upon another 5 percent on July 1, 1977 (*MEED,* Jan. 7, 1977:7). This power struggle was resolved in favor of the moderates, once the OPEC states which raised prices suddenly faced a drop in demand for their oil. The

National Iranian Oil Co. (NIOC) reported a drop of almost 50 percent in its oil sales during the week following the January conference, and during February it was reported that Iraqi crude was being sold at a 30 to 35 percent discount (*MEED,* Jan. 7, 1977:7, Feb. 18, 1977:22). The cancellation of the additional 5 percent price increase during the July, 1977, OPEC meeting not only indicated once again the extent to which oil producers are dependent upon Western markets, but it also underlined the extent of Saudi Arabia's hegemony within OPEC and OAPEC.

As I have suggested, the strident rhetoric of the radicals within OAPEC is less a testament to their ideological commitment to socialism than to their concern with development problems and their discontent with the conservative Arab states which dominate OPEC and OAPEC. It is not surprising, therefore, to find that Iraq and Libya actually increased the amount of oil sold during the 1973 boycott (Stork, 1974:225–226, 235). While the media in these countries accuse the United States of being an imperialist power, American firms continue to be awarded lucrative contracts in both Algeria and Iraq.[11] Despite Libya's lead among the Arab oil producers in nationalizing foreign oil companies, relations between Western multinationals and the Libyan government have been cordial (*New York Times,* Sept. 17, 1976).

The cleavage within OAPEC between conservatives and radicals suggests an additional hypothesis which reinforces the arguments regarding economic interdependence. The state apparatus in the conservative oil-producing nations has severely limited political participation to the tribally based royal families, the upper echelons of the religious hierarchy, a small number of influential merchant families, and technocrats. One of the most severe internal contradictions facing the conservative Arab oil-producing regimes is how to preserve their traditional monarchies in the face of attempts at rapid industrial development. Rapid socioeconomic change without any corresponding political change bodes ill for the conservative monarchies. These regimes have already been challenged by opposition emanating from the military, native technocrats in the oil industry, and the large numbers of Palestinians who are employed in these states (Halliday, 1975:78–79, 80; *Oman: A Class Analysis*, 1974:6).

The threat of the radical OAPEC states is as much political as it is economic since they all possess highly nationalistic republican regimes which provide an alternative political model for anti-monarchist elements in the conservative states. Furthermore, the radicals have been staunch

11. Contrast the stridency of "hawar ma'al-rafiq 'Adnan al-Hamdani hawla istratijiyat al-siyasa al-naftiya" [an interview with Comrade 'Adnan al-Hamdani, Secretary General of the Ba'th Party's Committee on Oil Affairs Concerning Oil Policy Strategy"] in *majallat al-naft wa-l-tanmiya* [Journal of Oil and Development] (1975:21–33) with the list of contracts awarded by Iraq to foreign firms in *MEED Annual Review* (1977:67–68). It should be noted that one of the most prominent of the foreign firms operating in Iraq is the Bechtel Corp. of San Francisco.

supporters of the Palestine Liberation Organization (PLO) which contains a large Marxist faction hostile to the conservative monarchies. Given the large number of Palestinians among the working classes and technocratic strata, the conservative oil producers are faced not only with potential domestic opposition but also with a Palestinian fifth-column within their borders.

Consequently, the conservative oil producers are even less willing to provoke a severe confrontation with the West. Saudi hesitancy following the outbreak of the Six Day War to institute an oil embargo against the United States brought protests from Palestinians in the oil fields. Similar protests have erupted in Kuwayt where Palestinians comprise over half of the total population (Stork, 1974:113; *MEED,* Sept. 3, 1976:3–4). Were another Arab-Israeli war to erupt, the conservative OAPEC regimes would face the unsettling prospect of having to choose between failure to initiate an oil boycott—thereby incurring the wrath of the domestic populace and Palestinian elements—and, by initiating a boycott, provoking the United States and its NATO allies to withdraw their military and technical personnel. Such a step on the part of the West would deal a serious blow to economic development in these countries and make the conservative monarchies more susceptible to internal subversion.

A third hypothesis relates to the conservatives' fears of increased Soviet penetration into the Middle East. Attempts have been made with varying degrees of success to woo present and former Soviet clients such as the PDRY, Somalia, Egypt, Syria, Sudan, and the PLO away from Soviet influence. Since Soviet influence has often been enhanced by increased political instability in the Middle East, and given that the Soviet Union is most influential in the radical states, it is in the interest of the conservative members of OAPEC to prevent renewed conflict with the West from erupting, for this would place them in a weaker position regarding the Soviet Union and its Arab allies.

Not only is the West essential to all OAPEC states economically, but it is essential to the conservative OAPEC states militarily if they are to retain their hegemony within OAPEC. This does not mean that the interests of Western countries and the conservative OAPEC states are identical. Hostility to Israel is still intense within these states (although recent indications are that Saudi Arabia would recognize Israel in the event of a comprehensive peace settlement) (*New York Times,* Mar. 10, 1978). The declining value of the United States dollar threatens to lead OPEC to raise oil prices or to shift from the dollar to another Western currency as the accepted basis of oil payments. However, the response of Saudi Arabia, the most powerful OPEC and OAPEC member—hence the key to OAPEC-Western relations, has been relatively moderate regarding its political and economic differences with the West. The Saudi regime is no longer speaking of an oil boycott in the event of a new Arab confrontation with Israel but rather of a significant increase in the price of a barrel of oil.[12] On economic issues, the Saudis tacitly supported the West in the North-South Dialogue by refusing to support radical demands for transfers of wealth from the prosperous industrialized nations to underdeveloped countries.[13] Saudi Arabia has committed some of its massive reserves to easing the budgetary deficits of those underdeveloped countries which are resource-poor and have suffered from the rapid increase in oil prices since 1973 (*MEED,* Mar. 4, 1977:39). By assuming a greater role in the international economic system, Saudi Arabia has indicated a willingness to relieve the West of having to commit its own resources to assure the continued health of the world economy. Having joined the elite among Western nations who exercise a dominant role in global economic affairs, the Saudis have also become more committed to assuring its continued health and stability, thus reinforcing their reluctance to provoke confrontation with the West. As the Saudis envisage their future role, it is to act as a powerful broker between the advanced industrialized countries of the West and the underdeveloped world (*New York Times,* Dec. 23, 1977).

SUMMARY

It has been argued that the increasing economic interdependence between the OAPEC states and the West will continue to reduce the possibilities of renewed conflict between oil producers and oil consumers such as occurred during the fall of 1973. It is predicted that all of the OAPEC states will become increasingly reluctant to resort to another oil embargo as their investment in programs of economic and military development continues to grow. The costs of a withdrawal of Western investment and technical expertise as a result of an embargo would deal a serious blow to such development programs. Similarly, it would leave many regimes, especially the conservative monarchies of the Arabian Peninsula, vulnerable to political challenges by radical elements within the Arab world and to the threat of increased Soviet influence. Thus, the hypothesis that economic interdependence will increase is reinforced by considerations relating to intra-OAPEC cleavages and the desire of the conservative producers to maintain their political and economic hegemony within the oil cartel. For the West, the disruption of oil supplies and trade with highly lucrative Arab markets would represent a political and economic disaster. As a result, the advanced

12. In 1977, the Saudis were talking of a $3 increase in the price of a barrel of oil.

13. In the North-South Dialogue, Saudi Arabia, and to a lesser extent, Iran opposed "OPEC hawks" such as Iraq and Nigeria at the conference. They supported the West on such issues as raising the question of oil supply and demand and in fighting to prevent the issues being moved from the forum of the conference to that of the United Nations (*MEES,* 1977:4).

industrialized countries are eager to avoid conflict with the Arab oil-producing nations and increasingly seek to placate them in their political and economic disputes with the West. Rather than solely emphasizing the potential for conflict between OAPEC and the West, greater attention should be given to the nature of the emerging alliance between its conservative members, especially Saudi Arabia, and the dominant Western power, the United States.

REFERENCES

BLAIR, JOHN
1976 *The Control of Oil.* New York: Pantheon Press.
The Gulf Committee
1974 *Oman: A Class Analysis.* London (No publisher is indicated in brochure.)
HALLIDAY, FRED
1975 *Arabia Without Sultans.* New York: Vintage Press.
HUREWITZ, J. C. (ed.)
1976 *Oil, the Arab-Israeli Dispute and the Industrial World. Boulder, Colo.:* Westview Press.
Majallat al-naft wa-l-tanmiya (Journal of Oil and Development)
1975 vol. 2, no. 2
Middle East Economic Digest
1976 (July, August)
1977 (February, July, August, September, November, December)
1978 (January, March)
Middle East Economic Digest Annual Review
1976 (December 31)
Middle East Economic Digest Special Reports
1976 Saudi Arabia (December)
1977a Libya (February 18)
1977b Sea Transport (March 25)
1977c Manpower and Recruitment (June 24)
Middle East Economic Survey
1977 (May, June)
Middle East Research and Information Project Reports
1976 No. 36 (April 1975)
RUSTOW, DANKWART and JOHN MUGNO
1976 *OPEC: Success and Prospects.* New York: New York University Press.

STORK, JOSEPH
1974 *Middle East Oil and the Energy Crisis*. New York: Monthly Review Press.

WELLS, DONALD
1976 *Saudi Arabian Development Strategy*. Washington, D.C.: American Enterprise Institute for Public Policy Research.

Who's Who in Saudi Arabia
1977 Jedda, Saudi Arabia and London: Tihama and Europa Publications.

The Oil Weapon

WILLIAM R. BROWN

In the international arena all states are expected to operate with the power at their disposal. This premise is also true of any combination of states, such as the NATO complex or that represented by "the Arabs." For a long time now, the United States and Israel have been able to manage the implications of Arab power without accommodating the Arabs' view of appropriate political relationships for the Middle East. Despite early prognostications about the evil things that the Arabs would do if they ever became powerful, the Western industrialized countries found that they could live with the influence that marketing arrangements after 1973 allowed Arab oil producers. We even became comfortable in the presence of the "oil weapon." Some observers now see oil as no more potent a political device than the spectre of Sorel's general strike that transfixed so many Frenchmen during the 1930s.

Much of this argument rests on the assertion that when the Arabs declared an oil embargo in 1973, it was ineffective, a conclusion that cannot be challenged. Then there is the assertion that petroleum is a matter of economics and that the Arab oil producers are not likely to sacrifice immediate income for some far fetched political purpose associated with the Palestinians. The implication is that a homeland for the Palestinians and the rescue of the Dome of the Rock from Israeli sovereignty are just not important enough in the overall scheme of things to cause the Saudis and Kuwaitis, not even Mu'ammar Qadhdhafi, to take the risks that would be involved in any effort to redress the situation.

Then there is another assertion: The oil glut beginning in 1981 proved that OPEC is dead—and so is any possibility of the Arabs using oil for political purposes. Rather than proceed with this argument in general

Reprinted by permission of the publisher, The Middle East Institute, from *The Middle East Journal,* Vol. 36, No. 3, Summer 1982.

terms, it is more fruitful to look at the details of petroleum marketing arrangements—to place the prospects for Arab action into a factual context. What does the information at our disposal tell us about the Arab capability to apply political pressure through oil? What is the ability of consuming countries to withstand that pressure? Particularly as we look ahead to the mid-1980s and beyond, we see the significance of these questions.

A total suspension of Arab oil production would be effective. Even those who deny OPEC power concede this point. But a suspension of production would be difficult to coordinate, and it would be so damaging to friend and foe alike (not to mention the Arabs themselves), that it is unlikely to occur unless out of cataclysmic happenstance. Clearly another selective embargo (without total suspension of production) would be self-defeating. But this prohibition does not preclude the Arabs from considering a policy that would selectively pressure the United States. What sort of policy, then, would the Arabs have to follow to edge toward the position in which their oil resources received greater, if not ultimate, consideration in a Middle East political settlement?

MARKETING DYNAMICS

More than one-half of the oil utilized by Western Europe (excluding Britain and Norway) and Japan must still be imported from producers who for eight years (1973–1981) exercised a cartel's control over petroleum supplies through the Organization of Petroleum Exporting Countries (OPEC). And as late as 1981 the United States depended upon imports for 35 per cent of its petroleum requirements, although only 20 per cent of our consumption was of OPEC origins. The changes that occurred in the petroleum market over these eight years were in price, not in the availability of supply. For the United States, the cost of imported oil increased from $2.40 a barrel in 1973 to $13 in 1976 and on to $37 in 1981.[1] This increase resulted in the usual market reaction to higher prices—consumption declined. The outcome was a glut of oil on the international market and a moderation of prices.

Beginning in 1981, and on into the first quarter of 1982, between two and three million barrels a day were being produced in excess of consumers' demand. This condition resulted from two factors. First, non-OPEC producers—principally Mexico, Britain, Norway and Egypt—entered the market with a total production that reached perhaps six million barrels a day. The Soviet Union, with some fluctuations in its supplies, brought 1.5 million barrels to the European market. But in addition, there was another

1. *Monthly Energy Review,* DOE/EIA-0035 (82/03), US Department of Energy, pp. 14, 76.

problem. OPEC producers had different marketing positions. And quite naturally, they adopted divergent marketing strategies.

Logic would tell us that those having limited reserves would wish to conserve any natural endowment, selling it only as their own demands for imports required that they generate foreign exchange. But there has been more to it than that. Indeed, the countries with limited economically exploitable reserves (Algeria, Nigeria, Indonesia, and to an extent Libya) for awhile attempted to sell as much as they could as fast as they could—but also at the highest possible price. All were spending at a high rate on consumption, economic development and arms, with imports constituting a major part of expenditures. They saw themselves as being in need of their oil earnings immediately.

In addition, the petroleum resources of these countries will be largely depleted by the year 2000. This is not the case, however, for the world as a whole. Total reserves will last much longer, even at our sometimes profligate rate of consumption. Conceptually Algeria, Nigeria and Libya saw their reserves as being large enough to require them to do whatever they could to retain their share of the market, particularly as upward movement in prices brought a noticeable reduction in consumption within the importing countries. None of these exporters seemed to be concerned with the possibility of high prices ultimately driving consumers to other sources of energy (principally synthetics produced from coal, oil shale, or tar sands). Their position was rational insofar as their reserves would be exhausted before a large-scale conversion to synthetics could take place. Long-range considerations did not exist. They felt compelled to produce at levels beyond their immediate foreign exchange needs. To cut back production and conserve could mean that any producer who followed this strategy might not be able to regain its market at some future date without drastically reducing prices. Such a move would defeat the very purpose for which each producer joined the cartel. They did fail to recognize, however, that high prices at the gasoline pump would allow higher cost producers to enter the market. After oil reached $16 a barrel, much of Mexico's oil became economically exploitable. The same occurred for oil from the North Sea and the Northern Slope of Alaska as prices climbed above $25. Thus, by driving up prices, Algeria, Nigeria, and Libya created their own competition and ultimately contributed to conditions under which supplies exceeded demand.

Standing in an entirely different position from these producers was Saudi Arabia. Holding 26 per cent of the world's economically exploitable reserves and in 1981 accounting for 40 per cent of the cartel's sales, the Saudis differed from their associates insofar as they could look ahead to high volume exports for as long as 50 years. Necessarily, theirs had to be a long-range strategy. They did not want to raise prices to a level where consuming countries would be compelled to institute crash synthetic fuel programs or engage in large scale conversion to oil substitutes—coal, nuclear power, hydroelectric, etc. In fact, price was a secondary consideration for the Saudis. Substitutes would come along eventually, and for the Saudis they could appear most advantageously at just about the time their capacity for large scale sales abroad had ended. At that point—perhaps in 2050—they would be left with enough oil in the ground for domestic consumption and for a respectable petrochemical industry. Thus, Saudi Arabia wished to maintain prices and production at a level that would maximize income but also keep consuming countries buying, even while these countries moved toward substitutes for OPEC oil at a moderate pace. While most observers of the petroleum scene focused their attention on the price of oil, the Saudis were thinking about long range marketing structures.

Somewhere between the two extremes of Saudi Arabia on one hand and Algeria, Libya and Nigeria on the other, were Iraq, the United Arab Emirates, Kuwait and possibly Iran. They had moderately large reserves and were not totally attracted by short run considerations. But neither could they be as patient as the Saudis. And even that was not the whole picture. The varying configurations of the oil producers' extensive gas resources also caused them to view production, price levels and future sales of oil in diverse ways.[2] The final mistake of the aggressive wing of OPEC occurred in 1979. The Iraq/Iran war took a potential 6 million barrels a day off the market as the warring countries hit one another's oil installations. OPEC used this critical situation to boost prices by a full 30 percent to $40 or better a barrel.

If prices had been permitted to increase further (beyond $40 a barrel), the entire picture would have changed again. Venezuela has only modest reserves (perhaps 10 per cent of Saudi Arabia's) and has favored the strategy of Algeria, Libya and Nigeria. But with its huge reserves of heavy oils (tars, sands, etc.) in the Oronoco Basin, it could assume a commanding position within OPEC if prices ever reached $80 a barrel. At that point Venezuela's oil would be a money maker and its viable reserves would be *twice* those of Saudi Arabia's. Surely Venezuela would like to see prices eventually reach a point where all this would come to pass.

Whereas OPEC holds 75 per cent of exploitable oil reserves at current prices, if the level ever reached $80, their hold would be reduced to 25 per cent. Saudi Arabia would lose much of its influence in OPEC and in the market as a whole. Talk of Saudi power would disappear. At higher prices, other types of high cost energy would also come into their own. The Saudis therefore had reason to work for price stability at a reasonable level.[3]

The upshot of these conditions was that each producer saw advantage in a different strategy. This situation explained the difficulty OPEC members

2. *Oil Market Trends,* October 7/14, 1981, Vol. 17, no. 15, SP59.

3. *Ibid.,* July 15/22, 1981, Vol. 17, no. 3, Arab Press Service, Nicosia, SP9-11. *Operations in Oil Diplomacy,* Vol. 2, no. 1, Arab Press Service, Nicosia, SP4-7.

had in 1981 and 1982 in arriving at a unified pricing and production policy. In political terms, it meant that oil was a tenuous weapon. Herein lies the truth of the naysayers who contend that oil could never be used against the supporters of Israel in order to achieve some settlement of the Palestinian problem that served the purposes of the Arabs.

Beginning in 1981, a sense of well-being among Western consumers was engendered by the quarreling within OPEC. As a measure to assure price stability, Saudi Arabia in 1979 chose to undercut OPEC producers who opted for spiraling prices ($40 or more) by increasing its production at one point to 10.3 million barrels daily and by insisting on selling oil at well below the price level to which some OPEC associates aspired. With the economies of the Western industrialized countries showing signs of sluggishness (i.e., growth in GNP far below that which governments sought), and with the conservation ethic forced upon the consumer by high prices, the international demand for oil declined in 1980–81 by 16 per cent. Most OPEC producers were required to sell less in 1980 than in 1979. Moreover, they saw the need to cut production further in 1981, and 1982 was even a worse year for them.

From a high of 31 million barrels a day produced in 1977, OPEC output fell by April 1983 to less than 14.5 million barrels. The appearance of British, Mexican, and even Soviet oil on the European market meant less dependence on OPEC for the industrialized countries. And many preferred to buy from these sources, even when prices were comparable. OPEC represented economic and political pressure. The natural inclination of consumers was to attempt to undercut this influence.

Pressure, therefore, was on the producers and not on the consumers of oil. Even with reductions in production, crude oil piled up in terminals at the rate of two to three million barrels a day. By the end of the first quarter of 1982, prices had declined by 25 per cent. By early 1983, they were down by 35 per cent.

How serious was the situation for those who would employ Arab power? Experts have calculated the minimum production figures (referred to as the red line) which each producing country required at existing prices to meet financial commitments. Each could, of course, survive on less, but at the sacrifice of development, armament or consumption expenditures. As flagging world demand carried production below the red line, strains developed among OPEC members over pricing agreements. Each hoped to see others bear the burden of sacrifice. There was some question whether even Arab producers, including the Saudis, would be interested in an oil strategy that took income for any period of time below the figure that could be obtained from red line production. As for non-Arab members of OPEC, below this level they would surely attempt to exploit any shortage created by an Arab oil cut back that had political purpose. Once expectations are built into a growing economy, severe social and political dislocations can occur from drastic reductions in expenditure. Good fortune can be quickly translated by a blessed populace into a natural right. And, indeed, some OPEC producers experienced privation in this respect.

The OPEC members have made numerous attempts to alleviate their plight. At a meeting in Geneva in May 1981, the cartel (excluding Saudi Arabia) responded to conditions by agreeing to reduce production by 10 per cent. Then, at an October meeting, Saudi Arabia acceded. It, too, would reduce production as a step to remove downward pressure on prices. At the same time, OPEC accepted a base price of $34 a barrel.

But purchases continued to slump. In 1982, exporting countries outside OPEC, led by Britain, started another round of price reductions. Effective price as reflected in noncontract purchases on the open market fell to $28. During the final months of 1981 and for the first part of 1982, Western countries drew from huge inventories that had been allowed to build up in excess of commercial requirements. At this point, OPEC sales fell below the red line. Finally, in March 1982 OPEC for the first time set a production quota—17.5 million barrels a day with Venezuela, Libya and the United Arab Emirates sharing the major cuts with Saudi Arabia.

Even with the drastic steps, the oil market did not respond. OPEC coalescence at times became a myth as Libya, Iran, and Nigeria sold quantities in excess of their quotas and below the established price. Finally, in March 1983, in the atmosphere of an international debt crisis that threatened the world's financial structure, OPEC made a valiant effort to preserve its place in the petroleum market. Quotas were reshuffled and prices were calculated on the basis of a $29 a barrel standard.

Minimum Production Levels[4]

(million barrels/day)

	Red Line	1979 Production	1981 Production	April 1983 Production
Algeria	0.9	1.15	0.80	0.70
Kuwait	1.2	2.50	1.15	0.60
Libya	0.7	2.09	1.14	1.10
United Arab Emirates	0.6	1.83	1.47	1.00
Iraq	2.5	3.47	1.00	.80
Iran	2.0	3.16	1.38	2.20
Saudi Arabia	6.0	9.53	9.85	4.50
Nigeria	1.8	2.30	1.43	1.00
Venezuela	1.5	2.35	2.10	2.00
Indonesia	1.4	1.59	1.60	1.25

4. *Oil Market Trends,* September 30/October 7, 1981, Vol. 17, no. 14, SP56; and March 24/31, 1982, Vol. 18, no. 12, SP51; and April 13/20, 1983, Vol. 20, no. 15, SP57; *Monthly Energy Review,* March 1982, pp. 88–89; *New York Times,* March 15, 1983, p. D-22.

From among all Arab producers, only Saudi Arabia had, or could ever hope to have, a position of sufficient strength in the oil trade to contemplate using its position for political purposes. In 1981 and 1982, we were to see verification of this assessment in the treatment Qadhdhafi suffered at the hands of Ronald Reagan. Nonetheless, petroleum clearly remained a factor in US thinking about the Middle East. Our preoccupation with Persian Gulf oil contributed to the Saudis' ability to acquire F-15's, Sidewinder missiles, and AWACS aircraft from the United States. It required the United States to work Saudi Arabia into its yearning for a Middle East "strategic consensus" against the Soviet Union. It contributed to Mr. Reagan's generous acknowledgment of Prince Fahd's peace plan, even while the plan was anathema to Israelis.

The source of all this concern was oil. It gave the Saudis a voice which others were prone to heed. In a way, the Saudis were already using the oil weapon. Any country with a $40 billion annual development program (accompanied by the commensurate need for costly imported Western technology) and with annual operations of a similar magnitude on the international financial market, was going to exercise influence. In these amounts—$75 to $90 billion a year in foreign transactions—we were no longer talking about money. We were talking about power. And we could be sure of one thing. However they might use the oil weapon, the Saudis were not likely to sacrifice the authority which their position provided. The Saudis were mindful that as the oil actually needed from Saudi Arabia declined below the red line, no margin was left for the use of the oil weapon.

THE AMERICAN SITUATION

Conditions on the international petroleum market would appear to justify the conclusion that the Saudis lost their capability to pressure the United States with oil. By the end of 1982, total petroleum products supplied in the US fell to 14.8 million barrels per day. About 8.6 million barrels were produced domestically, 4.7 million barrels were imported and the balance was taken from inventories—a short run phenomenon.

These data demonstrated that the Arabs by no means had a hold on the United States' oil supply. They showed that our dependence on imported oil was declining. In fact, imports for 1982 were 20 per cent below the previous year and a full 46 per cent below the 8.5 million barrels a day imported throughout much of 1977.

But petroleum security is difficult to assess simply from the daily averages of crude oil imports. Too many extraneous factors intervene. During 1981, the government's policy of developing a strategic petroleum reserve added 200,000 barrels a day to our imports in excess of consumption. This continued in 1982. Building the strategic reserve exaggerated the extent of dependence reflected in import figures. In the other direction, there is the management of inventories. As long as OPEC was periodically increasing prices, and the rest of the world followed suit, oil companies increased inventories. Each time prices increased, so did the value of the inventories. Market phenomena were causing the United States to import more than was needed for consumption. But after October 1981, as prices began to slip, this practice ceased. The cost to oil companies of holding these huge inventories was perhaps $3 billion annually. With no windfall profits, oil companies chose to reduce this burden. Market requirements began to be met partially from inventories. At one point, imports dropped to 3.7 million barrels daily—55 per cent of 1980 levels. By January 1, 1982, inventories had declined by 40 million barrels.[5] Theoretically our petroleum security declined accordingly.

Nevertheless, American inventories at the end of 1981 were high and the Strategic Petroleum Reserve was growing. Total stocks of all petroleum from the tankers already in harbor to the service stations in our neighborhoods constituted better than 1.4 billion barrels as compared to one billion barrels in 1973 when we first became aware of the oil weapon.[6] Even with a total withdrawal of Arab oil from the American market, and assuming that those amounts could not be replaced from other sources, the United States could continue levels of consumption identical to 1982 for four years. In 1979 we could have held out under similar circumstances only for about a year.

But the United States failed to develop a coherent approach to assuring its energy requirements. Supposedly the market mechanism could do the job. Despite the great deal of talk, during the nine years following the 1973 embargo, the United States did very little to find substitutes for oil. Whereas in 1973, petroleum products provided 46.7 per cent of our energy, in 1982 they accounted for 42.8 per cent. Petroleum consumption declined by 13 per cent while energy consumption itself decreased by five per cent during this period. Perhaps a greater disappointment was that an anticipated relief from dependence on imported oil never materialized. Domestic production did not respond to higher prices as the oil experts and one American president after another had claimed that it would. Even with prices at their highest, production in 1981 remained at the average of the previous eight years and below the level of 1973.[7] There was considerable huffing and puffing about coal liquification and the vast quantities of shale oil out West, but in an operational sense few changes occurred in the sources of energy. The United States had not moved vigorously to establish

5. *Operations in Oil Diplomacy,* December 9, 1981, Vol. 2, no. 6, SP50; *Oil Marketing Trends,* October 7/14, 1981, Vol. 17, no. 15, SP62, and *The New York Times,* December 4, 1981, p. D-6.
6. *Monthly Energy Review,* March 1982, p. 92.
7. *Ibid.,* March 1982, pp. 6, 89.

its independence from foreign oil supplies, even though it did substitute British and Mexican for OPEC oil. In fact, as prices declined early in 1982, American companies began to postpone and even cancel projects for extracting oil from the tar sands of Canada and the shale deposits of the Rocky Mountains.[8]

Moreover, signals from the oil experts contained mixed blessings. Generally they saw a continuation of the current situation insofar as the consumption of petroleum, and in particular imported petroleum, was not expected to increase before 1984. Some observers saw consumption remaining static throughout the 1980s.[9] While this could be interpreted as an optimistic note regarding reliance on OPEC and more pointedly on Arab producers, it was necessary to remember that this assessment depended on a relatively low level of economic activity that no one really wanted. Even under ideal marketing conditions, with the United States and its friends (associates in the International Energy Agency) having first call on oil from non-Arab exporters, better than three million barrels daily was required from Arab wells to meet our combined demand at the 1980 level of consumption.[10] As the United States, Western Europe, Japan and other countries of the Western alliance worked to improve overall economic performance, they unavoidably would increase their dependence on imported and most probably Arab oil.

U.S. Imports[11]
(daily average)

	b/d (000)		Per cent of Consumption**	
	1980	1982*	1980	1982
Saudi Arabia	1,257	577	7.4	3.8
Libya	552	28	3.2	0.2
Algeria	483	163	2.8	1.0
UAE	179	98	1.05	0.6
Iraq, Kuwait, Qatar	70	14	0.4	0.1
	2,541	880	14.86	5.8
Other OPEC	1,710	1,275	10.0	8.3
Non-OPEC	2,580	2,931	15.2	19.2

*Eleven months average.
**Products supplied.

8. *The Christian Science Monitor,* March 3, 1982, p. 1; *The New York Times,* April 5, 1982, p. D-1.
9. *The Wall Street Journal,* November 12, 1981, p. 8; *The New York Times,* January 11, 1982, p. D-1; *The Christian Science Monitor,* January 14, 1982, p. 11; January 15, 1982, p. 10.
10. Extrapolated from data on worldwide crude oil production and IEA consumption, *Monthly Energy Review,* February 1983, pp. 96, 97.
11. *Monthly Energy Review,* March 1982, pp. 34–35; February 1983, pp. 17, 33.

Any reduction in the use of petroleum has its costs. Smaller automobiles getting higher mileage and lower home temperatures are virtuous. But in addition, some economists suggest that since 1973, attempts to limit the use of energy have been partially responsible for our difficulty in increasing worker productivity. The conclusion to be reached? *The oil market in the early 1980s was depressed to a level that was not likely to prevail for long. More normal economic activity and consumption would almost assure that the oil weapon still had viability.*

FEATURES OF AN OIL STRATEGY

Until 1981, the "Oil Arabs" in their diplomacy relied upon the knowledge that they *possessed* a capability to exercise great power. While this approach has assured various considerations from the United States, it was inadequate for appreciably influencing the course of events related to Palestine and Israel. To have any real impact on this issue, the Oil Arabs would have to go a step further and show that they could actually *use* the political capability that oil supposedly provided. It was not that those who would engage in an oil strategy needed to bring the United States to its knees. They only had to create the impression that a shortage existed, or that it was about to occur. The oil weapon would amount to the application of many of the same principles that are employed in nuclear diplomacy. To begin with, the strategy would not depend on physical force. Its purpose would be to coerce the United States, without the open use of force, into doing something that we would otherwise not do. An oil strategy would depend upon the United States and the Oil Arabs having a certain community of interests. The Arabs' objective would be to create a situation in which these interests outweighed points of difference over the idea of Palestine.

In fact, the Arabs would hope to structure our motives, but not by making explicit and overt demands that in exchange for a secure oil supply we must take certain steps related directly to resolving the Palestinian problem. Rather, the Arabs' ploy would be to refigure sufficiently the conditions under which oil is marketed for the United States to feel compelled to act on its own.

With conditions that existed after 1980, the use of oil for political purposes was no longer plausible. But conditions could still change. And in the event of the return of a tight market, the position of the Saudis and the other Oil Arabs of the Gulf pointed to a particular type of strategy, i.e., a moderate and limited approach that would be meant to cause pain but not excessive damage. In fact, the initial steps for an oil strategy had already been taken by early 1982. The mechanisms for such a plan are:

1. marketing more oil through bilateral state trading arrangements and less through the multinational major oil companies,

2. requiring all shippers of petroleum to provide discharge certificates showing the destination of the shipments,
3. insisting that a prescribed portion of oil be shipped on the vessels of the exporting country, and
4. increasing the refining capacity of the exporting countries.

The Arabs' aim in mounting an oil strategy would be to establish greater control over the marketing of their petroleum. Saudi marketing practices have, in fact, always been consistent with this purpose. The multinational companies are a device that is largely beyond Arab control. In conjunction with an oil strategy, Arab suppliers, therefore, would have to move gradually away from Texaco, Socal, Exxon and Mobil while facilitating more sales through such organizations as the Saudis' own state trading corporation, Petromin. Eventually, Petromin might even achieve the capability of marketing oil for other Arabs.

The initial effect of this move would be to make world logistics of the petroleum trade less coherent insofar as the major oil companies have provided a large part of the communications that keep the free world's supply and demand in balance from one region or locale to another. Less oil in multinational channels, even while world supplies remain adequate, means less information about the trade and therefore less access. Uncertainty would develop, and uncertainty is the beginning of fear that supplies might be inadequate. Consequently, it is also the beginning of an oil strategy.

But during 1982, the Arabs lost much of the impact this factor could have provided. As OPEC discipline crumbled, more oil appeared on the spot market; i.e., daily trading usually from Rotterdam that does not occur under standing contracts. By March 1983, 20 per cent of the oil in international trade was being handled in this way. New channels of trade had developed. The multinationals were no longer providing all the communications that kept the market in balance. And more than half of the trade was outside OPEC channels. Thus, in all probability, neither OPEC nor Arab producers could ever again create confusion in the market and contribute to regional shortages simply by dealing directly with purchasers.

An oil strategy would not mean the total denial of supplies to the major oil companies. Generally, the trade would continue, but at a lower level. Some companies, such as Shell and Mobil, have acquired rights to various quantities of oil by making capital investments under joint ventures in Saudi Arabia, principally for refineries. These arrangements would continue under any circumstance.

The purpose of denying the United States a commanding position in the petroleum trade goes beyond the matter of access to source of supply. In the oil business these additional considerations are referred to as "downstream operations," i.e., all the functions after oil is pumped from the wellhead and placed into storage tanks in the producing country until it is actually put into autos at service stations in the consuming country. One innovation developed by exporting countries during 1980 was to require shippers to provide a discharge certificate. These could become an intimidating device if exports were certified only for specified destinations. Commercial intelligence could be gathered from various sources and bans could be threatened against ships or shippers that were detected engaging in diversion. It amounted to one more small mechanism for monitoring and controlling the trade.

The actual ownership of shipping was another matter. Several exporting countries hoped to develop fleets of their own and require that a portion of purchases be transported on these vessels. In this regard, Kuwait eventually intended to move 50 to 70 per cent of its exports on its own ships. Abu Dhabi talked of 100 per cent, while the Saudis looked forward to handling 40 per cent of their own transport. In 1980, OPEC countries transported 15 per cent of their own exports and by the mid 1980s this figure is expected to increase to 35–40 per cent.[12] In this move the producers' control over the trade would increase.

Shipping is closely tied to the plans of Arab producers for refining their own petroleum. By 1985 new refineries currently under construction were scheduled to increase Saudi capacity from 760 thousand to two million barrels of refined product a day. Depending on marketing conditions, this figure could increase to three million by 1990. Other Arab OPEC members were following the same course but at a less ambitious level than the Saudis. Total Arab refining operations stood at two million barrels in 1980 and would reach 4.9 million in 1985. The idea behind this development was that producers would eventually sell more petroleum products and less crude oil. Ideally, petroleum exporters would be able to enter into marketing arrangements that required purchasers to take a package including petroleum products and even petrochemicals—ethylene and fertilizer—if they also hoped to have access to crude oil. In effect, they would be adding value to a portion of their sales.

In 1983 about 40 per cent of the capacity of the free world's refineries were idle. The phenomenon developed because as crude oil prices reached astronomical levels, refining had become a losing proposition. As long as the multinational companies controlled the industry from the wellhead to the gasoline pump, this factor didn't matter. It was all part of the total cost of operations. But as the multinationals were forced to abandon the principal of vertical integration that allowed them to control the industry, they chose to shift the losses of refining to others. In the United States alone, 75 refineries, representing a third of capacity, were closed. Because of obso-

12. Fereidun Fesharaki, "Wide Impact Seen for OPEC's Refining Push," *Petroleum Intelligence Weekly,* Supplement (June 22, 1981), p. 6. *Downstream Trends,* November 25/December 2, Vol. 17, no. 22, Arab Press Service, Nicosia, SP85.

lescence, most of these would probably never be reopened. All this permitted the OPEC exporters who were developing refining capacity to shift a portion of their sales from crude oil to petroleum products.[13]

As the shift in the trade from crude oil to refined products occurred, exporters of petroleum established control over the market. If demand and supply again come into balance in the mid-1980s as some observers believed, they would be able to bargain over supplies of gasoline, distillates and residual fuels rather than just over crude oil. Their influence would be over the types of product that were available and not just over the destination of petroleum. The question of *who* buys would be extended to a matter of *who* buys *what.* In the process, consuming countries could become less secure. Imbalance in the supply of the various products could become more likely for reasons of market inefficiencies as well as being the result of political calculation.

Increased refining by producing countries also has an influence on shipping. By 1982, there was already an excess of capacity in tankers. Perhaps 25 per cent were not being used—or at least not being used efficiently. This was particularly true of the supertankers. There was not, however, a surplus of the type of tanker that carries products as opposed to crude oil. And it is in this area that OPEC countries, and particularly the Arab members, were building new fleets.[14] Thus, shipping was not simply to be a matter of competition among existing tankers or between the existing fleet, which was largely Western owned, and that which the Arabs were building. New types of tankers would be required and the world could expect the producers to build and use their own. The plans of the Saudis and Kuwaitis, therefore, were not just hypothetical projections. Once again, OPEC and Arab control over the trade would tighten in the process.

But the plans and practices of Arab oil producers raised a number of economic issues. With refinery operations being high cost in nature, why would Arab countries want to enter trade in refined products? It was obvious that purchases by consumers were shifting from crude oil to petroleum products. Thus those producers with products as well as crude oil to sell had some advantage in the highly competitive market of the early 1980s. And Kuwait went a step further. It purchased refineries and retail facilities in Europe, beginning in Benelux and then taking in Scandanavia. This assured a market for its crude oil. In fact, Kuwait in a small way was re-establishing a vertically integrated industry as it had been controlled by the multinational companies. In these new marketing practices we were to see the contradictions of political considerations and economic interests.

13. *Downstream Trends,* October 14/21, 1981, Vol. 17, no. 16, SP61; November 18/25, 1981, Vol. 17, no. 21, SP81-84; November 25/December 2, 1981, Vol. 17, no. 22, SP85; March 9/16, 1983, Vol. 20, no. 10, SP37. *Oil and Gas Journal,* December 29, 1980, pp. 121, 135. *Technology,* April 20, 1981, p. 113. *Oil Market Trends,* March 3/10, 1982, Vol. 18, no. 9, SP37.

14. Fesharaki, p. 6.

The decline of the West's refining capacity was probably the greatest threat to its petroleum security. A sign that economic interests were winning out was Kuwait's purchase of European refineries. Kuwait's refineries were, in effect, beyond its political control. This move made it less likely that Kuwait would ever participate in a political strategy that called for a selective boycott of any Western consumer.

Under conditions of (1) worldwide depressed economic activity, (2) decreased demand for energy, (3) increased competition from Mexico and Britain, and (4) a glut of petroleum on the world market, it would be impossible for Arab countries to create the conditions that could constrict oil shipments to the United States and Europe. Only in a robust international economy in which oil supply was in critical balance with demand could oil be introduced more directly into any political equation between the United States and the Arabs. These conditions could occur sometime after 1985, when the Arabs have greater command of "downstream operations." At that point, even a shortfall of one million barrels a day that could be covered for an extended period from the United States' Strategic Petroleum Reserve would create nervousness. It is only at the margin that the Arabs need cause tensions.

In a real sense, the Arabs probably do not have the capability to create an oil shortage in the United States. Even with an increase of economic activity, as Europe and the United States move out of recession, no one expects demand to press upon the OPEC production capacity of 34 million barrels a day. Through arrangements concluded under the Gulf Coordinating Council for unified oil policies, Saudi Arabia might be able to acquire necessary cooperation for an oil strategy from Kuwait, Qatar and the United Arab Emirates. But disciplining Libya, Iraq and Algeria would be difficult. Such non-Arab members of OPEC as Venezuela, Nigeria and Indonesia could be expected to go their way. Who can say what Iran might do? Then there is Mexican and North Sea oil to offset any Arab withholding of oil from the market.

But the success of an oil strategy would not simply rest on the amount of oil reaching the United States and Europe. The panics of 1973 and 1979 resulted largely from failings of the market mechanism—the *belief* that oil was not available. Here we are dealing with a psychological factor and the dynamics of American rather than Arab politics. The Arabs would not threaten a true limitation of supply quite so much as chaos. The United States would be susceptible to this suggestion insofar as the popular image of Arabs in our country identifies them with unruliness. The mere knowledge that the Arabs were reducing shipments, and had extended their control over the trade through their own refineries, ships and marketing channels would create anxieties. The purpose of an oil strategy would be to make supply sufficiently tight for the American public to become aware of the dangers that were involved. Herein lies the one source of pressure to which American leaders respond—the American public. Thereafter, the Arabs might conclude, the scene would be set for serious negotiations, first perhaps between the United States and the Saudis. The Arab purpose would be to get the United States to put the full force of its diplomacy behind a Palestinian settlement.

The oil weapon has a face other than the frightening countenance that is often placed upon it. As presented here, it is all part of rather ordinary marketing developments in which all oil exporters, and not just the Arabs, are engaged. Oil producers are beginning to create their own institutions. They no longer need to depend upon our trading channels, ships or refineries. The same trend is occurring in finance. By the end of 1981, the oil exporters were no longer a net provider of funds to Western banks. Their own private sector and their national banks had developed the capability to facilitate the international transactions resulting from oil production. Economic institutions mean economic (and political) muscle. At the very time analysts in consuming countries were gleeful over the troubles of OPEC, the cartel has been moving ahead with new controls that ultimately could tighten the vise on Western oil supplies, with or without the oil weapon. Moreover, having their own financial institutions frees the Arabs from some of the fears of US retaliation such as the freezing of Arab assets or a repeat on a larger scale of President Reagan's ordering American technicians to leave Libya—and then for US companies not to purchase Libya's oil. In this regard, the oil weapon becomes more viable.

A number of things could be said about the politics that can emerge from international petroleum transactions. Surely costs would be involved for those attempting to use the oil weapon. The difficulty for Saudi Arabia, with its existing marketing and political stance, would be that it was attempting to achieve too much with a single device. In the process, it could vitiate the impact of its own power. Through their position in OPEC, the Saudis have attempted to hold to pricing policies that serve their long range economic purpose. As part of this effort, they had hoped to work closely enough with the major multinational oil companies to preclude the industrialized countries either from shifting rapidly to investments in alternative energy or from drawing too much oil from non-Arab sources. In the latter, they have clearly failed.

The Saudis also have objectives in the areas of military supply and economic development which they hope to achieve with oil. We can group these two items under the heading of technological transfers. In each case, Saudi expectations are tied closely to the United States. In 1981, the United States agreed to provide Saudi Arabia with AWACS early warning radar protection and long range offensive capability for its F-15 fighters at what the Reagan Administration considered to be a substantial domestic cost. In satisfying a variety of Saudi requests, the US can argue that it establishes a claim to Saudi oil without changing its position on the Arab/Israeli settlement. Can, in fact, one implement be used by the Saudis to acquire arms, build a technologically advanced economy, assure an advantageous international price for petroleum *and* change the US position on Palestine? And if the achievement of these objectives is not possible simultaneously, are the Saudis prepared to make sacrifices in any or all of the other areas to reach political goals associated with Palestine?

The monumental nature of these questions is enough to send the Saudis scurrying for an alternative in their efforts to exercise influence over the resolution of the Arab/Israeli problem. But the alternatives are not very good. Destabilizing the international financial market with erratic movement of the $150 billion it has invested abroad would be difficult to achieve, self-defeating in nature, and not in keeping with the image the Saudis hope to project in the international community. Squeezing off imports of machinery and high technology from the United States has also been suggested as an option. Despite the sizeable nature of US exports to the Arab world ($9 billion to Saudi Arabia alone and $13.3 billion to all Arab oil exporters in 1982), many transactions are tied into long term development contracts with American firms. Changes in trading patterns could not be made in a dramatic way to seize the attention of the American public without violating contracts and seriously damaging the Saudi development effort. As for withholding new contracts and sales opportunities from American companies, any attempt to pressure a nation by denying it something which it does not yet have can only be expected to produce limited results. Suppliers would be hurt, but the prospect of creating the sort of public pressure with this tactic that would be necessary for a US policy change on Palestine is unlikely. Other than resorting to the oil weapon, therefore, it would appear that the Saudis' only option is to attempt to accommodate US interests and hope that we will reciprocate.

Assuming that the Saudis should ever decide to use oil more aggressively for political purposes, when and under what conditions might they be inclined to act? Admittedly, a fortuitous convergence of conditions would be required. Simply in a practical sense, the oil weapon as seen here would not be viable until 1985, when Arab refining and shipping capacities begin to fall into place. Even then, both the free world economy and demand for oil would have to be stronger than they were from 1981 through 1983. Parenthetically, it should be noted that the Saudis find Western economic recovery compatible with their interests on a number of counts, including the use of the oil weapon.

The Saudis actually need time. They are not yet fully accustomed to assuming a more active role in Middle East diplomacy than they occupied during the 1970s. The Fahd peace plan and the initiative to bring an end to the fighting in Lebanon in 1981 suggested that a new attitude was emerging. The Saudis seemed much more at ease in the forefront of politics than they had been five years earlier. But, then, after the Israeli invasion of Lebanon in June 1982, the Saudis took virtually no overt or positive role in the political affairs of the region.

Even with a new confidence in their capabilities, the Saudis would not be likely to move unless they were totally exasperated with Washington's procrastination as successive administrations attempt to manage rather than resolve the problem which Arabs see as being the result of Israeli obstinacy. Before using the oil weapon, the Saudis would have to be convinced that the United States government would not act unless it was compelled to do so. In effect, the Saudis would first have to conlude that the policy of attempting to move the United States with positive inducements—relatively moderate price increases, adequate supply and accommodating diplomacy—had failed. Even at that point, movement by the Saudis could not be strident. An unsophisticated and blatant use of oil would stiffen United States resolve and work to the benefit of Israel.

Finally, what are the ramifications of failure? Actually, neither the moderate Arabs nor the Americans have any alternative to getting along with the other. A controlled, disciplined, and low keyed strategy need not arouse either Arab or American passions in a way that would be permanently damaging to the relationship. Thus, the consequences of failure would be limited. Even if an oil strategy were not totally successful, it would show a new determination on the part of the Arabs that would probably have some influence on American behavior.

Essentially, in attempting to deny the United States and Europe their full petroleum requirements, the Saudis would be involved in pressuring a government that had been friendly toward them. The United States would, no doubt, find this disconcerting. When the full implications of having any other government in Riyadh are considered, however, it becomes apparent that the United States would have little choice but to tolerate the Saudi ploy—even while attempting to defeat it. Here is where we see the risks and irrationalities that have been likened to the strategies of nuclear diplomacy.

Pressing a friend is not without precedent. The United States and the European countries are constantly using similar tactics in their relations with one another. These are legitimate as long as they remain within limits. The basis for such action toward a friend is, in fact, that the relationship between the two is inextricable. Sometimes if direct action of the type contemplated in the oil strategy is not taken, the bitterness and frustration of inaction will set in. A relationship can deteriorate to the detriment of all concerned. Thus, the propinquity that characterizes the Saudi/US relationship would be critical to the strategy's success.

For the most part, the suggestions in this article have not been expressed in discussions of the Middle East, possibly because the mere stating of them makes it appear that whoever says such things is "siding" with the Arabs against the United States. The purpose of the article is not to revel in a hypothetical punishment of our country or to sound a cheerless warning.

Rather, it is to introduce an element of caution into the delights and even truculence associated with the independence from Arab perceptions that some observers now see for us in the glut that has characterized the petroleum market. It would appear prudent to remind ourselves that others still have the ability to influence our future, even our use of petroleum. We still live in a world of growing, not lessening interdependence.

PART TWO

THE MIDDLE EASTERN OIL PRODUCERS: PROBLEMS OF WEALTH WITHOUT DEVELOPMENT

Economic Development

As developed in Part I, a major purpose of this book is to analyze the reasons for and ramifications of the oil price revolution of the 1970s and its effects on the international economy. At the same time, however, the most wide-ranging and fundamental effects of this revolution have occurred in the oil-producing states themselves. The focus of Part II then is on the impact of oil wealth on the economies, societies and politics of the ten Middle Eastern states where petroleum and gas production form the principal component of the gross national product (GNP).

The central role of oil wealth in all these countries means that the tremendous changes brought about by the oil price revolution go far deeper than just a transformation of their oil sectors. The capital provided by oil revenues constitutes the primary determinant of each of these countries' economic structure and development potential. As a result, the ability to almost totally restructure the national economy has had direct and serious consequences on the composition of the society. In turn, this invariably results in political tensions as the basic sociopolitical framework of the country is fundamentally altered. In Iran, the destabilizing effects of this process helped bring down the Shah's regime and create a new and diametrically opposite political system. The social and economic changes in the states on the Arab side of the Gulf have been just as far-reaching, but so far the political disruption has been kept to a minimum.

Despite the tremendous rise in oil income during the 1970s and early 1980s, the Middle Eastern oil producers remain developing countries and part of the Third World. The amount of wealth transferred to OPEC since 1973–1974 has indeed been staggering and has allowed its thirteen member states to achieve a position of considerable economic power on the global scene. Nevertheless, it should be kept in mind that these states are in possession of only tiny, fragile and largely undeveloped economies. Their combined GNP of *ca.* $300 billion pales by comparison with just that of the U.S. at *ca.* $2200 billion.

For the Middle Eastern members of OPEC, oil is the principal—if not the only—major resource they possess. The capital it provides, while extremely important in the development process, is only one of many required inputs. Blessed only by relatively recent oil production, Saudi Arabia and the smaller Gulf states have had to begin economic development virtually from scratch. As Shaykh Ahmad Zaki Yamani, the Saudi Minister of Petroleum, pointed out a few years ago,

> At the risk of vexing you, I must maintain that we are still a poor country. Not only do we lack industry, agriculture; worse still, we lack manpower. We make our young people study and we send them to foreign universities, but it takes years to obtain a degree or a technical diploma. Meanwhile we have to import engineers, technicians, specialized workers that we don't know where to house because we lack hotels. To build hotels we need contractors, but the contractors themselves need hotels to live in. It's a vicious circle that exhausts us. Among other things we lack cement. We lack harbors because we lack the cement to build them. Last, but by no means least, we lack water. We haven't a single river, a single lake. We depend on rainfall alone. For 100 years, it has rained less and less frequently, for the last 25 hardly at all.[1]

Algeria, Iraq and Iran may have more resources to work with and have enjoyed a longer lead time, but they also face serious problems. Larger and more diverse populations make it more difficult to distribute the benefits of oil income. Even ambitious and well-planned development efforts run afoul of a myriad of bottlenecks. Essential skills and education among their people are still in short supply. The state must necessarily play a strong, leading role in the development process because of the weakness or absence of a viable private sector; this is true not only for the "socialist" economies of Algeria and Iraq but for the "traditional" systems of the Gulf as well.

The selections in this chapter begin with two case studies of how oil wealth has influenced economic growth and transformation in two very different states. Algeria, with a relatively large population of nearly 20,000,000, an extensive agricultural sector and heavy emphasis on industrialization, contrasts sharply with smaller Kuwait, inhabited by about one million people, of whom more than half are non-Kuwaitis, lacking any significant agriculture and possessing few other natural resources. J. R. Nellis first illustrates how the socialist ideology of Algeria has influenced its development planning and then Jacqueline Ismael outlines the evolution of Kuwait, a very atypical Third World country.

1. From an interview by Oriana Fallaci, *New York Times Magazine,* 14 September 1975; quoted in John Waterbury and Ragaei El Mallakh, *The Middle East in the Coming Decade: From Wellhead to Well-Being?* (New York: McGraw-Hill, 1978), pp. 75–76.

Following these studies, Iranian economist Jahangir Amuzegar takes a hard look at the mixed experience of OPEC and other oil-producing countries over the past decade of vastly increased oil revenues. He points out that in nearly all these countries, the "oil bonanza" has been a mixed blessing, as it has resulted in severe economic problems and increased sociopolitical tensions. Thus, he asserts, one must ask "not merely what the oil affluence has done *for* the petroleum exporters, but also what it has done *to* them."

Then John Townsend enumerates some of the severe difficulties in achieving the plans for industrialization in the Gulf. With the widespread lack of most resources, there is a real question whether these countries can ever realistically hope to establish viable economies before the oil runs out. Some observers have argued that they must necessarily remain *rentier* economies. Others suggest that they concentrate on creating a skilled corps of technocrats from among their population in order to supplant oil in future decades as a major foreign-exchange earner. Meanwhile, the sometimes helter-skelter development rush in the Gulf and Libya would be hopeless without the massive infusion of foreign workers at all levels—estimated at more than three million in the early 1980s. J. S. Birks and C. A. Sinclair summarize the impact that labor migration has had in the region and offer several scenarios for the future.

Bibliographic Sources for this Chapter:
14, 17L, 18, 23C, 23E, 25, 28, 33, 37, 41, 42, 65, 73, 95, 100, 102, 130, 132, 136P, 141, 142, 146, 149, 153G, 153H, 153I, 157, 160, 162, 169, 185, 186, 187, 189F, 189G, 191, 195, 197, 198, 199, 200, 207, 210, 214, 215, 216, 217, 219B, 219D, 219E, 219H, 220K, 220M, 224, 225, 226, 253, 258, 280, 281, 289, 291, 292, 295, 296D, 296L, 309C, 309E, 309H, 309L, 318, 341, 353C, 354, 377, 381, 384, 388D, 402, 408, 410, 411.

The Algerian Strategy of Development

J. R. NELLIS

Algeria is important for its wealth, for its size and location, for the dynamism and austerity of its leadership, and for its pretensions to socialism and leadership of the Third World. Clearly, an imposing list. Yet the Algerian approach to development is little known and insufficiently understood, at least in the English-speaking world. In France, on the contrary, and throughout *la francophonie,* Algerian movements and events are closely watched and intensely debated. Much of the controversy has concentrated on the question as to whether or not Algeria deserves its self-proclaimed status as a socialist state; [1] the celebrated *autogestion* effort of the 1960s has been thoroughly and carefully analysed,[2] and the nationalisations of foreign oil companies—as well as a few other salient economic enterprises—have received considerable attention.[3] It should be noted that the extensive French literature on post-1962 Algeria has focused on events up to 1971, though a few materials on more recent developments are beginning to emerge,[4] and that a small number of articles in English on

1. See, for example, G. Chaliand, *L'Algérie est-elle socialiste?* (Paris, 1964), and more recently and more vehemently, Kader Ammour et al., *La Voie algérienne: les contradictions d'un développement national* (Paris, 1974).
2. See Gérard Duprat, *Révolution et autogestion rurale en Algérie* (Paris, 1973); Claudine Chaulet, *La Mitidja autogerée* (Algiers, 1971); and Serge Koulytchizky, *L'Autogestion, l'homme et l'état: l'expérience algérienne* (Paris, 1974).
3. For an introduction, see Larbi Talha's 'L'Economie maghrébins depuis l'indépendance', in Centre de recherches et d'études sur les sociétés mediterranéennes, *Introduction à l'Afrique du Nord contemporaine* (Paris, 1975).
4. Good examples are the chapter by Bruno Etienne on 'Le Socialisme algérien', in ibid., and the article by Bernard Cubertafond on 'Réflexions sur la pratique politique algérienne', in *Maghreb/Machrek* (Paris), 69, 1975, pp. 25–32.

Extracted by permission of the publisher, Cambridge University Press, from J. R. Nellis, "Socialist Management in Algeria," *Journal of Modern African Studies,* Vol. 15, No. 4, 1977, pp. 529-536.

major Algerian programmes, such as *la révolution agraire,* have recently been published.[5]

Having genuflected to the literature, one can now make the point: neither in the extensive and generally excellent French literature on Algeria, nor in the few studies available in English, is there an adequate overall description of the strategy of development as it has been crystallised in the period 1971–6. This is more than regrettable; very significant alterations in both policy and style have been made by the Boumediène régime since 1971, and are continuing at the time of writing. Aimed at correcting partially this void, this exposé attempts to describe generally the evolution of Algeria's strategy since *le redressement*—the change of régime—in 1965.

In June 1965, Algeria's first Chief of State, Ahmed Ben Bella, was overthrown. Power was assumed by a Council of the Revolution, headed by then Colonel and now President Houari Boumediène.[6] The prime justification for this *coup d'état* was that the inefficiencies and the unplanned 'socialism of emotion' were leading the country to economic ruin. It was at first fairly widely assumed that the change of régime signified a clear victory for the elements to the right of Ben Bella and his supporters, notably the army, the middle- and upper-level peasantry and property owners, the most influential civil servants, and those heavily concerned with the traditional Islamic values and character of Algerian society. Boumediène insisted that it was inefficiency which had been renounced, not socialism *per se.* Speaking only a few months after the coup, he said:

> Socialism is not this incoherent collection of improvised measures and personal reactions that for three years gave the people only an erroneous idea of socialism. Socialism is a long and laborious process of construction that requires the elaboration and application of a comprehensive plan.[7]

There were observers who thought—and many who still believe—that such words were but lip-service to an impossible-to-renounce revolutionary ideal, deeply set in post-independence Algeria because of the populist nature of the anti-colonial struggle, and the at least partial success of the political mobilisation programmes of the Ben Bella period.

5. Some of these are: Hamid Ait Amara, 'The Agrarian Revolution', in *CERES* (Rome), 40, July–August 1974, P. J. Stewart, 'Algerian Peasantry at the Crossroads', Institute of Development Studies, University of Sussex, Discussion Paper No. 69, March 1975; Tony Smith, 'The Political and Economic Ambitions of Algerian Land Reform, 1962–1974', in *Middle East Journal* (Washington), 29, 3, 1975; John Waterbury, *Land, Man and Development in Algeria* (Hanover, N.H., Reports of the American Universities Field Staff, 1973), pt. III; and John Nellis, 'Workers' Participation in Algeria's Nationalized Industries: la gestion socialiste des entreprises', School of International Affairs, Carleton University, Ottawa, Occasional Paper No. 30, 1976.

6. Boumediène died on 27 December 1978. Chadli Benjadid was subsequently elected President (ed.).

7. From a speech made in November 1965; quoted by Waterbury, op. cit., p. 1.

It is certainly true that the new leaders came to power through the action of the army and the support of the more prosperous peasantry, and there is considerable evidence, from the early activities of the Boumediène régime, to support the view that there had been a sharp push to the centre, if not a counter-revolutionary full turn to the right. The *autogestion* sector, small but important beyond its size because of the tremendous emphasis it had received in the post-independence period, was at first ignored and allowed further to stagnate, and then revised into a mixture of centrally controlled state farms, and generally unregulated co-operatives. Some observers have argued that the only reason the self-management sector was not totally dismantled was simply because it 'could not be abolished without serious political repercussions'.[8] And on the early—1966 and 1967—occasions when leftist critics called for the rejuvenation of a truly participatory *autogestion* programme, or attacked the emerging state capitalist system, they too were ignored or more forcefully silenced. Centralisation and tightened bureaucratic control replaced the (admittedly largely theoretical) decentralised and localised decision-making arrangements, and officials of a technocratic, efficiency orientation began to replace those closely identified with Ben Bella's stress on spontaneity and mass mobilisation.

The Government justified such changes on the grounds that it was laying the basis for an efficient brand of participatory socialism, and in the more stable situation of 1968 Boumediène was able to argue as follows:

> We are for self-management, but a viable self-management that yields a profit, that results in an efficient organization of work and an increase in production. To liberate the worker is a revolutionary principle . . . but to produce is also a necessity.[9]

Critics countered with the notion that state capitalism and rigid centralism were being substituted for the previous emphasis on the workers and peasants.

At the outset, the Government had little ammunition with which to refute these attacks: it offered Algeria a collegial, faceless, form of leadership, and there was no dramatic immediate improvement in the overall level of administrative effectiveness. Eventually some elements of the student population, the trade union movement, and the intellectual left slid into overt hostility to the new state of affairs. By 1967, Charles Gallagher, an American observer, could write with deep pessimism:

> There is no party except for the facade . . . labour unrest has grown to the point where . . . urban workers must be counted among the active op-

8. David and Marina Ottaway, *Algeria: the politics of a socialist revolution* (Berkeley, 1970), p. 227.
9. Quoted in ibid., p. 270.

ponents of the regime . . . [There is a] lack of mutual comprehension found in relations between the government and the urban proletariat . . . [The 1965 coup has] produced the semi-stability of mediocrity . . . Given the complexity of the economic and social problems confronting the country, and the limited abilities of not only most of the leadership but also of many middle- and lower-level cadres, it would seem likely that matters will not improve for some time to come—and they may get worse before they get better.[10]

But 1967 was to prove the low point in the cycle for Algeria; since then the Boumediène régime improved greatly its internal situation, and rebuilt its regional and international reputation. The pessimism of Gallagher's statements proved unwarranted, for time and money have produced a competent and self-assured Government, which since 1968, and especially since 1971, has steadily widened the number and nature of its internal reforms, constructed a respected international position, and reattached itself to the outskirts of mainstream modern socialism.

It is worthwhile to present a general summary of how this process took place, as the ambiguities and retrograde steps in the 1965–71 period only become explainable in the light of later policies and undertakings. What appears to have happened after the coup is as follows: the new leaders recognised the essential fragility, indeed the nonexistence, of the state apparatus, and consciously chose as a first priority the strengthening of all bureaucratic structures. Past programmes based on a 'decentralist' or participatory ethic were not supported, or were eliminated, in the interests of building a powerful central apparatus. The party and the union were bureaucratised and became simple instruments for the execution of policies made above. The régime repressed dissent from those who complained of what they saw as a turn to the right, and of the failure to take action against the growth of a petty and medium bourgeoisie, groups which were profiting greatly under the new system. There were no true agrarian reforms during 1965–71, nor were meaningful restrictions placed on either Algerian private enterprises or the investments of the rapidly growing bureaucratic élite.

But nationalisation of foreign-owned business and industry took place starting with the banks, insurance companies, and some mining interests in 1966, moving slowly into the oil-gas complexes in 1968 and 1969,[11] and culminating with the take-over of the remaining French oil interests in 1971. With this action, the major sources of rent-producing natural resources were under Algerian control. At the same time, the leadership apparently

10. Charles F. Gallagher, 'A Note on the Maghreb', in I. William Zartman (ed.), *Man, State and Society in the Contemporary Maghreb* (New York, 1973), written originally in 1967, p. 12.

11. For a semi-official description of the steps and stages pursued since 1965, see the account of an address given by Professor Bouzidi of L'Institut des Sciences Économiques d'Alger in *El Moudjahid* (Algiers), 25 September 1975.

concluded that the structures of the state were now sufficiently strong (indeed, perhaps far too much so) to allow a re-emphasis on popular mobilisation and participation. These were the goals at which the régime had consistently said it was aiming, but, as noted, many observers had dismissed such statements as a mask for the continuation of the state-capitalist approach.

Now, the policies of the 1965–71 period *were* thoroughly state capitalist, and the single-minded pursuit of centralism and industrialisation by such means has left a legacy of rural stagnation and lack of capital, and has been responsible for the emergence of a comparatively wealthy, powerful, and ideologically 'untrustworthy' upper bureaucratic élite which is an important barrier to the construction of a more egalitarian society.[12] In spite of these difficulties, the actions of the post-1971 period have demonstrated the inadequacies of the view which dismissed the Boumediène régime as centrist, if not reactionary. Though there is always something suspicious in any argument which insists on taking steps to the rear in order to ensure future progress, this seems to have been the Algerian experience. In 1965 the country was simply not being governed, and the possibility existed that it might even disintegrate. The coup, and the resulting stress on basic governing and control mechanisms, changed this. It would seem that Boumediène and his closest collaborators did not jettison their belief in participatory socialism, but rather were sincere when they said that they were trying to create conditions in which such a brand of socialism could be created and sustained.

Today, as since 1965, the key Algerian hope remains that of using its present oil and gas wealth to finance state controlled and directed industrialisation. The dedication to a heavy industry approach, strongly reminiscent of former Soviet and East bloc orthodoxy, has recently come under considerable strain as a miserable record of agricultural production and high external prices have forced Algeria to spend over a quarter of its oil revenues simply to import essential foodstuffs.[13] Still, the official position remains that industrialisation is the 'sole remedy to the problems of under-development' and the only strategy 'capable of promoting economic and social development in poor countries'.[14] Or at least so stated an author-

12. For a critical account of Algeria's post-coup rural and agricultural policies, see Elbaki Hermassi, 'Periphery Revolutions: commitment and disenchantment', M.E.S.A. meeting, 1975. For a sympathetic view of the Algerian approach to socialist development which spells out the difficulties created by the emerging new class of bureaucrats, see the articles by Paul Balta in *Le Monde* (Paris), 28–30 December 1974; and for a detailed critique which rejects the claim that the bases are being laid for a socialist society, see Kader Ammour et al., op. cit.

13. I have seen cited, but cannot recall where, the figure of 30 per cent of Algeria's oil and gas revenues going to pay for food imports, mainly cereals. *The Quarterly Economic Review of Algeria* by *The Economist* (London), III, 1975, p. 3, reported that the 1975 wheat harvest was a poor 700,000 tons, leaving two million tons to be imported, thus giving some notion of the size of the problem.

14. 'L'Industrialisation: remède unique aux problèmes du sous-développement', in *Bulletin Économique. Algérie Presse Service* (Algiers), II, 1975, pp. 3–6.

itative pronouncement in June 1975. A slight modification was made in October that year when the results of yet another poor harvest became known, and an official document then ranked agriculture—for the first time—as equal in importance to the industrial sector.[15] But the Government allocates only 15 per cent of its total investment to agriculture, including irrigation, in contrast to about 70 per cent for industrial activities which, indeed, received even more in the late 1960s.[16] One can look for increased investment in and emphasis on agriculture, but probably not at the expense of the co-operative approach in the rural areas, or the continued dominance of the industrial strategy, to which the leadership is thoroughly committed.

Major contemporary problems include the comparatively limited oil deposits, combined with the usual shortages of skilled and trained manpower. Bureaucratic practices are everywhere very slow and tediously inefficient; it seems that *none* of the numerous state industries, save the giant oil and gas combine known as *Sonatrach,* operates at a profit.[17] As noted, agricultural productivity is declining. As for regional cooperation, socialist Algeria has luke-warm relations with free-market Tunisia, and a distinctly frigid attitude towards the Kingdom of Morocco, with which it almost went to war in early 1976 over the issue of the future of the former Spanish Sahara.

The continuing international economic slump is dangerous for Algeria, not only because of the inflated cost of the necessary imports and the decline in world growth necessary for the acceptance of its products, but even more serious is the closure of the migrant labour market in Europe, and especially in France. There are perhaps as many as one million Algerians, including both workers and their dependents, in Europe as a whole, a migration which has eased tremendously the internal employment demands. The importance of this safety valve is shown by the fact that in 1971 there were 480,000 unemployed in Algeria, of which 362,000 were in the urban areas: it is unlikely that the present situation is any different, and may be considerably worse. The oil and gas industries provide capital, but not very many jobs: only 35,887 as of 1973.[18] The effective closure of the European migrant market, combined with the high internal rate of unemployment *and* the capital intensive nature of the oil and gas (and many other heavy) enterprises, explains the insistence on widespread industrialisation in the shortest possible time. Algeria realised more clearly and earlier than most oil producers that its only hope for self-sustaining growth was the careful exploitation and nationally-directed investment of its finite

15. Quoted in the weekly economic supplement of *El Moudjahid,* 21 October 1975.
16. Ibid.
17. See the article signed by 'B.T.' on income division in *gestion socialiste* industries, which implies that few if any state enterprises are profitable, in *El Moudjahid,* 10 March 1976. The *Quarterly Economic Review of Algeria,* op. cit. p. 4, stated flatly that the state corporations 'other than *Sonatrach*' are running deficits.
18. See Djilali Sari, 'L'Évolution de l'emploi en Algérie', in *Maghreb/Machrek,* 69, 1975, pp. 42–9.

present resources. The Government is convinced that if an industrial base is not laid by the time the mineral resources run down, then the country will quickly slide back to a permanent state of secondary dependence. To repeat, low agricultural productivity and expensive agricultural imports are increasingly bothersome and may lead to a redeployment of resources and investment, but industrialisation is likely to remain as the foundation of Algerian policy.

Centrally planned and state controlled industrialisation requires—according to post-1971 Algerian thinking—a skilled, dedicated, and especially a mobilised populace which can only be achieved through a variety of educational, participation, and distribution programmes. The most important of these are: (i) efforts to 'decentralise, deconcentrate, and democratise' the political system which have resulted in the election of consultative 'popular assemblies' in the 676 *communes* and 15 *départements*—and in February 1977 a long-standing promise by the President was fulfilled with the election of a national assembly; (ii) *la révolution agraire,* a well advanced plan designed to limit the size of landholdings and to give viable farms to the landless and poor; and (iii) *la gestion socialiste* in the industrial and service sector. Related developments also underway include a national debate over a recent (1976) and thoroughly socialist constitution, *la charte nationale,*[19] as well as attempts to rejuvenate the party, to collectivise and improve the lot of the rural pastoralists, and to encourage voluntary, 'consciousness-raising' labour among students and urban workers. In a related vein considerable emphasis is being placed on the Arabisation of the entire educational system and the reform of the national university.[20] These are the 'grand institutions' of the Algerian state: 'the central supports for the building of socialism'.[21] They are designed almost expressly to counterbalance the expanded and powerful administrative structure. The régime is now seeking to avoid being suffocated by a sterile and unresponsive bureaucracy and—conversely—being overwhelmed by the excessive and unproductive demands of an unskilled population, eager for self-expression and self-control after more than 20 years of war and austerity.

Alternatively, of course, it could be contended that the régime has been reluctantly forced to re-emphasise mobilisation and decentralisation because of public demand,[22] or simply as revolutionary 'window-dressing' in order to mask the perpetuation of the centralised state-capitalist approach. This is a powerful argument which fits neatly the observed facts, especially

19. Front de Libération Nationale, *Projet de Charte Nationale* (Algiers, 1976).

20. Two detailed and highly critical articles by Azzedine Chabane on University reform were published by *El Moudjahid,* 24 and 25 October 1975.

21. *La Charte Nationale,* p. 11.

22. In the spring of 1976 the intense national debate over *La Charte Nationale* revealed the extent and depth of popular discontent with many programmes and the public style of the régime. Significantly, the top leadership appeared to be well regarded.

the hesitations and caution as regards mobilisation. But I reject such a view, mainly on the grounds that if the Government had been insincere it would hardly have erected such an elaborate and consciousness-raising series of mobilisation structures; insincerity could have been more easily and less dangerously accomplished. It is more realistic, if more perplexing, to regard the Boumediène régime as embarked upon a public struggle with the forces it itself has created: a powerful bureaucracy, an entrenched peasantry, a people weary of economic and political austerity. There being no easy explanation available, I must simply assert—rather than show—that the stress by the national leaders on popular mobilisation is a sincere reflection of their inmost ideological convictions. Their caution stems not from insincerity but rather from a knowledge of the fragility of the nation and the limitations of the population. The mobilisation programmes are regarded, rather paternalistically, as schools within which the inhabitants can be socialised to progressive norms and behaviour.

Postscript - 1983

Houari Boumediène died, of natural causes, at the end of 1978. He was replaced by Colonel Chadli Benjedid. This choice reflected the power of the army as the ultimate arbiter in the Algerian polity; it was also a pragmatic compromise which avoided the choice as successor of either a militant socialist or an economic liberalizer. Though the new leadership immediately announced its intention to pursue the socialist policies of its predecessor, in fact there has been a considerable shift of emphasis in the development strategy in general, and operational practice in particular.

First, it is now openly recognized that the industrialization policy is not working, that the state-run industrial sector continues to generate neither profits nor sufficient employment, and that the previously used rationale—that one must give the infant industries a certain but unspecified amount of time to master operational efficiency—is untenable. Fifteen years of public sector industrial experience have generated some modest productivity improvements, but they are now widely judged as insufficient to justify a continued high level of industrial investment. Thus, recent reforms have:

- broken down some of the largest state corporations, including the famous SONATRACH, into several smaller units, in hopes that more direct and effective managerial control can be applied;
- suggested a lightening of interventionist administrative procedures, and their replacement by market mechanisms and freer pricing systems;
- halted expansion in several industries, and called for the completion of previously started but never completed projects; and
- acknowledged the desperate need to increase investment in housing and social infrastructure.

Second, it is now obvious that the agricultural sector is performing disastrously. Food self-sufficiency has fallen to 40 percent of requirements, versus 70 percent a decade ago. The annual average growth rate of agricultural production in the period 1960–1979 was a meager 0.5 percent, to be matched against a 3.05 percent annual average growth rate of population in the same period. The authorities continue to blame poor weather for the country's agricultural difficulties, but have begun to show an awareness that excessive and incompetent management, and the continued stagnation of production in the *autogestion* sector, are contributing factors which should be, but have not yet been, directly faced.

Third, the participatory reforms instituted in the late Boumediène period have not bourne their expected fruit. While there are a large number—more than ever—of representative institutions at the local, regional and national levels, in the workplace and on the farm, the society continues to be thoroughly regulated and controlled by the top political leadership, working through the omnipresent, omnipotent and omni-incompetent bureaucracy. There is thus now a pervasive malaise in Algeria, as the regime seems incapable of implementing the socialist/participatory aspirations of the mid-1970s, but is equally incapable of relaxing state control. The result is a bit of the worst of both worlds, confirming the view of the orthodox left that Algeria is not now and never has been in transition to "true" socialism, and confirming the view of the liberal marginalist that state intervention cannot create development.

The overall point is that the Algerian development strategy is now less audacious, more pragmatic, less stridently socialistic, and more concerned with productivity gains—still within the context of an interventionist, statist system.

The Economic Transformation of Kuwait

JACQUELINE S. ISMAEL

The emergence of Kuwait as a major oil supplier to the industrial nations radically transformed the economic infrastructure of Kuwait. The traditional seafaring industries, already depressed in the interwar period, were essentially obliterated by the early fifties by the dynamic expansion of the oil sector. In 1948, for example, only 82 ships went pearling; and by 1955 only eleven. Similarly, the last Kuwaiti deepsea dhow was purchased by the Office of Education in 1954 for preservation as an historical memorial. Dhow construction, maritime commerce, and pearling, then, already marginalized in the interwar period by the forces of production of the industrial world, were essentially extinguished as productive sectors by the new integration.

In the post-oil era, the character of Kuwait's economic infrastructure was transformed from production of surplus value to consumption of surplus products. Government expenditure on development was the principal mechanism facilitating this transformation. At its core was the transfer of public revenue to the private sector. Within the structure of peripheral capitalism, development took the form of the expansion of the consumer market for the commodities of the industrial world. In the fifties and early sixties, both the physical infrastructure of the affluent society and the bureaucratic infrastructure of the welfare state were created from the huge oil revenues flowing into the nation. Both facilitated private accumulation from public revenue and expropriation to the world centers of capitalism.

The physical transformation of Kuwait City from a sun-baked adobe town, four miles long on its desert side, to a modern metropolis of the most contemporary design and ostentatious architecture, 25 miles long on its desert side by 1963, was one of the initial mechanisms of private accumula-

From Jacqueline S. Ismael, *Kuwait: Social Change in Historical Perspective* (Syracuse University Press, 1982), pp. 102-11, by permission of the publisher.

tion-expropriation of public funds. As early as 1950, foreign consultants were brought in to draw up a Master Development Plan for Kuwait City, and its suburbs, identifying the sites for public institutions, industrial areas, ports, and residential areas.[1] In 1952, a Development Board was established to co-ordinate construction activity, especially the development of a public utilities infrastructure. The Board devised a program of development projects for the period 1952/53–1956/57 at the cost of 91.5 million Kuwaiti dinars (KD). Expenditure on development projects between 1956 and 1965/66 amounted to KD 345 million.[2]

The relationship between public expenditure and private accumulation-expropriation was evident in the land acquisition and construction program. Within the city, the government bought up huge tracts of land for public projects and resold the improved surplus at about four percent of its cost to the treasury, setting off lucrative land speculation. In addition, the government-owned banking institutions made substantial advances to merchants investing in land. Outside the city, in the once public domain of the desert, those who had insight into expansion plans fenced in and claimed ownership of huge tracts of the barren sand. The government then purchased these lands at handsome prices to accommodate expansion. According to the report of the mission of the International Bank for Reconstruction and Development (IBRD) sent to Kuwait in 1961 and 1963:

> The Government buys land at highly inflated prices for development projects and for resale to private buyers. Land purchases amounted to between KD 40 million and KD 60 million in most recent years. Whatever the political or developmental justifications for this practice, the prices fixed by the Government for these transactions and the small amount thus far collected on the resale of the land make the public land transactions a rather indiscriminate and inequitable way of distributing the oil revenues. In addition, probably the largest share of these funds are invested abroad, so that the land purchase program fails to accomplish its main objective of invigorating the Kuwait economy.[3]

Fakhry Shehab further observed:

> Enormous private fortunes were amassed by both selling to, and buying from, the state. It has been estimated that between 1957 and 1962 close to $840,000,000 of public money was spent on land.
>
> This huge expenditure would have been justified on the grounds that it engendered economic activity and diffused a large portion of the new

1. The original plan was superseded in 1954 by a new plan for the development and expansion of Kuwait City. See *The Economic Development of Kuwait,* pp. 28–30.
2. The Planning Board, *The First Five Year Development Plan,* 1967/68–1971/72 (Kuwait: al-Rissala Press, 1968), pp. 21–22.
3. *The Economic Development of Kuwait,* p. 4.

wealth, were it not that in fact only a limited amount was piped into the local economy. By far the larger part was remitted abroad either directly, or indirectly through the banking system.[4]

In spite of IBRD recommendations in its report, *The Economic Development of Kuwait* (p. 89) that "the purchasing of land at high prices in excess of development needs" as a mechanism for disbursing public funds to the private sector be replaced by more economically productive mechanisms, this continued to be an important component of development expenditure, as reflected in Table 1.

Another example that demonstrates the government's essentially paternal relationship to private accumulation occurred in 1977. Following a considerable expansion of stock market speculation throughout 1975 and 1976, there appeared a slump in the market in 1977, indicated both by a 30.6 percent decrease in the volume of traded shares compared to 1976 and an overall average decline of 38 percent in the price of shares. In an effort to offset the considerable losses experienced by speculators as a result of the slump, in December 1977 the Government undertook to purchase any quantity of shares that might be offered for sale at prices it prescribed.[5] By February 1978, it was reported that the gesture cost the State KD 125 million.[6]

The government's efforts to modernize the City of Kuwait resulted in a construction boom, particularly in the period 1952 to 1965. Foreign planning consultants, architects, engineers, construction firms, and labor planned and created a city with the best material and technologies the industrial world could supply.[7] In contrast to the land acquisition program, however, government outlays in this period to create social overhead capital did generate considerable economic activity. In addition to a great many public buildings, commercial centers, apartment blocks, and suburban community projects built in the period, the following were also constructed:

1. 176 government schools and 32 private schools.
2. 8 hospitals, 2 sanatoria, 37 dispensaries and health centers, 148 school dispensaries and 9 centers for preventive medicine.
3. 1,100 kilometers of paved roads.
4. A number of electric power stations and an expansive network for

4. Fakhry Shehab, "Kuwait: A Super-Affluent Society," *Foreign Affairs* 42, no. 3 (April 1964):469–470.

5. Central Bank of Kuwait, *Economic Report for 1977,* pp. 67–72.

6. "Kuwait," *Financial Times Survey* (London), February 27, 1978, p. 14.

7. In 1961, for example, the IBRD mission noted that about half the inbound tonnage of the Port of Kuwait was material for the construction program. *The Economic Development of Kuwait,* p. 110. For a comprehensive critique of the development of modern Kuwait City, see Saba George Shiber, *The Kuwait Urbanization* (Kuwait: Government Press, 1964).

distribution and street lighting laid; between 1956 and 1965, installed capacity increased from 30,000 kwh to 370,000 kwh.

5. Water desalinization plants with a capacity of producing 15 million gallons per day; in addition, development of underground brackish and sweet water resources producing a total of 23 million gallons per day.
6. The creation of Kuwait Airways and the development of support and maintenance facilities.
7. The expansion of port facilities.
8. The creation by the Ministry of Public Works of several industries supplying construction materials.[8]

The development of the bureaucratic infrastructure of the welfare state was also related to the expansion of the consumer market for the commodities of the industrial world. In this case the target was to increase the consumer capacity of the population by raising its standard of living. Two mechanisms have been utilized to achieve this: through the expansion of social services, and in the form of wages to public sector employees. This program was facilitated through the transformation of Kuwait into the welfare state—the total service society with almost every human need from the cradle to the grave serviced by institutional arrangements. Health, education, welfare, housing were transformed from the private family sector to the public service sector. Between 1962/63 and 1978/79 expenditure on public services increased from KD 27.4 million (Government of Kuwait, *Annual Statistical Abstract* [ASA], 1976:182) to KD 291.5 million (ASA, 1979:302).

Corresponding to the growth of public services was the increasingly bureaucratic organization of society. From the simple structure of a ruling family that served as the government administration in the pre-oil era, by 1979 the government was composed of seventeen ministries. As early as 1955, 55.6 percent of the labor force was employed by the government. In 1978, the civil service comprised 133,248 employees, about 44 percent of the labor force and representing approximately one civil servant for every ten persons in the population.

The civil service is actually overstaffed, however, by the policy of guaranteeing to all Kuwaiti citizens a job in the public sector. This has been a major mechanism for increasing the standard of living of the population. Commenting on the administrative effects of overstaffing, the IBRD mission to Kuwait observed:

> Even a cursory passage through administrative offices reveals great segmentation, showing little groups of two to five persons performing a few moves in an administrative process. This fragmentation seems unneces-

8. Mahmud Bajat Sinan, *al-Kuwayt: Zahrat al-Khalij al-Arabi* (Kuwait: Blossom of the Arab Gulf) (Beirut: Matabi Dar al-Kashshab, n.d.), p. 174.

sary in what is after all a small State and administration. . . . There is a profusion of forms and ledgers, often individually well designed and always beautifully printed, which have to be filled out laboriously, in many cases with data that already are easily accessible in other forms and ledgers. At the same time, important data are seldom available. The amount of work done by each person or section is often small because there sometimes simply is not enough work to fill the day. . . . There are, of course, many exceptions to this general image, but a classified civil service of about 36,000 members serving a population of about 350,000 speaks for itself.[9]

The increasing standard of living of the Kuwaiti population as a result of liberal government welfare and employment policies is reflected in the increasing level of consumer demand for goods and services. In 1975, average per capita income in Kuwait was $11,365, one of the highest in the world. The corresponding growth in the level of consumption is indicated by the shift in living space. While in 1957, there were 2.1 persons per room in Kuwaiti dwellings, by 1965 this had decreased to 1.9; and by 1970 to 1.5. Correspondingly, the rooms per dwelling reflected a steady increase from 3.3 in 1957, 3.5 in 1965, to 4.2 in 1970 (ASA, 1975:335–36). The increasing standard of consumption is also indicated by private car ownership. In 1965, there were 0.18 cars for every person in Kuwait 15 years or over, or 0.70 cars per household; by 1970, this had increased to 0.25 cars for every person 15 years or older, or 0.92 cars per household; and by 1975, to 0.35 cars for every person 15 years or older, or 1.3 cars per household (ASA, 1976:277). In fact, on every indicator of standard of living, the population of Kuwait improves yearly its consumption capacity for luxury commodities.

Table 1. State Expenditure on Property Acquisition and Revenue from Sales of State Owned Property, 1964/65–1971/72 (in KD millions)

	Expenditure on Property Acquisition	Percent of Development Expenditures	Property Sales Revenue
1964/65	KD 45.0	50.0%	KD 2.5
1965/66	29.2	38.2%	1.5
1966/67	10.0	12.3%	2.0
1967/68	30.0	33.7%	1.5
1968/69	17.0	29.5%	1.5
1969/70	10.0	14.8%	1.8
1970/71	25.0	33.0%	1.7
1971/72	29.7	33.0%	1.6

Source: Data for 1964/65–1967/68 from *Statistical Abstract,* 1968, pp. 113–15; data from 1968/69–1971/72 from *Statistical Abstract,* 1972, pp. 182–85.

9. *The Economic Development of Kuwait,* p. 45.

Table 2 reflects that the economic infrastructure of Kuwait is increasingly dependent upon oil. The calculation of gross domestic product is based on the value of all goods and services produced in Kuwait for final use. Nevertheless the calculation of GDP at market prices introduces some error. Had the impact of inflation been taken into account, the real growth between 1970 and 1979 would have been considerably less than the figures indicate. For example, price indices rose from 117.5 in 1973 to 175.6 in 1979 (1972 = base year).

Table 2 does not by itself reflect the role of oil revenue in the generation of GDP. Data on GDP uses provided an indication of the pattern of consumption and development expenditure, and the roles of the private and public sectors in social and economic development. However, such data are only available for 1977 but may be used as an approximate indicator, as reflected in Table 3.

As Table 3 indicates, expenditure on exports represented 75.1 percent of total expenditure on GDP, while total private and government consumption and capital formation represented 41 percent and 26 percent respectively. The ratio of private consumption expenditure to total consumption expenditure amounted to 62 percent in 1977. The high ratio of expenditure on imports to GDP (45.3 percent), and the ratio of import expenditure to consumption and capital formation expenditures (67.7 percent), reflect how greatly the Kuwaiti economy depends on the outside world to satisfy its needs for goods and services. It also provides an indicator of the role of public expenditure in generating private sector economic activity. Private enterprise in Kuwait, in fact, is fundamentally based on public revenue transfer to the private sector.

Government transfers to the private sector occur through three main mechanisms—current expenditures, which provide the basis of purchasing power of the population and sustain the increasing level of imports, land purchases (discussed earlier) and capital expenditure which together provide the basis for the direct transformation of public revenue to private capital. Table 4 provides a summary of these for the period 1973/74 to 1979/80.

Table 4 shows that wages and salaries represent the largest single item of current expenditure. General expenses, which represent outlays on goods and services, constituted the second major item in current expenditure. The relative importance of these expenditures decreased from 59.1 percent of the total in 1973/74 to 49.1 percent in the 1979/80 budget. However, this was a result of substantial increase in capital expenditures rather than cutbacks. The absolute increases in wages and salaries and general expenses reflect inflation as well as substantial real expansion in expenditure on goods and services.

Domestic transfers consist of a number of items, the most important of which are the exceptional allowances for retirees, social allowances, price

Table 2. Gross Domestic Product (at Current Prices), by Industrial Origin, 1970, 1975, and 1979 (KD million)

	1970 KD	Percent	1975 KD	Percent	1979* KD	Percent
Oil Sector	579.5	57.2	2,363.3	70.0	4,650.3	72.3
Non-Oil Sector						
Agriculture and Fisheries	3.0	0.3	8.8	0.3	14.4	0.2
Manufacturing	45.1	4.5	196.7	5.8	349.6	5.4
Electricity, Gas, and Water	7.2	0.7	13.3	0.4	26.3	0.4
Construction	28.7	2.8	70.0	2.1	160.0	2.5
Wholesale and Retail Trade	81.0	8.0	211.3	6.2	331.6	5.2
Transport, Storage, and Communications	26.8	2.6	55.0	1.6	84.7	1.3
Financial Institutions	32.6	3.2	59.2	1.7	173.9	2.7
Insurance	0.6	0.1	6.0	0.2	13.3	0.2
Others	208.9	20.6	393.9	11.7	626.4	9.8
Total	1,013.4	100.0	3,377.5	100.0	6,430.5	100.0

*Central Bank of Kuwait estimates.

Source: Central Bank of Kuwait, *Economic Report for 1979,* pp. 28–29.

Table 3. Gross Domestic Product Uses, 1977

	Amount (000)	Percent
Private Final Consumption	986,205	25.4
Governmental Final Consumption	604,289	15.6
Total Capital Formation	1,006,671	25.9
Change in Stock	130,000	3.3
Exports	2,918,000	75.1
Imports	−1,760,000	−45.3
Expenditure on GDP	3,885,165	100.0

Source: Central Bank of Kuwait, *Economic Report for 1979,* p. 30.

subsidies for some basic commodities, assistance to the General Housing Authority and some other outlays to social affairs. The relative importance of these outlays remained consistent, reflecting the pressures of inflation on dependent groups.

Capital expenditure consists of direct development expenditure, loans to, and participation in domestic institutions. Direct development expenditure represented a major portion of total domestic expenditure over the period 1973/74 to 1979/80. The relative importance of this item increased from 15.7 percent to 29.1 percent over the period. However, a high portion of this increase was accounted for by the rise in world prices of construction materials, equipment, and labor. The development expenditure primarily represents allocations to the development of electricity, water, transport, communications, trade, and housing. In addition to the development expenditure, the Government provides financial assistance to domestic companies and institutions. This assistance is given either in the form of loans at low rates of interest, or through direct participation when the private sector is short of funds.

Over the post-oil period, development in Kuwait has been purchased at the cost of increasing dependence upon specialization in the world division of labor as an oil producer. In this period, dependence has been transformed from a specific to a universal characteristic of Kuwait's structure by policies that tie private and public consumption levels directly to oil revenues. Even the development of productive forces that has occurred in manufacturing, energy, and water production and transport and communications infrastructures has served to increase dependence rather than decrease it by increasing dependence upon externally produced technologies, raising consumer absorption capacities for imported luxury commodities and reducing the capacity of labor, to develop autonomously interdependent rather than dependent productive forces.

Table 4. Summary of Domestic Expenditures, 1973/74, 1976/77, and 1979/80 (KD million)

	1973/74		1976/77		Budget 1979/80	
	KD	Percent	KD	Percent	KD	Percent
Current Domestic Expenditure						
Wages and Salaries	197.5	44.2	368.2	26.0	645.4	28.8
General Expenses	66.4	14.9	314.3	22.2	453.1	20.3
Domestic Transfers and Unclassified Payments	17.0	3.8	63.5	4.5	97.9	4.4
Land Purchases	19.4	4.3	108.4	7.6	330.0	14.7
Development	70.2	15.7	353.1	24.9	650.9	29.1
Local Loans	13.7	3.1	35.2	2.5	—	—
Contributions to:						
Shares in Local Companies	45.8	10.3	51.6	3.6	—	—
Government Financial Institutions	16.6	3.7	123.5	8.7	60.0	2.7
Total	446.6	100.0	1,417.8	100.0	2,237.3	100.0

Source: Central Bank of Kuwait, *Economic Report for 1980,* p. 50.

Oil Wealth: A Very Mixed Blessing

JAHANGIR AMUZEGAR

For nearly a decade now, the spotlighted fortunes of some oil-exporting countries have been a focus of global interest. Raw materials exporters among the less-developed countries have dreamed of being able one day to establish a producers' association similar to the Organization of Petroleum Exporting Countries (OPEC). The industrial powers have helplessly watched their political clout and economic affluence dwindle before the kingdom of oil. The capital-short and aspiring Third World planners have kept telling themselves (and each other) that if only they had this black gold, the magical *élan vital* for their economic takeoff would be close at hand.

With the significant jump in the price of oil in 1973–74, and the projected transfer of wealth to the oil exporters, the dawn of prosperity and progress for the petroleum-rich countries was widely and uncritically presaged. OPEC's immense and growing revenues were exponentially projected not only to improve balances of external payments, but also to help finance domestic development and defense needs. The Midas touch of oil was expected to reduce domestic costs of living through larger supply of (oil-financed) imports. Massive oil receipts were widely believed not only to enlarge and extend domestic goods and services, but to make budgetary deficits fiscally extinct. Vast new opportunities for wise domestic investments were to guarantee gainful employment for the rising labor force. Rapid, all-around economic prosperity was thought to be achievable without forced saving or painful belt tightening. Material prosperity and progress, in turn, were hopefully believed to lead to more social cohesion, greater political stability, a meaningful industrial democracy, and an eventual advanced-country status.

The outcome, as even the dullest observer knows by now, has been astonishingly less rosy and far more checkered. Prosperity has been achieved at the cost of mystifying new sociopolitical tensions in some countries. In others, progress, although palpably evidenced, has put in motion destabilizing internal forces. Except for a dwindling few countries, initially large balance-of-payments surpluses have turned into deficits. Skyrocketing imports, while contributing to both higher standards of living and greater productive capacity, have oddly enough exacerbated domestic inflation. Budgetary gaps in most countries, while partly the result of expanded public services, have been further aggravated. The sweet dreams about sociopolitical integrity, middle-class solidarity, and genuine technological progress still remain to be fulfilled.

In short, many oil exporters now seem to think, and some even publicly admit, that the "oil bonanza" has not been a clear or unmitigated blessing for them. A disenchanted statesman among OPEC insiders goes as far as to say that history may show that the oil-exporting countries "have gained the least, or lost the most, from the discovery and development of their resources."[1]

Even if a more sympathetic evaluation of the past record, or a brighter outcome of still future events, should significantly blunt this rather pessimistic assessment, the oil exporters' manifestly enormous social, political and economic problems cannot be summarily dismissed as temporary aberrations. For most of the oil-rich countries, the tradition-shattering changes, the political uncertainties, the security risks, the economic bottlenecks, and the structural distortions are probably greater today than in the pre-1973 era. The question being asked now is not merely what the oil affluence has done *for* the petroleum exporters, but also what it has done *to* them!

II

The purpose of this brief review is to examine the experience of some major oil countries over the past several years, and to relate that experience to what (with the benefit of some hindsight) might be regarded as the inexorable dynamic of an oil-reliant economy.[2]

Some of the troublesome and unpropitious consequences that have attended development strategies in the oil-rich countries during the period from 1973 to 1982 have been so similar in form and substance as to rationally rule out mere circumstantial coincidence. When such different

1. Ali A. Attiga, "Economic Development of Oil Producing Countries," *OPEC Bulletin,* November 1981, p. 7.

2. The inner dynamic of an oil-rich economy discussed here connotes neither a Marxian (dialectic) nor a Rostowian (deterministic) approach to stages of development. It speaks only of the universality of certain emerging problems which were neither anticipated nor supposed to occur—but in fact did occur—within a diverse group of petroleum-exporting countries.

economies as Algeria and Venezuela, Ecuador and Saudi Arabia, Gabon and Norway, Indonesia and Libya, Kuwait and Nigeria, Iran and Mexico end up facing nearly identical economic problems, there must be reasons other than happenstance.

These countries differ from one another in many respects—in area, size of population, natural resource endowment (including oil reserves), ethnic and religious origins, stage of economic development, standard of living, type of government, international association, and a host of other social and cultural values. In area, they range from Kuwait with only 18 thousand square kilometers to Algeria with nearly 2.4 million. In size of population, they rank from slightly over a million for Gabon to upward of 146 million for Indonesia. In volume of oil reserves, Saudi Arabia is reportedly endowed with 178 billion barrels of crude, compared to Ecuador with less than 2.5 billion. Some of the so-called low-absorbing countries in the group (e.g., Kuwait and Saudi Arabia) are dependent on oil for more than four-fifths of their government revenues, while such dependence in others (e.g., Mexico and Norway) is about one-fifth. Several countries in the group—the "high absorbers"—are endowed in varying degrees with other raw materials, a large supply of arable land, and a moderate to good industrial base (e.g., Algeria, Indonesia, Iran, Nigeria, Norway, Venezuela). While the non-oil sector accounts for as little as 28 percent in Saudi Arabia, it amounts to as much as 67 percent in Indonesia and Nigeria.

In ethnic and religious composition, some countries are Muslim Arabs; others are non-Arabs with a Muslim majority; others still are non-Arab Christians of different cultural and historical heritages. The stage of economic progress is highly advanced in Norway, and moderately forward in Mexico and Venezuela, but largely nascent in Saudi Arabia. Per capita gross national product (1980) ranges from $430 a year for Indonesia and $1,010 for Nigeria to $12,650 for Norway, and nearly $20,000 for Kuwait. Political regimes vary from a conservative theocracy in Saudi Arabia to a highly advanced democracy in Norway—with a full spectrum of several party and nonparty systems in between. Some countries are members of OPEC; others are not. In short, there are probably greater heterogeneities in this group than there are in any other identifiable group of nations.

In the face of such outstanding physical diversities and inequality of natural and technological endowments, the one common binding element among these countries has been oil wealth. They all have been blessed with the shared prospect of receiving revenues from petroleum exports in a most fortuitous form which economists call *rent*.[3] They have also shared the aim

3. Economic *rent* is defined as a special, and fiscally speaking unearned, reward to the least-cost producers of a commodity in temporary short supply. Thus, due to the relatively low costs of extracting a barrel of crude in some oil-rich countries compared to the expenses of oil extraction elsewhere (or the costs of alternative sources of energy), the low-cost producers are able to anticipate and reap "windfall" profits from the sale of their oil.

of using the oil proceeds to achieve three targets: raise the living standards of the present generation; follow a domestic development strategy which can ensure the welfare of future generations; and reduce dependence on oil through domestic diversification and/or external investments.

III

That the receipt of massive revenues for oil-rent taxation has been a clear financial boon to the economies of major oil exporters is a matter of consensus. Whether or not this oil wealth has put the recipients on the path to sustainable growth, prosperity, and progress, however, still remains to be answered. On the positive plane, there have been a number of integrative forces leading toward economic growth and material prosperity.

First, the quadrupling of oil prices in 1973–74, and corresponding revenues accruing directly to the state treasuries, helped finance increased public spending, and gave a thunderous boost to domestic economic activities. The immediate expansion of government expenditure varied from an annual rate of 26 percent in Algeria to 120 percent in Nigeria and 140 percent in Iran. Fiscal outlays in some of these and other oil-exporting countries were two to five times larger than the corresponding annual averages during the previous five years, 1968–1973.[4] A large part of public expenditure went for investments in infrastructure, social service facilities, and productive capacity expansion. The ratio of investment to gross domestic product (GDP)—a good measure of national capital formation—more than trebled in Saudi Arabia, and doubled in Iran, Kuwait, Nigeria and Venezuela. Five years after the first oil price rise, 9 of the 13 OPEC members were devoting between one-fourth and one-half of their GDP to domestic investment—a feat scarcely matched by most of the developed or developing nations.

Growing public expenditure and investment caused, and was followed by, expansion of credit to the private sector. Annual private credit expansion in most countries in the five years after 1973 was more than twice the figure for the previous five years. Expansion of public and private demand, in turn, led to the growth of real GDP for all countries during most of the 1973–79 period. The average annual growth rate varied from less than 6 percent for Algeria and Venezuela to about 7.5 percent for Indonesia and Nigeria and over 11 percent for Saudi Arabia. More impressive were the rates of growth of non-oil GDP, which averaged about nine percent for all OPEC members, and as high as 13 percent for Kuwait and 15 percent for Saudi Arabia. The impact of oil was thus positive, extensive and enduring.

The second integrative force released by the oil affluence manifested itself in the government's ability to enhance public welfare through a series

4. Figures are based on data from *International Financial Statistics,* International Monetary Fund, Washington, D.C.

of measures such as expanded and improved public goods and services; increased labor employment particularly in the public sector; enlarged supply of consumer necessities (including food and fuel) at subsidized prices; higher national standards of living through larger domestic output and sizable imports; and, not least, reduced tax burdens. In some of the Arab countries in the Persian Gulf, national commitment to the creation or expansion of a state welfare system followed, and replaced, tribal and paternal attitudes and customs. In others (pre-1979 Iran in particular), the creation of a welfare state was modeled after some of the more "progressive" states of Northern Europe, and became an end in itself.

In the Gulf countries, free medical services for all ages, free education at all levels, housing and home sites at below-market prices and on easy credit terms, highly subsidized basic services and essential items, generous retirement pensions at a relatively early age, and the right to government employment at last resort became some of the outstanding manifestations of instant economic well-being. In these and other oil-exporting countries, not only was the physical landscape (particularly in large cities) changed beyond recognition, but the standard of living for a large portion of the population reached unprecedentedly high levels. Private consumption—a clear indication of improved material welfare—rose at the median annual rate of at least seven percent for all major oil-exporting countries in the period from 1970 to 1980, almost twice as fast as the 1960–70 rate, and almost twice as much as in low-income developing countries. For some countries (e.g., Saudi Arabia, Kuwait and Libya) the average annual growth was around 19 percent. For Nigeria, the yearly rate of acceleration was five times faster than the decade before the oil price rise.[5] Crowning the plethora of welfare measures, and higher consumption levels in most countries, there was a reduction or elimination of taxes on income, excises, imports and inheritance.

Third, there was a strong impetus to the rise of the middle class. The sheer increase in wealth, and the seemingly unstoppable dynamism of economic growth, clearly paved the way for the emergence and solidification of new middle-income strata in society—civil servants, entrepreneurs, professionals, medium-sized shopkeepers, skilled technicians and the like. The existence and growth of such a constituency is historically regarded as the main rudder of stability in the inherently destabilizing process of economic development—what Joseph Schumpeter called "creative destruction." Although this role has not been played unfailingly in all oil-exporting countries, the growing middle class has nevertheless been a progressive

5. The increase in *average* per capita private consumption, however, masks the pattern of internal income redistribution among different social groups, and between individuals in the same group. Although difficult to quantify due to lack of statistical data, it is often argued that the new oil incomes have shown a great measure of maldistribution in many oil-based countries.

element in discarding some of the traditional growth-inhibiting factors in society, and in championing certain modernization causes. In particular, this class is potentially capable of blending together agreement on an identifiable national purpose, as well as the nature and scope of an internally relevant form of government.

Finally, public ownership of oil reserves in almost all major oil-exporting countries, combined with a captive export market where increasing volumes of oil could be sold, frequently at sharply higher prices, and paid for in hard currencies, gave national economic planners a rare and unconstrained opportunity to shape national economic destinies. Freed from stringent and back-breaking requirements for mobilizing domestic savings or obtaining investment capital abroad, national authorities had extensive and flexible options to put together a set of economically rational, socially unifying, and politically acceptable policies for the use of oil proceeds. The inevitably dominant role of the government in the economic development process had the potential of serving as a very stabilizing force.

IV

These beneficial aspects of the oil price rise for the major oil exporters are thus undeniable. What has been perhaps less clearly acknowledged or agreed upon are the cost factors. As time has passed, a number of unforeseen socioeconomic setbacks and reversals have emerged. These vigorous and persistent negative forces, initially affecting the economic arena, have subsequently engulfed the entire national superstructure in a chain reaction.

In the World Bank's tactfully guarded language, few of the oil-exporting countries have yet found "a development strategy that promises not only industrial growth, but improved rural development, expanded employment and redistribution of incomes, the provision of basic services to the poor and limitation of population growth."[6]

Foreign correspondents and press commentators have been harsher and more blunt in their assessments. The oil-exporting countries have been frequently criticized (although not always fairly or objectively) for misguided economic policies, megalomania, bureaucratic bungling, waste, inefficiency—and worse. None has come out with a clean bill of economic health. For example:

- *Algeria's* economic performance, under its own brand of socialism, has been blamed for its slow pace of industrial progress despite one of the highest investment/GDP ratios in the world; "abysmally low" produc-

6. *World Development Report, 1981,* Washington: The World Bank, 1981, p. 114.

tivity of its state corporations; imbalances in intersectoral priorities; and other shortcomings.[7]

- *Ecuador* has been scolded by its own President Osvaldo for the "extravagance of the 1970s," with agriculture declining, much of the new oil wealth "squandered," and forecasts of lean times ahead.[8]
- *Indonesia*—blessed with "endless resources"—is portrayed as suffering from chronic and high unemployment; immodest inflation; swollen bureaucracy; "skyrocketing" subsidies; "galloping" fuel consumption; and other "deep" economic troubles.[9]
- *Iran's* modernization under the Shah has been termed a blunder with "disastrous consequences" because of unduly activist government intervention in the economy; high wages paid to workers in nationalized and subsidized industries; crowding out of private sector investments; and the creation of a frustrated educated class.[10]
- In *Mexico,* oil has been considered responsible for bringing in massive economic imbalances—hyperinflation; a stagnation in tourism and non-oil exports; the highest balance-of-payments deficits in the country's history; one of the largest external debts for any developing country; towering interest rates; an explosion of consumerism; and a reduction of real purchasing power for "ordinary" Mexicans.[11]
- *Nigeria* has been faulted for an uneven pattern of sectoral growth; a large domestic resource gap despite $30 billion in annual income from oil; a snail's-pace growth in the standard of living of the masses; larger dependence on imports (even on food which was once exported); periodic food shortages; badly deteriorating public services; emphasis on the production of non-traded goods combined with heavy restrictions on external trade and exchange; a double-digit inflation rate; one of the highest annual population growths in the world; and declining non-oil exports.[12]

Even *Saudi Arabia,* which seems to have had better success with its development efforts than many others, has been subjected to rumblings about skewed income distribution between Saudi citizens and guest workers; the torrid pace of real estate speculation; clashes between concepts of modernization, industrialization and materialism on the one hand, and traditional Islamic fundamentalism on the other; concern over a permanent

7. *Financial Times,* December 24, 1981; and *The Economist,* January 18, 1982.
8. *Reuters,* January 4, 1982; and *The Wall Street Journal,* April 22, 1981.
9. *The New York Times,* February 8, 1981; and *The Wall Street Journal,* January 20, 1982.
10. *The Wall Street Journal,* August 25, 1981.
11. *The New York Times,* April 6, and December 24, 1981, and February 20, 1982; *The Wall Street Journal,* August 13, 1981, and January 28, 1982; and *The Washington Post,* January 2, 1982.
12. *The Times* (London), March 16, 1981; *The New York Times,* July 23, 1981; and *The Christian Science Monitor,* October 8, 1981.

population of immigrant workers; eroding traditions of seniority, hospitality and thrift; and waste.[13]

Judgments on *Venezuela*—Latin America's richest country—have run along similar lines. Although the Venezuelan economy is still perhaps in better overall condition than those of other Latin American countries, there are criticisms that oil riches have failed to "trickle down" despite more than 20 years of democratic government. A nearly ninefold increase in oil incomes in six years has not produced a better life for the average Venezuelan, but led to economic dislocation and steep inflation; virtually all the grandiose schemes have fallen short of expectations; out-of-control bureaucratic expenditures have surpassed the oil income with the national debt climbing upward; the country has faced the first trade deficits in 50 years; unemployment has reached a two-digit figure; and state industrial losses have been on the rise. In the words of its own elder statesman, Perez Alfonso, nothing was better in 1980 than in 1974—not even the growth rate.[14]

And *Norway*—a country with a long tradition of democratic and efficient economic management, the advantages of a late start in the oil game, and every chance to avoid the oil trap—has not escaped the oil curse. While Norway's plight has been dubbed the "disease of the rich," and its problems portrayed as "golden ones," the dilemma created by the oil wealth has apparently been both real and painful. By various accounts, the newfound oil wealth has priced some traditional industries out of world markets due to high wages paid to skilled workers and the effects on domestic prices. The traditional manufacturing industries have faced spiraling costs and dwindling productivity; their exports have stagnated in the face of an appreciating krone. Oil riches have been used to subsidize consumption, job creation and decaying industries. Inflation has been kept partly in check only by a long wage and price freeze, and subsequently ran at a double-digit rate. Industrial production in 1981 was only at the 1973 level. And there has been a population shift toward the Oslo region.[15]

V

Such assessments of individual countries may be regarded by their authorities as too harsh, unjustified or at best one-sided. Yet they often reflect factual information supplied by the countries themselves. Certainly the

13. *The New York Times,* March 1, and March 23, 1981; *The Wall Street Journal,* June 2, 1981; *The Washington Post,* January 3, 1982; and *The Economist,* February 13, 1982.

14. *The Wall Street Journal,* February 3, 1981; *The New York Times,* August 3, 1981; and *The Washington Post,* October 7, 1981. Similar indictments have been leveled against Libya. See J. A. Allan, *Libya: The Experience of Oil,* London: Croom Helm, 1981, Ch. 10.

15. *The Christian Science Monitor,* March 31, 1981; *The Economist,* April 18, 1981; *Financial Times,* December 4, 1981 and *The Wall Street Journal,* January 13, 1982.

relatively hard figures alone—on inflation, budgetary gaps, external imbalances, rising dependence on oil, etc.—would be enough to call for an explanation as to what went wrong. A brief glance at the pattern of developments since 1973 in the major oil-exporting countries may offer some clues.

As already noted, the immediate and almost simultaneous result of the 1973–74 oil price rise, for the producing countries, was a sharp increase in public investment—in infrastructure, basic services, and industries with long gestation periods. There was also a substantial expansion of credit to the private sector. In consequence, investment, wage levels and consumer demand shot up, with domestic production lagging far behind, in the producers' largely underdeveloped economies with their many supply rigidities.

Thus, internal price levels began to rise steeply. The consumer index, which rose at an average annual rate of only 0.6 percent in Kuwait during the 1960–70 period, accelerated at an annual rate of over 18 percent in 1970–80; for Nigeria, the figures were 2.6 percent versus 19 percent; for Iraq, 1.7 percent and 14 percent. In Iran, the index climbed from an average of about five percent in the 1968–73 period to more than 15 percent in 1974–79. Venezuelan price levels went up eightfold within a decade, and inflation in Mexico rose to a rate of 30 percent in 1981. Even in Norway, the inflation rate during the 1970–80 decade was twice the 1960–70 rate. Among the low-absorbing surplus group, Saudi Arabia had the highest rate—an average of 24 percent a year, with a peak of 35 percent in 1975, compared to less than five percent a year from 1968 to 1973.

To meet rocketing demand for development, defense and consumption needs, the authorities opened wide the import gates. An extensive array of capital and consumer goods soon found their way into the activity centers of the oil-exporting lands. For example, foreign purchases during 1970–80 grew at an average annual rate of 20 percent in Nigeria and 35 percent in Saudi Arabia. Compared to annual imports in the 1960–70 decade, the lowest growth ratio was still three times higher in the case of Saudi Arabia, and as much as six times more for Indonesia, 13 times larger for Nigeria and 15 times bigger for Algeria.[16] While a substantial part of these imports were in the category of capital goods and defense needs, consumer items too kept their distinct share.

In theory, and under normal circumstances, such unrequited imports (i.e., paid for by extra incomes from higher export prices and more favorable terms of trade) should have served as a buffer against immoderate price hikes. In the case of the oil economies, however, the salutary effects of increased imports (and additional supplies) were substantially offset by

16. All figures are based on data from *World Development Report, 1981,* Washington: The World Bank, 1981.

two attending unwelcome factors. Import prices in the industrial countries were themselves on the rise due to internal inflation (partly blamed on higher fuel costs) and because of worldwide shortages of certain items.[17] Moreover, a series of impervious physical and administrative bottlenecks (e.g., port congestion, overworked transport, poor distribution systems, and inadequate experience in inventory management) played havoc with the planned deliveries and utilization of imported capital and consumer goods. The results were admittedly much waste and more inflation, particularly in the urban centers.

Rapidly rising prices, in addition to triggering familiar imbalances in the economy (i.e., unintended income redistribution, unproductive speculation, commodity hoarding, capital flight and other social ills), went hand in hand with other similar distortions. First, wages and salaries in the expanding public sector (and among growing public enterprises) accelerated upward beyond modest gains in productivity. With the tacit approval if not actual encouragement of the state, private sector wages and salaries soon followed the trend. Thus, wages rose at an average of 30 percent in Iran in 1974–75 alone; for certain special categories (e.g., industrial and construction workers) the 1975–76 wage index went up as much as 48 percent. In the mid-1970s, labor remuneration rose at an average annual rate of nearly 19 percent in Venezuela; in Mexico, urban minimum wages went up at an average annual rate of 27 percent—exceeding both productivity gains and cost-of-living increases. In Kuwait, wages rose by an average of about 25 percent in the period from 1973 to 1979; in Saudi Arabia weekly pay rose at an annual average rate of 15 percent in the 1976–80 period and as high as 35 percent in 1977. In all other oil exporters the heightened demand for housing, construction projects, and non-tradable goods and services was reflected in a drastic rise in local wages (particularly for skilled and semi-skilled groups)—with the private sector having to follow high public sector rates. In some countries around the Persian Gulf, manpower shortages and spiraling wage hikes were somewhat eased through immigration. In others, higher wages added to rising production costs.

At the same time, prices of imported goods and those of the nationalized and "modern" industrial enterprises were kept down by increasing subsidies out of the national budget.[18] In Indonesia, subsidies grew 15 times

17. According to data published by the International Monetary Fund, the export unit value index of the industrial countries—a rough measure of import costs to OPEC members—increased by nearly 25 percent in 1974 and 11.5 percent in 1975, with the average for the 1970–79 decade reaching 12 percent as compared to a mere 1.2 percent for 1960–69. See *International Financial Statistics,* Sup. No. 2, Washington: IMF, 1981. Some OPEC officials and neutral observers have charged that prices of certain specialty goods and services destined for OPEC were deliberately jacked up by foreign suppliers. See, for example, Walter J. Levy, "The Years that the Locust Hath Eaten," *Foreign Affairs,* Winter 1978/79, p. 299, and Robert Graham, *Iran,* New York: St. Martin's Press, 1978, p. 85.

18. The Saudi government reportedly pays farmers more than three times the world price for wheat, while subsidizing half the costs of agricultural machinery and fertilizers, in addition

larger in six years. In the six surplus countries (i.e., Iraq, Kuwait, Libya, Qatar, Saudi Arabia and the United Arab Emirates), food, fuel, water, electricity and other basic necessities were heavily subsidized. Notably cheap fuel prices in a number of oil exporters (e.g., Indonesia, Ecuador) encouraged domestic consumption to the point of jeopardizing the growth and indeed the continuation of petroleum exports beyond 1990.[19] In others, the subsidized new industries became increasingly inward-oriented and unable to compete in world markets.

Furthermore, welfare benefits and entitlements were increased and broadened generously—with scant regard to their snowballing effects. Stark warnings came later. The relative stability of nominal oil prices after 1975—in the face of worldwide inflation—coupled with the paucity of non-oil taxes and climbing public expenditures in a majority of oil-exporting countries, pushed budgetary deficits to historically high levels in some countries (e.g., Nigeria) and to an uncomfortable magnitude from time to time in others (e.g., Algeria, Indonesia, Iran, Mexico, Norway and Venezuela). Even such low-absorbing countries as Libya and Saudi Arabia experienced budgetary deficits in some years between 1974 and 1982. And with the exception of six relatively small countries (e.g., Iraq, Kuwait, Libya, Qatar, Saudi Arabia and the United Arab Emirates) all other major oil-exporting countries ran into periodic or protracted balance-of-payments deficits during the 1973–82 period.

Still further, the exchange rate was allowed to rise effectively and to remain overvalued. By keeping prices of foreign goods and services relatively cheap in local currencies, the authorities sometimes managed to keep domestic inflation temporarily or partially suppressed. But relatively high and stubborn inflation throughout the 1970s caused the effective exchange rate (i.e., the nominal rate adjusted for local rates of inflation compared to those of trading partners) to appreciate by various degrees among some oil exporters. The 1973–78 effective appreciation was about 50 percent for Nigeria, 70 percent for Indonesia, 40 percent for Gabon and 25 percent for Ecuador.

In other countries (e.g., Saudi Arabia) the exchange rate was officially adjusted upward periodically against the U.S. dollar. Only Indonesia and Mexico reduced their exchange rates to improve competitive positions. In other countries, the rate was linked to a weighted average of currencies of major trading partners. To the extent, however, that local inflation gener-

---

to interest-free loans and free gas for electric generators. In the industrial sector, incentives include free land leases, no-interest credit up to 15 years, tax holidays and custom-free imports. See *The New York Times,* January 24, 1982, and *The Economist,* February 6, 1982.

19. Interestingly enough, subsidized domestic prices of petroleum products substantially below world prices have not only been a major drain on national budgets; they have been contrary to the very advice given by OPEC leaders to the Western industrial countries in favor of energy conservation.

ally surpassed the prevailing rates abroad, the nominal exchange rates adversely affected non-oil exports. Instead of keeping pace or increasing relative to oil exports, the percentage of these non-fuel items in total merchandise exports dropped after 1973 in almost all oil-exporting nations.

Finally, troubles became overpowering when domestic infrastructure (e.g., port capacity, warehousing space, urban dwellings, transport, power supplies, building materials, etc.) gave way, as they did in almost all the nonindustrialized producing countries. Racked by spiraling costs, and ravaged by frequent breakdowns of incapacitated facilities, many governments were faced with a situation that called for immediate remedies. The authorities' reaction to the urgent physical and financial needs of the overheated economy was reflected in their resort to price controls, anti-profiteering campaigns, forced sale of privately "hoarded" merchandise, customs auctions, importation of foreign labor, foreign borrowing, contracting out of many public projects to foreign firms on a "turn-key" basis, etc. With the exception of a few small Persian Gulf states which followed liberal trade and exchange policies, most other major oil exporters were forced to exercise routine controls on prices and incomes. Most also kept domestic interest rates on deposits and loans at lower levels than the prevailing world market rates—for different reasons, but mostly in the hope of keeping inflation down. In certain open economies like Iran, Kuwait, Mexico and Venezuela, these interest differentials gave rise to considerable flights of capital.

VI

The cost of economic progress in the major oil-exporting countries has thus been neither small in magnitude nor limited in scope. World Bank data show that despite the average annual growth rates of domestic investment in many oil economies during the 1970–80 decade running more than three times faster than similar rates in the period from 1960 to 1970, these countries managed to increase their overall annual GDP growth in the 1970s by no more than one-third over their respective rates in the 1960s.[20] Almost all countries were also faced with a decline in agricultural activity, and an increased dependence on petroleum as a source of sustained livelihood and continued growth. For some countries, growing budgetary deficits, rising external debt, and internal unemployment were additional costs.

The question is why? Why did things not work out as expected? Why have so many different countries been so uniformly trapped not only by severe economic problems but some also by internal political ferment, social tension, identity crisis, and an overall malaise? Has there been coin-

20. Part of the surprisingly high capital-output ratios was due to the concentration of public investments on low-yielding infrastructures or long-gestating heavy industries. But there were also other reasons.

cidental bad luck? Or have there been identifiable causative forces at work?

The answers are frustratingly elusive. Only certain corroborating clues may perhaps offer plausible explanations for the oil countries' *embarras de richesse*—though still at the risk of some oversimplification.[21] At the top of the list, one may place what the French call a "lyrical illusion," i.e., an exhilarating state of near euphoria where state planners and the rank and file alike were disingenuously confident that oil money would solve all problems. Having considered the shortage of foreign exchange in much of the postwar period as the most critical single impediment to domestic economic development, national authorities saw their sudden affluence as the key to meeting all their needs. The preamble to Iran's Fifth Plan before the 1979 revolution epitomizes this supreme confidence. Based on the assumption that foreign exchange was no longer a constraint, the new plan was regarded as the "spearhead of one of the country's most brilliant and significant transformations [toward] the period of Great Civilization" when the country was to become one of the world's largest industrial powers.

Similarly ambitious outlooks and aspirations prevailed in Algeria, Ecuador, Mexico, Nigeria, Venezuela and others. Planned expenditures in almost all major oil exporters were geared to projected oil revenues rather than domestic absorptive capacity. Making the task of planning and implementation many times more difficult was the paucity of basic data, the dearth of experienced planners, the acute shortage of skilled work force, and inadequacies of bureaucratic superstructure. Blind confidence masked stark realities.

Second, where development and current outlays were not totally out of line with a country's potential absorptive capacity, the subsequent behavior of oil exports and prices made a hash out of confident expectations. Many post-1974 plans in the oil countries were based on the assumption of a steady rise in the volume and price of oil exports. However, the small nominal increase (and an actual real decline) in crude prices between 1975 and 1979, followed by a gradual decline in world demand for oil (and particularly OPEC oil) in 1980–82 drastically cut crude exports in some countries. Weakening spot market prices led to a series of official price cuts by many exporters (and in the case of some oil exporters to even below the benchmark price). Plummeting oil revenues, in turn, delivered devastating blows to the launching or completion of certain major projects.[22] As a

21. In a broad sense, the experiences of the major oil exporters have not been too different from those of other countries undergoing commodity booms in the past. In fact, some historical parallels can be found with discovery and development of precious metals in North and South America. What makes the case of oil somewhat *sui generis* is the prevalence of *similar* problems in *dissimilar* countries at the *same* time, and the seeming disregard of the obvious and experienced risks.

22. In Nigeria, for example, a mammoth $14-billion liquefied natural gas project had to be postponed. In Iran, all the six giant projects which constituted the core of the Sixth Plan fell by the wayside after the revolution. Retrenchments occurred also in Venezuela, Mexico and even Kuwait. See *The New York Times,* February 20 and 22, 1982.

result, the resource balance in many countries turned negative, and resort was made to borrowing from international capital markets. Severe contractionary policies were also introduced in nearly all high-absorbing countries in the late 1970s. In the process of this deflationary adjustment, private investment was seriously "crowded out" by public sector outlays.

Third, a basic inhibitive element seems to have been the inherent character of oil extraction and export. Unlike other dominant single industries (e.g., automobiles, housing, shipbuilding, or arms) which draw their inputs of land, labor and capital from a wide variety of other smaller industries, and in turn stimulate and evoke a wider range of productive and entrepreneurial activities, oil offers few such backward and forward linkages. Thus, in spite of major progress made in expanding domestic use of crude in meeting local demand for oil derivatives and for export, petroleum remained a highly insulated and technologically advanced industry with little direct spillover into other economic sectors. Forward linkage projects such as petrochemicals, aluminum and steel, which were being developed as an integral part of development strategies, were deliberately created to fill the void.

Fourth, a highly crucial seed of subsequent troubles in the less-developed members of the group appears to have been an uncritical identification of the "development process" with industrialization, and the equation of industrialization with modernization. From Algeria to Venezuela and from Gabon to Saudi Arabia, highest priority in public investment (or subsidies to the private sector) was given to a technologically advanced group of industrial schemes.

Advanced Western countries' politico-economic clout was generally attributed to their military/industrial might. Excessive reliance was also placed on the once fashionable theory, propounded in the 1950s by Albert Hirschman in particular, that deliberate tilting of the economy in favor of strategically selected industries is the best way to get an underdeveloped country off dead center.[23] State planners in nearly all countries accordingly opted for heavy industrialization as a major development priority. The belief was that a big industrial push would stimulate a cluster of other activities whose benefits would soon "trickle down" to all lower social strata. For the oil-producing countries, however, the result in the 1970s was a much-regretted neglect of agriculture and rural development—with many dire consequences. The depopulation of rural areas not only reduced farm output, but contributed to the catastrophic overcrowding of urban centers and the explosive onset of social tensions.

Closely related to the preference for industrialization over agriculture, and protected import substitution over competitive export promotion, has

23. A. O. Hirschman, *The Strategy of Economic Development,* New Haven: Yale University Press, 1958, p. 36.

been the choice of capital-intensive industries as the focus of public investment. Almost all countries bent on industrialization (read: modernization, progress) have preferred projects such as direct-conversion steel, petrochemicals, heavy machine tools, metallurgicals and fully mechanized state farming. This phenomenon, dubbed Pharaohism by the critics, has meant the introduction of advanced technology in countries with little or no tradition in such sophisticated undertakings. In Saudi Arabia and a few smaller and sparsely populated Gulf states, this strategy may still prove to be appropriate, but the labor-saving character of such projects makes them more questionable in richly populated countries like Iran, Nigeria and Mexico.

Even in countries with relatively small populations per square kilometer of land (e.g., Algeria, Saudi Arabia, Venezuela), the tightly "encapsulated" and imported technology imposed heavy burdens on the meager stock of skilled workers, technicians, engineers, teachers and managers. As the skilled labor shortage became more acute, poaching and pirating among various public and private enterprises tended to increase.[24] National planners have tried to justify their strategy by arguing that modern equipment is both more cost-effective and a better source of training modern skilled workers. But in practice, things have turned out differently: equipment has often not been properly used or maintained, and local unskilled labor, instead of being properly trained, has often been replaced by foreigners.

Fifth, many governments, inadvertently or by design, have assumed an increasingly dominant role in the economy without adequate administrative and bureaucratic capabilities. There has been an inherent ambiguity in their pursuit of the double-barreled aim of nationalization of a large sector of the economy and, at the same time, creation of a welfare state. As rural migration quickened, a growing number of unemployed (and unemployable) migrant workers from farm areas thus became wards of the state—helping to increase overstaffing of public enterprises and inefficiency in public services. A staggering chunk of the public budget was subsequently diverted from productive investment to the support of a European-style welfare system without the benefits of *prior* industrialization and adequate government machinery. Under irresistible political pressures from a largely poor, frustrated, and impatient population demanding to receive its share of the oil revenues for a better life *now,* most government leaders simply could not afford to sit on the oil money and avoid spending it in one way or another.

Sixth, there has been a tendency toward a "maximalist" approach to domestic investment and public spending. Rooted in Hirschman's contest-

24. In Iran, with less than one-third of the per capita income of some European countries in 1978, some run-of-the-mill engineers, bankers and bilingual secretaries received several times higher salaries than their Western counterparts.

able "big push" theory, state planners have tried to hitch their economic wagons to unreachable stars. The justification has been disarmingly simple: without a broad vision, lofty goals and hard-to-reach targets, the resources and energies of a nation cannot be optimally mobilized.[25] The outcome of this bold assertive philosophy, however, has invariably been a hyperboom. For a number of political, psychological and security reasons, most governments—particularly in the high-absorbing countries—seem to have opted for spending almost all of their new wealth at home instead of placing it abroad for slower and more orderly domestic investment at a later date.[26] An impatience with the pace of progress has led national planners to go for highest-growth scenarios about which they themselves must have harbored some doubts.

Under the pressures of unmeetable demands by the public and private sectors, the fragile domestic productive structure thus became intolerably strained. And when the warning signals flashed, the authorities' response was inescapably political, and their measures often expedient. Many hurried and inadequately planned public undertakings, lacking experienced managers and skilled workers (and overburdened with politically vocal but economically underproductive labor) are now judged by outside observers to be much costlier than anticipated. They have surely accomplished much less than expected, or than could have been achieved in a calmer and more resilient atmosphere.

Seventh, in order to combat raging inflation, the authorities somewhat wishfully resorted to the sort of "over-the-counter" remedies which helped relieve the symptoms temporarily, but like all such medication, made the removal of underlying causes perhaps more difficult. The benign policy of reducing high costs of living for the poorer strata in society (and spreading oil benefits to the widest possible segment of the population) was implemented through public subsidies to high-cost domestic producers and inflation-ridden imports. The threefold objectives of these subsidies, i.e., helping low-income consumers, reducing costs of inputs in other economic activities, and protecting the new "infant" industries against foreign competition, were both popular and plausible. But these galloping outlays—easily financed while government's oil revenues were on the rise in 1974–75—turned into major burdens on the budget in the subsequent slack periods, 1976–78 and 1981–82.

Meanwhile, the control of inflation was undercut by the decision to allow wages and salaries to rise for public servants and urban workers—a policy

25. Western observers point to Saudi Arabia's $250-billion Third Plan (1980–85), compared to its $18-billion First Plan (1971–75); Iran's near doubling of its Fifth Plan (1973–78) from $36 to $69 billion is another frequently cited example. See "Filling a Void," *The Economist,* February 13, 1982.

26. One clear exception has been Kuwait, which with the benefit of an early start and severe physical limitations for internal development, has chosen consistently to invest a large part of its oil revenues abroad.

designed to mitigate inflationary hardships, secure industrial peace and insure political loyalty. These increases necessarily spread to small-scale industries, non-oil exports and import-substituting activities, so that increased overall inflation canceled out the possible benefits of official or de facto currency devaluations in some countries.

Moreover, higher wages in the private sector were no help in reducing unemployment; in some high-absorbing countries (e.g., Indonesia, Mexico) the jobless rate reportedly remained perhaps as high as 30 percent of the labor force. Resort to a cornucopia of "quick fixes," like price controls, credit and interest-rate ceilings, import licensing, private investment regulations, etc., was psychologically effective for a brief period, but proved both socially divisive and economically counterproductive at a later stage in many countries.

Eighth, incontestably the massive influx of foreign exchange revenue from oil caused domestic currencies to appreciate officially, or in real terms, through domestic inflation. Oil-inflated and overvalued exchange rates were often allowed to prevail mainly in the hope of reducing inflation through cheapening imports.[27] As a result, exporters of traditional items found themselves at a serious competitive disadvantage both in world markets, and in meeting prices of substitute imports. In consequence, almost all major countries saw the shares of agriculture and manufacturing (so-called directly productive sectors, which are at the mercy of foreign competition) shift downward in favor of government bureaucracy, domestic and external trade, and "luxury" construction (which are non-traded goods and thus immune to exchange rate changes). Despite the earnest pledge to reduce dependence on oil as a fountainhead of economic life and the locomotive of domestic development, the share of this sector in GDP increased in almost all countries.

Finally, mention must be made of the divisive social impact of the new "oil psychology"—variously termed petromania, quick-money fever, or the catch-as-catch-can syndrome—in nearly all countries. On the one hand, those groups seized by what was called the "lyrical illusion" have been both eager for progress and impatient with its slow pace. On the other hand, in countries where deeply ingrained patterns of behavior had not been altered for centuries, there has been intense resistance to change by other groups. The reformist middle class and others caught up in the oil boom have tended to identify cultural traditions with poverty, suffering, inferiority and weakness. Traditionalists (or religious fundamentalists), however, consider these same traditions as the source of inner strength and prideful identity. Hence, petromania has triggered what many observers see as an inter-

27. Mexico's second official devaluation of the peso by 30 percent in 1982, on top of the 40 percent in 1976, has often been cited as a prime example. See *The New York Times,* February 19 and 22, 1982; and *The Wall Street Journal,* February 22, 1982.

minable struggle between privileged, "Westernized," progressive elites and cultural hard-liners.

In its most controversial aspect, the new "oil psychology" is portrayed (and deplored) as the materialization of traditional ethos; a ruthless search for wealth driven by frantic lusting after money status; and above all a decline in the cardinal virtues of self-reliance, both national and personal. And the sense of conflict is fed by the contrast between genuine entrepreneurs and less genteel manipulators, the rise of a *rentier* society, welfare handouts, and an ever-widening gap in income between rural and urban populations.[28]

These social divisions—impossible to quantify—have not only played a part in the economic problems of most of the oil-exporting countries; they may in the end become (as is already the case in Iran) the basis of political convulsions.

VII

From an economic standpoint alone, the near uniformity of problems faced by the major oil-exporting countries in the aftermath of their new affluence seems to point to a dynamic momentum which has clearly defied political, cultural and geographic differences. As each bit of progress has opened up a new Pandora's box of fresh problems, the oil-exporting authorities have met similar challenges with like responses. The new oil wealth seems to have imposed its own model on all oil exporters. Given an unavoidable political decision to use the oil money for further development and diversification of the economy, and in the absence of a strictly coordinated planning process, development expenditures proposed by various public agencies have invariably exceeded anticipated revenues. Even with "brutal" pruning of the budgetary requests, national authorities in most countries have found no alternative but to expand (sometimes reluctantly) oil production and exports to meet the "irreducible" minimum budgets. Larger exports at higher prices, in turn, have whetted the agencies' appetite for increased appropriations. And the circles have turned vicious.

The bewildering similarity of the problems faced by oil exporters of different stripes and status, and the uncommon uniformity of responses and reactions shown to these problems, lead one to believe that an oil-generated economy must possess an inner dynamic of its own. This rather mystical belief is reinforced when one notes that both the integrative and the negative forces discussed above seem to have been at work not only in the countries that were suddenly and unexpectedly faced with more money than they knew what to do with (i.e., most OPEC members), but also in

28. See A. A. Attiga, "How Oil Revenues Can Destroy a Country," *Petroleum Intelligence Weekly,* Special Supplement, October 19, 1981; and Peter Kilby, "What Oil Wealth Did to Nigeria," *The Wall Street Journal,* November 25, 1981.

such countries as Mexico,[29] Norway and even the United Kindgom[30] which were latecomers to the scene, evidently well aware of certain risks, and publicly committed to avoiding the pitfalls common to other oil exporters.

Thus, all of them—developing or developed, large or small, rich or poor, northern or southern—seem to have encountered the inherent collision between the requirements of economic rationality (i.e., productive investments dictated by cost-benefit calculations) and the exigencies of social justice (i.e., a welfare state mandated by the availability of easy money). In trying to combine these conflicting objectives, some equity has been achieved at the expense of efficiency: the expansion of the welfare constituency through larger segments of society has unavoidably made fewer resources available for genuine economic growth. Some efficiency also has been obtained at the sacrifice of oil economic ways and mainstays: traditional values, non-oil exports and self-sufficiency have been threatened. All in all, the "oil bonanza" has not meant impeccable bliss for any.

Are there lessons for the future in what has happened? With oil prices and production levels now dropping rapidly, in what some see as a continuing trend, what will happen to the economies and political stability of the oil-producing countries, especially those for which oil is the centerpiece of their economic structures?

Firm predictions are impossible. But three fundamental points are worth stressing.

To begin with, in almost all oil countries—even those with a thriving private sector—people's dependence on the state, as a nucleus of authority, assistance and support, has palpably increased. The state has become not only the focus of power but the initiator, promoter and monitor of all major economic activities—without being adequately prepared to assume these responsibilities. A major restructuring and reinforcing of government capacity is essential.

Second, governments themselves have become alarmingly dependent on oil revenues not only for long-term national economic development but for current consumption. Increased national commitment to elaborate and expensive development plans has increasingly bound the leadership to the maintenance, if not expansion, of oil production and export—even in the

29. As a seasoned observer reports, when Mexico discovered large new oil reserves in the late 1970s, it vowed to avoid the "mistakes" of other energy exporters, i.e., "disastrous political results" in Iran, Venezuela's reliance on food imports, Nigeria's port congestion, and Saudi Arabia's enterprises overrun by foreign technicians. And yet, "almost fatefully," Mexico also became addicted to oil. See *The New York Times,* February 20, 1982.

30. A recent commentary on the British economic situation finds it "hard to believe" that the British people have benefited from North Sea oil—in view of current output, investment and living standards which are at their lowest in years, with record unemployment and interest rates, and double-digit inflation. It concludes that the government has thus far "frittered away" most of the benefits of North Sea oil, and shows every sign of continuing to do so. See *The Times* (London), December 14, 1981, p. 15.

face of declining world demand. And a majority of increasingly articulate urban populations has become addicted to heightened consumerism. In most cases, national priorities must now be reordered.

Third, and most crucially, oil revenues have themselves become dependent, to a growing degree, on factors outside the control of the governments of the producing countries. With demand now below available supply at the official price, the key determinants are the pace of economic activity and growth in the outside world, oil conservation and increased efficiency in its use, and the supply and price of alternative sources of energy. Domestic monetary, fiscal and exchange policies have always been influenced by developments abroad, particularly for oil countries with substantial foreign assets; now these policies have been, to a troublesome extent, placed at the mercy of extraterritorial decisions, actions and trends.

Thus, governments that have come to be held responsible for the national destiny now face new and unforeseen difficulties. Even in the days of boom, their troubles were legion. How they cope with a continuing situation of stagnation, or worse, in the world oil market is an issue in which not only the oil producers but their major customers have an enormous stake.

Problems Confronting the Establishment of a Heavy Industrial Base in the Arab Gulf

JOHN TOWNSEND

INTRODUCTION

The objective of this article is to analyse the broad strategy of the oil-producing nations of the Arab Gulf for the establishment of a heavy industrial base and to interpret the major constraints standing in the way of a rapid transformation of the Arab Gulf into an industrial region. It is particularly concerned with markets and prices, with management, with skills and with labour. A detailed discussion of every heavy industrial project in Saudi Arabia, Kuwait, Iraq, Bahrain, Qatar and the United Arab Emirates will not be provided, but an indicative list of the major heavy industrial projects in operation, proposed, planned or under construction, in the general area of the Arab Gulf may be found appended to this article in its original place of appearance. This list has two important qualifications: first, it makes no mention of industry in Iran. Even though much of Iranian industry is under a cloud of uncertainty at the moment, it must never be forgotten that industry in the Arab countries of the Gulf is likely to be in competition with Iranian industry over the course of the next two decades. The second qualification is that, although great care has been taken in compiling the list, it is not necessarily comprehensive nor does it give an up-to-the-minute status report on individual projects. Whereas much publicity is given to projects when they are first mooted, or when a feasibility study is commissioned, very little is said when projects are cancelled or deferred.

It must be stressed that a number of heavy industrial projects, for example, the Arab Ship Repair Yard in Bahrain, the ALBA aluminum smelter in Bahrain, the fertiliser and steel plants in Qatar, and many refineries and smaller industries are operating successfully and making a profit. Most of these projects have had the inevitable start-up problems and management

Excerpted by permission of the publisher, Croom Helm Ltd., from Tim Niblock, ed., *Social and Economic Development in the Arab Gulf,* 1980.

problems. In some cases, these problems have been serious, but in no case has there been a failure to solve the problems and overcome the difficulties. It is only to be expected that industrial projects incorporating advanced technology will have problems during the implementation stage. Attention will not be concentrated on these problems.

Rather, the concern is with understanding the basic environment of the Arab Gulf and with examining the industrial development strategies of the Gulf states as they seek to master their environment and to harness their basic resources in a way hitherto undreamed of by their peoples. Policies for industrialisation, being radical in the geographic and social environment, have inevitably brought into the open contradictions and constraints, some of which will doubtless be mastered. Others are so fundamental that they could either emasculate plans for the establishment of a heavy industrial base or, being resolved, change the basic character of the societies of the states of the Arab Gulf.

Plans for broadening the economic base of any oil-producing nation in the Arab Gulf start with a simple and impeccable statement of principle: to invest oil revenues today in such a way that tomorrow, when oil revenues may no longer be important, they will not be missed. More precisely, governments have set out to spend oil revenues in such a way that modern science, technology and management may be applied to all resources in the Gulf countries, including resources in the agricultural sector, so that the over-all productivity of these resources should be permanently raised. In other words, they are setting out to create an industrial base. The resources which are plentiful are capital, energy, raw materials for petrochemical industries, and certain building materials. The resources which are extremely scarce are water and labour.

MAKING OPTIMUM USE OF EXISTING RESOURCES

The most prominent reason given to justify industrialisation as a means to ensure long-term prosperity in the Gulf is that in this way the best use can be made of existing resources and the value of raw materials can be enhanced. Let us now consider what is involved in seeking to achieve such objectives. The most abundant raw material in the countries of the Arab Gulf, after crude oil, is natural gas. This gas, which is also a cheap energy source, is the basic input for capital-intensive petrochemical plants which produce ethylene and ethylene by-products. Ethylene is the basic feedstock for the international petrochemical industry, and Saudi Arabian plans call for the production of two million tons of ethylene equivalent by the early 1980s, all intended for export. Iraq, Abu Dhabi and possibly others are also planning ethylene plants and would intend to export production.

Natural gas is also a fuel for metallurgical industries, such as aluminum and iron ore smelting. The value added in each case is likely to justify the import of the necessary ores. The principal market for the aluminum ingots produced in the Arab Gulf is a world market, although local and regional markets can be expected to take an increasing share of production. The production of steel bars and billets in the Arab Gulf is intended for local and regional markets. In addition to the aluminum smelter planned for Saudi Arabia, there is already a smelter operating in Bahrain. Dubai's smelter was due to start production in 1981 and Iraq's at about the same time. As far as steel is concerned, the Qatar steel mill started operating in August 1978, and Iraq's mill in January 1979. Qatar's mill is intended to serve a regional market, Iraq's a local market only.

In a sense, the strategic location of the Arab Gulf countries *vis-à-vis* world trade routes, and especially tanker routes, is an economic resource. Dry docks and ship repair yards form an important element in the Gulf countries' plans for creating a heavy industrial base. The market for tanker and ship repairs is, of course, international. The governments of the Arab Gulf countries, with the exception of Oman which is not a member of OAPEC, are all shareholders in the Arab Ship Repair Yard (ASRY) in Bahrain. Dubai's dry dock and ship repair yard complex was formally opened in February 1979. The Dubai yard was built in direct competition with the Bahrain yard and has a theoretical capacity of three times the latter.

These examples, which are not intended to be exhaustive, reveal the major problems confronting the governments of the countries of the Arab Gulf as they try to establish a heavy industrial base. The examples point to the problem of markets and of prices and, allied to these, the problems which could arise from regional duplication. With these problems must be considered questions of imported technology, management and manpower generally.

Let us start with the question of markets. As far as the export-oriented industries are concerned, there is no reason why Arab Gulf producers of ethylene, ethylene-equivalent products and aluminum ingots should not ultimately sell all their products profitably on world markets. Aluminium Bahrain (ALBA), for example, established over seven years ago, is now making money after a long period of losses. This profitable operation is due mostly to increasing demand on world markets for aluminium, but a managerial shake-up and a determined attack on production costs by management have also contributed to this favourable situation.

ALBA is a joint venture in which the government of Bahrain has held 77.9 per cent of the equity, with foreign companies with international reputations in the aluminium business holding the balance. The Saudi Arabian Basic Industries Corporation (SABIC) has this year agreed to buy 20 per cent of the equity in ALBA from the government of Bahrain, so that Bahrain's experience can benefit SABIC in Saudi Arabia's plans for a smelter. Dubai's Shaikh Rashid, the moving spirit behind his Emirate's

Dubai Aluminium Company (DUBAL), believes that the present buoyancy in world markets for aluminium will continue into the 1980s when the DUBAL smelter comes into operation. At that time, Iraq's smelter should also be competing for a share of the market. The problem is obvious: no one can predict with confidence the likely state of the world market for aluminium throughout the 1980s, and four Arab Gulf producers will be competing with each other for a share of this market.

There is no reason why these four producers should not each win and keep a market share. This will require the best in professional production and marketing management, with a tough cost discipline. Production costs in the Gulf countries tend to be high compared with the international competition because distance and a harsh physical environment add a significant factor both to initial plant construction and to plant maintenance costs. Local labour shortages tend also to put up production costs, and distance can add a factor to distribution costs. None of these problems is incapable of solution but, it should be stressed, the solution lies in good management. Any temptation on the part of Arab Gulf governments to subsidise production costs to maintain competitiveness may give short-term benefits, but could not be compatible with the longer-term establishment of a viable heavy industrial base.

The joint venture as a means of establishing heavy industry is much favoured by countries such as Saudi Arabia, Bahrain, Qatar and the United Arab Emirates. The theory is simple: a joint venture permits an Arab Gulf country to maintain control of its heavy industries whilst obtaining from the joint venture partner the management, technology and access to markets deemed essential for successful heavy industrial projects. Equity participation by the joint venture partner, up to a maximum of 49 per cent, is favoured rather than a straightforward management contract because it is believed that a foreign joint venture partner who is prepared to stake his own money in a project is, in effect, locked in and hence will be more likely to exert every effort to ensure the profitability and the success of the project.

Foreign partners are possibly less happy with the concept. The Saudi Arabian steel joint venture with West Germany's Korf Stahl A.G. took a considerable time to negotiate. Korf wanted only a management contract, whereas SABIC wanted them to take 49 per cent of the equity. Korf finally agreed to take a share in the equity of the project after the Saudi-West German Joint Economic Commission agreed to a scheme for guaranteeing private West German investment in the Kingdom.

There is an apparent lack of confidence by potential foreign joint venture partners in the viability of Arab Gulf heavy industrial projects. This lack of confidence would appear to be behind the reluctance of international companies to finalise joint venture agreements which have been on the drawing boards for some years. One which seems now almost certain to go ahead is a Saudi-Japanese joint venture to build a 310,000-tons-a-year ethylene plant. Originally conceived as a joint venture between SABIC and Mitsubishi, the project now will be between SABIC and a Japanese company to be known as the Saudi Petrochemicals Development Company, a consortium led by Mitsubishi and including no fewer than 54 Japanese companies. The role of the Japanese Ministry of International Trade and Industry has not been formally announced, but it seems probable that the Ministry is offering some form of guarantee to the Japanese investors. In short, some foreign joint venture partners seem to be seeking guarantees from their own governments before they are prepared to invest in heavy industry in the Arab Gulf.

This reluctance by foreigners to invest is not a major hurdle, though it may require a certain amount of rethinking of policies on the part of Arab Gulf governments. The government of Iraq does not favour joint ventures. The ASRY shipyard in Bahrain has no foreign capital invested in it, nor does the Dubai dry dock project. ASRY in Bahrain is managed by a Portuguese shipyard operator under a ten-year management contract, and the Dubai yard will have a similar arrangement with another foreign shipyard manager.

Just as difficulty in finalising joint venture agreements with foreign investors should not be an insurmountable problem in the way of establishing a viable heavy industrial base in the Arab Gulf, nor should the apparent duplication of projects be allowed to become a stumbling block. Much is talked about regional co-operation in the Arab Gulf. Many institutions have been set up to further this cooperation, for example, the Gulf Organisation for Industrial Consultancy, but the plain fact is that competition is more apparent than co-operation, in spite of regional organisations like Gulf Air and ASRY.

ASRY is worth examining in a little more detail, as it is an example of regional co-operation which risks being made less viable by the potential competition from the Dubai dockyard. Saudi Arabia, Kuwait, Bahrain, Qatar and the United Arab Emirates are equal major shareholders in ASRY, with Iraq and the Libyan Jamahiriya each having a small shareholding. Although the shareholders do expect some financial return on their investment, a commercial return is less important for them initially than the fact of an Arab-owned shipyard which can act as a catalyst for wider industrial expansion in the region of the Arab Gulf. Their immediate commercial objective was to establish an efficient dockyard operation which would attract international shipowners because of the price and the quality of the work carried out. In this objective they have succeeded and the ASRY yard has a full order book. Yet this comparatively satisfactory state of affairs could be upset by competition from the Dubai dockyard, which is equipped to carry out the same work as the Bahrain yard and which has three dry dock basins to Bahrain's one. The ASRY management is deeply concerned at the prospect of regional over-capacity and has indicated that it is ready to work out a formula for co-operation with the Dubai yard.

The question of regional co-ordination of industrial plans is not as simple as it looks. Apart from the obvious political considerations, there is a fear among some Arab Gulf governments that state capitalism, which they have been obliged to adopt to get heavy industries launched and which inevitably tends to be less efficient than the private enterprise which is at the heart of Arab Gulf successful business, would naturally be strengthened at the expense of private enterprise and hence of over-all commercial competitiveness. The question of state as opposed to private sector ownership of Arab Gulf heavy industries is not an issue at the present time, but it is likely to be one uppermost in the minds of Arab Gulf leaders over the course of the next few years as heavy industry competes with government for the best Arab Gulf minds and abilities.

ASSOCIATED PROBLEMS IN INDUSTRIALISATION

The dimensions of the Arab Gulf's likely dependence on foreign labour as a result of industrialisation are discussed in detail by Birks and Sinclair in the following selection. Here it will simply be stressed that this dependence would hold serious political and social implications for the states of the Arab Gulf. Local governments face a dilemma: they do not want to have such large numbers of foreigners, but at the same time the present and future economic health of their communities would seem to depend on foreigners.

Apprehension for a future where Gulf nationals constitute a steadily diminishing proportion of the total population is one factor causing some Gulf citizens to question whether heavy industrialisation is really desirable. There is an increasing tendency to listen to voices which contrast the present pattern of development unfavourably with the past. Such voices contend that the traditional communities of the Arab Gulf were not really underdeveloped before the oil era brought such sweeping changes in its wake. The traditional communities were certainly poor but, it is argued, they had great moral strengths, were balanced and had simple effective rules of social justice which were understood and accepted by all. Each community was recognised as being the best judge of its own interests, and communal rules at once permitted communal wishes to be optimised and precluded much interference by rulers and officials in communal activities. The organisation of each community's institutions, economic structures, markets and administration was simple but in general adequate for the needs of the members of the community. With very little contact with the outside world, the obvious differences between the life patterns of the Arab Gulf communities and the outside world were scarcely noticed and, if noticed, were not valued. There was no deeply felt need to 'modernise' or to 'catch up'; these are concepts which have been imported from the industrialised nations, along with a taste for a wide range of consumer goods.

The policy prescription following from this attitude is that each Arab Gulf nation should produce just enough oil to generate revenue for its own immediate needs. Such a policy would have added advantages, so it is argued, in that it would put up the price of oil and conserve the depleting resource. From this point of view, the question is not whether the purchase of modern capital-intensive technology to create an industrial sector will be effective, but rather whether the attempt should be made at all. This attitude must be counted among the problems confronting the establishment of a heavy industrial base in the Arab Gulf.

International Labor Migration in the Gulf

J. S. BIRKS AND C. A. SINCLAIR

The prevalence of unemployment and underemployment amongst the relatively large workforces of the capital-poor states (which average 3.4 million workers) ensured a steady, spontaneous transfer of labour from the capital-poor to the capital-rich states within the Arab world. This began in the 1950s.[1] However, the demand for labour in the oil-exporting states reached a new peak after 1973, in terms both of the quantities and of the quality of the labour required. The scale of migration was transformed as a consequence. Migrant workers soon represented a large proportion of the workforces of the capital-rich states. In some sectors of the economy requiring specialised skills, immigrants came to comprise virtually the entire workforce.[2]

Even in the largest economy and labour market—that of Saudi Arabia—expatriate workers now account for more than 40 per cent of the economically active. The countries with smaller economies, Kuwait and the UAE, rely on migrant labour to the extent that 69 per cent and 85 per cent respectively of their labour forces are non-national. Qatar, the smallest economy, depends upon a labour force in which more than four out of every five workers are expatriates.[3]

1. For an introduction to this, see J. S. Birks and C. A. Sinclair, 'The International Migration Project: An Enquiry into the Middle East Labour Market', *International Migration Review* (New York, Spring 1979).

2. Throughout this paper, only migrants for employment are considered. Refugees, those migrating for educational purposes, tourists, and those crossing borders for other reasons such as pilgrimage are excluded. Normally, figures quoted refer to migrant workers, excluding dependants. Where dependants are included, they are mentioned specifically.

3. J. S. Birks and C. A. Sinclair, *Nature and Process of Labour Importing: The Arabian Gulf States of Kuwait, Bahrain, Qatar and the United Arab Emirates*, mimeographed World Employment Programme Working Paper, WEP 2-26/WP 30 (ILO, Geneva, August 1978).

Excerpted by permission of the publisher, Croom Helm Ltd., from J. S. Birks and C. A. Sinclair, "Economic and Social Implications of Current Development in the Arab Gulf," Tim Niblock, ed., *Social and Economic Development in the Arab Gulf,* 1980.

The large majority of migrant workers in 1975 were Arabs. In the principal countries of employment, Arabs accounted for 75 per cent of all migrants, Asians 18 per cent, Europeans 2 per cent and 'others' about 5 per cent. Overall, almost half of all migrants were working in Saudi Arabia. Libya, the UAE and Kuwait together accounted for 48 per cent. Bahrain and Qatar, with their smaller economies, provided employment for 5 per cent of all migrant workers in 1975.

Most Arab migrants work in Saudi Arabia (57 per cent); this is only to be expected in view of the Kingdom's central position in the region, physical size, enormous oil revenues, commensurate objectives and political inclinations. Closer inspection of the characteristics of Arab migrants reveals that, in 1975, Egypt, Jordan and the Yemen Arab Republic together accounted for 74 per cent of the total migrant workforce. Most Egyptian migrants in 1975 were working in Libya, though 95,000 (24 per cent) were in Saudi Arabia. Egypt's proximity to Libya is the chief explanation for their substantial presence there. Almost all Yemeni migrants were working in Saudi Arabia: the attraction of the high wages offered in the neighbouring country evidently accounted for this. Jordanian and Palestinian migrants were working mainly in Saudi Arabia, but also in Kuwait, Libya and the UAE.

By 1975, some Asian labour was being used in the Arab world, but on a small scale. Only in the UAE and the Sultanate of Oman did Asian labour make a substantial contribution to the labour force in 1975, when the majority (57 per cent) of Asian workers in the Arab world worked in the UAE.

Indians and Pakistanis have long been involved in economic development in the Middle East. As traders and entrepreneurs they had existed, in small numbers, under the traditional economic order in the Peninsula. In World War II they were introduced in substantial numbers as part of the British war effort in the region. Subsequently, Indians and Pakistanis continued to make a contribution to the labour market, becoming associated with certain tasks and established in particular countries, such as Kuwait, Saudi Arabia and Bahrain. Until 1973, though, the numbers of workers in the Arab world from the Asian sub-continent were small, and their participation informal and spontaneous.

THE NEW SCALE OF ECONOMIC DEVELOPMENT IN 1975 TRANSFORMS MIGRANT LABOUR PATTERNS

In 1975 several aspects of the Middle Eastern labour market were changing. The volume of demand for labour soared as development plans, redrawn on a new scale after the 1973 oil price increase, were implemented. Targets embodied in them were exceeded. The increased labour demand engendered by this rapid development of the capital-rich states occurred just when Iranian and Iraqi migrant workers in the Gulf states were returning home. Iran and Iraq, which had previously supplied a significant amount of labour to Saudi Arabia and the Gulf states, began to develop their own economies, attracting home large numbers of their nationals.

The expansion in labour demand and the contemporaneous withdrawal of part of the supply increased markedly the pressure upon the capital-poor Arab states to supply yet more manpower. However, by 1975, a relatively high proportion of all potential migrants in the Arab labour-exporting countries was already working abroad.

Indeed, the proportion of the national labour forces of Jordan (East Bank), the Sultanate of Oman and the Yemen Arab Republic which was employed abroad was close to one-third in 1975. In Syria, Egypt and the Sudan the number of migrants abroad, though forming a smaller proportion of their domestic workforces, was also close to the potential number of workers who were prepared or able to migrate.[4]

Nevertheless, during 1975 and 1976, when the opportunities for migrant workers in oil-rich states became even more attractive, yet more workers departed from the capital-poor states. However, the rather limited number of extra workers drawn from the Arab supplying states in 1975/76 fell well short of the burgeoning demands of the oil-rich states. As a result, the rapidly industrialising wealthy states had to turn elsewhere for additional supplies of labour.

THE CHANGING PERSPECTIVE OF THE MIGRANT-RECEIVING COUNTRIES OF THE GULF: THE EMPLOYMENT OF INDIANS AND PAKISTANIS

The shortage of Arab labour in the region meant that the oil exporters had to broaden their areas of labour recruitment. They turned, in their search for more manpower, to Asia, in particular to the Indian sub-continent and the countries of the south-east of the continent.

The four Gulf states of Kuwait, Bahrain, Qatar and the UAE had already experienced a marked shift in the composition of the labour market before 1975. Between 1970 and 1975, the proportion of jobs occupied by expatriate Arabs fell by 17 per cent. The Asians' share of the workforce rose from a quarter to a half of the total, their number rising from 83,900 to 247,700. These Asians were mainly Indians and Pakistanis. The increased participation of Indians and Pakistanis was both a cause and a consequence of the formalisation of labour supplies from the sub-continent through a series of recruitment and manpower-supplying agencies.

4. J. S. Birks and C. A. Sinclair, *Human Capital on the Nile, Development and Migration in the Arab Republic of Egypt and the Republic of the Sudan,* World Employment Programme Research Working Paper, WEP 2-26/WP 29 (ILO, Geneva, 1978).

The recruiting agents, private sector entrepreneurs from Asia, drew profit from the migration of so many for employment. Representatives of agents, living in the Gulf states, identified employers' labour requirements and specified the skill levels, trades and numbers needed. The requirements could be met quickly as the agents in India, Pakistan and Sri Lanka had ready lists of would-be migrants according to qualifications and availability. These applicants would be trade-tested before being despatched to the country of employment.

Therefore, in the mid-1970s, just when Arab labour was becoming short in supply, expensive and of less predictable quality (as the potential maximum level of Arab exports was neared), Indian and Pakistani labour of known skill and capability became available quickly and easily. This regularisation of the Asian-Arabian labour flow by agents, who could above all supply labour speedily, encouraged the rapid increase in utilisation of Indian and Pakistani labour in the capital-rich states during the period of their most frenetic development.

In 1977 and 1978, however, both Indian and Pakistani governments intervened in this system of labour recruitment. Intervention stemmed mainly from the abuse of human rights suffered by employees recruited through labour agencies. It is also of note, though, that some concern had been expressed in Pakistan about shortages of certain types of labour, ostensibly brought about by the scale of labour exporting.

The constraint upon the supply of Indian and Pakistani labour resulting from government intervention has added strength to the Oriental connection, to which this paper now turns.

THE PERSPECTIVE OF THE GULF STATES AND SAUDI ARABIA CONTINUES TO EVOLVE: THE ORIENTAL CONNECTION

Formalisation of labour supplies to the capital-rich states has been taken to a new extreme by countries which have recently become, and now feature large in the Arab region as, suppliers of labour. The countries of South-East Asia—referred to here as the Orient, and including Indonesia, Korea, Malaysia, Nepal, the Philippines, Taiwan and Thailand—have, since 1973/74, come to account for a rapidly increasing share of the labour market. In 1970 very few migrants of Oriental origin were working in the capital-rich states of the Arab Gulf; by 1975 they numbered some 15,000 workers. In 1979, the number of workers from the Far East in the capital-rich states exceeded 160,000.

The utilisation of Oriental labour is the result of the coincidence of the varied social and economic programmes pursued by the governments of the Gulf states. Disenchantment with the wider cultural and social implications of hosting large migrant populations in these Gulf states is growing. The indigenes of the Gulf states are becoming concerned at the possibility of being 'swamped' by the immigrant communities associated with the migrant labour forces. The implications of, for example, Qatari nationals comprising so small a proportion of the total population of their own country are great, and little evaluated. Apart from fears of numerical domination, there is concern amongst Arab policy makers in the Gulf states over the considerable cultural impact which expatriates have on the Arab indigenes and their customs. Shaikh Sultan Bin-Mohammed al Qasimi, Ruler of Sharjah, observed: 'they [migrant workers] have caused problems, and have had an effect on life, on traditions, on everything.'[5] In Kuwait, the rights of various classes of 'national' are openly debated. Although less overt in the other capital-rich states, these considerations feature increasingly in decision making, and are assuming significance in the shaping of future economic development and the labour markets.

These concerns are separate from (though related to) the economic costs of supporting growing expatriate communities. The establishment of nuclear family groups amongst the immigrant communities is a major aspect of their growing political motivation and strength. As immigrant families become established permanently, so their household heads become concerned to secure rights similar to those of nationals.

This growing disenchantment in the Gulf states with hosting such large numbers of migrants appears to present planners with a conflicting aim to the underlying thrust of economic and social policy in the capital-rich states of the Middle East—rapid economic growth based upon industrialisation. How, given the limited human capital in the capital-rich states, can rapid economic development continue without their populations becoming totally outnumbered in their own country?

ENCLAVE DEVELOPMENT IN THE CAPITAL-RICH STATES

The compromise chosen by the rulers of the oil-exporting states has been to separate physically new industrial areas (upon which future growth and development will be based) from existing urban areas. In this way, it is hoped that what are perceived to be the undesirable aspects of labour importing (the close and prolonged contact between indigenes and immigrants and the economic cost of supporting expatriates within the country of employment) will be minimised. Examples of such 'enclave' industrial areas are becoming numerous in the capital-rich countries of the Arab world: Shuaiba (Kuwait), Umm Said (Qatar), Jebel Ali (Dubai), Ruwais (Abu Dhabi), Yenbo and Jubail (Saudi Arabia) were all built on 'desert' sites away from major centres of population.

In order to limit the costs of infrastructure provision and to facilitate the 'containment' of the immigrant populations at these large industrial sites,

5. *Middle East Economic Digest* (15 December 1978), p. 43.

there is a clear trend toward the recruitment of single, unaccompanied males to build and subsequently also to run the plants. This enables the industrial areas to be operated on a 'work camp' basis, with only a minimum amount of services provided for the all-male population and labour force. Thus social and infrastructural overheads are minimised, and the growth of the expatriate population is kept to a minimum by the exclusion of the migrant labourers' families, who remain at home in the country of origin of labour. Such work camps, in which the men who operate these enclaves live, resemble the normal camps associated with the construction industry. However, the camps associated with enclaves are to be operated on a long-term, rather than just a temporary, basis. Workers in enclaves mingle little with the indigenes and the established communities in the major urban centres. The cultural impact of the large expatriate labour forces is therefore minimised.

The countries providing the labour for this enclave form of development have been, almost without exception, Oriental. These South-East Asian companies have taken on large-scale contracts and supplied workforces of all skill levels prepared to live and operate in these work camp conditions. The camps are provided by the contractors themselves as part of the project. The Arab capital-rich governments have therefore discharged their responsibility for the provision of infrastructure for these immigrant workforces.[6]

One of the earliest major enclave contracts which set the pattern was the dry dock in Bahrain.[7] South Koreans provided the labour force and expertise for the $30 million contract. The Koreans were entirely self-sufficient in housing, which they built themselves. Food and recreational facilities were flown direct from Korea. After completion of the dock on time, the Koreans all left Bahrain. In their departure, the Koreans appeased one of the major fears of the governments of the capital-rich countries—the dallying of workers after the completion of the contract for which they were brought in. It is this process by which the more permanent communities of immigrants are built up. The departure of the Koreans contrasted very favourably, in the minds of the Bahrainis, with the aftermath of the construction of the aluminium plant for ALBA, when large numbers of migrants who had been employed building the plant remained. The aluminium plant was not constructed as an enclave.

A similar policy of enclave containment of labour forces has been adopted for the industrial development of Jebel Ali, Dubai. Isolation is achieved by siting the development away from Dubai city. The chief reason

6. J. S. Birks and C. A. Sinclair, *Saudi Arabia and the Libyan Arab Jamahiriya, Key Countries of Employment,* mimeographed World Employment Programme Working Paper, WEP 2-26/WP 35 (ILO, Geneva, Summer 1979).

7. J. S. Birks and C. A. Sinclair, *The State of Bahrain, Country Case Study,* International Migration Project (Durham, 1978).

militating in favour of enclave development here is the sheer volume of labour required for Jebel Ali. Figures of estimated project employment in 1985 are equivalent to the entire population of Dubai Emirate in 1977. If this scale of demand for labour were to be met in the open market of the Emirates, it would be disruptive of the whole economy and would make arrivals of labour hard to monitor and control. In short, the UAE's economy and labour market would be severely distorted if developments on the scale of Jebel Ali were not isolated from the labour market at large.

Oriental companies are supplying the self-contained workforce to construct the large-scale plant and infrastructure of Jebel Ali. The site will continue to be run as an enclave after the completion of construction, being operated by contracts similar to those issued for its construction phase. It will probably never be thought desirable to integrate the Jebel Ali developments into the wide economy of Dubai or the Emirates as a whole.[8]

Although less advanced at the present, the Ruwais development (Abu Dhabi) will be built and operated on a similar basis. The Yenbo and Jubail enclaves in Saudi Arabia are also being built by Orientals with enclave-type contracts. Saudi Arabian manpower will not be available to operate these industrial areas (let alone construct them, a task which is almost entirely in the hands of the Orientals), so Yenbo and Jubail will continue as enclaves, operated by Far Eastern labour which will not integrate with the Saudi national populations.

Yet Orientals are employed within the Arab world not only as enclave labour forces. Companies from South-East Asia have tendered successfully and are meeting deadlines on non-enclave contracts; the provision of infrastructure in Riyadh, for example, is being effected by Oriental firms, as are major hotel and office building contracts in other capital-rich states.

Orientals will probably continue to be employed in the capital-rich countries by choice instead of Arab labour. Not only is Oriental labour more cost-effective; not only do Oriental companies offer very attractive 'enclave packages' to capital-rich countries, minimising difficulties for indigenous planners; the Orientals, by virtue of their preparedness to live in enclave developments, and return home after contracts are complete, offer overriding social advantages as a labour supply for which knowledge of Arabic is not essential.

CONCLUSIONS

It might be thought remarkable that Asians and Orientals will come to dominate the labour markets of the capital-rich states when widespread unemployment exists in Egypt and Morocco. Considerations of equity and Arab solidarity should militate in favour of using this unemployed Arab

8. J. S. Birks and C. A. Sinclair, *The United Arab Emirates, Country Case Study*, International Migration Project (Durham, 1978).

labour before that of other regions. Indeed, this view is embodied in the labour law of the capital-rich states.

Yet in reality these considerations are of little relevance: whatever the overt desire of governments, development is executed through private contractors and establishments. These executors of development are concerned primarily with market factors. In the labour market, Asian and Oriental manpower is more cost-effective in the majority of cases than is Arab labour. Thus the former is preferred. Employment of Orientals in enclaves—where a large part of the future labour demand is found—falls within the overt (and now declared) policy aims of the governments of the Gulf states.

The evidence suggests that the Gulf states have little intention of abandoning their development aims; indeed, they have only a limited capacity to restrain the rate of growth. Their dependence on migrant workers is pervasive and extensive, and will inevitably increase.

In 1975 indigenous labour accounted for 50 per cent of employment in the capital-rich states; by 1985 this will have fallen to 41 per cent. In 1975 Arabs (indigenes and expatriates) absorbed 88 percent of all jobs. By 1985, it is likely that the Arab share will fall to 65 per cent. It is the implications of the Arab share of the market falling and of nationals becoming an ever-smaller minority in the capital-rich states, especially the Gulf states, that cause intermittent attempts to restrain development.

The fear is that, as nationals become a smaller minority, their national identity and culture will be lost. These concerns have some empirical basis: by 1985 the number of Asian and Oriental workers could have risen to as many as 1,600,000 from their 1975 total of 292,000. These figures relate to migrant workers: the settling of even a proportion of these workers with their families will create yet another challenge to the social fabric of the Gulf and other capital-rich Arab states.

Social Change

Iran provides a good example of how rapid economic change can create severe dislocations within a society. The pressures generated in this manner were one major contribution to the outbreak of the revolution. Pervasive social change is something that all the Middle Eastern oil states have experienced. In part, this is inescapably disruptive. In other ways, however, its effects may be ameliorated by successful adjustments; change is, after all, constant in every society. The last few decades have gradually seen the development of national identities in the Middle East out of local, sectarian or tribal ones. In the Gulf, the ruling families have evolved out of the strongest tribal clans and lay claim to legitimacy as symbols of national unity, by providing continuity in a sea of change.

Nevertheless, there are major problems caused by oil-fed economic change. In the larger countries, there is often massive unemployment in the midst of great demand for workers—nationals frequently lack the necessary skills or background for wage-sector jobs. The local population often disdains menial work, leaving it for foreigners to do. The rush to the cities, where the boom is most evident, results in the disruption of family and village ties. It may also signal an abandonment—or at least a weakening—of traditional values and patterns of authority in the headlong rush to modernization. All too often, the consequence is the anomie of urbanization: modern-looking and Western-oriented cities are inhabited by still largely traditional populations. As expected, social change affects the younger members of society the most. They are the first of their families to become literate, to travel or study outside their country or even village, and to accept positions within the government bureaucracy. In a part of the world where age commands respect, ministers, directors and governors are frequently under the age of thirty.

Within all these countries, a social cleavage comes into existence with the appearance of Westernized elites, who have adopted foreign lifestyles (*e.g.*

nuclear-family housing and 9-to-5 jobs), social values (*e.g.* drinking of alcohol), and political goals. The bulk of the population, however, remains heavily traditional. Middle Eastern society is characterized by a sharp dichotomy between the elites and the masses—the middle classes are still very small and their impact on society limited. But modernization does not only touch the elites; it also has an impact on the masses. Traditionally, there are two major kinds of lifestyles: that of the settled population or peasants, and that of the nomads or Bedouin. Both have been equally turned upside-down by the changes of recent years. While the average American's conception of a Middle Easterner may be that of a camel-riding nomad, the peasant is far more typical of most countries; only in Saudi Arabia and the smaller states of the Gulf have the nomads predominated.

Bedouin lives have been transformed even more drastically than those of the settled population. The economic transformation of their countries has often dictated sedentarization, which has been simultaneously encouraged by the state and even sought by the Bedouin themselves. But settlement has caused its own problems, as Sa'id Salman Abu 'Adhirah explains in his contribution below. The end result, for peasants as well as the Bedouin, is often very mixed feelings. Their dilemma was eloquently expressed by the Saudi Deputy Minister of Planning, Faysal al-Bashir, in an interview with CBS News Correspondent Ed Bradley.[1]

BASHIR: When I look back at it now, I don't believe anyone planned it. I just happen to love education. My father tried many times to stop me from going higher, because to him the fifth grade, or even the third grade, was the highest degree anyone can master. He was illiterate. He's still illiterate. My mother was illiterate and is still illiterate. Therefore, I don't blame them. But in my case, it happened; it was really an accident of history, an accident of circumstance. I just loved education. I did it, so to speak, my way. And I was really just like a particle in the sea, at the mercy of waves in that sea. And this is what happened to me. When I look back at it, sometimes I am very sad.

BRADLEY: Sad? Why?

BASHIR: Because, by the standard of my tribes, I am certainly considered to be, up to five years ago, one of the biggest failures they ever met.

BRADLEY: Why?

BASHIR: Because they thought that I'll be staying beside them to help them, in a way to be their leader. And you find me sad through using two yardsticks of judgment. When I see some of the tribesmen, I just start remembering those days. I am a romanticist. I think it's a beautiful life

1. CBS Reports, "The Saudis," originally broadcast on Tuesday, 21 October 1980.

when everything is going very well. When spring is good, grass all over the desert, sheep are running, you feel a sense of community. You can touch each one around you. You don't feel the impersonal life of the city. That was the beauty of desert life. That was the beauty of nomadic life. But when the spring is no good, when there is not enough rain, nomadic life is one of the hardest and shiftiest lives you can imagine, physically as well as mentally. The survival of the fittest does exist in the desert. You must be strong. You must be sharp in your judgment, because nature does force the human being to behave that way. I really don't regret leaving it, but I do feel sad sometimes when I remember that sense of community. When you feel really secure, everyone is your cousin or a brother or a relative, despite the fact you have no blood relationship to him. They are around. You feel together. And this is the sadness sometimes.

The only real comfort I get right now, is when I notice most young kids are following the path I did without knowing who did it, really, who took it, whether it's Faysal al-Bashir or what have you. They are doing what I did and following that path, and I believe that I made the right decision.

The intertwined processes of sedentarization, detribalization and urbanization may be seen most clearly in Saudi Arabia and the smaller Gulf states. The latter especially are perhaps best characterized as city-states: glittering metropolises of 100,000 to more than a million people, mushrooming in sites occupied by mud-brick villages only a few decades ago and dwarfing in both population and importance their small traditional hinterlands. These cities display all the handiwork of modern urban planning, complete with towering office and apartment blocks, ring roads, broad divided thoroughfares, and frequent lush parks—but with a few unique twists. Manama (capital of Bahrain) retains much of its old *suq* (market) atmosphere, even if the principal items for sale often seem to be portable tape decks and American jeans. Dubai (the second-largest city of the United Arab Emirates) sits astride "the Creek," which can be crossed either by a modern highway tunnel or by small wooden passenger ferries. Muscat (capital of Oman) sits in a small bowl surrounded by black volcanic peaks, its waterfront commanded by two picturesque Portuguese forts. The impression that the streets of these cities are heavily crowded with Pakistanis and Indians, along with other foreigners, provides evidence of another factor in the rapid social change these states are undergoing.

While physical transformation may be most complete in the Gulf emirates and Saudi Arabia, change has had fundamental impact elsewhere among the Middle Eastern oil states as well. In the second and third selections for this chapter, Dale Eickelman and Mary Elaine Hegland provide thoughtful case studies of how "progress" has changed the lives of peasants of the Gulf. In the case of Oman, change has been more gradual and subtle, even though its long-term effects may be just as detrimental to the composition of "traditional society." In Iran, the aftershocks of revolution have made villagers more activist and demanding; the events that have rocked the cities of Iran in the last few years have been accompanied by alterations in the countryside as well.

Bibliographic Sources for this Chapter:
4, 17J, 17K, 23F, 36, 37, 41, 42, 72, 89, 95, 102, 103, 105, 113, 130, 132, 133, 134, 142, 151, 153F, 157, 161, 162, 191, 199, 202, 207, 216, 217, 219F, 219G, 220E, 220I, 220J, 225, 226, 230, 260, 280, 288, 292, 296G, 305, 309A, 309C, 309M, 381, 388B, 411, 412.

Sedenterisation and Settlement of the Bedouin

SA'ĪD SALMĀN ABŪ 'ĀDHIRAH

I do not know the exact date when projects for sedenterisation of the Bedouin started in the Arab countries, but do not think such projects existed twenty years ago. It was in 1962 that they started in Jordan, and this was followed by three other Arab countries, including Saudi Arabia. Similar projects existed in Egypt but their aim was not only to settle Bedouin but to reclaim land, and only a few Bedouin were involved. Sedenterisation projects—which killed the Bedouin heritage and eradicated Bedouin tribal society—seemed at the beginning to be for the welfare of the Bedouin as they entailed attractive promises and brought wide hopes for a transfer from the life of hardship and misery to an undreamt of life of happiness and comfort.

Pioneer projects were set up to convince them of the promises made by the officials. Land was tilled and divided up into farms upon which houses were built and distributed to some of the leading families (*Dhawū kalimah masmū'ah*) close to the ruling circles.

In principle these pioneer projects were very successful, particularly during the first five years when the farmers or Bedouin settlers started to reap the fruits of these plans through the sale of their crops, particularly lucerne and vegetables, and through the wages the settler families received from the Government in return for their work in these agricultural units.

The State, as represented in the Ministry of Agriculture, was responsible for organising all agricultural operations and its officials worked unstintingly, hastening to offer aid and advice in difficulties, great or small. The only part played by the Bedouin was to divert channeled water from one basin to another, on one day in every two or three weeks. It was in fact excellent work and an unrivalled opportunity, especially since all agricultural operations were mechanised, whether it be cultivating, harvesting or

Reprinted by permission from *Arabian Studies,* Vol. 4, 1978.

transport. The administration of the project stipulated that the Bedouin work for a period of two years as a labourer, following which he would, as I have said, receive an agricultural unit. A large number of Bedouin joined these projects as labourers, in the hope of acquiring possession of the agricultural units. They left their families in the desert, their natural habitat, and completed two years of training and practice. Then commenced the second phase, the distribution of the units. The poorer Bedouin were set aside and the agricultural units handed over to the Shaykhs and their sons, or to persons of rank and authority close to the responsible officials or supported by a senior official. The Administration, however, far from immediately dashing the hopes of the Bedouin, even added to its promises to them, maintaining that they would receive the agricultural units at the first opportunity, and that they would only have to wait a few months until the land was reclaimed and distributed along with houses constructed for them. They waited months—then years.

They waited on with disaster looming ahead of them. With the sad heart of a Bedouin who witnessed the tragedy, actually living with it as an official in hope and expectation, I tell of the tragedy of the destruction of the life of the Bedouin, their customs, and their traditions inherited from their fathers and fore-fathers.

Firstly, relying on the promises made by the State, the Bedouin took to selling their camels and sheep in preparation for the move to settled life. Thus they lost their greatest asset which had been their pride, and which at the same time ensured the supply of their needs in the way of food, drink, transport, even wealth. In this way they lost all means of existence of the Bādiyah.

Secondly, when, as a result of this they abandoned the desert with tears in their eyes, they stuck close by the projects, settling by them in the expectation of receiving their agricultural units. Some continued to live in 'houses of hair' (tents), then some built a simple house of two or three rooms in places allocated to them by the State, while others built such houses at their own expense, but these were few in number. The majority borrowed the money from the Agricultural Loans Foundation (*Mu'assasat al-Aqrād al-Zirā'iyyah*) which welcomed them with open doors.

During this phase (the phase of transfer of the Bedouin from the desert to the settlement projects), the State realised its inability to continue these projects because of the immense costs involved, as well as on account of a succession of falls in aid extended by the rich to the poorer countries and, furthermore, the soaring prices of building materials in all countries of the world. The State had no alternative but to announce its inability to absorb and settle all these people.

This was a tremendous shock that dashed all their hopes to the ground. It was not only a tragedy—it was disaster.

The Bedouin, having lost their assets, their sole wealth, found that it was hard to regain them. A Bedouin in his tent is hospitable so he would maintain the obligations of hospitality in his house from his ever diminishing income, for now he had no camels producing offspring or sheep at pasture.

Social relations among the Bedouin declined and weakened, for the Bedouin began to think twice about visiting a friend who now was living in a stone house with a closed door upon which he had to knock for his friend to answer. This knocking at doors is not liked by the Bedouin who does not know what is going on in this house with closed doors—and no one might be at home. This man or that was accustomed to go to his friend's 'house of hair' which was open by day or by night, and he was able to see at a glance who was inside, it being the duty of those in the tent to receive him hospitably whatever the circumstances be.

Even if it were a woman in the tent she could perform the duties of entertaining the guest, and the guest could personally slaughter the animal for his meal, as well as preparing Arab coffee himself. These were the traditional ways.

But his friend has now come to live in a closed house, and the Bedouin is shy of sitting with a woman in a closed house like this. In such case the guest would be obliged to go to another house or to go away altogether.

Traditions of hospitality themselves have changed, for, instead of the animals customarily slaughtered in honour of and to entertain the guests, it was at times difficult to obtain these animals and, against his will, the Bedouin had to resort to killing chickens, or buy meat sold in the market. This caused awkwardness and embarrassment to both the host and his guest.

Times for gathering together to take coffee, customarily the hours of morning, noon, evening and night, came to an end, for every Bedouin became preoccupied with his own circumstances, his new life and how to organise it.

Then disposition and respect dwindled, for the Bedouin started to sound out the opinion of his daughter by one or other means, perhaps through her mother, regarding the person she would consent to marry. The authority and right of a cousin to the hand of his paternal uncle's daughter has diminished—in fact almost completely disappeared (though I personally am opposed to this form of marriage, but I refer to it to give a clearer picture of the kind of change which took place in customs and traditions).

Contact with the cities resulted in the adoption of bad habits unknown to the Bedouin all their lives. They learned about wine and heard obscene language. Their ways of bringing up their male children and daughters altered—and there is a vast difference between the child brought up in the 'house of hair' in a desert and the child born and growing up in a house amongst sedentary companions.

MIGRATION IN REVERSE

As a result of this a migration in reverse has commenced, and a return to the desert began when it was all too late and conditions had already so deteriorated that the Bedouin has now to build up himself anew.

In this article I do not attach blame to the State nor do I wish to arouse suspicions about its good faith. To tell the truth is always hard, and I do not want to go to jail as a result of writing this article, for I am a Bedouin who lives in the desert stretching from the Gulf to the Ocean.

What I want to show and to stress is that it is a duty and a wise course to revive Bedouin life through aid to the Bedouin in the place where they were assured of a livelihood—namely by digging wells for them in grazing areas where they always used to pasture their animals. They should be further aided by the provision of medical treatment through mobile medical dispensaries which would visit them regularly, even if only once a week. Medical veterinary services should also be extended to their animals. Mobile schools should be sent to them to provide elementary classes—after which pupils would go to boarding schools.

It is my belief that such a system as this would form the basis, initially, for the transfer of the Bedouin from the life of the Bādiyah to settled life, if it is proposed to persist in the policy of sedenterisation of the Bedouin.

But—I must ask a question. Why should the Bedouin be settled? Do they constitute a force to be feared if they remain together? Or are they an evil society that must be destroyed? These are questions to which I find no answer.

I mourn the Bedouin in their new life. Farewell to our noble old traditions—I have loved you as a child and as a young man. What can I do but bid you farewell, you life of the Bādiyah? My heart is bitter with sorrow for you and my sole consolation is that books and histories have preserved some of our traditions and customs. Some day it will be said there was once upon a time a Bedouin way of life which began to vanish in the last years of the twentieth century.

Omani Village: The Meaning of Oil

DALE F. EICKELMAN

The oil states of the Arab peninsula have all experienced stunningly rapid transformations with the advent of oil, but even in this unusual context Oman is exceptional. Oil was only discovered there in commercial quantities in 1964 and first exported in 1967; substantial programs of economic and social development got under way only after 1970. Prior to 1970, Oman was as isolated in the Middle Eastern context as was Tibet in Southeast Asia. Transport to the interior was still primarily by mule and camel; rice, cooking oil and inexpensive textiles constituted the principal regularly available imports. Hospitals, motorable roads and schools were virtually nonexistent and visitors to the country were rare. So when oil revenues were applied to development, the changes brought about were all the more striking because of their late start and because Omanis, many of whom had worked in neighboring countries in the 1950s and 1960s, had more concrete visions of their country's future than were available to countries that had begun development in earlier decades.

Oman like other oil states is actively seeking to reduce its dependency upon oil, but to date 95% of government revenues are derived from its export. Given the technical requirements of oil production, only a small fraction of Oman's estimated 900,000 inhabitants are directly employed by oil-related industries. The major oil company has only 4,000 employees, of whom 2,700 are Omanis. An equivalent number of Omanis are employed in directly related industries. Although roughly 70% of Omanis continue to live in the small villages and towns of the interior and the coastal plains, oil has also profoundly affected their lives.

Some aspects of these transformations are easy to discern, such as stunning increases in the number of hospitals, clinics, schools, modern housing,

A condensed version of this article appeared in *Aramco World Magazine,* Vol. 34, No. 3 (May-June 1983).

paved roads, employment in the public and private sectors, and the number of Omanis who *commute* from small towns and villages for work in the modern sector elsewhere in Oman and the nearby United Arab Emirates. Yet there are other more profound aspects of this great transformation difficult for outsiders and even for Omanis themselves to grasp because they are qualitative ones not readily amenable to enumeration: What exactly do the massive increases in revenue and expectations mean to most Omanis? How have the dramatic material changes affected the daily lives of Omanis and their notions of work, family, society and community? What sort of future do Omanis envisage for themselves and for their children?

Oman is a highly diverse country of more than 1,200 towns and villages, so the answers to these and related questions are necessarily complex. The way to begin is to look beyond statistics and abstract notions and seek to understand the daily life in villages where the "great transformation" made possible by oil is just beginning to be felt.

Wadi Ghul is such a village. It is a cluster of some forty households and 250 inhabitants situated in a narrow canyon 14 kilometers away from Hamra, a small provincial town in Oman's interior. There are two other small settlements nearby; together with Wadi Ghul, the total population of the three villages is no more than 900. All the inhabitants of these villages are members of the Abriyin tribe, one of the major tribes of the interior. The tribe's shaykhs or leaders reside in Hamra, which is also the nearest administrative center. Almost every town in Oman is characterized by its relatively small scale and Hamra, with 2,400 inhabitants, all but 200 of whom are Abriyin, is no exception. The residents of Wadi Ghul and its neighboring villages have a strong sense of regional community. They were originally client groups of the Abriyin and paid the shaykhs in Hamra a small annual tribute in exchange for their protection. Now that intertribal fighting is only a distant memory of older men, villagers think of themselves as Abriyin pure and simple and stress more their common identity as Omanis.

As late as 1978 travel between Wadi Ghul and Hamra was necessarily by foot and by mule. Sayf Said, a villager in his 20s who now is a soldier with the Omani army, told me that until recently he like other villagers measured distance in time, not in kilometers. By this measure, Hamra was four hours distant by mule. Today pickup trucks owned by some villagers, especially those with sons in the army, now leave almost every morning for Hamra and the neighboring market centers of Bahla and Nizwa, and Hamra is little more than half an hour away. The distance between Hamra and the outside world similarly has been shortened with the completion of a graded road in 1976 and a paved one in 1979. As a result, villagers now have regular access to the clinic and other government facilities in Hamra, although regular schooling is still out of reach. The new road also means that in addition to Hamra, villagers today can choose to sell their produce and herds in the slightly more distant markets of Nizwa and Bahla. Muleteers still ply their trade between Hamra and villages inaccessible by road, but the government road network is rapidly expanding to all but the most inaccessible villages.

Like most Omani villages, Wadi Ghul is built near the ruins of an older settlement and its size is limited in large part by the scarcity of water and suitable agricultural land. High on the side of the narrow canyon in which the village is situated are the remains of fortified watchtowers, some of them hundreds of years old, from which the entire canyon and its approaches can be seen. One of these watchtowers is so situated that buckets could be lowered into a reliable underground water supply. Now that intertribal fighting is a thing of the past, these towers have fallen into disrepair.

Wadi Ghul has none of the obvious outward signs of the wealth of Hamra such as electricity (since late 1982), television aerials, stores carrying frozen goods, or the clinic, sanitation services and schools of the provincial capital. The consequences of new wealth in Wadi Ghul show primarily in its houses and the presence of several pickup trucks parked where the graded road ends about half a kilometer away from the village. There is also a small store open several hours daily. The newest houses are made with cement rather than with the stones laboriously collected in the region and broken to size. The manual labor for traditional house construction is no longer available because so many of the village's young men work elsewhere in the army, as drivers or in other unskilled tasks. Indicative of the villagers' concern with religion and with presenting their best "face" to the outside world, the first buildings to be made of cement were their new mosque and a *sabla,* or guest house, where visitors are honored with the traditional hospitality of coffee and dates.

Despite the superficially timeless look of the settlement, none of the older stone houses is more than thirty years old. Because of the earlier threat of intertribal raiding, villagers lived in Wadi Ghul canyon only during those parts of the year when their presence was required for planting and harvesting. Caves were used as dwellings, and for the rest of the year they lived in one of the other, larger nearby villages. Gradually, more permanent houses were constructed, the first ones at a higher, easily defended location, and the present ones closer to the irrigated fields and orchards. The stones of the first houses were reused to construct houses in Wadi Ghul's present location, so little remains of the first settlement except the now-abandoned mosque.

Households in Wadi Ghul are grouped by kinship clusters. One of these is the Awlad Srur, the children of Srur. Srur significantly is not a distant ancestor. Although villagers say that they can trace their ancestors back for generations, in practice they rarely indicate ties more than two generations deep. Srur is the deceased father of three elderly brothers who presently together with their married sons constitute fourteen separate households that collectively maintain a small guest house and share the expenses of entertaining visitors.

Because of the exercised preference of marrying cousins and other close relatives, there are complex bonds of kinship and marriage among all inhabitants of Wadi Ghul. In fact, some of the women told my wife that they found it difficult to imagine how a women could marry a "stranger" not of the same extended family or an immediate neighbor. Each villager is related to the others in a multiple of ways and can trace ties of self through their fathers, mothers, husbands and wives, and brothers and sisters. With the sole exception of a woman who was originally from a village two kilometers away, marriage for the other women meant little more than a move of several hundred yards from the home of their father to that of their husband.

This clustering of relatives has practical advantages. Although each household manages its own expenses, meals are communally prepared and women ordinarily work together at their daily tasks: carrying water, cooking, tending children, sewing and helping with various herding and agricultural activities.

The new wealth is reflected in both new house construction and in all house interiors. House interiors, even of wealthier villagers, tend to be sparsely furnished with rugs and a few cushions. Machine-made imported rugs are beginning to replace the beautiful, locally made Omani rugs (*saha*) made by villagers themselves. In the women's work areas, usually a small building separate from sleeping quarters and used also for cooking, many of the cooking pots and utensils are imports as well. So are the ubiquitous thermos jugs and cups of coffee, usually from China. Plastic cans for water replace the less portable clay vessels of a decade ago, and most households now possess a sewing machine. Women sew their own clothes and those of their children. Almost all men now have their clothes made by tailors in local markets.

Wadi Ghul, like other Omani villages of the interior, looks as if it is self-sufficient, but this impression vanishes quickly when one talks with villagers, who say that they are not now and were not in the past. In the past, many were compelled to emigrate to neighboring countries; one of the oldest villagers once was a pearl diver in Bahrain. In the late 1940s and 1950s, many villagers worked in Saudi Arabia, and since the 1960s in nearby Gulf states. Land and water ownership also indicates an intricate dependence upon the outside world. Most of the minute parcels of land, some no more than two meters long and one wide, are locally owned and bought and sold only within the community. Yet now as in the past, some land and water rights were transferred out of necessity to shaykhs of Hamra to raise cash in years of particular scarcity, and repurchased whenever possible.

Income from agriculture, herding and, until its recent local demise in 1981, the related craft activity of weaving saddle bags and rugs fluctuates widely from year to year. As elsewhere throughout northern Oman, rainfall is negligible and irregular. Even when rain falls occasionally in torrents, it is almost without value unless it falls in the mountains, where it seeps into the earth and can be channeled into underground conduits of water called *falajs,* gently sloping underground tunnels which conduct water by gravity flow from hillsides to lands suitable for cultivation. Every Omani village of the interior is formed around at least one *falaj* system; a few have more than one. The rights and obligations that have grown up around the operation of these vital irrigation systems form the backbone of the local sense of community just as do the daily cycle of five prayers in which all village men and women participate and other shared community activities.

Every household has some income from agriculture and herding, but no household is either self-subsistent or self-sufficient. Dates and limes are the principal crops, and a few vegetables such as onions, radishes, garlic, and cucumbers, and cantaloupes and watermelons are also grown. A decade ago tomatoes were also successfully introduced. These vegetables are planted in the shade of the date palms. Of all this produce, only limes are sent in good years for sale in local markets. Yet now as in the past, villagers rarely have sufficient produce for their own needs. In years of water scarcity or of poor yields due to crop diseases, the villagers of Wadi Ghul are compelled to buy even dates from markets in neighboring towns. In good years, falling agricultural prices restrict the potential for profit in selling to local markets.

Part of the food consumed is of local origin, such as the thin, waferlike bread called *rkhal* made from date palm fronds. Chickens, who wander freely around the house compounds, are a source of eggs and meat, and sheep and goats are slaughtered on major occasions such as feast days. For as long as anyone can remember, rice and coffee, both imports, have been staples. They were carried to inner Oman in the past by mule and camel caravans from Muscat. Now, with improved transportation, these goods and tins of pineapples, tomato paste, anchovies, and other easily transportable items are available. Oranges and apples, now a fairly regular part of the diet, were such luxuries in the recent past that they were kept under lock and key for honored guests.

Women stay close to the village and its fields most of the time. If no men are present when guests arrive, women will serve them coffee and dates, in accord with the conventions of hospitality, but will quickly withdraw when men are present. Men are responsible for herding further afield and for marketing. In summers they take their herds of sheep and goats to high mountain pastures, sometimes remaining away for weeks. In the colder winter months, shepherds and their herds return to the lower valleys of the Wadi Ghul region.

Even more than agriculture, herding ties the villagers of Wadi Ghul to the outside world in complex ways. Herds are expensive and their wellbeing depends largely upon fluctuating seasonal conditions and the skill of individual shepherds. Bad weather, insufficient forage or disease can suddenly decimate herds, and few villagers have sufficient capital to withstand a major catastrophe and reconstitute a sizable herd. In the past, much of the capital for herds came from the wealthier shaykhs of Hamra, although now that there are alternative lucrative investments in trading and construction, investment in livestock is rapidly tapering off and the size of herds has sharply declined. In this sense, the incorporation of Oman into the modern economy has had a profound impact upon local patterns of exchange, although some of the slack in investment has been taken up by family members who have joined the army and who contribute part of their cash income to relatives who remain in the village.

Improved transport in recent years gives villagers a choice of markets in which to sell their animals. Almost daily someone from Wadi Ghul goes to each of the three major nearby markets and villagers naturally choose to sell their animals in the location offering the best price. In the past there was no effective alternative to Hamra itself. Men and women are involved in separate and intricate visiting patterns; the constant exchange of visits insures a sharing of information concerning economic and other "news."

Another local source of income related to herding has been until recently the weaving of saddle bags and Omani rugs. For the obvious reason that saddlebags are becoming a thing of the past as transport becomes motorized, their manufacture is declining rapidly. The decline of rug weaving is more complex. There are two color combinations of rugs. One pattern is brown and white, using only the natural wool of animals. The other is red and black, with the red thread prepared by dying white wool with madder, a rich crimson dye from India which is purchased in the Hamra market.

In the past, many of these rugs were locally used, although current fashion now is to display machine-made imports in guest rooms. In the last decade, rugs were destined primarily for the markets of Nizwa, Bahla and Muscat. Occasionally townsmen commissioned a rug, but most ended up in the hands of auctioneers, who for a modest fee which never exceeded 10% of the value of the rug (currently about $70 each) saw to their eventual sale.

Wool for rugs and saddlebags was obtained by shearing the sheep and goats of each household's herds. This was usually done after the colder winter months, when herds were in the vicinity of Wadi Ghul, and sometimes late in the summer as well. The wool was then spun into yarn, the slowest part of the entire process. Shepherds, women and older children spun whenever they had spare time between their chores and the demands of hospitality. Each complete ball of yarn, called a *kibba,* took about four days to complete, and nine balls of yarn were needed for a complete rug.

Making the rug itself has been until recently an exclusively masculine activity and is done on a simple two-beam wooden loom which can be set up on the ground and easily carried from place to place. Relatives passed the craft to each other, and households that engaged in rugmaking made anywhere between ten and fifteen rugs annually.

There is a continued demand for Omani rugs, now in large part destined for Oman's growing foreign community. The demand has caused prices for Omani rugs to treble between 1980 and 1982. Nonetheless, rugmaking together with other traditional crafts is rapidly disappearing in face of the more attractive alternatives of schooling and wage labor. In Hamra, which itself had no modern schools until 1972, virtually all school-age children of both sexes now attend both primary and intermediate schools, and many children in villages closer to Hamra do so as well. As schooling spreads, the participation of youth in the traditional activities of agriculture, herding and related crafts drops markedly. Because of the high price of rugs, the Omani government has had limited success with a pilot project in a coastal Omani village to encourage women to take up rugmaking as men abandon the craft. Yet for the time being, rugmaking has disappeared as a craft in Wadi Ghul and many other villages in the interior. In 1980 six young shepherds in the village were still engaged in weaving. On a return visit in 1982, I found that they had all joined the army and had not been replaced by other villagers.

Now that Wadi Ghul has a road, younger villagers say that a primary school is their most urgent requirement. Until a decade ago, the only way to become literate was to attend the outdoor Quranic school shared by Wadi Ghul and two other nearby villages. Here children from the ages of roughly six through fifteen gather under the shade of a date palm to learn to chant the Quran. Children often attend for several years if they are not needed for agricultural or herding activities. Unfortunately, as many of the villagers remarked, schools such as theirs do not regularly teach reading or writing. It was not unusual to see students who had attended the school for up to three years being unable to write their names or simple phrases in Arabic.

The lack of a school does not mean that village youth remain illiterate. In remote villages such as Wadi Ghul, the army serves a major educational function. Every recruit who joins the army is required to attend literacy classes until he possesses a basic literacy. Because of advanced teaching techniques and an almost exclusive use of Omani instructors, the army has a remarkably successful program in which most recruits achieve basic literacy within four months. The most able recruits are encouraged to proceed to more advanced levels of education. Of even greater significance, these programs inculcate notions of citizenship and skills which enable soldiers effectively to represent their village to local officials and outside administrators.

Because of the impact of schooling on their own lives, recruits often try to teach their younger brothers and sisters the rudiments of reading and writing during their monthly leaves, and seek ways to persuade the government to build a school in their village.

What does the future hold for a village like Wadi Ghul? Perhaps soldiers like Sayf Said, most aware of what the village is now and what it might become, are best able to convey a notion of the future. One Friday afternoon, when many of the younger men on leave from the army were present, ten of us sat outside in the shade of one of the guest houses and discussed the futures open to the village. I emphasize futures, for Sayf and the others were aware that their actions for the first time were likely to shape what would become of their village.

Unlike some youth in other countries of the Middle East, those of Wadi Ghul are not yet very attracted to the "bright lights" of Muscat or large cities elsewhere. Their first thought continues to be their villages, where they have their wives and families. Most marry young, and many have children before they are twenty. They are aware that they can never expect to live on agriculture and herding alone, and recognize the reluctance of their sons and younger brothers to engage in these activities. The impact of the decision of youth not to engage in these activities once schooling is widespread will be significant. For Wadi Ghul, the great transformation of schooling has yet to come.

When it does, my friends argued that it will not result in a progressive abandonment of their village. For many decades, the men of Wadi Ghul have had to emigrate for part of their working lives in order to support their homes and their relatives. As employment opportunities have opened up in their own country, work elsewhere has ceased to imply seasonal migrations or long absences of many years. Work in Oman itself means frequently weekly, and in almost all cases monthly visits to their village. Approximately 50 men from Wadi Ghul, ranging in age from their late teens to their early thirties, work elsewhere in the country. Unlike their elders, they can return home frequently to their wives and families. In their judgment, the increased opportunities for work in Oman itself has given their village a new lease on life.

The strong sense of family and community is likely to keep communities such as Wadi Ghul intact for the time being, even if their local economy is marginal and must be sustained by remittances from villagers working elsewhere in the country. The "good life" as seen in Wadi Ghul does not consist of expensive luxuries. Even when money is available, there is little desire for expensive clothes and furnishings. Villagers invest in basics: pickup trucks to facilitate travel between work and village, wristwatches (a necessity in the modern army), sewing machines, concrete houses, and an improved diet. They find it inconceivable to live in a situation in which neighbors are not kin. Wadi Ghul and villages like it have long been subject to strenuous economic vicissitudes. For them, the "great transformation" brought about by oil is a fortunate circumstance because it frees them from the economic insecurity of the recent past. Oil's great transformation has permitted substantial improvements in the standard of life and soon will mean access to the benefits of modern education. The economic and educational changes are seen as means to improve the quality of life in the villages themselves, not abandonment of villages for new styles of life. Omanis are justifiably proud of their past, place great value upon ties of family and village, and consciously recognize the strength of their social and cultural traditions as they face a future of challenge and promise.

One Village in the Iranian Revolution

MARY ELAINE HEGLAND

In Iran, during the summer and fall months of 1979, one repeatedly heard reports of conflict over land, of peasants evicting their landlords to take over and farm the land themselves. One such uprising took place in Aliabad,[1] not far from the southwestern provincial capital of Shiraz. This article addresses two main questions: What were the circumstances and processes which led to this uprising? Why, in Aliabad, as in many other villages, did an uprising occur when it did, several months after the Iranian Revolution of February 11, 1979?

Encouraging the development of a local revolution in Aliabad were three main factors. First, the villagers were filled with a bitter but suppressed resentment towards the landlord because of a history of conflict over land and water and his brutal treatment of those who opposed him. Secondly, contrary to expectations, the new government did not support the supremacy of the landlord's family, the Askaris, with sanctions and force. Thirdly, the dissidents effectively utilized slogans, symbols, and types of revolutionary activity developed during the course of the Iranian Revolution, as well as the expressed attitudes of Imam Khomeini and new government policies to legitimize their local revolutionary activity, thereby gaining cooperation from authorities in Shiraz.

1. The name of the village and names of persons have been changed in the interests of privacy. This article is based on an 18-month period of research in Iran, extending from June 1978 until December 1979, which was made possible by a fellowship granted by the Social Science Research Council and the American Council of Learned Societies. The conclusions, opinions, and other statements in this article are those of the author. A great debt is owed many kind and open-minded Iranians who offered friendship and assistance.

Reprinted by permission by the publisher, Middle East Research & Information Project, from *MERIP Reports,* No. 87, May 1980, pp. 7-12, *MERIP Reports,* Box 1247, New York, NY 10025.

A HISTORY OF CONFLICT

Three major incidents in the recent past had aroused great antipathy towards Seyyid Ibn Ali Askari, who lived in nearby Shiraz, but owned a third of the land in this village of about 4,000 inhabitants, and towards Seyyid Yakub Askari, his brother living in the village. The first incident coincided with the overthrow in 1953 of Prime Minister Mossedeq and the return of the Shah to Iran. Under the leadership of several progressive religious figures and educators in Aliabad, most of the population had supported Mossedeq and travelled into Shiraz to demonstrate in his favor (much as they participated in pro-Khomeini demonstrations in late 1978 and early 1979). In this early period Seyyid Yakub, his relatives, and other allies took the part of the Shah, for they profited from their connections with his government. After the coup, Seyyid Yakub's henchmen attacked the homes of the pro-Mossedeq leaders in the middle of the night and beat them severely. At least one of the leaders was then exiled from the village for a number of years; others were punished as well.

A second set of events that contributed to the reservoir of popular anger grew out of a peasant sharecroppers' strike in the late 1950s. This was significant because of its effect on the distribution of land following the 1962 land reform. The peasant sharecroppers declared their share of the crop to be insufficient. Seyyid Yakub Askari, then representative of the former landlord, was forced to hire labor to farm the land himself. At the end of this year, neither the former landlord nor many of the peasants were yet willing to relent. Seyyid Yakub announced that all previous sharecropping agreements were null and void and rights to agricultural land were available to any and all for the asking. Some peasant sharecroppers had by then found jobs elsewhere, and others, too stubborn and resentful of the former landlord and his representative, abandoned their traditional agricultural rights. In the end, Seyyid Yakub and many of his relatives and supporters took over full rights to the agricultural land in dispute. Several traders, absent for long periods of time due to their work and who thus often maintained a neutral position in village affairs, also took advantage of this opportunity. Because land reform stipulated that land should go to those presently possessing agricultural rights, those former sharecroppers who had forfeited their rights after the strike lost out and did not receive any land; while many men who had been traders or landless shortly before land reform, took possession of land.[2]

2. Before the strike some 150 to 200 household heads, or about 40–50 percent in the village of 3,000 persons, held sharecropping rights. Most other men were in trading at various levels, a vocation many people considered to be preferable. Peasant sharecroppers were assumed to be the luckless, less clever and less able persons who could not command the relatively greater resources necessary to go into trading. Many people combined the two occupations, and there was some shifting between them as well. In the past as well as today most men were engaged in a variety of economic activities in an attempt to make ends meet or to get ahead. There were

The process of land reform, more than any other aspect of peasant-landlord relations, aroused the greatest outrage and deepest resentment. First, because many families had abandoned their traditional farming rights during the strike or because their agricultural activities were not economically viable, they received no land whatsoever or such a small area of land that they considered it useless and refused to accept it. Secondly, as the former landlord was an in-law of the Shah, he had received advance warning of the impending land reform. He quickly sold half of the village to friends there, although apparently this did not become public knowledge until later. Seyyid Ibn Ali Askari bought two-thirds of the land for sale and one-third was purchased by Seyyid Yakub and some 20 others. As a result, only half of the village land was distributed among the persons holding agricultural rights; they felt it was their right to receive all of the land. Thirdly, the best land—flat, irrigatable, and close to the village—went to the new small landowners. Only the dry, hilly land further from the village was left for those with agricultural rights. Fourthly, many of those peasants who had kept their agricultural rights after the strike feared that the land reform official seeking the dimensions of their land was doing so for the purpose of taking it from them. They reported an acreage less than what they were actually farming, and then received less land than was their right. Other people owning rights to land but working in other jobs or trading outside of the village were absent during the few days allowed for registering for land reform and did not receive land.

Directing all of their fury about these discrepancies in land distribution at Seyyid Ibn Ali and his family, the villagers refused to sign the required papers and violence broke out between the two factions. The villagers succeeded in routing the landlord faction from the village. But Seyyid Ibn

---

frequent examples of socio-economic mobility. After the strike the number of sharecroppers declined; only about 100 men received land during land reform. Such land was generally not sufficient to support a family. Ten to 20 of these small owners supplemented the meager income from their land by working as agricultural laborers. The others worked outside the village, as did most other village men. Only about 10 percent earned the greater part of their income by working in the village as agricultural or construction laborers.

As trading opportunities decreased and the amount of capital required for a successful enterprise increased, the number of men in trading pursuits declined drastically. Ex-traders as well as ex-farmers entered factories or commuted to Shiraz to work in construction or other jobs.

The village upper classes relied on one or more of the following economic bases: (1) opium smuggling; (2) ownership of land bought from the former landlord before land reform; (3) trading. Several men had managed to continue in trading and maintained prosperous shops in the village or even commuted to their shops in Shiraz.

Before the strike, the landlord took two-thirds of the *ābī* (irrigated) crop and one-third of the *bash kārī* (dry land) crop. Peasant sharecroppers were responsible for providing human and animal labor, seed, and equipment. In addition, they had to provide labor for maintaining irrigation streams and ditches. A year or so after the strike a delegation of a new set of sharecroppers, now including Seyyid Yakub himself and prosperous traders, opium dealers and other well-to-do relatives and supporters of Seyyid Yakub, visited the landlord to confer with him regarding the division of the crop. They persuaded him to lower the share to one-half the *ābī* and one-fifth the *bash kārī*.

Ali used his period of absence from the village to rally support from among his contacts with government officials. He even succeeded in obtaining a letter from the Shah, according to some reports.

After some months he triumphantly re-entered the village, accompanied by several truck loads of armed gendarmes. Three days of virtual martial law followed. Men and boys hid in terror. Women didn't dare go outside of their homes. All those seized were tortured and beaten, held a few days within the village or the gendarmerie station several kilometers away, and then taken to prison in Shiraz. Young boys were hung upside down and beaten on the bare soles of their feet for a day or more, until they revealed the whereabouts of older brothers and fathers. Four or five persons, reportedly, died as a result of beatings and torture received. At least 27 men were imprisoned for about six months. Following their release they moved, as others had already done, to other villages to avoid further retribution from Seyyid Ibn Ali and his family. The wrath of the villagers was reduced to a smoldering resentment.

Seyyid Ibn Ali's advantageous connections with government officials and those in authority were inherited from the former landlord, who was from a nationally prominent family. In addition, through his heavy involvement in opium trading and processing, he had contacts with heads of government offices in Shiraz, many of whom used opium. With others he had ties based on bribery that related to his smuggling activities. He also discovered another method of widening his connections in political circles—using the land he had bought before land reform. He sold a large area of land, which actually had been appropriated to holders of agricultural rights under land reform, and used the money to bribe personnel in the court system. He sold sections of land for a moderate price to the chief of police in Shiraz, to the head of the court system, and to several other influential persons. He continued to encroach for his own use on areas of land and water resources reserved for the peasants. He antagonized the other small landowners by doing likewise with their land, until he controlled an estimated two-thirds of the village land, instead of the one-third which he had "legally" purchased.

After land reform, agriculture in Aliabad declined drastically. Relatively little land or water was left to the agriculturalists after Seyyid Ibn Ali confiscated these resources, often selling them to Shirazis for villas or factories. The villagers blamed this decline, and the fact that subsequently most men were forced to either move into Shiraz to work or to commute to jobs in the city, on Seyyid Ibn Ali and his supporters. But there was no recourse open to them, and they experienced years of helpless fury.

POLITICAL EXPECTATIONS

Given this suppressed hatred of the people of Aliabad, it is rather remarkable that, after the fall of the Shah's regime which propped up Seyyid

Ibn Ali as a village dictator, they did not immediately challenge the supremacy of the Askaris. Their failure to do so is related to perceptions about the impact which the revolution and the new government would have on political structure and process within the village.

During the course of the revolution, Seyyid Yakub and his allies were outspoken in their opposition to Imam Khomeini and in their support of the Shah. Seyyid Ibn Ali, on the other hand, was politically astute. He correctly sensed the probable outcome of the struggle at an early date and professed his devotion to Imam Khomeini. The sons of both Seyyid Yakub and Seyyid Ibn Ali as well as of their allies were active in the revolution. Indeed, Seyyid Muslim, the 27-year-old son of Seyyid Yakub, was a main organizer of revolutionary activities. A highly successful entrepreneur dealing in the lucrative real estate business, leasing gardens, and also in trade, he was not tied to an employer and was free to organize the village youth and bus them into Shiraz for daily demonstrations. In addition, he was an experienced administrator. The villagers were quite accustomed, however grudgingly, to looking to him as a leader. Finally, as both his father and he were on excellent terms with the gendarmes, he did not fear punishment for his revolutionary activities. Although Seyyid Yakub was responsible for giving the names of those active in the revolution to the gendarmerie head, he would hardly have given them the name of his own son.

Upon the success of the revolution on February 11, 1979, all expected the new authorities to reward those who had helped bring about their victory. Should the Khomeini forces prevail, as they eventually did, Seyyid Muslim and the other young Askari men were expected to be able to protect their elders from summary revolutionary justice. Similarly, should the Pahlavi regime retain power, the older Askaris were prepared to shield their sons from any official retribution. In either case, assumed the villagers, the Askaris had situated themselves to maintain their local power and influence, and few anticipated any practical change in the political structure even in the event of a Khomeini victory.

Reflecting the common assumption that Imam Khomeini's political role and performance would be similar to those of the Shah, one peasant proudly told me that he was old enough to remember five shahs—including the present one. Many people expected a secret service organization, similar to the Shah's SAVAK, in any Khomeini regime, and that the same type of political organization of exchanging favors, would be operative. All power and authority would continue to emanate from the central government and the central figure of Imam Khomeini. They assumed that this power would come down to the local level, available for the Askaris to run the village. Seyyid Ibn Ali Askari would no longer be the power intermediary with Seyyid Yakub Askari serving as administrator; now the visiting *mullah* from Qum would delegate authority to Seyyid Muslim Askari for local administration.

Political activity was based on these assumptions for several months after the revolution. Thus, although no one paid the customary and required visit of congratulations to Seyyid Yakub on New Year's Day (March 21, 1979), many people did call on the visiting *mullah.* Even persons from other villages came to the *mullah* with their problems and disputes, as previously people had gone to Seyyid Yakub for assistance. By taking gifts to the *mullah,* villagers hoped to get on his good side and thus be granted special consideration in future times of trouble, just as in the past they had taken trouble to be on Seyyid Yakub's good side. The *mullah* worked closely with Seyyid Muslim, as the most influential local political figure, and was a frequent guest at his home. Seyyid Muslim was in charge of organizing the village men to guard the village every night. When the *mullah* was provided with rations from the Shiraz authorities to give the night sentries, he handed them over to Seyyid Muslim for distribution.

Seyyid Yakub and his allies, to gain the good graces of the new authorities, visited the mosque and were verbally chastised by a religious figure from Shiraz for failing to support the Imam Khomeini. They expected all would be forgiven with an apology, as had been customary when joining forces with the victorious contender in a political struggle.

Gradually, however, both the *mullah* and the people realized the situation had changed drastically. The *mullah* did not have the backing to enforce his decisions, nor to help people with problems and resolve conflicts. Dissatisfied with the lack of obedience to his rulings, the *mullah* eventually left. The gendarmes, fearing revenge for their past brutal treatment of villagers, either left the area or stayed at their post but refused to involve themselves in village affairs. The *pasdaran* (Revolutionary Guards) from Shiraz were unorganized, inexperienced, unwilling to jeopardize their popularity, and too few in numbers to have an effect on security in the region. Unsure of what the new policies would be, personnel in government offices were reluctant to take action of any kind. Religious leaders were fearful of taking initiatives which might lead to further dissension and strife and certainly were unprepared to champion such an unpopular family as the Askaris, even if some members *had* participated in the revolution. Minor attacks against Askari property went unpunished. In May, a young man actually assaulted Seyyid Muslim. The gendarmes failed to make an immediate appearance and arrest the young man; the Askaris, for their part, speedily and cheerfully patched up the quarrel!

The villagers noticed this reluctance on the part of the authorities to back the Askaris and became aware that more local initiative and activity were permissible under the new conditions. They began to organize, gradually overcoming their fear and gaining new confidence in their cause and devotion to it, much as had happened among the Iranian population as a whole during the course of the national revolution. During this local process, the symbols, political categories, and activities developed during the course of the revolution were used by activists to legitimize their actions for themselves, for potential supporters and for government authorities.

SYMBOLIC WEAPONS

Shortly after the revolution the educators and religious figures who had been enemies of the Askaris since the time of Mossadeq resumed their efforts to regain peasants' land confiscated by Seyyid Ibn Ali. They filed complaints with the many government offices in Shiraz and even traveled to Tehran to confer with officials of the land reform offices, but with no results. The changed perceptions and attitudes, though, gained momentum over the summer months of 1979.

The reformist and revolutionary emphases in Shi'a Islam took on local meaning. This vocation to struggle against injustice and oppression could apply to situations other than ridding the country of the Shah. Villagers began discussing the past exploits and characteristics of the notorious Seyyid Ibn Ali in the vocabulary of the victorious revolutionaries: *zolm* (oppression), *zur* (force), *taqouti* (decadent, idolatrous), *zede enqelab* (anti-revolutionary), *estabadad* (tyranny), *savaki* (SAVAK agent), *mofsed* (corruptor), and *roushveh* (bribe).

Just as mosques had been centers in national revolutionary activity, the Aliabad mosque became a forum used by the activists to arouse sentiment against Seyyid Ibn Ali. Every Friday during July of 1979, the mosque was the scene of inflammatory speeches by visitors from Shiraz and local leaders denouncing Seyyid Ibn Ali. Although in the past, mosque activities had been completely regulated and financed by the Askaris, the former rulers were all too aware of the change in their position, and did not even attempt to enter the mosque during those months.

The villagers were not content to emphasize the Shi'a injunction to struggle against unjust leaders; They also expounded on its socialistic aspects. Men and women would comment positively on the simple life led by Imam Ali, and then mutter that it wasn't Islamic that some people should have hundreds of acres of land while others had none whatsoever, that some people should own whole apartment buildings in Shiraz (as did Seyyid Ibn Ali), while others had not even a home to their name. Through discussion and education, the anti-Seyyid Ibn Ali alliance increased in numbers over the summer. Then, on October 2, 1979, Seyyid Ibn Ali made a move calculated to win him the sympathy of Shiraz's religious hierarchy; but it backfired, further unifying the opposition to him and mobilizing it into action.

Accompanied by a group of four Revolutionary Guards and engineers, Seyyid Ibn Ali arrived in Aliabad prepared to divide some 40,000 meters of his land among 40 *mostazefin* (poor and oppressed). Upon learning that the landlord planned to give away land actually belonging to the peasants to so-called *mostazefin* (who were in fact his own relatives and a few influential persons whom he hoped to buy off), the peasants and landless protested. The *pasdaran* and engineers, innocent of contention for the land, stopped their work and asked the protestors to explain their case to the court the following Saturday. Early Saturday morning, October 6, several hundred irate men left in buses, pick-ups, taxis and private cars to chant in front of the governor's office and the court and to swear out more than a hundred complaints against Seyyid Ibn Ali, including four accusations of murder. Seyyid Ibn Ali was arrested the same day. For several days in a row the men of Aliabad commuted into Shiraz to shout slogans and chants in front of government offices and homes of religious figures and to add to the file against Seyyid Ibn Ali. Seyyid Muslim organized the Askaris for a competitive show of force, but they were badly outnumbered and outshouted.

In hopes that Seyyid Ibn Ali would be able to reach a compromise with the peasants, the authorities let him out of jail on October 10. Accompanied by his relatives and other supporters in a noisy caravan of cars, Seyyid Ibn Ali came to Aliabad with the aim of pacifying his enemies. But the population reacted with outrage, crowding into the village square and even shouting in anger at a jeepful of normally revered Guardsmen sent to preserve the peace: "You think we're still back in the Shah's time so that you should stick up for tyranny?"

At six the next morning the mosque loudspeaker broadcast the message to gather for another trip to Shiraz: "Don't be afraid! Fight for your rights!" Several hundred men repeated their trek to Shiraz to shout chants in front of the governor's office. The leaders spoke with the governor's representative, explaining why the people of Aliabad wanted Seyyid Ibn Ali put back in prison. The landlord was again seized, this time in the process of arranging to leave the country!

With this second imprisonment, more people in Aliabad, who at first had been reluctant to defy Seyyid Ibn Ali openly, allowed themselves to be persuaded of their Islamic duty and joined forces with the opposition. The Shiraz Court eventually acquiesced to continuous pressure and granted the activists explicit, written permission to farm those areas of Seyyid Ibn Ali's land which he himself had failed to plant. (Allowing peasants to sow land not currently farmed was a general policy designed to encourage agricultural productivity.) Armed with this document and optimistic about the final outcome of the dispute over land, the leaders announced that on November 1 and 2, all married natives of Aliabad who wished to receive land should register at the mosque. With those natives who were currently living in Shiraz also eligible, the signatories totaled some 800 men. On Friday morning, November 2, crowds of men gathered at the mosque and marched out to a section of land previously controlled by Seyyid Ibn Ali. In the forefront was the green flag of Islam. Two groups chanted the alternate phrases of a revolutionary couplet, as was customary in the large national marches and demonstrations against the Shah. At the site, the green flag of Islam was planted in the ground and men took turns sowing by hand, while the elated onlookers shouted *"Allah-u akbar"* and other inspiring slogans.

Since it was late into the winter wheat season, the activists decided that rather than waste precious time dividing the land up at this point, they would farm the land jointly this year. Ten men were hired to be responsible for agricultural work, and each new shareholder was taxed 400 *tumans* to defray expenses. All shareholders were expected to participate in the necessary communal activities, such as cleaning out the irrigation ditch.

Another incident two weeks later aroused the Aliabad population again and further undermined the position of the Askaris. Gendarmes and revolutionary guardsmen caught the Askaris in the act of firing rifles and throwing rocks from the roof of Seyyid Yakub's home upon unarmed crowds below. Seyyid Yakub, his sons, Seyyid Ibn Ali's sons, and other allies were arrested. They were subsequently released, but warned by authorities in Shiraz not to return to the village. Virtually exiled and fearful for their safety, they remained in hiding. Other supporters did not dare return to the village and stayed with friends or relatives in Shiraz.

Taasua and *Ashura,* the eve and the day of the anniversary of Imam Hussein's martyrdom, had in 1978 been the turning point of the Iranian Revolution. Proclaiming that it is more important to follow Imam Hussein's example by being prepared to die in the struggle against injustice and repression than it is to weep for his demise, Imam Khomeini urged Iranians to defy the orders of the Shah's government and take to the streets. He instructed them not to weep and beat their chests and to lash their backs with chains, a traditional manifestation of grief for the death of Imam Hussein, but to exhalt in his spirit of courage and altruism, to follow in his path and to be prepared to die for their faith and for the freedom of succeeding generations. For the first time, all over the country, thousands upon thousands braved the military to scream in terrible anger, "Down with the Shah." This marked the beginning of the struggle-to-the-death with the Pahlavi regime.

For the people of Aliabad, commemoration of *Taasua* and *Ashura* in 1979 was a celebration of the fact that the newly understood revolutionary meaning of Shi'a Islam was alive in their own village. By planning and organizing the commemorations themselves, the activists effectively demonstrated that Imam Hussein and his martyrdom belong to those who fight for justice. It is not the property of the wealthy and powerful to legitimize their supremacy, as had been the case in the past. Then, the Askaris had organized, regulated, and financed *Ashura* activities, as befitted their *seyyid* status. This year the Askaris did not even dare make an appearance.

Seyyid Ibn Ali had managed to bribe his way out of prison the day before

Taasua. This provided the activists with an opportunity to put into immediate practice their contention that the struggle against injustice is more important than mourning the death of Imam Hussein. Upon learning of Seyyid Ibn Ali's escape, men and women abruptly halted the mourning rituals and went to the nearby gendarmerie office to complain .When they were informed that the matter was out of the jurisdiction of this branch office, the men continued on to Shiraz to demonstrate in front of the homes of religious leaders and influential persons. They did not return until they were assured that the landlord would again be imprisoned. By this time, 10 o'clock at night, it was too late for the customary afternoon and evening ritual of *Taasua*. But people assured one another that it did not matter; the more important obligation was to go after Seyyid Ibn Ali and bring him to justice.

By the time of my departure from Aliabad in December 1979, the landlord still had not been apprehended. The pressure on government officials to continue the search for him went on, but several more thoughtful persons confided that it would be better if he fled the country. People would begin to forget, and his relatives and supporters, if not personally guilty of unjust acts, could be assimilated into the community again. If there are no drastic reversals of the new situation in Aliabad, the gratification of the villagers at freeing themselves from the power of the village tyrant and his cohorts should mean considerable loyalty in the future to Imam Khomeini and the Islamic Republic.

Political Evolution

It is obvious, of course, that extensive socioeconomic change does not occur in a vacuum. This is especially true when it is of the magnitude experienced by these ten countries since 1974. A major effect of the kinds of change outlined in previous chapters has been to put considerable pressure on existing political systems. In Iran, these pressures were a major contribution to the unstable political climate which developed under the Pahlavi regime. It should be remembered, however, that the ramifications of sudden oil wealth in Iran were only one cause of the Iranian revolution. As Eric Rouleau shows in his following contribution, the revolution grew out of a combination of factors, most of which were unique to the Iranian case. While the Middle Eastern oil producing countries may have oil, socioeconomic underdevelopment and geographical proximity in common, they nevertheless vary considerably in social composition, historical background and political experience. Consequently, assumptions such as that the Iranian experience will inevitably provide the example for other producing states must be treated with caution.

While some of the states under discussion are monarchies, as was Iran, they are only remotely comparable to the Shah's regime. The patrimonial governments of the Arabian Peninsula (Saudi Arabia, Kuwait, Bahrain, Qatar, the United Arab Emirates and Oman) are firmly rooted in traditional sociopolitical milieux. To date, their ruling families remain closely linked to the process of creating nation-states out of tribal environments, as shown below by J. E. Peterson. There is little of the megalomania and paranoia of the Iranian imperial court; attempts are made to retain "democratic" aspects of the traditional way of governing such as the *majlis* or daily court where any Saudi may approach the king directly with a problem or complaint. Rulers work hand-in-hand with religious leaders, rather than oppose them.

Even though the governments of these states are changing rapidly, expanding and assuming new functions, rule is still by consensus and not by repression. Major decisions are reached by consensus within the ruling family and not by dictatorial fiat; Abbas Kelidar demonstrates this process in his study of succession in Saudi Arabia. The problem of unemployment which plagued Iran is nonexistent on the Arab side of the Gulf; instead, great numbers of expatriate workers, both skilled and unskilled, must be imported to fulfill development plans. Opportunities for any bright and educated national, regardless of background, remain almost limitless; thus, non-members of ruling families frequently occupy important government positions and participate directly in the decision-making process.

Of course, the ability of these political systems to retain legitimacy has been helped by their small and relatively homogenous populations. This has made widespread distribution of oil wealth much easier and apparent, economic bottlenecks more surmountable, and the myth of the nation as one large family more sustainable. But these states are fast approaching a threshhold, where traditional ways of exercising authority must soon accommodate modern political institutions. While the older generation may still rely on the *majlis,* younger generations may place more weight on formal channels of participation, such as consultative or legislative assemblies. The key to the survival of these monarchies seems to be adaptability and willingness to welcome greater participation in more and more areas of political life.

Other Middle Eastern producers, of course, are far from being monarchies. Algeria and Iraq have both undergone the experience of colonization and possess revolutionary governments guided by secularized, socialist ideologies. The Iraqi political system, dominated by the Ba'th Party, the Tikriti clique and Saddam Husayn, is dissected and analyzed below by Phebe Marr. Husayn's hold over the country remains firm despite several years of mostly losing battles with Iran.

At the other end of the Middle East, the socialist orientation of Algeria has already been indicated by the examination of its economy. Its leftist ideological stance has been dominant since the ruling Front de Libération Nationale (FLN) wrested independence from France after an eight-year struggle. The nationalist credentials established by the FLN, the rooting of its Arab socialist ideology and the cohesiveness of its leadership since the deposition of President Ahmad Ben Bella in 1965, have all combined to produce a stable regime in Algiers with the ability to weather continued economic difficulties. The willingness of the country's politicians to work within the existing system and the strength of Algeria's political institutions was further demonstrated in late 1978 when the death of President Houari Boumediene was followed by a strongly fought but nonviolent contest for power and the eventual succession of dark-horse candidate Chadli Benjadid.

Nearby Libya has become a unique case. The revolution of September

1969, which toppled a corrupt and anachronistic monarchy and established a military-dominated republic, followed an almost-classic Middle Eastern pattern. The revolution's young leader, Mu'ammar al-Qadhdhafi, seemed to be driven by a predictable mixture of Nasirism, socialism and Islamic fervor. Gradually, however, the path of Libya's politics began to diverge radically from that of other revolutionary regimes. While most Westerners regard Qadhdhafi only as a supporter of liberation movements and international terrorism and as a quixotic adventurer in neighboring countries, he has also pursued a drastic experiment in the redirection of internal Libyan politics and society. People's committees were created in an attempt to bypass bureaucracy and directly involve the country's population in the government. The ruling Revolutionary Command Council and the cabinet gave up their authority to the General People's Congress. Qadhdhafi resigned his titles to become "Leader of the Revolution" and to write the *Green Book,* his outline of a utopian order.

As might be expected, the net result of the Libyan experiment has been confusing. While there seems to be considerable participation in the system at all levels, Qadhdhafi retains firm control over the country's ultimate course. The gross economic and social inequalities prevailing in 1969 have largely disappeared and such measures as the nationalization of rental properties (including homes and apartments) and nearly all commercial establishments have helped to thoroughly distribute Libya's oil income throughout all levels. At the same time, the continuous introduction of these and other measures has unsettled much of the population and the inevitable inefficiency and waste have been made possible only by the full exploitation of Libya's oil. Lisa Anderson includes background on the domestic politics of Libya along with her examination of Libya's foreign policy in the next chapter.

This chapter continues with Eric Rouleau's study of the causes of and the course of the Iranian revolution, and concludes with James Bill's analysis of the structure of the emerging political system several years into the era of the Islamic Republic of Iran. Although his analysis ends in 1981, the situation remained essentially unchanged into early 1983. The key to Iran's future remained the question of who would succeed Khomeini. (See Appendix E for a chronology of the Iranian revolution.)

Bibliographic Sources for this Chapter:
1, 3, 10, 12, 17, 21, 22, 23F, 24, 35, 37, 38, 39, 41, 42, 61, 62, 69, 84, 95, 98, 100, 101, 102, 104, 105, 114, 121, 122, 123, 131, 132, 140, 147, 152, 153, 154, 157, 158, 159, 161, 163, 165, 170, 172, 183, 185, 191, 199, 203, 209, 211, 212, 216, 217, 219C, 219I, 219J, 219K, 219L, 220C, 220E, 220F, 224, 225, 226, 239, 243, 244, 246, 248, 250, 257, 259, 261, 265, 270, 275, 276, 280, 282, 283, 285, 286, 296E, 296F, 311B, 319, 327, 338, 353B, 353J, 364, 381, 388A, 390, 401, 409, 412.

Tribes and Politics in Eastern Arabia

J. E. PETERSON

In the short space of a century or so, eastern Arabia—particularly the region now comprising Bahrayn, Qatar, the United Arab Emirates and the Sultanate of Oman—has undergone a fundamental evolution in political authority. At the beginning of this period, the principal political unit was the tribe, decentralized and egalitarian in nature. Through a combination of largely external factors, this environment was drastically transformed with increasing rapidity. Eventually, the present situation was reached whereby a number of "nation states," or, perhaps in some cases more accurately, "city states," have come to exercise ultimate political control. Despite some difference of opinion over who should govern in these states, there is little denial of their legitimacy as sovereign states. Equally evident is the reliance on political institutions in the Western mold. The growth of these political modifications has meant a parallel eclipse in the influence and power of the tribes.

This process of transformation was necessarily centered around first the consolidation and then the formalization of leadership institutions. Traditionally, tribal leadership was vested in the shaykh, originally an honorific title of respect carrying the connotations of respected elder and wiseman. The shaykh also came to be an elected leader in the sense of *primus inter pares.* [1] Over the years, certain eastern Arabian shaykhs acquired additional power and prestige as a result of being recognized by the British as Trucial

1. In reference to the classical Bedouin society of Arabia, Werner Caskel has written that "In this tribal organism there was no official leader, let alone a hierarchy. A leader can acquire a position of any official character only by being appointed to, or confirmed in, his office by a non-Bedouin power; otherwise he is only *primus inter pares.* His authority is usually inherited, but it is sometimes won by his own efforts." "The Bedouinization of Arabia," *American Anthropologist,* 56, No. 2, Pt. 2; *Memoir* No. 76 (April 1954): 37.

Reprinted by permission of the publisher, The Middle East Institute, from *The Middle East Journal,* Vol. 31, No. 3, Summer 1977.

Shaykhs. Eventually, they adopted the title of *ḥākim* (ruler) and finally, at independence, the title of *amīr,* combining the attributes of ruler, commander and prince. This process, however, benefitted only certain tribal leaders. Many of the remaining shaykhs found their authority diminishing and, in the near future, they may see their separate leadership status disappear entirely.

Of equal importance in the transformation of eastern Arabia has been the dichotomy between *badū* and *ḥaḍar,* and the gradual preponderance of the latter. The term *badū* bears the associated meanings of Bedouin, nomadic, rural, of the desert, whereas *ḥaḍar* denotes a civilized region, the settled population, town dwellers. Throughout the span of Middle Eastern history, the *badū* and *ḥaḍar* populations have maintained an unstable relationship characterized by mutual need and mutual mistrust, each dependent on the other while maintaining claims of superiority over the other.[2] In eastern Arabia, *ḥaḍar* culture has been concentrated in Oman while the origins and customs of Gulf society have tended to come from the *badū* tradition. The historical development of the two areas demonstrates this basic distinction. The Gulf states blossomed as distinct political entities only relatively recently, whereas the acknowledgement of a legitimate central authority in Oman is a long established tradition. While the culture of the Gulf largely emanates from the desert, Omani culture has grown out of a long history of continuous cultivation, predating the arrival of the Arab tribes and complemented by the country's ancient tradition as a secure fastness for the Ibāḍī sect of Islam against a wide range of both Islamic and European invaders.

Despite this conceptual distinction between Oman and the Gulf, tribalism has served as a fundamental political force in both areas. The degree of cohesion within the tribe varies widely. At one end of the spectrum are strongly knit tribes under the leadership of a *tamīma,* or paramount shaykh, who exercises political authority over all sections of the tribe. At the other extreme, a tribe may actually consist of nothing more than a loose federation of autonomous subsections. Although a tribe may be either settled (*ḥaḍar*) or nomadic (*badū*), the settled tribes are more likely to possess a strong central organization whereas the nomadic tribes generally are divided into independent family units. This observation is complicated, however, by the existence of a number of tribes which contain both settled and nomadic sections.

Equally important is the realization that tribal politics is constantly in a state of change. Formation of new tribes is not unusual, as is illustrated by the example of the Balūsh tribe, composed of ethnic Baluchis originally

2. The classic exposition of this dichotomy was given by Ibn Khaldūn in *The Muqaddimah,* translated by Franz Rosenthal (New York: Pantheon Books, 1958). The complementarity of the two cultures is the focus of Cynthia Nelson, ed., *The Desert and the Sown: Nomads in the Wider Society* (Berkeley: University of California Press, 1973).

from Baluchistan, along the Iran and Pakistan coast, who migrated inland to the Ẓāhirah (Dhāhirah) province of Oman and there formed a tribal unit analogous to their Arab neighbors. Another example is that of the Āl Bū Shāmis of the Buraymī region, who have only recently divorced themselves from their parent Nu'aym tribe. The reverse situation, that is, the disintegration of an existing tribe, is also not unknown, a case in point being the Banī Riyām of the mountainous interior of Oman. Following the flight of the tribe's *tamīma* in 1959 after the failure of a revolt against the Sulṭān, the Banī Riyām lost their position as one of the most important tribes in Oman and the tradition of strong leadership by the Nabāhina clan gave way to the virtual independence of a dozen or more subdivisions.

Change may also occur within a tribe. Under the leadership of a particularly forceful shaykh, a comparatively small or weak tribe with little control over its subdivisions may strengthen its internal organization and increase both its influence and the size of its *dīra* (range). The reverse side of the coin is the damage done to a tribe's existing position by weak leadership. Besides resulting in diminished status and perhaps reduced territory *vis-à-vis* neighboring tribes, the weak leader faces replacement by a close relative; alternatively, the reins of tribal leadership may be usurped by a rival clan.

Although a major impact of tribalism on the modern states of the area has been the evolution of ruling families out of tribal leaders, there have been other profound effects as well. A number of essential concepts incorporated into eastern Arabian governments are based directly on tribal custom. Among these ideas is that of the *majlis,* a public session whereby the individual citizen is granted personal access to the Ruler and has an opportunity for immediate redress of his grievances. Another is the *shūrā,* the process of consultation with tribal or community notables, an idea which has been formally incorporated into most of the area's governments through provisions for consultative or legislative assemblies. The incorporation of these traditional practices in the governments of the area has contributed to a sense of continuity in the midst of rapid change.

EVOLUTION OF TRIBAL RÉGIMES IN THE GULF

In tracing the political development of the various Gulf states, it seems useful to divide the period into three somewhat arbitrary phases: (1) the rise of certain powerful tribes to political prominence, with the accompanying establishment of nuclear settlements along the coast; (2) the accrual of autocratic powers by the shaykhs of these tribes; and (3) the eventual appearance of territorial states in a constitutional framework. The first step, the concentration of power and the creation of nuclear settlements, was accomplished by the more cohesive tribes, exhibiting strong, central leadership and shifting from a nomadic or semi-nomadic existence to a more settled environment. Their moves were undertaken for a variety of reasons, and the underlying compulsions for the shift, as well as the process itself, have yet to be adequately studied. However, it is probable that changes in occupational pursuits, such as from the emphasis on herding or date cultivation, to fishing and pearling or increased opportunities for trade, played a central part. In a sense, these moves reflected the age-old pattern of tribal migrations in the Arabian Peninsula, as tribes were forced to seek security and independence in new locations as a result of pressure from other tribes.[3]

The establishment of coastal bases at al-Manāma, al-Dawḥa (Doha), Abu Dhabi, Dubay, *etc.*, was a significant departure for a number of reasons beyond that of simple physical relocation. The act of selecting a permanent geographical center resulted in an identification between the tribe and its settlement. Furthermore, this small shift from *badū* to *ḥaḍar* culture required a concomitant increase in reliance on stronger political authority. In this milieu, the eventual result was an intensification of the concept underlying the ceremony of *mubāya'a*—one denoting homage, allegiance and acknowledgement as sovereign—to the tribal leader. This, then, was an early step in the transformation of leadership along a spectrum from tribal shaykh to protector of a territorial entity to *ḥākim* to *amīr*. Concurrently, the shaykhly clans of the predominant tribes of the nuclear settlements evolved into aristocratic families, eventually providing the ruling families in each of the amirates.[4]

At the same time, the predominant shaykhs began to assume increasing responsibility over all members of the tribe, even though some tribesmen were physically separated from the rest of the tribe. In addition, they sought to gain support for their predominant position by strengthening existing tribal alliances into permanent confederations, a course skillfully pursued by the shaykhs of Abu Dhabi.[5] Finally, the shift to coastal residences opened the way to greater contact with the outside world—which may have had an additional effect in promoting an exchange of ideas and

3. The odyssey of the Āl Khalīfa, from Kuwayt to the Qatar Peninsula to Bahrayn in the eighteenth century provides a pertinent example. See Ahmad M. Abu Hakima, *History of Eastern Arabia 1750–1800* (Beirut: Khayat's, 1965).

4. The pivotal rôle played by these families is brought out by Peter Lienhardt in "The Authority of Shaykhs in the Gulf: An Essay in nineteenth-century history," *Arabian Studies,* Vol. 2 (1975): 63–64.

5. The shaykhs were from the Āl Bū Falāḥ section of the Banī Yās tribe, whose main center had been in al-Līwā' (al-Jiwā) oasis to the southwest of Abu Dhabi town. The settlements of al-Līwā' have long been shared peacefully by the Banī Yās and the Manāṣīr. To the east and around al-Buraymī oasis, Āl Bū Falāḥ influence was maintained by close ties with the Ẓawāhir (Dhawāhir). A third pillar of support for the Abu Dhabi shaykhs was provided by the widespread 'Awāmir tribe. See J. B. Kelly, "A Prevalence of Furies: Tribes, Politics, and Religion in Oman and Trucial Oman," in Derek Hopwood, ed., *The Arabian Peninsula: Society and Politics* (London: George Allen and Unwin, 1972), pp. 137–141.

weakening the hold of traditional beliefs and customs.[6] It should be pointed out that there are notable exceptions to this generalized view of the development of coastal settlements. One is the long tradition of seafaring pursued at such ports as Ra's al-Khayma (formerly Julfār) and Khawr Fakkān—the Arab navigator for Vasco da Gama's initial voyage across the Indian Ocean is thought to have been from Ra's al-Khayma.[7] A second example is the agricultural settlements of northern Bahrayn Island inhabited by the Baḥārina.[8] Nevertheless, the distinctions between these minority traditions and the more recent settlement patterns outlined above became increasingly blurred with the coming of the Europeans and subsequent Arab reaction to this development.

The second stage in the political development of the Gulf, the concentration of political power in the hands of a few shaykhs, owes a great deal to the activity of the British in the Gulf over the last 150 years. Britain's supremacy over the southern shore of the Gulf was first manifested by the imposition of maritime restrictions on the Arab population, beginning in the early nineteenth century, and directed against Arab "piracy" (as perceived by Britain) and slave trading. These restrictions eventually reached the point of preventing Arab warfare-by-sea, which was formally renounced in the Treaty of Perpetual Maritime Peace, signed in 1853. The network of truces engineered by this document came as a result of British perceptions that the shaykhs of the nuclear settlements—*i.e.*, the "Trucial Shaykhs"—exercised political authority over all residents of their settlements. In other words, their position was considered to be closer to *ḥākim* than the traditional conception of a tribal shaykh.[9] Despite having achieved recognition of a sort by the British, the Trucial Shaykhs received no benefit from the trucial relationship, as regarded their internal position, until much later. Among the indigenous population, the status of a Trucial Shaykh continued to depend on the vigor he displayed and the skill he could show in maintaining tribal support.

The third stage, the development of a territorial state, began with the quickening of British involvement with the Arab littoral. The Trucial Shaykhs were held responsible for all members of their physical or tribal communities, reinforcing the transition from *primus inter pares* to autocrat. The establishment of British protection over these settlements in the early

6. The cultural impact of the outside world on the southern shore of the Gulf in a later period is covered in M. Morsy Abdullah, "Changes in the Economy and Political Attitudes, and the Development of Culture on the Coast of Oman Between 1900 and 1940," *Arabian Studies,* Vol. 2 (1975): 167–178.

7. Gabriel Ferrand, "Shihāb al-Dīn Aḥmad b. Madjid," *Encyclopaedia of Islam,* 1st ed., Vol. 4, pp. 362–368.

8. The Baḥārina, of mixed Arab and Persian stock, apparently are the original inhabitants of Bahrayn. They are Shī'ī Muslims, unlike most of the Arabs of the Gulf who are Sunnī.

9. The Muwaḥḥidūn (Wahhābīs) of central Arabia also made this assumption regarding the rôle of the Trucial Shaykhs. See Lienhardt, "The Authority of Shaykhs in the Gulf," p. 70.

1890s not only formalized the British position *vis-à-vis* the Trucial Shaykhs, but also bestowed a legal status on the concept of "shaykhdom." At this stage, the term denoted an embryonic territorial entity geographically centered on the nuclear settlement but under the political control of an autocratic tribal shaykh who exercised a measure of responsibility over the *dīras* of the tribes owing allegiance to him. The territorial sovereignty thus implied in turn reinforced the legitimacy of the Trucial Shaykhs and their families. Nevertheless, despite the emergence of the concept of shaykhdom, consensus on which settlements constituted shaykhdoms, particularly along the Trucial Coast, was not reached until quite recently. The town of Kalbā was considered to be a trucial shaykhdom from 1936 until 1951, and Ra's al-Khayma and al-Fujayra, two of the seven members of the United Arab Emirates, did not achieve trucial status until 1921 and 1952 respectively.

Final impetus for consideration of the shaykhdoms as full fledged states came as a result of the drive for oil and the attendant need for delineation of boundaries. Petroleum concessions were obtained for al-Bahrayn and Qatar in the early 1930s and exploration agreements were secured for various shaykhdoms of the Trucial Coast soon after. All of these and subsequent contracts were negotiated between the various companies on the one hand, and the Trucial Shaykhs on the other, under the watchful eye of the British government. The granting of concessions by the Rulers was done under the assumption by both parties that the Trucial Shaykhs as a group held sovereignty over all territory lying between the Gulf coast and the borders of the Sultanate of Muscat and Oman, even though the limit of each shaykhdom's territory was still undetermined. The vagueness surrounding boundary questions would undoubtedly remain vague today had not the possibility of oil deposits in undemarcated regions required the drawing of precise boundaries. The most notable example of impediments to the solution of this thorny problem was the longstanding dispute between Saudi Arabia and the British, acting on behalf of the shaykhdoms, over the Qatar and Abu Dhabi borders with Saudi Arabia.[10]

By the 1950s, increasing expectations of oil in the northeastern part of the Rub' al-Khālī desert, along with various other factors, resulted in the extension of Saudi control over part of al-Buraymī oasis, thus precipitating

10. This dispute dated back to the period of Ottoman ascendancy in Arabia at the turn of the century. The "Blue Line" of 1913 was the first attempt to delineate the border between the then-Ottoman Najd and the Gulf states. Saudi claims were set forward in the "Red Line" of 1935; the British presented counter-proposals later that year in the "Green Line" and then the "Riyadh Line." For greater detail on these proposals, see Great Britain, *Arbitration Concerning Buraimi and the Common Frontier Between Abu Dhabi and Sa'ūdi Arabia, Memorial submitted by the Government of the United Kingdom of Great Britain and Northern Ireland* (2 vols.; London: HMSO, 1955), and Saudi Arabia, *Memorial of the Government of Saudi Arabia, Arbitration for the Settlement of the Territorial Dispute Between Muscat and Abu Dhabi on One Side and Saudi Arabia on the Other* (3 vols.; Cairo: 1955).

the protracted "Buraymī crisis" which pitted Saudi Arabia against Abu Dhabi and the Sultanate of Muscat and Oman, and, by extension, some American individuals against some British individuals. Although the expulsion of the Saudis by the British-officered Trucial Oman Scouts in 1955, following the breakdown of an arbitration tribunal meeting in Geneva, imposed a *de facto* solution to that particular boundary question, a formal agreement on the Saudi-Qatari border was not reached until 1965, and not until 1974 on the Saudi-Abu Dhābi border. Saudi Arabia and Oman, however, have yet to agree on a *de jure* border. The difficulty of drawing boundaries in the midst of a maze of mountain enclaves in the Trucial Coast has proved particularly troublesome. Although claims over many disputed areas were settled by a British official, Julian Walker, others remain.[11]

Despite the dilemma over boundaries, the oil companies went ahead with drilling operations and eventually brought oil onstream in most of the shaykhdoms. The wealth provided by this newly tapped resource provided the fortunate Rulers of the so-blessed shaykhdoms with ample means with which to consolidate their internal positions, as well as to advance their prestige at the expense of the Rulers of non-oil-producing shaykhdoms. Even before the revenue from oil exports appeared, the modest concession payments enabled Rulers to expand the traditional practice of largesse in furthering their prestige and reputations. With the commencement of payments on oil exports, the Rulers embarked on ambitious programs of socioeconomic development. For the first time, basic services were provided, such as roads, housing, education and health care, which gained the respect of the population and added functions to the institution of the Ruler. This newly acquired wealth provided the means for expanding what had been rudimentary government on a personal level into a more complex edifice. Functional government departments were established, buildings raised, services expanded and expatriates hired to advise, administer and train local personnel to eventually staff their own bureaucracy.

As a result of their having taken advantage of the opportunities presented for change and evolution, the states of the lower Gulf achieved full independence in 1971 without suffering the instability and disintegration which some had feared after Britain's announcement of withdrawal in 1968. Both Bahrayn and Qatar emerged as full fledged members of the Arab League and the United Nations with few fundamental changes apart from their assumption of complete responsibility for defense and foreign affairs. The Rulers of the Trucial Coast, however, exchanged British restrictions on their sovereignty for collective restrictions as embodied in the federation of the United Arab Emirates.[12] Apart from the novelty of

11. One such unresolved territorial dispute received considerable attention in the area in the late summer of 1976 when Sharjah announced plans to build a multi-million-dollar shopping complex, the Charles de Gaulle center, on land also claimed by Dubay.

12. A detailed examination of the progress of these states in the first years of their indepen-

the UAE in the Peninsula as an experiment in power-sharing, it was notable in formalizing the necessary adjustments in relative status between the Rulers of the various shaykhdoms, largely caused by the discovery of oil: whereas once the Qawāsim of the northern settlements had enjoyed the power and respect due to the Trucial Coast's premier ruling family, the arrangements of 1971 clearly favored the Āl Nuhayyān of Abu Dhabi and the Āl Maktūm of Dubay, since the two families garnered the Presidency, the Vice-Presidency and the most important cabinet positions in the new federation. The UAE was also important in driving home the realization that the individual Rulers and their mini-states could not go their separate ways but needed to coöperate in the larger political entity provided by the UAE.[13]

THE TRADITIONAL TRIBAL SYSTEM OF INTERIOR OMAN

Having dealt with the evolution of the Gulf states, it remains necessary to outline the development of Oman along substantially different lines. Oman's experience has been shaped largely by two factors: the rôle of Ibāḍism in creating a sense of a distinct Omani community *(miṣr)*;[14] and by the dichotomous nature of the Ghāfirī and Hināwī tribal confederations. Although Ibāḍism arose in al-Basrah as a moderate variant of the Khārijī sect of Islam, Omanis were heavily involved in its early development and the first Ibāḍī state, under the leadership of an Imām, was established in Oman in the eighth century A.D. Thus, the Omani heartland not only acquired a distinct religious and communal identification but it was also provided with a tradition of a central political institution or quasi-government. The focal point of this system was the Imām, whose office was modelled on the pattern of the tribal shaykh and included the concept of *primus inter pares*. The resultant democratic, centrifugal nature of the Imamate gave considerable latitude and independence to individual tribes.[15]

As a result of this decentralized religio-political system, the history of Oman in the Islamic period was characterized by a series of historical

---

dence is to be found in John Duke Anthony, *The Arab States of the Lower Gulf: People, Politics, Petroleum* (Washington: The Middle East Institute, 1975).

13. The Qāsimī shaykhdom of Ra's al-Khayma remained aloof from joining the UAE at its inception in hopes of a fortuitous oil strike in its territory. This independent stance lasted only three months, however, when no oil proved forthcoming and Ra's al-Khayma was obliged to join the federation without obtaining any of the conditions it had previously set out for membership. For a detailed analysis of the formation of the UAE, see John Duke Anthony, "The Union of Arab Amirates," *Middle East Journal,* 26, No. 2 (1972): 271–287.

14. "...by which term may be understood the territorial expression of the community...." J. C. Wilkinson, "The Ibāḍī Imāma," *Bulletin of the School of Oriental and African Studies,* 39, Pt. 3 (1976): 537.

15. A good exposition of Ibāḍī tenets as they apply to Oman is contained in *ibid.,* pp. 535–551.

cycles. First, the tribes were united behind a vigorous Imām. Following his death, a secularized dynasty of Imāms emerged and Omani influence was extended overseas, principally to East Africa. Eventually, the dynasty collapsed in civil wars over the succession to the office of Imām, and finally a new cycle began with the election of a strongly religious figure as Imām.[16] This pattern was finally altered in the nineteenth century as a result of increased British activity in Oman.

The Āl Bū Sa'īd dynasty, established in the mid-eighteenth century, had followed this pattern by forsaking its interest in the tribal politics of the interior for a maritime empire. Following the death in 1856 of the family's most capable Ruler, Sa'īd bin Sulṭān, the Āl Bū Sa'īd gradually lost control of all Oman except a narrow coastal strip. Their claims to legitimate leadership over the country were largely forfeited when the office of Imām was discarded in 1784. By the early twentieth century, a definite schism had emerged with the British supported Āl Bū Sa'īd Sulṭāns on the coast and a renascent and tribally dominated Imamate in the interior. This state of affairs was formalized by the Agreement of al-Sīb in 1920, which was negotiated by the British Political Agent and signed by representatives of the Sulṭān and important tribal figures. By its terms, the existing autonomous nature of inner Oman was recognized by the Sulṭān and the interior was left to a tribal balance of power system under the loose supervision of the Imām.

The key to the working of this tribal system in the interior was the Ghāfirī-Hināwī dichotomy. Although based to some extent on north Arab/south Arab lines and on the distinction between Ibāḍīs and Sunnīs, the Ghāfirī and Hināwī confederations essentially grew out of alliances forged during the civil wars of the early eighteenth century. As a result of these alliances, all Omani tribes became identified with one of the two sides. If the inhabitants of the upper part of a valley or town were Ghāfirī, then the populace of the lower part of the valley or town would generally be Hināwī. Tribes with traditional feuds generally gravitated to opposite sides of the division. Since identification with one confederation was not absolute, tribes could and did switch allegiance. As a result of this fluidity in alliances, it required an exceptionally skillful individual to secure the support of tribes from both the Ghāfirī and Hināwī factions in order to claim

16. "It is easy to see that the Ibadi politico-religious ideology is an impractical basis for the permanent development of a state. It automatically develops a cycle which encompasses its own downfall. As the country is united so does its wealth and prosperity increase and the religious ideal weaken; the leadership becomes the prerogative of a single group and degenerates into temporal power *(salṭanah)*. There ensues a struggle for power in which tribal *'asabiyah* is brought into play and every potential weakness in the country exploited until full-scale civil war is the outcome. The situation is usually resolved by one or more of the parties calling in an outside power, normally with disastrous results for the Omanis in general." J. C. Wilkinson, "The Origins of the Omani State," in Derek Hopwood, ed., *The Arabian Peninsula: Society and Politics* (London: George Allen and Unwin, 1972), pp. 78–79.

leadership over inner Oman. Furthermore, the key to coöperation between the two confederations lay in the religious appeal: only the symbol of the Imamate could unite the majority of Omani tribes.

Between 1920 and 1954, these conditions held: inner Oman functioned as a relatively cohesive unit at peace with the coastal Sultanate, under a triumvirate composed of the Imām and the Ghāfirī and Hināwī leaders. But in 1954, the death of the old Imām and the subsequent alliance between the new Imām, his ambitious brother and the equally ambitious Ghāfirī leader, plus covert Saudi assistance, combined to trigger a chain of events that brought an end to the Imamate. By early 1959, all of Oman had been brought under the Sulṭān's control. The office of Imām lay vacant and all the tribes paid allegiance to the Sulṭān.

TRIBAL COMPONENTS IN THE DEVELOPMENT OF MODERN STATES

In eastern Arabia, even though the state has replaced the tribe as the primary political unit, it still relies heavily on various tribal components. The most obvious and important of these is the ruling family, whose political position is absolute.[17] The family provides both the Ruler and the Heir Apparent, who generally also serves as the Deputy Ruler. In addition, other family members hold the major ministerial portfolios and command of the military forces partly to retain ultimate power within the family and partly to satisfy the ambitions of close relatives who otherwise might be tempted to instigate *coups d'état.* In 1982, the Āl Khalīfa held seven out of the 16 ministerial positions in Bahrayn, while the Āl Thānī held 11 of the 16 posts in Qatar.[18] The allocation of cabinet posts in the United Arab Emirates has been particularly ticklish, since there are a number of ruling families to take into account. The ruling families of the seven shaykhdoms had members in 12 of the 28 posts announced in December 1973, and in 10 of the 23 posts in the cabinet during 1982. The most important posts belong to the Āl Nuhayyān of Abu Dhabi and the Āl Maktūm of Dubay, including Prime Minister and Deputy Prime Minister. In Oman, close relatives of the Sulṭān—known as the Āl Sa'īd—occupied six of the 22 cabinet posts, while other members of the ruling family hold an additional three posts. Furthermore, many of the more than 30 *wālīs* (local representatives of the

17. Emile Nakhleh has formulated the concept of "urban tribalism" in explaining the rôle of these ruling families. "Legitimacy, authority and the allocation of values within the political system reside in an all powerful ruling family which views the governing process as a synthesis of a well-entrenched tribal tradition and a functionally limited form of public administration." Letter to the editor in the *Middle East Journal,* 27, No. 2 (1973): 273. He incorporated this concept in his study, *Bahrain: Political Development in a Modernizing Society* (Lexington, Mass.: Lexington Books, 1976).

18. Complete lists of all these cabinets are to be found in *The Middle East and North Africa, 1982–83* (London: Europa Publications, 1982).

Sulṭān) belong to distant branches of the Āl Bū Sa'īd. Besides holding political supremacy, the ruling families of the area also enjoy high status due to their relative wealth, education and, in recent years, commercial ties.

Complementing the family élite is a second élite group composed of the shaykhly clans from other major tribes in the state. The allegiance of this group to the state and the Ruler has been secured through subsidies provided by the Ruler, intermarriage with the ruling family, and the distribution of choice government posts, particularly as ministers or ambassadors. Certain of these clans have also been quick to take advantage of the increased economic activity of the last several decades in establishing commercial activities and obtaining franchises for foreign products. The 'Aṭīya family of Qatar is representative of this group. In the mid-1970s, one member of the family was Minister of Public Works, another was second in command of the army, while others served in the diplomatic corps. The family is tied to the Āl Thānī through various marriages and is also part of the merchant community. In several states, representatives from the major tribes and merchant families were allocated seats in consultative assemblies. Although these institutions provided their members with a certain amount of additional prestige, their full political and institutional potential has yet to be realized. Most of the 50 members of the shortlived National Consultative Council in Abu Dhabi were tribal representatives.

Particular tribes have frequently been tapped as a source of armed retainers, and *badū* tribesmen are often employed as night watchmen. The situation in Oman is more complex, as the coöptation of representatives from major tribes of the interior of Oman in the Sultanate government dates only from the *coup d'état* in 1970. Previously, nearly all authority was kept firmly in the hands of the Sulṭān who entrusted little to his small, mixed cadre of officials. This position was reversed by the more pluralistic régime after 1970.

EROSION OF TRIBAL SOCIETY

As the pace of the development of modern political units and the process of socioeconomic transformation quickened, tribalism as a political, social and economic force in the societies of eastern Arabia declined. In the past, occupational patterns were closely linked to the tribe, in the type of activity pursued as well as the location. These patterns were regulated by a seasonal cycle followed by various semi-nomadic tribes of the Gulf. The winter occupations of nomadic herding and the harvesting of date palms in tribal oases were left behind in the summer as the tribesmen moved to the coast for employment in fishing or pearling.

The early presence of oil companies in the area tended to complement this cycle, as casual laborers, truckdrivers and guards were recruited from the tribes in whose *dīras* the exploration for oil was being conducted. These new job opportunities were not viewed by the tribesmen as different in kind from the supplemental employment in which they had been engaged during earlier pearling seasons, although they were more lucrative. Many tribesmen saw oil company employment as merely a short-term means of gaining extra money which could be invested in livestock or palm trees. Even later, when concessionary payments to the Trucial Shaykhs allowed a modest expansion in employment, the tribesmen who drifted to the towns rarely remained there longer than a few months, by this time possibly investing their surplus earnings in the purchase of a taxi rather than a few palms.[19]

Significant departure from traditional practices only came following the discovery of oil and its export. The mushrooming of nuclear settlements as urban centers of an increasingly Western nature resulted in more and more opportunities for permanent employment. Changing values caused more tribesmen to retain jobs on a permanent basis, since this afforded them such advantages as education for their children and the benefits of modern health care. The newfound oil wealth trickled down to all segments of the indigenous population, especially through government land purchases and provisions for free education and inexpensive housing. The result was an exodus by many tribesmen from the interior oases to the nuclear settlements and replacement of the tribe by the family as the primary social unit. As education became widespread, its rôle in expanding an awareness of the outside world grew. For the first time, the women of the tribes left the interior and joined their male relatives in the towns: their social horizons were broadened, the opportunity for education extended, and a crack in the door toward employment and social mobility was opened.

One example of the profound changes in tribal politics wrought by oil company activities is that of the Durū‘ of Oman, a *badū* tribe occupying a large *dīra* between the Rub‘ al-Khālī desert and the settled areas of Oman's al-Ẓāhira province.[20] In common with most *badū* tribes, tribal organization among the Durū‘ (singular, Dir‘ī) has been extremely loose in the past. However, after Petroleum Development (Oman) (PDO) began drilling operations in Dir‘ī territory in the mid-1950s, the tribe's political system began to resemble the local equivalent of a labor union. All local labor used by company parties had to be recruited from the Durū‘ and the shaykhs of the premier clan, the Mahāmīd, were employed as Arabian equivalents of "shop stewards," and used as "labor supervisors" on drilling sites to prevent wildcat strikes. Internal power struggles in the Maḥāmīd clan resulted in PDO's introduction of a system whereby three of the more prominent

19. One example of this process, as it developed in the state of Abu Dhabi, has been ably described in Frauke Heard-Bey, "Development Anomalies in the Beduin Oases of al-Liwa," *Asian Affairs,* 61 (N.S. 5), Pt. 3 (October 1974): 272–286.

20. The following discussion of the Durū‘ is largely based on information contained in a mimeographed study produced *ca.* 1971 by Petroleum Development (Oman) Ltd.

shaykhs each served three-month periods in rotation as "duty shaykh." By 1970, this system proved ineffective and the rôle of the "duty shaykh" was limited to outlying parties while their place elsewhere was taken over by an "employees' representative committee."

The growth of government authority and services has also contributed to the erosion of tribalism by usurping functions previously performed by the tribes. Gradually, the position of the various Trucial Shaykhs grew more autocratic. Once simply the strongest tribal shaykhs in their immediate areas, they eventually acquired complete political authority over all the tribes, as well as the non-tribal population, residing in a shaykhdom. Government agencies were created to assume functions previously filled by tribal leaders. Courts were established and legal codes written, reducing the role of *'urf* (tribal law), with its emphasis on collective responsibility, in favor of Western legal precepts, with their stress on individual responsibility. In most Gulf states, this expansion of the government apparatus first consisted of additions in departments and advisors directly subordinate to the Ruler, although the last decade has seen the introduction of constitutional modifications, such as a Council of Ministers and/or a legislative/consultative assembly in all the states of the region.[21]

Slowly the influence of the non-Trucial Shaykhs declined. Their traditional rôle as intermediaries between the Ruler and the members of the shaykh's tribe was superseded by the establishment of government agencies in direct contact with the people, such as courts, municipalities, hospitals and health clinics, schools, and programs for roadbuilding and agricultural improvement. Although this trend operated also in Oman, particularly after 1970, the situation there was complicated by the existence of what had long been an autonomous region in the interior. As a result of the schism between the coast and the interior, the power of the tribes of inner Oman persisted until later than was the case in the Gulf. Nevertheless, the events which led to the dissolution of the Imamate in the 1950s and restoration of Sultanate control over the interior set in motion the same forces which resulted in the decline of tribal influence in the Gulf.

The reunification of Oman came in late 1955 when the armed forces of the Sultanate occupied the major towns of the interior. Government authority was vested in the person of a sort of "superwali" who represented the Sulṭān in dealings with the various tribal shaykhs. In the summer of 1957, the leader of the Ghāfirī confederation and the last Imām's brother, along with the erstwhile Imām, staged a brief insurrection. They managed to gain temporary control over much of the central portion of Oman before being driven up onto the Jabal al-Akhḍar mountain by a combined British and Sultanate force and then were finally driven from Oman in early 1959.

21. For a discussion of the development of this process in Abu Dhabi, see Frauke Heard-Bey, "The Gulf States and Oman in Transition," *Asian Affairs,* 59 (N.S. 3), Pt. 1 (February 1972): 19–20.

As a result of this revolt, the Sulṭān, acting through his *wālīs* in all the sizable towns of central Oman, exercised direct control over all the tribes except for those with *tamīmas* who had remained loyal to him. The principal tribe involved in the rebellion, the Banī Riyām, suffered the destruction of its principal settlement and the imposition of a military governor over its mountain villages. The untrained and ill-equipped levies which had characterized the Sulṭān's military were reorganized into the Sultan's Armed Forces (SAF) under the leadership of a commander seconded from the British Army, and military camps were established at key locations throughout the interior. The new army also assumed responsibility for the collection of taxes from that area for the first time in nearly a half-century. Another indication of the Sulṭān's increased control was his ability to issue orders to all the shaykhs of the interior and have them obeyed without question.

The only tribes to retain some measure of autonomy were those with leaders who had declared their loyalty to the Sulṭān previous to, during and after the revolt of 1957–1959. Principal among these leaders was Aḥmad bin Muḥammad al-Ḥārithī. Aḥmad's career is well worth examining in some detail since his rise and fall illustrates the transformation of tribal politics in Oman over the last 30 years, as well as the general decline in tribal power throughout eastern Arabia. Under Aḥmad's great-grandfather and grandfather, the Ḥirth had become the principal tribe of the Hināwī confederation and the Ḥārithī *tamīmas* the most important tribal figures in the country. When Aḥmad's father became *tamīma* of the tribe in 1946, his weak leadership and continued illness resulted in a loss of prestige for the tribe. The father's death two years later set off a succession struggle between Aḥmad and Aḥmad's uncle, Ṣāliḥ bin 'Īsā. When Ṣāliḥ emerged victorious and was generally recognized as *tamīma,* Aḥmad went to see the Sulṭān and offered him his support.

This alliance was tested a few years later when Ṣāliḥ joined with the forces of the short-lived Imamate of 1954–1955 and subsequently left for exile in Egypt. The loyalty shown by Aḥmad to the Sulṭān at this time earned him the position of *wālī* of Nizwā, the most important town in inner Oman. His further expression of loyalty in 1957 gave him a position of unquestioned power over the province of al-Sharqīya, in which the Ḥirth tribe was situated, as well as considerable influence over much of the rest of the interior. As a result of his association with the Sulṭān, albeit as a subordinate, Aḥmad was able to create a personal aura of power that had not been equalled in the interior since the middle of the eighteenth century, a particularly remarkable achievement in light of the ebbing position of the rest of eastern Arabia's tribal leaders during this period.

Nevertheless, Aḥmad was either unable or unwilling to comprehend the fundamental shift in political power following the 1970 *coup d'état.* The new régime was no longer dependent on the fealty and support of tribal figures.

Nearly all the financial reserves accumulating from several years of unspent oil revenues were expended on a rapid program of economic expansion—although a considerable amount went to fighting the guerrilla war in the southern province of Dhufar—and the base of participation in the government was significantly broadened. The old situation of an isolated Sulṭān requiring all government business to be conducted through him and entrusting very little substantive authority to his few inadequate minions, was abruptly reversed by the expansion of government machinery and widening of the state's political outlook following the 1970 *coup*. Key posts were filled by men from merchant families, educated Omanis returning from exile under the old Sulṭān, and representatives of various other minority groups.

Tribal organization and tribal confederations were no longer a major factor in Omani national politics. Although one or two ministerial positions were reserved for representatives of interior tribes, the shaykhs were bypassed as government services were extended directly to the population and individuals were recruited for government employment on their personal merit and not necessarily through membership in any particular tribe. The few remaining *tamīmas* were forced to accept the supremacy of the government in all matters, retaining authority only in internal tribal affairs. When Aḥmad bin Muḥammad al-Ḥārithī refused to acknowledge the complete sovereignty of the Sulṭān, he was quickly put under house arrest in Muscat, where he remained through the 1970s.[22]

With Aḥmad's demotion, the last significant bastion of tribal autonomy in eastern Arabia fell. Henceforth, the only tribes which were able to conduct their internal business relatively free from government supervision were the few completely *badū* tribes of the desert. Even there, the impact of oil had transformed daily life and drawn tribesmen away to the coastal settlements. Clearly, tribalism as a factor in the political life of the states of eastern Arabia is declining in importance. The vastly increased resources available to the area's Rulers has allowed them to consolidate and delegate political power in an unprecedented manner. Shaykhly clans have generally been able to retain influence only through lateral movement into such fields as commerce and real estate. The road to prosperity for the average tribesman has required less dependence on tribal connections and more on non-ascriptive criteria.

Only the ruling families continue to play an important, central political rôle. Yet, in their evolution into autocratic dynasties, the tribal orientation of these families may be fading. Although radical elements in the Peninsula have sought to enlist tribal elements in subversion against the existing régimes, any successful pressure for broader political participation is likely

22. See J. E. Peterson, *Oman in the Twentieth Century: Political Foundations of an Emerging State* (London: Croom Helm; New York: Barnes and Noble, 1978), pp. 118 and 122.

to come from other elements of society, such as professional groups, social clubs and organized labor.[23] Nevertheless, the position of the ruling families will undoubtedly remain paramount for some time to come, particularly since they constitute an important link in the area's transition from tribal societies to modern states.

23. The social development necessary for the appearance of these groups has so far been largely limited to Bahrayn, owing to its longer period of oil income. Even there, however, the ruling family has been reluctant to share political power, as evidenced by the events of August 1975 when the National Assembly was dissolved indefinitely. See Nakhleh, *Bahrain*.

The Problem of Succession in Saudi Arabia

A. R. KELIDAR

Saudi Arabia is governed by a monarchy where the King enjoys almost absolute power under the all-embracing religious law of Islamic Shari'a. Under this system the King is the Head of State, his own prime minister, the religious spiritual leader of the community of Muslims, and the paramount chief of a tribal confederation of national magnitude. He appoints his ministers, provincial governors, and other officials, who are directly responsible to him. Though a cabinet was formed in the late 1950s, this is presided over by the King or one of his deputy prime ministers, usually the Crown Prince. This arrangement was started by the late King Faisal when he was Crown Prince to his brother King Saud, only to set in motion a process that eventually led to the deposition of Saud as King and the assumption of Faisal to the throne in 1964. The politics of the succession in 1964 or again in 1975 when King Faisal was gunned down by a mentally deranged member of the Saudi ruling house has not altered in any noticeable way the nature of political rule. It has remained authoritarian, with power concentrated in the hands of the King or his Crown Prince. There is no public opposition to government policy, no system of political parties or independent interest groups, and no elected representatives. Politics remain traditional, autocratic, highly personal and extremely centralized, and rather unaffected by the economic transformation which the country is undergoing.

The King and his princely class of ministers are supported by a limited number of expatriate Arabs who supervise the economic transformation of the country in the wake of the oil boom. However, they are being replaced by qualified Saudi nationals with considerable encouragement from certain reformist and forward looking members of the ruling house. The expa-

Reprinted by permission of the publisher, Royal Society for Asian Affairs, from *Asian Affairs*, Vol. 65 (N.S. 9), Pt. 1, February 1978.

triates are motivated by pecuniary gain as the sole reward for their expertise. As they are on short contracts, they expect to be changed and the Saudis regard them as expendable. They are disfranchised by virtue of their alien origins, and cannot develop a vested interest or a power base in the country. They do not seek political participation in the political process, and have been unable to seek political accommodation on a permanent basis. As such they are implicitly trusted as whatever political significance they may enjoy in the system is dependent on the nature of the relationship they establish with the King or his ministers.

There is little doubt that the system has proved to be remarkably stable and resilient—the smooth transition in March 1975 following the assassination of King Faisal indicated the degree of unity and political stability which could be regarded as the accomplishment of the late king following the years of political dissension and the wasteful largesse of Faisal's predecessor. It has shown that when faced with that kind of tragic adversity the House of Saud is capable of withstanding the challenge of closing ranks to manifest the cohesion required to ensure its survival. The strength of the régime lies in the large numbers of the ruling family, estimated to be between 3,000 and 5,000 princes, its collective interest in self-preservation, and its tribal affiliations through marriage. Ibn Saud, the founder of the present kingdom, who set out from his refuge in Kuwait in 1902 with a small band of relatives and followers to reconquer his ancestral capital of Riyadh, adhered to the age-old Arabian practice of marrying into the tribes to cement the alliance he made with them. In this respect he would have made Henry VIII cover his face in shame in terms of marriages, divorces, abandonments and procreation. His direct male descendants number something like 400 princes. The evolution of a vested interest in the existing order as a result of the oil boom must have contributed to the smooth and peaceful transfer of political power to King Khalid. His immediate acknowledgement by the other members of the royal family as their new King was to emphasize the importance of continuity—he being the second most senior of Ibn Saud's surviving sons, as well as having been named as Crown Prince by his predecessor.

Yet in Arabia the process of succession is not traditionally by primogeniture. Neither Arab society nor Muslim civilization managed to work out rules and regulations, institutions or formulae for political succession. The problem bedevilled Muslim rule throughout the centuries, and in the modern nation-states of the Muslim world the search for a new basis of legitimate rule has not been conclusive. In the tribal tradition of Arabia, rule passes to the oldest surviving and generally the most accepted male relative—be he an uncle, brother or cousin. It goes without saying that the incompetent rarely survives for long. In the short history of Saudi Arabia there have been four occasions where the process of succession could be observed. In the first instance it was after the death of Ibn Saud in 1952. He, it seems, impressed upon his sons that they should not challenge the right of his eldest surviving son Saud to succeed him. They did not, until the very existence of the Kingdom was threatened by a combination of internal subversion and sedition aided and abetted by Nasser's quest for pan-Arab leadership and the Egyptian military presence in the Yemen. As a result a struggle for power ensued which was resolved by Faisal's assumption of the throne in 1964. There was no doubt of the genuine reluctance of Faisal to defy the wishes of his dead father or to force his brother to abdicate. He tried to share power with the King on the basis of King and Premier but the experiment was a total failure when Saud, especially after his visit to Egypt in 1964 to attend the summit conference of the Arab Heads of State, wanted to assert his position of primacy. His express desire to assume his full powers as King precipitated a crisis in the Saudi ruling house, and a meeting of some one hundred princes and sixty-five leading Islamic religious leaders was held at Faisal's behest to consider the situation. The meeting resolved that Saud should be asked to abdicate in favour of Faisal. Faisal's considered response when told of the princes' deliberations was "We do not depose a King until all form of persuasion has been used." In fact Saud agreed to abdicate because he was told in no uncertain terms that if he did not he would be deposed and exiled to live in poverty as all his possessions would be expropriated, or he could live at home under permanent house arrest. Being the fun-loving King he was—he decided to go. His legacy for Saudi Arabia, oddly as it may seem today, was virtual bankruptcy. It was Faisal who set the kingdom on the road to recovery. What was significant was that the document proclaiming Faisal as the new King was signed only by sixty-eight indicating their importance in the Saudi household.

Soon after Faisal had taken over, Khalid was appointed Crown Prince when Muhammad, the oldest surviving son of Ibn Saud, either declined the role or was passed over because of his close association with Saud while King. The selection of Khalid—an old and sick man even then; a man who enjoyed being a desert squire with an immense love for racing camels and hunting hawks; a man more familiar with the hospital surgical ward than the corridors of power—ensured that Faisal would not be challenged by his Crown Prince. It also meant that the Crown Prince would not exercise the same authority or wield the power that Faisal had when he was Crown Prince. The position of Khalid was completely overshadowed by the King who assumed the same role as the father had in the administration of the kingdom for fifty years as well as by that of the second deputy prime minister—the King being the Premier and the Crown Prince as his first deputy. The second deputy prime minister Prince Fahad, the sixth son of Ibn Saud, became Crown Prince and first deputy prime minister to Khalid in 1975. Although Fahad was rather keen to separate the premiership from the position of the Crown Prince under Faisal in order to establish his claim to the eventual succession as well as to enlarge the scope of his responsibilities vis-à-vis the King, Faisal was rather reluctant to concede that prerogative to him. Disinterested as Khalid was in the affairs of State, and content to play the role of a figure-head, the political processes of decision-taking and policy-making were left to Fahad who emerged as the strong man of the Saudi régime. Fahad enjoyed the same privileges as Faisal had when Crown Prince to Saud, but without the acrimony associated with the latter. (Ed. note. Fahad succeeded to the throne without incident upon Khalid's death on 13 June 1982.)

This recurring anomalous practice is likely to continue and may well become a destabilizing factor in the politics of succession and the peaceful manner by which the transfer of political power has taken place. There are several other factors of considerable importance which will influence any future course of action that the Saudis may follow. To begin with, it is misleading if not dangerous to think of the Saudi ruling house as a monolithic institution bound all the time by a consensus that will please all. There have been sufficient indications to show that there are divisions, rivalry, dissension and jealousy motivated by political or personal differences that could precipitate a power struggle within the ruling group. The defection of Prince Talal and his small band of princely supporters to the cause of radical notions of Arab nationalism in 1962 and his advocacy for the establishment of "a democratic and constitutional form of government" may have been premature, but it was an early manifestation of the divergence in the political outlook among the Saudi rulers. The call appreciated by Nasser of Egypt fell on deaf ears in Saudi Arabia, and even Talal was to abandon it when threatened with total impoverishment in exile. Moreover King Faisal, himself, was not murdered by a disgruntled Asiri or a dissident Hasawi but a member of his own family. To dismiss these manifestations as worthless because they have not managed to introduce radical alterations to the system of government will be foolhardy. The permanent cohesion of 3,000 people with their divergent interests, diverse political aspirations and ambitions is difficult to maintain especially when the country is undergoing such a massive economic transformation. Such economic development is bound to have some drastic social and political repercussions. (Ed. note. Fahad succeeded Khalid as King on the latter's death in 1982, after this article was written. Consequently many of the individuals discussed below have since assumed different positions.)

The apprenticeship which Prince Fahad had served under his guide and mentor King Faisal secured him a predominant position in the Saudi royal hierarchy which made it possible to accede to the throne. Fahad's status as the future King of Saudi Arabia was further fortified by the fact that he is a member of a group of princes generally known as the "Sudairi seven" all of whom being the progeny of a mother who hailed from the important tribe of the same name. Fahad's mother was a favourite wife of Ibn Saud whose own mother was also a Sudairi. In true Arabian fashion the children of a favourite wife enjoy preferential treatment from the father, and therefore from the outset the seven Sudairis must have held a notable place in the Saudi ruling house and society. Their position is so strong and entrenched that many of their supporters (members of the clan) who had expected to be appointed to the cabinet in 1975, by virtue of ability and long service to the State, had to be overlooked for fear of creating too much resentment among the other princes. Nevertheless there exists considerable opposition and resentment of the Sudairi pre-eminence particularly as they control all the important government departments.

Before acceding to the throne, Fahad was the Crown Prince and first Deputy Prime Minister. His younger brothers held the following posts: Sultan was Minister of Defence, Nayif was Minister for the Interior and Turki was Deputy Defence Minister; Salman was Governor of Riyadh, the capital. The other two of the seven are Ahmad and Abdul Rahman, one of whom was the Deputy Governor of Mecca—the capital of Hijaz. Fahad has made sure therefore that his own clan are in charge of the internal security forces as well as all branches of the armed forces. He has also arranged that when one of the brothers relinquishes some of his responsibilities as had been done in the case of the Ministry of the Interior a sibling would take it over. No doubt the same thing will happen if Sultan were to give up the Ministry of Defence; his sibling deputy will assume the post.

These precautionary moves on the part of the Sudairi princes are not without reason, but must be an expression of the clan led by the Crown Prince that he intends to pre-empt any challenge to his authority, particularly as the faction opposed to the Sudairi hegemony is led by Prince Abdullah, second Deputy Prime Minister and Commander of the National Guards, a reorganized and much better armed tribal force than its forerunner—the White Army. Since the Sudairis are called so after their mother the Abdullah group seems to have spoken for their father. They have been dubbed as the Abdul Aziz faction. Abdullah as second Deputy Prime Minister under Khalid, the same position occupied by Fahad under Faisal, staked his own claim to become Crown Prince when Fahad succeeded Khalid. The Sudairis, conscious of the potential challenge posed by Abdullah's claim, seem to have suggested that he should relinquish his command of the National Guards if he wanted to be acknowledged as the Crown Prince, pointing out the precedent set up by Fahad when he abandoned the Interior Ministry upon his elevation to be the Crown Prince. Abdullah, however, was reluctant to leave his position so vulnerable by giving up his power base in the realm, particularly as he knows that Fahad had handed his portfolio to a Sudairi sibling. (Ed. note. In the reshuffling after Khalid's death, King Fahad assumed the position of Prime Minister, Prince Abdullah became First Deputy Prime Minister and remained Commander of the National Guard, and Prince Sultan became Second Deputy

Prime Minister and remained Minister of Defense. Abdullah was also named Crown Prince, as expected.)

The two factions hold different views and entertain divergent aspirations which may have direct relevance to the political and economic options facing the Kingdom. These would have been completely peripheral, indeed parochial to the outside world but for the recent development which took place as a direct result of the 1973 war in the Middle East and the quadrupling of oil prices, and their drastic effect on the world economy generally and Middle East politics in particular.

The energy crisis has catapulted Saudi Arabia into unprecedented prominence, and bestowed on it wealth and influence that made the kingdom not so much of an anachronism but more of an anomaly in the modern world. By virtue of its oil resources this desert kingdom has become the richest country in the world. Possession of about one-quarter of the free world's petroleum reserves and its dominant position in OPEC has given Saudi Arabia a voice in regional and world affairs wholly disproportionate to a population of merely 5 million. This is not to say that the kingdom prior to the energy crisis did not play an important role in determining the nature and the direction of Arab politics. There are examples to be cited where certain developments in the politics of the region could be ascribed to Saudi influence and even machination. Before the world had heard of the coined phrase petro-dollars or begun to worry about the future of the North–South dialogue, the civil war in Lebanon, or President Sadat's trip to Jerusalem where he accorded the Jewish State fulsome recognition—matters with which the Saudis are inextricably linked—the Saudis were instrumental in transforming the Arab League from an institution devised as a vehicle for the promotion of Arab union into an important forum where several independent sovereign States were vying with each other for the protection of their territorial integrity and national interest. Saudi diplomatic and other resources have been constantly deployed to prevent union between the Arab countries. The case of the proposed union between Iraq and Syria in the late 1940s is an example where Saudi effort put an end to the proposal. The downfall of the union between Egypt and Syria (into the United Arab Republic in 1958) was sought by the Saudis by all available means including the payment of a bribe to the tune of £2 million to have the President of the UAR, Nasser, killed.

At present Saudi Arabia plays a more crucial role in Arab politics as all the confrontation States in the Arab–Israeli dispute (i.e. Egypt, Syria and Jordan) are dependent on financial subvention from the oil producing countries. These States have come to realize that it will be extremely difficult for them to wage a successful war against Israel without the deployment of the oil weapon and the neutralization of American support for Israel. Sadat's dramatic and historic trip to Israel may be explained by the fact that he possessed neither the control over oil supplies nor the leverage in Washington which the Saudis have. Sadat, therefore, found himself in the unfortunate position of having to fight a war in which his military and other political capabilities were undermined by not only Egypt's dire economic plight but also Saudi reluctance to impose an oil embargo should an outbreak of hostilities be resumed. The interdependence between Saudi economic development and the economics of the West and their drive against communist encroachment either in Europe or the Middle East would not incline the Saudis to jeopardize their own position by undermining the political, social and economic fabric of the Western World. "We know," Shaykh Zaki Yamani once remarked, "that if your economy falls we fall with you."

These factors as well as the gigantic strides undertaken by the Saudi political leadership to transform the country from a backward semi-nomadic and ignorant society into a modern, forward-looking and highly industrialized nation-state are likely to influence the course as well as the personalities involved in Saudi politics. The huge expenditure of $142 billion on a five-year development plan is bound to create fundamental changes in outlook, attitude and traditional values in a country where such vast development suffers all the constraints except the financial. The country does not possess more than a rudimentary infrastructure in relation to the government's ambitions and human resources. The task of directing the administration, whose most obvious manifestations are chronic inefficiency aggravated by the merry-go-around of bureaucratic red-tape and graft, is commanded by the few men at the top.

The emergence of such a Westernized group commonly described as the intelligentsia in third world countries will in due course become a factor of direct relevance to the problem of political succession. No part of the ruling élite could afford the alienation of this segment of society, especially when it is undergoing such rapid changes. Men equipped with the technological and scientific skills—the product of modern training and education—are of paramount importance to the new society which the leaders are attempting to create. People like Shaykh Zaki Yamani, Minister for Oil, and Hisham Nazir, Minister of Industry, are only the vanguard of this social group. It will be easy to absorb and integrate them into the system as long as their number is small. Moreover, it will be done without endangering the structure of political power as the exclusive preserve of the Saudi princes. But as the membership of this group increases, as it is bound to do so through the spread and expansion of educational and other services, the demand for power-sharing will be intensified. Their admission to the political process will inevitably dilute the political authority of the Saudi ruling family and give them a voice in determining the nature of the arrangement and the manner of selection or election of the rulers. Their exclusion may create a bottleneck effect on an entrenched and inward-looking oligarchy which will seek to preserve its prerogatives. The intelligentsia, however, will become an aspiring political élite demanding political accommodation on institutionalized basis in the form of a constitutional representative system of government.

Such demands may be manipulated and utilized by members of the ruling house in their struggle for power. Prince Talal's campaign for a democratic constitutional government failed in the early 1960s because it did not find echoes in Saudi society. The situation may not be the same when the next Saudi prince takes up the challenge. There is little doubt the demand for such radical change in the system of government will outrage the religious fundamentalists of the puritanical Wahhabi movement on whose tenets the State was set up. In 1962 Prince Abdullah dismissed Talal's demand for a constitution in the following terms: "Talal," he said, "alleges that there is no constitution in Saudi Arabia which safeguards democratic freedoms. But Talal knows full well that Saudi Arabia has a constitution inspired by God and not drawn up by man. I do not believe there is any Arab who believes that the Koran contains a single loophole which permits an injustice to be done. All laws and regulations in Saudi Arabia are inspired by the Koran, and Saudi Arabia is proud to have such a constitution."*

It is clear that Abdullah, the leader of the Abdul Aziz faction, derives his power in the Saudi establishment not only from the 20,000-strong force of tribesmen but also as the spokesman of the conservative traditional forces in the country. His constituency of support and influence lies in the direct and close liaison he maintains with the desert tribes of Najd which is historically the heartland of the Saud family. He also exerts considerable sway over the widespread and powerful Shammar tribes of Northern Arabia from whom his own mother originated. Fahad on the other hand is more progressive and reformist, aware and conscious of the need for administrative modernization. His powerful position in the Saudi political structure is based on the control of the armed forces, the police and the frontier forces. Fahad has also been concerned with the important contribution the intelligentsia could make to the stability and future prosperity of Saudi Arabia. In an obvious attempt to court their support, he made a statement several years ago which somewhat resembled the demand made by Talal some fifteen years earlier, in which he declared his intention to set up a consultative Council for the Kingdom which was to include all the dynamic young people who would contribute their ideas and participate actively in the administration of the State. The fact that this Council has not been set up yet may provide an indication of the opposition that such proposals would encounter in the Saudi State, despite the fact that Fahad had made it clear that the members would not be elected but appointed.

However, the generation gap is of significance in this respect and even Fahad may find it difficult to countenance such fundamental changes in the

*Al-Hayat 18 August 1962, cited in *Faisal King of Saudi Arabia* by Gerald de Gaury, p. 107, London, 1967.

government of the country. But it is not so with the new generation of Western-educated young princes who may have more in common in terms of their expectations for the future of the Kingdom with members of the intelligentsia than with their peers and elders in the ruling house.

Such a group is emerging gradually and its potentials in providing future political leaders is beyond doubt. The champion of this group is Prince Saud al-Faisal, the son of the late King and currently the Foreign Minister of the Kingdom. The young prince bears all the hallmarks of a future King. He is known to advocate reform more staunchly than his uncle Fahad. He has called for the training and employment in government service of an élite core of educated technocrats and administrators to enable Saudi Arabia to make the transition from a tribal society to a modern State. The young technicians and bureaucrats are promoted to occupy key posts in the administration. Prince Saud is seeking the accommodation of these groups in the political system to prevent their alienation in order to buttress the State so that Saudi Arabia may withstand the strains and tribulations of the challenge of the modern age. It is for this reason that Fahad has found in Saud a welcome ally against the conservatives. However, Saud with a formidable power base of his own may come to the conclusion that his uncle is too traditional for the time. Under Saud, the formation of a third faction can be observed which may hold the future of Saudi Arabia in its grasp. This clan is made up of the eight sons of the late King, and what they do is as important as who they are in view of the patron-client relationship: Saud, Foreign Minister; Abdullah, a poet and businessman; Muhammad, former Minister of Irrigation (responsible for the iceberg project); Khalid, Governor of Asir province; Abdul Rahman, Commander of an armoured brigade; Saad is the Deputy Director of Petroman—the Saudi National Oil Company; Bandar is an officer in the Air Force and Turki is assistant to the man who is in charge of Saudi Intelligence, who happens to be their maternal uncle. A dazzling array of talents by any standard, but in the case of Saudi Arabia where political institutions are lacking and in the absence of an accepted formula to decide the succession, they could certainly crown or dethrone kings.

Iraq: Sociopolitical Developments

PHEBE MARR

Iraq entered the 1980s governed by the regime of the Arab Ba'th Socialist Party (ABSP), which came to power in July 1968, and which has now been ruling longer than any other regime since the fall of the monarch in 1958. Until the war with Iran, the regime had brought the country a measure of stability, a constantly expanding development program, and a large measure of beneficial social change; this is all the more impressive since Iraq is notoriously difficult to govern. In the decade following 1958 there were four different changes of regime and even more attempted coups. The longevity of the Ba'th regime has been due to a number of factors, among them the party's institutional development, its draconian measures against opposition, and an economic boom which, until recently, enabled it to alleviate discontent. The war with Iran, now in its third year, has slowed economic development and is putting this stability to a test. Thus far, Iraq has borne up well, but the regime is facing a number of internal and external challenges which may make its stability more difficult to maintain.

BACKGROUND

The Arab Ba'th Socialist Party (ABSP) now governing Iraq was founded in Damascus in 1947 by two Syrians, one of whom, Michel 'Aflaq, is now titular head of the Pan-Arab National Command of the party in Baghdad.[1] The Ba'th party is committed to secular Arab nationalism: to the unity of all Arab states; and to a revitalization of Arab society (Ba'th means "renaissance" or "resurrection") through scientific socialism, the content of

1. The command is pan-Arab in that it consists of Arabs from countries other than Iraq, yet it functions from its Baghdad base, and is very much under the control of the Iraqi Regional Command. The legitimacy of the command is not accepted by the Syrian branch of the party.

Reprinted by permission of the publisher from *AEI Foreign Policy and Defense Review*, Vol. 2, Nos. 3 and 4, copyright by American Enterprise Institute for Public Policy Research.

which has varied over time. As 'Aflaq's position indicates, the party theoretically has a pan-Arab organization with regional branches in various Arab countries but, in practice, since 1966 the party has been split into rival branches, one of which governs in Syria, the other in Baghdad.

The current Iraqi Ba'th regime came to power in July 1968 through two coups which followed in rapid succession—the first engineered by the military, the second by the Ba'th itself. Initially, five officers—three Ba'thist and two non-Ba'thist—governed through a Revolutionary Command Council (RCC) headed by Ahmad Hasan al-Bakr, although the civilian wing of the party, led by Saddam Husayn, played an important role behind the scenes. Since then the Ba'th party, and in particular the civilian wing, has consolidated its power and established a one-party state. The army officers who engineered the coup have been replaced largely by civilians; the Ba'th party has coopted all important posts in the state; and, with a few exceptions, a new generation of party leaders has taken charge.

The development of party institutions and their control over the organs of state have been rapid and impressive, as can be seen by three main developments. First, the party has broadened its base of support by enlarging its membership: although figures are not published, it has been estimated that party membership grew from 5,000 in 1968 to 50,000 regular members in 1978, with possibly 500,000 dependable followers.[2] This is confirmed by another source that estimates that the party's popular militia, a hard core of loyalists, numbered about 50,000 prior to the war.[3] Party recruits are carefully screened and go through a long period of probation and indoctrination before becoming full members, thus assuring the regime of a loyal cadre on which it can rely for support. Second, the party has gradually taken over the leadership of numerous popular and professional organizations, including the General Federation of Workers' Unions, as well as peasant, student, and women's organizations. Third, the party has moved to consolidate its position with rival political groups by forming a National Front. In 1976 this consisted of the Ba'th; the Iraq Communist Party (ICP), subsequently removed, several Kurdish groups supporting the government; and a few independents. The Ba'th has maintained a monopoly of power within the National Front.

The evolution of governmental institutions has proceeded more slowly. In 1970 a temporary constitution made the RCC the sole legislative authority in the state and vested considerable executive power in a president elected from, and by, the RCC. Ministers were to function merely as department heads. At the same time, the party's Regional Command was designated the major policy-making body of the state. The subsequent appointment of key Regional Command members to the RCC, as well as

2. *The London Economist,* June 24–30, 1978, pp. 78–79.

3. Majid Khadduri, *Socialist Iraq: A Study in Iraqi Politics Since 1968* (Washington, D.C.: The Middle East Institute, 1978), p. 44.

to important ministerial positions, ensured Ba'th control over the government. The enlargement of the RCC in 1977 to include all Regional Command members went even further in this direction, but it also made the RCC an unwieldy body for decision making and allowed power to gravitate into the hands of a smaller group within the Council. In 1980, in an effort to legitimate its rule and broaden its base of support, the regime took the step of creating a long-promised National Assembly. In a carefully controlled election held that year, the Ba'th party, not surprisingly, won an overwhelming majority of seats.

Despite this impressive facade of institutionalization, a definite countertrend has developed within the party, particularly since 1975—the concentration of power in the hands of one man, Saddam Husayn, and the concomitant personalization of decision making. Husayn has been an active participant in the party since his high school days when he was involved in the abortive assassination attempt in 1959 on General Qasim, then prime minister of Iraq. Most of his early adulthood was spent in clandestine party activity. He owes his early rise to power more to the patronage of al-Bakr, however, than to any other factor, and until 1978, the two formed a complementary partnership. Al-Bakr was the senior "father" figure in the party, a man with experience and a public reputation, whose contacts in the army proved invaluable in keeping the military in line. Husayn was the man with youth, energy, and dynamism, who championed the civilian wing of the party. With the support of al-Bakr, Husayn succeeded in removing most of his rivals within the Regional Command throughout this period to emerge as the strong man of the regime, and by 1978 all lines of authority led to his office. Meetings of the RCC and the Regional Command were called more to ratify decisions taken by Husayn and his coterie than to discuss and to make policy. Al-Bakr's resignation from the presidency in 1979 put Husayn in an unparalleled position of power.

Along with this trend, Husayn came to rely on a small group of men who, like himself, had grown up in Tikrit, a small town located on the Tigris River north of Baghdad. In the late 1970s, the Tikritis included four members of the RCC and several key ministers and high-ranking officers, many of them related to one another. Husayn, for example, is a cousin of Adnan Khairallah Talfah, his minister of defense, and is married to Talfah's sister. The leading figure in the intelligence service is Husayn's half-brother. While close family and local ties are typical of Iraqi society, the Tikritis are resented by younger party members who expect to make their way up the party ladder without benefit of relatives.

POLITICAL DYNAMICS IN THE REGIME

Although Husayn has managed to concentrate power in his own hands, he does not have a complete monopoly on it, and must consider balancing forces at top party levels to retain his preeminence.

Personal rivalries and competition for power are often difficult to identify because of the regime's secrecy. The problems, however, can be identified. They include potential jealousy of Husayn by some of the older party members, who regard themselves as his equal in party service; resentment of the Tikriti "clique"; and dislike by some of the newer Regional Command members of the concentration of power in Husayn's hands and the gradual eclipse of their own participation in decision making.

There are also differing ideological tendencies between the left and the right. The left puts more emphasis on internal socialist transformation and on the alliance with the Soviet Union, while the right takes a more nationalistic and pan-Arab line, including a hard line on the Kurds and the Shi'ite Muslims.

Husayn has shown himself to be pragmatic, flexible, and when necessary, ruthless in balancing these groups. He has consistently clipped the wings of both the left and the right, only to coopt their positions once their members have been removed from office. Several episodes illustrate his modus operandi. The first was his engineering of al-Bakr's resignation and his own succession to the presidency in July 1979. While this development had been widely expected, it occurred at a moment of crisis for the regime—an alleged coup attempt within the party, and mounting troubles with revolutionary Iran. Yet the succession was accomplished with smoothness and an aura of voluntarism which gave Husayn a measure of legitimacy. Meanwhile, to compensate for the loss of al-Bakr's stabilizing influence in the military, Husayn assumed the rank of a general in the armed forces in 1977 and has appointed his cousin minister of defense. It may be assumed that he has also staffed the upper levels of command with his personal supporters. However, some of these moves were unpopular in the army, and it is doubtful that he can manage the military as well as his predecessor.

The second episode is the alleged coup attempted by a number of party regulars in July 1979 and the trial and executions which followed. Five of those executed were Regional Command members. It is difficult to identify all the motives behind the attempt, or even to determine if anything as organized as a coup attempt occurred, yet it is certain that Husayn perceived a threat to his leadership and took severe action to nip it in the bud. According to one official statement the opposition was multifaceted, encompassing a variety of motives, but all the perpetrators were united against the leadership, that is, against Husayn.[4] This information suggests not only that personal motives played a leading role in the alleged coup, but also that ideological factors may have been at work. The fact that Syria was privately implicated by Iraqi officials suggests that pan-Arab sentiments may have been involved. There may also have been left-wing dissatisfaction over Husayn's treatment of the ICP and over his coolness toward the Soviet

4. Quoted in *Weltwoche Report,* September 12, 1979, p. 45.

Union. The fact that three of the five Regional Command members implicated in the plot were Shi'a Muslims points to Shi'ite dissatisfaction with the regime's handling of their situation. Whatever the causes of dissatisfaction, Saddam Husayn inaugurated his presidency by indicating that he would tolerate no opposition to his leadership within the party.

The third episode illustrates his willingness to accommodate potential opposition elements provided they do not threaten his preeminent position. In the summer of 1982, after the onset of the war and in the face of a need to propitiate Shi'a opposition, the Regional Command of the party was drastically reshuffled; of the seven new members a majority were Shi'a.

Whatever dissensions exist within the party, however, they are less serious than the traditional opposition from outside the party, particularly that of the Shi'a and the Kurds, both exacerbated by the way.

THE ELEMENTS OF OPPOSITION

Much of the opposition to the party is unrelated to the Ba'th per se, but springs from ethnic and sectarian groups which have been inadequately integrated into the state, and which have a long history of antigovernment activity predating the Ba'th ascendancy.

The Shi'a. As the Iran-Iraq war has shown, one of the most worrisome of these elements of opposition is that represented by the Shi'a or, more specifically, certain Shi'ite elements. The Arab Shi'a, who inhabit the delta of the Tigris and Euphrates south of Baghdad, constitute a little over half the population, yet they have, traditionally, been dominated by the much smaller Arab Sunni community. The Arab Sunnis, constituting about a third of the population, and situated in the cities and towns along the Tigris and Euphrates rivers north and east of Baghdad, have had a disproportionate number of members in the top levels of the bureaucracy and in the army officer corps. On the other hand, a substantial portion of the soldiers are drawn from the Shi'ite south, a potentially destabilizing factor in the event that the army is asked to support a cause unpopular with this group.

Opposition to the government comes from several mutually antagonistic Shi'ite groups. The conservative religious elements are the most significant. This group follows its natural leaders, the *mujtahids,* religious scholars who often number their followers in the thousands.

The religious leaders have behind them the support of several important institutions and an informal network of personal relations which make their movement difficult to penetrate. They have charge of a number of theological seminaries, mianly those in Najaf, in Karbala, and in Kadhimain (a suburb of Baghdad), where Shi'a from all over the world, but especially from Iran, spend years of study. Ties with students who return to Iran are strong. The religious leaders also have informal sources of income through contributions from their followers.

At the opposite end of the political spectrum, some Shi'a support the radical left. In 1973, the largest single group in the Central Committee of the ICP, comprising eight out of twenty-three persons, was Shi'ite.[5] The left has appeal to the Shi'a, as well as to other minorities, because of their concentration on social reform at home and their relative neglect of Arab nationalism. While the religious leaders and the Shi'ite Marxists are poles apart in philosophy, they are united in opposition to the essentially Sunni nationalist government.

Shi'ite opposition has generally taken a passive form, but it has reached flashpoint several times in the past decade. In 1974, five Shi'a religious leaders were executed for opposition to the state.[6] In February 1977 riots broke out during a Shi'ite procession in Najaf and the army was called out as a result. More serious than the riot in the minds of the Ba'th party leaders was the fact that a number of younger Shi'ite party members as well as police and security men sided with the rioters rather than with the government. Shi'ite riots occurred again in Baghdad and in Najaf in the spring of 1979.

The Iranian revolution, the emerging sectarian character of the government there and the Iran-Iraq war, are also creating significant problems for the Ba'th. The Iranian revolution has instilled in the Shi'ite community in Iraq a sense of identity and pride that it has hitherto rarely possessed. The Shi'a may now be expected to demand their rights more openly, more frequently, and more successfully.

The Shi'a in Iraq, however, exhibit several weaknesses which make a revolutionary movement such as that of Iran unlikely to develop. They are not organized for opposition, nor are they unified; moreover, they inhabit the flat plains of the delta where a guerrilla operation (such as that mounted by the Kurds in the mountains) is impossible. They also face a well-organized and determined government. The fact that few Shi'a have thus far shown open opposition to the war with Iran indicates that the Arab nationalist regime in Baghdad has more to offer them than the religious government in Tehran.

The Kurds. The Kurds, like the Shi'a, have a long and continuous history of opposition to the central government predating the Ba'th ascendancy. Constituting 15 to 20 percent of the population, the Kurds inhabit the mountainous northeast region of the country and have ties with their fellow Kurds in at least three neighboring states—Iran, Turkey, and Syria. The basis of Kurdish opposition is ethnic and national, rather than religious, since they, like the dominant Arab group, are Sunni. Unlike the Shi'a, the Kurds have mounted armed (and largely tribal) rebellions against the government in the 1920s, the 1930s, and during World War II.

5. Hanna Batatu, *The Old Social Classes and the Revolutionary Movements in Iraq* (Princeton, N.J.: Princeton University Press, 1978), p. 1107.
6. *Al-Hawadith* (Beirut), December 20, 1974, p. 20.

Kurdish nationalism began in earnest after the revolution of 1958 when Abd al-Karim Qasim, then prime minister and strong man of the first revolutionary regime, invited Mulla Mustafa Barzani, a redoubtable tribal fighter, back to Iraq from a long exile in the Soviet Union. After disagreements with Qasim, Barzani and the Kurdish Democratic Party (KDP), organized in 1946, took up arms against the central government in 1961. The fighting escalated into a full-scale guerrilla war which lasted intermittently until 1975, and may not yet be entirely over.

The current regime has dealt with the Kurdish situation in a more forceful and imaginative way than did its predecessors, and seems to have gone further toward achieving a permanent solution. In 1970 the Ba'th concluded a cease-fire with Barzani which provided for the granting of Kurdish autonomy within the framework of an Iraqi republic, Kurdish participation in the central government, the appointment of a Kurdish vice-president, and a larger share of development expenditure. Autonomy was to be preceded by a census which would determine the area in which the Kurds had a majority. Unfortunately, the cease-fire broke down because of distrust on both sides, and on March 11, 1974, the government unilaterally promulgated the autonomy law. Barzani, who disagreed with its provisions, resumed military action.

The central government now made rapid advances in the plains and in and around the large Kurdish cities, but, in spite of heavy bombardment, did not succeed in obliterating Barzani's movement in the mountain areas. Iran's involvement with the Kurds had meanwhile raised the crisis to the level of a potential regional conflict. At this point, both countries drew back from the brink and settled their differences—but at the expense of the Kurds. In March 1975, the two sides signed an agreement, with Algeria mediating the dispute. In return for the recognition by Iraq of the claim by the shah of Iran to a new boundary on the Shatt al-Arab in the south, Iran agreed to stop aid to the Kurds. Barzani took refuge in Iran and later in the United States (where he subsequently died) while Iran took in about 200,000 Kurdish refugees.

The Kurdish provinces in Iraq now constitute an autonomous region with an eighty-member legislative council and a smaller executive council. The Iraqi head of state has the right to dismiss the legislative council, thus rendering the autonomy more apparent than real. The legislative council has control over such domestic affairs as education, public works, and agriculture; the central government controls foreign affairs, oil, and defense. The oil areas have been specifically excluded from the autonomous zone. A number of progovernment Kurds have meanwhile been put into ministerial positions in the central government, and a Kurdish vice-president holds office.

By late 1976, most of the 200,000 Kurdish refugees in Iran had returned to Iraq, and the government had attempted to secure the border with Iran.

A large number of Kurds were resettled away from the borders.[7] On the more positive side, the government has begun a vigorous development program in the north and concentrated on agricultural schemes and land reform, which are bound to weaken the landed interests that traditionally supported Barzani.

While these events have greatly weakened the Kurdish movement, they have not ended it. Guerrilla activity still takes place inside and outside Iraq's frontiers. In June 1975 the National Union of Kurdistan (NUK) was founded by Jalal Talabani and some former leaders of the KDP. This group is said to have about 3,000 fighters whose political sympathies range from Marxist to the liberal democratic. Until recently the NUK has operated out of Syria and with Syrian help. A more important group, however, is the reconstituted KDP under the leadership of Mas'ud Barzani, the son of Mulla Mustafa, whose views are more nationalist than leftist. A group of his forces has reentered Iraq to spearhead the opposition inside the country. Both these opposition movements must, however, contend with a considerable number of Kurds who see their economic, if not their political, future in cooperation with the central government.

The resurgence of Kurdish nationalism in Iran has created both opportunities and difficulties for the Kurdish leadership in Iraq. While the Kurds in Iraq sympathize with the Iranian Kurds they cannot afford to alienate the Iranian government by supporting their kinsmen too directly, first, because of the Iraqi Kurds who remain in Iran and, second, because of their critical need for an open border. Thus, the KDP has tended to side with Khomeini in his war against Iraq. Meanwhile, the central government in Baghdad is supporting anti-Khomeini Kurds in Iran to weaken both the Iranian government and Iraq's own Kurdish opposition.

Given these splits in the Kurdish movement, it is difficult to see the Kurds mounting a serious attack on the government of Iraq at present. However, renewed guerrilla activity would slow down the development of the autonomous area and weaken the government's hold over it. If Kurdish activity escalates, it would surely cause disagreements within the regime on how to deal with it and, in addition, undermine the popularity of Husayn, who has staked much of his reputation on having "solved" the Kurdish problem. Should his solution unravel, so too would much of his legitimacy.

The Iraq Communist Party (ICP). A third element of opposition lies in the ICP, which finally broke with the regime in 1978. Next to the Ba'th itself the ICP is the best organized political party in Iraq, with a history going back to the 1930s. Its chief stronghold is among the intelligentsia (especially the students and teachers) and the workers, particularly those in such key sectors as the oil industry and the transport system. The ICP has always had

7. Chris Kutschera, "La Fiction de l'Autonomie au Kurdistan Irakien," *Le Monde Diplomatique* (Paris), August 6, 1977, pp. 6–7.

success in calling out and manipulating the masses in the streets, and, as previously indicated, it also has support among the Kurds and the Shi'a.

There is a long history of animosity between the ICP and the Ba'th dating from Qasim's regime, when bloody clashes occurred between the two parties. In 1963, when the Ba'th first came to power, they ruthlessly tracked the Communists down and eliminated many of them; by 1968, however, the Ba'th themselves had moved further to the left, and in 1974 they formed a front with the ICP. The alliance was, nevertheless, an uneasy one and in 1977, the situation of the ICP worsened when a growing coolness developed between the Ba'th and the Soviet Union. In 1978 the Baghdad government abruptly announced the execution of twenty-one Communists who were accused of having attempted to recruit members in the armed forces, such recruitment being illegal for any party but the Ba'th. In 1979, when more Communists were arrested, the Communist members in the cabinet resigned; the party leadership went into exile, and from Beirut and elsewhere, they have since published attacks on the regime. The Communists, most of whose supporters are now underground, thus constitute another element of open opposition to the regime.

While the ICP can cause difficulties for the Ba'th, particularly in helping to foment riots and demonstrations against the regime, there are serious limitations to its power. The party, though well organized, is small, and its support throughout the country has never been widespread, mainly because it runs into deep-seated opposition not only from conservative religious elements who dislike its atheism, but also from Arab nationalists who regard it as an appendage of the Soviet Union.

The Military. Little is known about political sentiments among the officer corps, but as the execution of the Communists indicates, the regime does not feel entirely certain of its loyalty. In general the officer corps is dominated by Arab Sunnis from whom the Ba'th draw much support. The military have been well taken care of by the regime in terms of salaries and other benefits, as well as through budgetary outlays on equipment and training. Most officers would appear to support the Arab nationalist (but not the leftist) aims of the regime. Certainly they would favor maintaining national unity against any disruption from the Shi'a or the Kurds.

The Ba'th use a number of mechanisms to keep the military under control. The party's Military Bureau, dominated by civilians, undertakes a thorough program of indoctrination in the army while party cadres function in all units, particularly at the upper levels, and report on the first sign of dissidence. Despite this control, however, the party has periodically had to conduct purges of army officers for presumed and attempted coups, or because officers have opposed government policies. In the past, dissatisfaction in the officer corps has focused on several issues, including the failure to settle the Kurdish problem. The key issue today, however, is the conduct of the war against Iran and the relative success of Iraq's defense.

Thus far, despite defeats and loss of territory originally taken from Iran, the loyalty of the army appears to be holding up well.

The Educated Middle Class. Opposition of varying degrees can also be identified among elements of the educated middle class, although it must be emphasized that the middle class as a whole is not in opposition; the regime finds considerable support among educated Iraqis for large parts of its program, if not for all of its methods and its goals.

The first source of this dissatisfaction is economic. The oil boom has created demands for more and for better consumer goods and for an improved system of distribution. Much of the middle class puts the blame for economic failures on the rigid socialist ideology of the regime and on too much bureaucratic red tape. Despite the regime's antipathy toward capitalism, it has not been able to prevent the rise of a new entrepreneurial class comprising the contractors and the skilled laborers, who command high prices for their services. This factor is resented by salaried professionals on fixed incomes. There are also complaints that wealth finds its way into the pockets of party and government officials. Although Iraq has done a better job than most Gulf states in moderating inflation, this, too, has taken a toll on the salaried middle class.

A second source of dissatisfaction is professional in nature. There are a number of indications that the regime has encountered conflicts between its demands for technical capacity and its requirements for political loyalty. One is the constant turnover in ministerial posts dealing with development; these rotate frequently between party loyalists with little advanced training or experience and highly trained professionals, who have usually been educated in the West. Another indication is the sizable number of well-trained Iraqis who are now working in Gulf countries and elsewhere in the Middle East. This dissatisfaction does not surface in active opposition but is passively helping to slow down the pace and the quality of development. Low rates of productivity in the economy were sufficiently worrisome to the party in 1976 and 1977 to occasion a series of meetings and seminars on the subject addressed by Husayn himself. In April 1978, a large-scale reshuffle of over 1,000 bureaucrats took place in the interest of efficiency.[8]

The educated middle class also desires greater political freedom, a sentiment it shares with the Shi'a, the Kurds, the Communists, and other groups. It must be emphasized, however, that as far as can be gathered, the regime has considerable support from the lower-middle class, the urban workers, and much of the rural population. All of these groups have benefited from the land-reform program of the regime, from its free health and education services, from food and housing subsidies, and other welfare benefits. These groups are far less interested in political freedom than in guaranteed employment and a rising standard of living. The regime is

8. *An-Nahar Weekly* (Paris), April 29, 1978.

banking on its economic development program and its egalitarian orientation in the distribution of wealth to keep the support of such elements. In this way it hopes to win the bulk of the population to its side and to weaken traditional loyalties to sectarian and ethnic identifications. Thus far, despite the stresses of the war, the regime's assessment appears to be correct.

EXTERNAL CONSTRAINTS ON POLICY

In addition to the potential opposition forces just cited, there are outside constraints operating on the formation of domestic policy in Iraq. At the moment the most critical is Iran, but Syria is also troublesome; both have the potential to stir up opposition within Iraq.

The problems with Iran are clear. Iran's revolution, in one blow, destabilized Iraq's entire eastern frontier, a factor which ultimately led to war. In the north, Iraq can expect more trouble with the Kurds if the war is not settled. In the south, the effects of the Iranian revolution on the Shi'a have already been described. The fact that Iraq went to war with Iran, largely in reaction to Iran's attempt to spread its revolution to the Shi'a in Iraq, illustrates the depth of the regime's fears.

The revival of discord between Iraq and Syria in the wake of the alleged coup attempt of 1979 may also prove troublesome, particularly in view of Syria's recent isolation on the Arab front, its involvement in Lebanon, and its recent reliance on the U.S.S.R.—all unpopular in Baghdad. Iraq's feud with Syria has in the past been exacerbated by conflicts of national interest, in particular Iraq's construction of a pipeline to carry oil from Kirkuk to the Gulf, bypassing Syria and depriving her of revenues, and Syria's completion of a large dam on the upper Euphrates which draws off water normally used by Iraqis to irrigate their fields downstream. These issues could be activated again.

In the past Iraq has also had uneasy relations with the conservative pro-Western states of the Gulf, which are suspicious both of its socialism and of its revolutionary activities. Recently, however, Iraq's relations with these countries have warmed; the war with Iran has cemented them rather than eroded them. Indeed, Iraq has become dependent on the Gulf states for income and support and can be expected to pursue a policy of cooperation, rather than confrontation, with the conservative Arab states of the Gulf for some time to come. This was made clear in a speech given on February 8, 1980, in which Husayn made a commitment to Arab solidarity, neutrality, and nonalignment.[9]

Iraq's main external constraint, however, lies in its dependence on the U.S.S.R. for weapons, for military personnel, and to a lesser extent for

9. Speech by Saddam Husayn, February 8, 1980, cited in *Foreign Broadcast Information Service,* February 11, 1980, pp. E1–E4.

technology. There are, however, limitations on Soviet leverage. For some time, Iraq has been attempting to diversify its arms supplies and has also shifted away from the purchase of Soviet and other Eastern bloc technology. Complaints about the ineffectiveness of Soviet material and workmanship have led to more contracts with the West. In 1976, 51 percent of the value of Iraq's imports came from EEC countries and from North America; 11 percent from the Socialist bloc. Of this total 21 percent came from West Germany alone; 13 percent from Japan, 5 percent from the United States, and a mere 2 percent from the U.S.S.R. Trade between the United States and Iraq rose from $210 million in 1977 to $846 million in 1982.

FUTURE PROSPECTS

Future prospects for Iraq depend very much on the outcome of the ongoing struggle with Iran. However, as of mid-1983, Iraq's political and social structure, with a few minor adjustments, has been able to withstand military defeats, substantial casualties, and a long and costly war, without internal upheaval and disruption. Moreover, in defending its own territory, the military performed well. Under the impetus of the war, however, Iraq has reached a point where its future direction, under the present regime, will tend more toward personal authoritarian rule, rather than toward a one-party system. Although party institutions will remain, they will be less important in under-pinning Husayn's government. He is likely to depend increasingly on a trusted group of supporters at the top of the structure while broadening support at the base through carefully controlled use of mass organizations and the new National Assembly. If Husayn remains in power, and the presumption at present writing is that he will do so, it may be expected that he will continue his policy of fragmenting and suppressing his opposition, of balancing forces within the leadership, and of continuing to use economic development and support from the lower and lower-middle classes as a basis for legitimacy. He can also be expected to continue a judicious balancing of contracts between East and West to maintain a degree of economic independence, and he may even loosen up on economic —but not political—controls. He will undoubtedly have to make gestures to conservative religious sentiment, especially to that of the Shi'a. He is currently emphasizing Arab nationalism to counter sectarianism, and he is moving out of his isolation on the Arab front, which will help consolidate his support with Arab Sunni and Arab nationalist elements.

Husayn nevertheless faces challenges from various elements that will test his leadership in the future, not the least his unpopularity for having plunged Iraq into an unsuccessful war. His canny sense of timing and his proven ability to manipulate the Iraqi political environment should enable him to survive the war, as he has survived crises in the past. If he should fail to do so, however, the most likely source of change would come from within the party itself, probably from a group dissatisfied with Husayn's leadership on personal or on ideological grounds. It is also possible that a non-Ba'thist group within the military could replace the regime. Whether military or Ba'thist, such a group is likely to be nationalist in its external policy and, particularly in the case of the military, pan-Arab in orientation; a coup of this sort is most likely to occur over dissatisfaction with Husayn's management of the war, or the need to remove him as a quid pro quo for a settlement, rather than over the direction of his policy. Indeed, Husayn's current leaning toward moderation and his rapprochement with the Gulf states is not likely to meet with opposition from the center of the party now in control. While less repressive, a military regime is likely to be less stable than the Ba'th, since it would probably open the door to constant military intervention, as was indeed the case in the past.

The replacement of the present regime by a more leftist and anti-Western group, from either inside or outside the party, is also possible. The leftists, clearly the losers in recent power struggles, are undoubtedly dissatisfied with Husayn's recent shifts. However, a more leftist group, if it should succeed in removing Husayn, would be least likely to maintain itself in power for long, since it is not likely to command widespread support among any of the potential opponents described above, except the ICP, unless it received substantial support from the U.S.S.R. There would be opposition from the conservative religious elements now in the ascendancy in the region, as well as from ethnic, sectarian, and Arab nationalist groups.

The prospect of a popular Shi'ite movement, perhaps in alignment with other groups, is a tantalizing one, in view of the war and Shi'ite opposition in Iraq. However, as events have shown, it is far less likely in Iraq than in Iran. The Shi'ite *ulama* have not attained the dominant position in Iraq that they achieved in Iran and are closely watched by the government. Moreover, a Shi'ite-led movement would be opposed by a stronger Sunni minority, as well as by a secular government. Many Shi'a, despite dissatisfaction with the present government, do not want to see disintegration of the state. It is, moreover, the Shi'ite areas, particularly in the south, which were the recipients of economic improvements before the war, although the damage to the Basra area has now eliminated that advantage. Above all, the example of instability, stagnation, and a revolution gone awry in Iran, have had a sobering effect in Iraq, giving the Shi'ite opposition second thoughts about disrupting the current stability and the benefits of future economic development. This is what the current regime is banking on, and its future policy is likely to reflect this assessment.

Khomeini's Iran

ERIC ROULEAU

From the very beginning of the Iranian Revolution, the West—and particularly the United States—seems to have been struck by a peculiar sort of political blindness. The first signs of revolt passed unnoticed. The explosions of rage in the spring of 1978, first in Tabriz and then in Qum, were attributed to "obscurantist mullahs" hostile to the Shah's agrarian reform. The immense demonstrations by millions of Iranians, as well as the strikes in the administrations, factories, schools, universities and oil fields which paralyzed the state and in the last analysis caused the monarch's inglorious departure, were attributed to the "fanaticism" of the Iranian people. How could it have been otherwise, it was asked at the time, since the population was following a reactionary old cleric in revolt against a man who had devoted his entire life to modernizing his country?

Rare were those who suggested that modernity is not necessarily synonymous with progress or well-being, or that the concepts of economic development current in the West—where quick material gain is often the only valid criterion—does not necessarily correspond to the true needs and interests of developing nations. Rarer still were those who pointed out the pitfalls of labeling an entire people fanatics simply because they were virtually unanimous in expressing their will. Even a study of Iranian history and pyschology would have revealed that the Iranians, while believers, are at the same time one of the least observant and most tolerant in the region. A mosaic of ethnic groups and religious communities, Iran has nonetheless been the scene of an astonishingly small number of strictly religious conflicts over the past two centuries.

If such is the case, how could Islam have played the role of prime mover in the Iranian revolution? To ask this question is to forget that other

Reprinted by permission of *Foreign Affairs,* Fall 1980. Copyright 1980 by the Council on Foreign Relations, Inc.

religions in other times and places have also provided an ideological dimension to political movements, and on occasion have been used to form states. The Italian preacher Girolamo Savonarola (1452–1498) rose against the French invaders before establishing a regime in Florence which was at once theocratic and democratic. Jean Calvin (1509–1564), acting in the name of the Reformation, burned his religious and political opponents alive in the theocratic republic he headed in Geneva, but his ethic—which among other things glorified work and justified interest-bearing loans—contributed to the rise of capitalism, the development of political democracy and Western cultural values. Oliver Cromwell (1599–1658) and the Puritans swept away the English royalty to establish the "Republic of the Middle Classes." These three men serve as examples to illustrate the revolutionary role religion has played at one time or another in the West. All occupy prominent places in our history books, yet each was reviled by his contemporaries as cruel, sectarian, fanatical.

In seeking to evaluate Imam Khomeini's movement, Westerners, and particularly Europeans, have remembered only the French Revolution, a comparison which was comforting to those who judged the Iranian Revolution retrograde. Hadn't the French clergy, allied to the nobility and Louis XVI, opposed the forces of progress by opposing the demands of the Third Estate? And wasn't the Iranian Revolution led, organized and oriented by clerics incensed by reform? It is true that the Catholic Church is not, generally speaking, known as a liberating force and that in our societies the notion of secularism is inseparable from that of democracy.

In any case, despite any parallels or analogies which might legitimately be established, the Iranian revolution is like no other. It is *sui generis*. Hence, the difficulty of understanding its day-to-day evolution.

II

The rise of the Shi'ite clergy is rooted in the history and ideology of Shi'ism. Ever since the disappearance of the Twelfth Imam in the ninth century, no temporal power has been legitimate or equitable in its eyes. Justice will not reign in the Muslim community until the "great occultation" (as the Twelfth Imam's disappearance is termed) ends with his return to earth (the resurrection).[1] Every ruler—or ruling party—is by definition a usurper, since by the very act of ruling he is substituting himself for the Twelfth Imam, the only one with the right to execute the divine will. The Shi'ite clergy's natural tendency has thus been to contest the authority and actions of the various dynasties that have governed Persia, and especially to contest the royal power every time it tried to open the country to foreign

1. It is interesting to note that when the *ulemas* (doctors of Islamic law) demanded at the beginning of this century the election of a parliament, they called it the "house of justice" *(adalatkhaneh)*.

influences apt to "pervert" Islam or introduce customs contrary to Muslim culture and tradition.

Thus, by the beginning of the nineteenth century, Shi'ism emerged as a kind of early anti-imperialist movement. In 1826, the *ulemas* declared a holy war against Russia. Three years later they had the members of an official delegation from St. Petersburg assassinated. They brought about the cancellation of the incredible monopoly for the exploitation of mines, forests, railroads, banks, customs and telegraphic communications granted to Baron Julius de Reuter in 1872. Their 1891 prohibition on tobacco consumption—largely observed by the population—led to the withdrawal of the tobacco monopoly accorded the previous year to a certain Mr. Talbot. Part of the clergy actively participated in the 1906 revolution aimed at establishing a constitutional regime. They did so not in the name of democracy—a "Western" notion abhorred even then—but to better control a royal power which favored European penetration. It was for the same reason that an important segment of the clergy under Ayatollah Kashani sided with Prime Minister Mohammed Mossadeq in 1951 when the latter nationalized the Anglo-Iranian Oil Company's interests in Iran.

The three favorite themes of the militant clergy—foreign domination, despotism, injustice—were precisely the evils suffered by the Iranian people under the reign of Mohammed Reza Shah. The Shah's agrarian reform benefited only a minority of the peasants, who, in any case, were soon taken over by big companies engaged in large-scale industrialized agriculture. The massive importing of agricultural goods, especially wheat from the United States—coupled with the absence or inadequacy of protective tariffs—contributed to the ruin of countless small farmers, aggravated rural unemployment and swelled the migration to the cities.

The Shah's modernization program—which created less an authentic development than a consumer society for privileged elites—quickly enriched the members of the royal family and the court, the entrepreneurs (almost all subcontractors for the large Western firms), the powerful merchants, the importers of spare parts and consumer goods, the speculators fostered by an unbridled capitalism worthy of the nineteenth century. On the other hand, those who suffered were legion: the small manufacturers and craftsmen squeezed by foreign competition, the workers (albeit well paid), the rapidly expanding middle classes, the millions of wage earners whose buying power was being eroded by a galloping inflation (over 50 percent in the two years preceding the fall of the monarchy).

The recession, which hit Iran as of 1976, increased the regime's unpopularity. Slumping oil sales and rising costs in imported materials forced the Shah to reduce considerably the credits allocated to development, giving rise to disillusionment commensurate with the grandiose hopes elicited by the oil boom of 1973–74. The austerity measures adopted seemed all the more unjustified in that the Shah continued to sink billions of dollars into useless military hardware, mainly from the United States, which piled up in his arsenals.

The middle classes wanted a constitutional system which would assure them the material security and political stability threatened by the arbitrary nature of the regime. The entire population demanded an end to the terror perpetrated by SAVAK, the formidable secret police. During the 37 years of Mohammed Reza Shah's reign, over a half million people are estimated to have been arrested, imprisoned or detained, briefly or for longer periods. Thousands of opponents or suspected opponents were packed off to the special courts; thousands were the victims of summary executions or assassinations, or died under the systematic practice of torture.

Most Iranians readily equated the regime which oppressed them with the United States, accused of having restored Mohammed Reza Shah to his throne in 1953 and keeping him there through the many forms of aid—political, economic, military and police—it supplied. It was notorious that the CIA worked closely with SAVAK, and that the Pentagon helped equip and train the imperial forces and advised them on a regular basis. American firms "pumped" the petrodollars out of Iran in exchange for needless armaments, industrial products and consumer goods. In the eyes of many Iranians, these "sales" were just another means of looting their country's resources.

III

In light of the above, one can more readily understand the dual nature of the uprising of 1978–79—dual in that it was directed as much against "American imperialism" as against the despotism of the Shah, considered two sides of the same coin. Largely spontaneous, the movement could very well have been taken in hand by Marxists or nationalists of the Mossadeq stamp. But the Shah's repression had pulverized the secular parties, traditional and otherwise. The National Front which came to prominence during the revolution was nothing but a collection of harmless notables—self-proclaimed heirs of Mossadeq who had somehow lost his prestige along the way. The leftist parties, such as the People's Mujahidin (progressive Muslims) or the People's Fedayeen (Marxist-Leninist), had concentrated too much on urban guerrilla activities and suffered too many human losses to be able to play a political role of far-reaching consequence. The Shi'ite clergy was thus a timely force, offering—in addition to its mobilizing ideology—the leadership and structure which was to assure the success of an enterprise that would otherwise have been doomed to failure.

The role of the clergy in Iranian society is considerable. Far more than the Sunni sheikh who is generally appointed and paid by the state, the mullah draws his livelihood from contributions of the faithful, with whom he shares prosperity or poverty, joys and sorrows. In the mosque or at the traditional gatherings he attends in private homes, virtually any issue of concern to the community is discussed; religion and politics are inseparable in Islam as in Judaism. The mullah can thus be a friend, confidant, adviser or guide for his flock, and acts as a moral support in times of adversity.

Using the sometimes considerable funds it collected (up to a fifth of the revenues of the faithful), the clergy established medical clinics for the needy, schools, and other social and philanthropic institutions during the Shah's reign. The preachers spoke out against social injustice, moral decay and corruption (understood to refer to that of the palace and its American advisers). They called for Islam's return to its roots as a way of defending national virtue and identity against the rape by Western technology. A good many men of the cloth experienced the Shah's prisons. Among the higher clergy, Ayatollah Taleghani, who died a few years ago, Ayatollah Mountazeri, considered to be the successor to Imam Khomeini, and Ayatollah Muhtazeri Rafsandjani, elected President of the parliament in July, suffered torture at the hands of SAVAK agents.

Imam Khomeini, who found refuge in the holy city of Najaf in Iraq, remained the distant symbol of this diffuse resistance. The political vacuum made him successively the standard bearer, catalyst, and finally the guide of the revolution. As a result of the positions he took during his 15-year exile, he appeared to embody the aspirations of various categories of the population. Shortly before his expulsion from Iran in 1964, he had become famous by publicly condemning both the Shah's violations of the Constitution and the granting of extraterritoriality to American civilian and military advisers and their dependents. "You have extirpated the very roots of our independence," he wrote from Najaf in 1967 to Prime Minister Hoveida.

This text, like so many others, was recorded and circulated clandestinely throughout city and countryside on tapes. In them, the Imam designated the United States as "the head of the imperialist serpent," accused the international monopolies of "looting" Iran's resources, complained of "domination" by foreign capital and the squandering of public funds in massive arms purchases. During the sumptuous festivities at Persepolis in 1971 marking the 2500th anniversary of the monarchy, he denounced the "insolent luxury" in which the Shah and his family lived. From the outbreak of the uprising in January of 1978, he incited the people to pursue their struggle until the fall of Mohammed Reza Shah. Disregarding many of his advisers, he insisted that the battle be waged without recourse to arms, and never tired of repeating that the Shi'ite faith would prevail over brute force. His calculation proved well founded: a year later, the imperial army—the most powerful in the Middle East after that of Israel—succumbed to "bare-handed revolutionaries" who had sacrificed tens of thousands of martyrs while achieving their victory.

IV

The second phase of the revolution opened shortly after Imam Khomeini's return to Iran in February 1979. It marked the beginning of the "anarchy," as the West called it—that classic phenomenon wherein the "sacred union" achieved around a common goal (in this case, the overthrow of the monarchy) explodes under the pressure of the centrifugal forces reflecting the interests, aspirations and political leanings of diverse groups of the population. Life more or less returned to normal. The ethnic minorities of the Shah's empire—Kurds, Arabs, Baluchis, Turkomans and so on—were all the more pressing in their demands for self-determination in that most of them did not belong to the Shi'ite community. The Marxists—orthodox communists, Maoists, Trotskyites, and so on, who had marched in the demonstrations under the green banner of Islam while chanting the name of Allah—now set up shop on their own under the insignia of the hammer and sickle. The middle classes formed in the Western school raised the standard of secularism, advocating the maintenance of a liberal (social democratic) economy and the establishment of a parliamentary system based on the Western model. The Muslims themselves were divided among rightist, centrist and leftist factions, each under the patronage of an ayatollah of like persuasion. Finally, despite its apparent allegiance to Imam Khomeini, the Shi'ite clergy was just as divided as the secular political world which indeed it more or less faithfully reflected.

It has often been written that Imam Khomeini is an intransigent man who, by ideological conviction as well as by temperament, brooks no half-measures. Hadn't he rejected all offers of compromise with the Shah, despite the insistence of many of his followers? And again, hadn't he refused, in the immediate wake of the monarchy's collapse, the proposals for a government of national unity which would surely have reduced the futile and costly disorders which ensued? But Imam Khomeini does not identify himself with Iran as General de Gaulle did with France. As the head of Shi'ite Islam, he could not and cannot conclude "tactical alliances" with forces contesting or—worse still—opposing the dogmas he is duty-bound to defend.

This does not alter the fact that the "guide of the Islamic Revolution" has shown himself as capable of maneuvers or even ruses as any secular politician when he considers such indirect means necessary to achieve critical objectives. The appointment of Mehdi Bazargan as head of the provisional government on February 5, 1979, was perhaps one of these instances. Otherwise, why would Imam Khomeini have chosen a man whose ideas were so far removed from his own? It is true that Bazargan had been imprisoned four times for opposition to the Shah and that he had supported Khomeini in exile. But by his own admission he was no revolutionary. He had gone to Imam Khomeini's retreat at Neauphle-le-Chateau to plead the case of a "step-by-step" policy, which implied the maintenance of the monarchy. In short, he advocated a path similar to that taken by Shahpur Bakhtiar, a friend whom he never wanted to condemn as a "counter-revolutionary" or "traitor."

The government put together by Mr. Bazargan was strongly tinged with conservatism. His ministers belonged to the moderate wing of the National Front and included the Mossadeqists so despised by Khomeini, former officers of the imperial army, and middle-class businessmen who had grown rich under the Shah. As of his first days in office, he publicly expressed his disgust with the institutions born of the revolution: the "Komitehs," the Islamic militias, the revolutionary courts dispensing high-handed justice on royalist dignitaries and SAVAK torturers. The summary trials and hasty executions repelled him. Mehdi Bazargan was a man of order: he wanted to conserve the instruments of the old regime—army, police, state administration—keeping purges to a minimum so as not to lose "estimable and irreplaceable elements." A worthy representative of the "bazaar"—a term widely used to designate the merchants and middle classes—the head of the provisional government also did not intend to alter the bases of the economy: it was under some pressure that he decreed the nationalization of the banks, insurance companies and large industrial complexes. He also tried, unsuccessfully, to resist the "unauthorized" occupation of the great agricultural estates by landless peasants and the election of "workers' councils" to take charge of the factories.

Mehdi Bazargan is also a pious and practicing Muslim, but he never shared Imam Khomeini's view that "Islam should take precedence over Iran." He complained of the insidious infiltration of "ignorant and arrogant" mullahs into the administration. He voted in favor of the "Islamic Republic," but shortly before his resignation he confided to the Italian journalist Oriana Fallaci that he feared a "dictatorship of the clergy."[2]

Two "mortal sins" precipitated Bazargan's fall: he opposed the Islamic constitution drawn up by the "assembly of experts" (composed almost entirely of clergymen), and he strove to normalize relations with the United States, Imam Khomeini's "Great Satan." It was on the Imam's orders that he reluctantly broke off diplomatic ties with Egypt after the peace treaty was signed with Israel in March 1979. He barely protested when the Shah was admitted to a New York hospital on October 22 and did not consider it useful to publicly demand the former monarch's extradition. A week later, he was shown on Iranian television engaged in cordial conversation in Algiers with Carter's National Security Adviser Zbigniew Brzezinski in order to obtain, among other things, a resumption in the supply of matériel and spare parts for the Iranian army. That was the reason—or the pretext—invoked for the occupation of the American Embassy and the seizure of its

2. "Everybody Wants To Be Boss," *The New York Times*, October 28, 1979, p. 66.

diplomats four days later by some four hundred "Islamic students following the Imam's line."

V

Imam Khomeini—who apparently had been informed that some such action was to be taken—had thus forced Mr. Bazargan to resign. Why had he waited nine months to do so? Hadn't he known from the very beginning that his prime minister was neither a "revolutionary" nor an enemy of the "Great Satan"?

Everything seems to indicate that the choice of Mr. Bazargan as prime minister was dictated by tactical considerations at a time when the Imam, recently returned to Iran, feared that the situation could slip from his control. The imperial army was still intact (and was not to collapse until February 12, one week after the provisional government was formed), as were the police force, the gendarmerie, the state administration. The Marxist parties (particularly the People's Fedayeen) and the People's Mujahidin were armed to the teeth and sought to outflank the Islamic movement, which was powerful at the popular level but completely lacking in the structures needed to exercise power (the Komitehs and Islamic militias were institutionalized only later). The middle classes, influential in the economy and in the government machinery, could easily slide into a dangerous passivity or seesaw among the moderate parties of the National Front.

Only Mehdi Bazargan seemed capable of juggling all these perils. The "bazaar"—Muslim and relatively conservative—was devoted to him. The leftist formations (especially the Mujahidin) thought they could count on his liberalism to secure a legal status in the young republic. A number of high-ranking officers—some of whom had secretly negotiated a compromise with him—had confidence that he could assure the army a smooth transition from empire to republic. In appointing him prime minister, Imam Khomeini thus hoped to buy time, the time needed to establish a "true revolutionary government." But this did not mean he gave him any real power. Mr. Bazargan was permitted to make endless declarations and protests over radio and television, but the essential decisions came from the clergy-dominated Council of the Revolution. It was Mr. Bazargan himself who, in the interview with Oriana Fallaci, best described his situation: "They've put a knife in my hand, but it's a knife with only a handle; others are holding the blade."

The seizure of American diplomats as hostages in Tehran sounded the death knell of the Bazargan government. It also brought into the open the conflict between Iran and the United States, which had lain dormant since the collapse of the monarchy. Imam Khomeini had not forgiven the seven Presidents who occupied the White House during the past 30 years for their unfailing support of the bloodthirsty dictator most Iranians saw in Mohammed Reza Pahlavi. Nor was the Imam prepared to forget that Mr. Carter had tried to save the Pahlavi dynasty to the bitter end. The ex-Shah's admission to the United States in October 1979 confirmed him in his suspicions that the American Administration was seeking to reestablish the fallen monarch on his throne. The American gesture, praised as "humanitarian" and "courageous" in the West, was perceived in Tehran as a new and intolerable provocation. As former Foreign Minister Sadegh Ghotbzadeh remarked to this writer: "It is as if Franco's Spain had offered to treat Hitler for cancer immediately after the Second World War." The only merit of the image is that it reflects the frustration and anger felt by the Iranians. The United States had seriously underestimated their potential reaction, just as it had seriously underestimated the profound motivations of the revolution itself.

In the view of certain observers, the Carter Administration committed the same error before and after the hostage crisis: it believed it could clear up the Iranian-American dispute by dealing with the "moderate" leadership in Tehran—first with Mr. Bazargan and his ministers, then with President Bani-Sadr and Mr. Ghotbzadeh. Washington believed this to be the most accessible route, but in reality it was blocked. It is true that the new chief of state and his foreign minister disapproved of the Islamic students' occupation of the American Embassy and favored normalization with the United States. But it was precisely these inclinations that undermined their credit with Imam Khomeini, at the same time weakening them in the face of their adversaries, particularly within the clergy.

In the last analysis, the American government could have played the "moderate" card successfully if it had been prepared to make enough concessions to give the moderates credibility in the eyes of Iranian public opinion. President Bani-Sadr had dropped his demand for extradition of the Shah. In exchange, he asked that the United States admit its responsibility in the crimes committed by the Shah and that it undertake not to interfere in Iran's internal affairs again. It is not to be excluded that Bani-Sadr could have delivered had his demands been met. At the very least, the climate would have changed significantly enough to open the way for the release of the hostages. But the White House did not understand that a price had to be paid for normalization. President Carter wouldn't even dissociate himself from the CIA-led coup of 1953 which restored Mohammed Reza Shah to power. If it is true that a great power cannot permit itself to be "humiliated" in such a fashion, one can ask oneself how a compromise could have been concluded without satisfying at least the minimal demands of the other side. Surely no one expected to obtain the hostages' release with nothing given in return.

It was clear from the outset that Imam Khomeini was going to utilize the widespread anti-American feelings to mobilize the population under his banner. Popular discontent—due mainly to economic problems and the duality of power—had reached alarming proportions by October 1979. The government (along with the secular nationalists and most of the leftist parties) was preparing to resist the adoption of the draft Islamic constitution then being drawn up. The occupation of the American Embassy thus came at the perfect time, pushing divisive factors to the background and uniting the people against the Shah's protectors. The same mechanism applied after the rescue mission of April 25, 1980: the U.S. attempt provided Khomeini with tangible evidence that the principal danger threatening the Iranian people was foreign intervention.

VI

The third phase of the revolution—following those of the uprising against the imperial regime and the founding of the Islamic Republic under a provisional government—was opened by Mr. Bazargan's resignation. Immediately after the seizure of the hostages, Imam Khomeini labeled the new phase the "second revolution." The patriarch of Qum had clearly decided that his faithful prime minister had outgrown his usefulness: the imperial army, decapitated and purged by the Islamic Komitehs, no longer represented a danger for the republic; the large industrialists who had controlled the economic power had been dislodged from their positions by various measures, including massive nationalizations. All that remained was to neutralize the representatives of the middle classes, the "moderate nationalists" who had been reviled publicly as "Westernized liberals" ready to "compromise with imperialism."

The task of ferreting out this new enemy fell to the Islamic students. They were doubly popular in the country at the time: they had defied all-powerful America—a role magnified by the importance conferred upon them by the American mass media and Administration—and they projected themselves as being at once the conscience and the instrument of the revolutionary movement. They were thus in a position to proceed, under the discreet protection of Imam Khomeini, with the elimination of the opponents of the Islamic regime.

Using more or less convincing documents—but impressive insofar as they came from secret archives of the U.S. Embassy in Tehran—the Islamic students succeeded in having numerous politicians or parties who passed for pro-Western in the Iranian Revolutionary context arrested or discredited. Mr. Amir Entezam, successively Deputy Premier and Ambassador to Stockholm under Bazargan, was detained as a "CIA agent." Accused of the same infamy, Hassan Nazih, former director of National Iranian Oil Company, and Rahmatollah Moghadam-Maraghi, head of the small Radical Party, both of whom had supported if not actually fomented the December 1979 rebellion in the Azerbaijani capital of Tabriz, were obliged to flee the country. Ayatollah Shariatmadari, a rightist liberal in whose name the Azerbaijani sedition had erupted, was reduced to silence after "compromising documents" were mysteriously circulated in Tehran. The Muslim People's Republican Party, led by Shariatmadari's son, was obliged to close its doors after dozens of its members were executed or arrested. A number of elected deputies lost their parliamentary seats, again thanks to documents from the U.S. Embassy, making it possible to charge them with "intelligence with the enemy." Men such as Khosrow Kashgai, chief of the powerful tribe of the same name, and Admiral Ahmed Madani, a moderate who won more than two million votes in the first presidential elections, were driven from the political scene. The list of the Islamic students' victims, too long to enumerate, testifies to their decisive role in the "second revolution." It also demonstrated the futility of Mr. Carter's efforts to end the Embassy occupation without being prepared to make Imam Khomeini any substantial offer.

The hostage affair rapidly became the principal cleavage separating Right from Left in domestic politics: all those who supported the Islamic students were in the good camp, on the side of anti-imperialism and the revolution, all the others could safely be labeled pro-Western counter-revolutionaries. In light of this, it is hardly surprising if few dared publicly to take a stand against the Embassy's occupation. The People's Mujahidin would have preferred the alternative of officially repudiating the some 900 agreements and contracts it claimed still tied Iran to the United States. Still, it saluted the November 4, 1979 event, although it later came to criticize the behavior of the Islamic students. The People's Fedayeen, who had ridiculed the superficial nature of Imam Khomeini's anti-imperialism in the early months, threw its entire weight behind the Embassy's occupiers after a few much remarked-upon hesitations. The moderate parties and public figures for the most part remained silent or adopted an attitude sufficiently ambiguous so as not to attract accusations of defending the United States. Only Mr. Bani-Sadr had the courage—or the imprudence—to state, as of November 6, that he disapproved of the seizure of hostages both on Islamic moral grounds and in the interests of the revolution. He lost no occasion to reiterate this opposition, and his relations with Imam Khomeini steadily deteriorated.

Paradoxically, Bani-Sadr—who had so vigorously fought Bazargan as head of the provisional government—steered himself on a course strongly resembling the one which led the former premier to his ruin. He thus laid himself wide open to his opponents, mainly the Islamic Republican Party and, above all, its President, Ayatollah Beheshti, who wasted no time relegating him to the dishonorable "liberal" camp. Predictably, they criticized him for his conciliatory attitude toward the United States, but also for his "illusions" concerning the possibility of close cooperation between Iran on the one hand and Europe and Japan on the other. Even his insistence on denouncing the Soviet intervention in Afghanistan was considered suspect, despite the fact that it was in keeping with the "neither East nor West" dogma of the militant clergy. In so doing, wasn't he trying to justify an indispensable normalization with the West in order to better cope with the peril from the East?

At the domestic level, Bani-Sadr turned out to be as much a man of order as Bazargan. Immediately after his election as President of the Republic on January 25, 1979, he announced his intention of dissolving the Islamic committees and the "guardians of the revolution" (the militias) as soon as he had reorganized the armed forces, gendarmerie and the police. Two weeks later, he obtained the release of Mr. Minashi, the Minister of National Guidance who had been arrested as a CIA agent at the request of the Islamic students. Bani-Sadr later tried to protect other moderate figures such as Admiral Ahmed Madani and Khosrow Kashgai, but failed. Opposed to political violence on principle, Bani-Sadr declared on July 27: "Denunciations, slander, torture, violence, massacres, prisons, are nothing more than manifestations of a Stalinist society."

A good Muslim but marked by French culture, as was Mr. Bazargan, Bani-Sadr entertained mixed feelings about the Shi'ite clergy. At the "assembly of experts" formed to draw up the new constitution, he surprised friends and enemies by vigorously protesting against the wide powers proposed for the fagih (the religious guide of the state, presently Imam Khomeini). Despite his denials, he was later accused of having abstained from the vote on the constitutional article dealing with the fagih's powers, even though these were reduced compared to those provided for in the initial draft.

Whatever the case, Bani-Sadr clearly took a stand for the separation of powers and the non-interference of the clergy in affairs of state, to the point of deriding "the Richelieus and Mazarins who crowd the Iranian political scene." Just after his election to the presidency, he told this writer that he owed his "victory to the people," before adding that he thanked "the lower clergy for its support." The higher clergy, for him, is that which supports the Islamic Republic Party of Ayatollah Beheshti, his bitter enemy.

On the morrow of his election, Bani-Sadr proclaimed Ayatollah Beheshti "politically dead." His optimism did not seem unfounded at the time. Ayatollah Beheshti had just suffered three important setbacks: he had wanted to be a candidate in the presidential elections, but Imam Khomeini had forbidden religious leaders to seek this office. Next, he had put forward the candidacy of Jalaleddine Farsi, but Farsi was disqualified on account of his Afghan origins. Finally, he had backed the candidacy of Hassan Habibi, who got a mere ten percent of the vote against Bani-Sadr's 70 percent.

So when Bani-Sadr took office in January he thought he had won the final round. He asked Imam Khomeini's permission to form a second provisional government to last until the election of parliament. Such a measure would have enabled him to edge out the Revolutionary Council—principally made up of Ayatollah Beheshti's friends—and to carry on with the political and economic reforms of his choice. But Imam Khomeini refused to grant his request, and Bani-Sadr had to be satisfied with the limited "privilege" of presiding over the Revolutionary Council's meetings. Ayatollah Beheshti used his dominant position within the Council to thwart all of Bani-Sadr's initiatives and to build his own power base. He succeeded in getting the majority of the parliamentary seats attributed to his party in March 1980. Next, he got himself appointed head of the supreme court. Having thus wrested control of the judiciary and the legislature, he set out to capture the executive power. As a means of stripping the President of the last vestiges of his authority, reducing him to a mere figurehead, he demanded that the President's nominee for prime minister receive the prior endorsement of his party.

President Bani-Sadr resisted at every step. He contested the results of the legislative elections, but Imam Khomeini judged them "fair." He opposed new purges Beheshti's party called for in the administration and the various branches of the security forces, but he was obliged to yield when the Imam launched his appeal for a "cultural revolution" aimed at ridding the state of "counterrevolutionaries" and "Westernized liberals." In any case, the radicalization of the Islamic Republic had become inevitable after the American rescue mission on April 25, which highlighted the passivity or, as some claim, the complicity of the armed forces. In the six weeks following the American expedition, seven distinct plots were uncovered and several hundred officers arrested. (Ed. note. The hostages were finally released on 20 January 1981, only hours before President Carter left office. Abol-Hásan Bani-Sadr was dismissed as Iran's president and went into exile in mid-June 1981, only a few days before a bomb killed Ayatollah Beheshti and 71 other leading members of the Islamic Republican Party.)

VII

The dangers threatening the Islamic Republic by the summer of 1980 were real enough. So was the overall deterioration of the political, economic and social situation. This journalist, in chance encounters with all strata of society—well-off and poor, urban and rural—heard nothing but complaints. The new agrarian reform decreed the preceding winter had still not been applied in many regions of the country. Compared to the year preceding the revolution, state investments in agriculture had dropped by 20 percent, those in industry by 50 percent. Deprived of capital, spare parts (a consequence of the Western embargo) and competent managers, the factories of the public sector were operating on the average at less than half their capacity and for the most part were showing serious deficits.

The precise number of unemployed is unknown, but estimates vary between two and four million out of an active population of 11.5 million. The state pays a modest subsistence allowance to 800,000 of them, the others living by their wits or on the resources of their families. The inflation rate, officially estimated at 25 to 30 percent, in fact is believed to approach the level which set the middle classes against the Shah: 50 percent or more for the well-off housewife's consumer basket. Rents, which had dropped considerably after the fall of the monarchy, are almost as high today as they were in 1977–78. Prices of certain food items such as fresh vegetables, flour, meat and fruit have soared to unprecedented levels.

The middle classes suffer more than others from the rising cost of living because of their consumption patterns. Wage earners, like government employees (1.4 million, including members of the armed forces), have seen their buying power seriously reduced not only by inflation but also by the halving of higher salaries and the elimination of overtime and various benefits—measures which were adopted with an eye both to fairness and economy. Indeed, the state has been running at a deficit since the reduction of oil exports. The total revenues of the government, $21 billion in 1980, were not even sufficient to pay the civil servants. Unable to sell more oil, the government was obliged to take out loans or dip into its reserves, reduced to some $7 billion since the Iranian assets in American banks (about $8 billion) were frozen.

The economic situation has thus sharply deteriorated since the fall of the Bazargan government, despite the instructions Imam Khomeini issued immediately afterwards, enjoining the authorities to "assure the well-being of the *mostazefin*" (the disinherited). The Imam refrained from appointing a new government and thus gave free rein to the Council of the Revolution and the high dignitaries of the clergy who had his confidence. It was at that point that mullahs were appointed everywhere—in the armed forces, the police, the administration, in large industrial complexes and in almost all the ministries—to supervise and check up on the secular authorities. Anarchy grew as a result, and, as one highly placed official remarked to this writer, "no one was obeying anyone anymore."

The state administration was virtually paralyzed by the inertia—voluntary or otherwise—of civil servants, many of whom no doubt consciously sought to undermine the regime. Strikes multiplied in state factories and public services. Daily cuts in electricity and running water antagonized the inhabitants of Tehran and other cities. The *pasdarans* (Islamic militias), loyal to rival decision centers, most often acted as they pleased. Certain army units refused to fight the Kurdish insurgents. During the summer the state was in an advanced state of decomposition while a series of military coups were uncovered, with ramifications in the political arena. Most serious of all was the possibility that popular discontent would be directed at the clergy, who would be held responsible for the deterioration of the situation.

Against this background Imam Khomeini's declaration of July 21, 1980, is more readily understood: "None of the present ministers is a revolutionary . . . and if the next government resembles this one, we can give the Islamic Republic up for lost, for we will be beaten." One also understands the significance of the massive purges, the arrests, the collective executions, the attacks against the offices of political parties, both Right and Left—those very parties that would benefit from the collapse of the Islamic Republic.

The moderate nationalists, principal targets of the "cultural revolution" unleashed in July, are considered the most dangerous. Indeed, their influence is pervasive in the economic centers, the state administration, the armed forces, the universities—in short, those areas traditionally dominated by the middle and upper middle classes, by the "Westernized" elites hostile to clerical power and socioeconomic upheavals. Their discontent—diffuse within the country because in the circumstances there is no figure to voice it—has been given focus in opposition groups abroad. Two daily broadcasts, one in the name of General Oveissi, the other in the name of Shahpur Bakhtiar, are beamed into Iran from "clandestine" radios in Iraqi territory and are very widely listened to by the population. It is no secret that Oveissi and Bakhtiar, among others, maintain close relations with the tribes and with civil and military groups—groups which, given more favorable circumstances, could possibly bring down the regime.

Despite the significant strides made by the Left these past months, its prospects are not bright—at least not in the foreseeable future. Unlike the Right, its social base is relatively limited. The Islamic movement, symbolized in the person of Imam Khomeini, has not lost the support of the less favored classes despite the disappointment and frustrations accumulating since the founding of the republic. The workers have not forgotten that their salaries in most cases have doubled since the revolution. Peasants receive generous subsidies—an average of 50 percent higher than they received under the Shah. Those who do not yet own land have not lost hope of benefiting from future agrarian reforms. As a result of the abundant rainfalls in the winter and spring of 1979–1980, the harvests were expected to be 20 to 30 percent greater. Above all, the *mostazefin* now feel they are living in "their own" republic, protected by Imam Khomeini.

Nonetheless, the People's Mujahidin, a Muslim movement with a very radical ideology, managed within a few months to become a mass party, particularly feared by Imam Khomeini, to judge from his denunciations of "Islamo-Marxists." In June 1980, the head of the movement, Massud Radjavi, was able to assemble 150,000 followers and sympathizers in a public meeting which had not been announced by the mass media. The prestige of the Mujahidin is high, both because of the armed struggle it waged against the Shah, and because of the positions—resolutely favorable to the poorest class—it has adopted ever since the birth of the Republic.

Nonetheless, the youth and inexperience of its cadres, coupled with the political errors they have committed, limit their possibilities of action, at least for the time being.

The Tudeh (communist) Party is smaller and has a much more limited following, but its influence has on several occasions proved far more decisive. Its cadres, generally trained in Eastern Europe, are of high quality. Perfectly informed about the international situation, capable of clearly formulating strategy and tactics and of applying them with unwavering discipline, they have been able to take advantage of the anti-American tide that has engulfed Iran in order to make friends and allies, even within the clergy. It has been noted that Imam Khomeini has never attacked the Tudeh by name—at least not as of mid-August 1980. When he denounces the communists, whom he sometimes calls the "American Left," he is especially aiming at the People's Fedayeen (Marxist-Leninist) which took up arms against the authorities in Kurdistan and elsewhere. His seeming indulgence for the orthodox communists can be explained by at least three facts: they have consistently supported him ever since his departure for exile in 1964 (their "clandestine" radio, probably located in East Germany, regularly broadcast his declarations recorded in Iraq); they are determined, like him, to "extirpate the very roots of American imperialism" in Iran; they have offered their help in defending and consolidating the Islamic Republic.

Two events occurring in rapid succession in June 1980 increased the Tudeh's margin of maneuver: the People's Fedayeen and the Kurdish Democratic Party (KDP) were both shaken by internal dissent. After long and arduous debates, the majority of the Fedayeen's leadership decided to adopt an "anti-imperialist" political line practically identical to Tudeh's. The same took place within the KDP, triggering the schism of the minority faction, which accused the majority (led by Abdel Rahman Ghassemlou) of cooperating with Iraq and NATO. The importance of these divisions is that they strengthened the Tudeh's authority and hence its ability to come to the rescue of a tottering republic.

Imam Khomeini certainly does not wish to be dependent on any non-Islamic group, particularly a Marxist one. His behavior from the very beginning of his political struggle in the 1960s shows his intention of reserving the monopoly on power for the "authentic Muslims." Whether or not he will be able to achieve this goal remains to be seen. The Islamic Republican Party, on which he seems to want to lean, is not a structured mass party, but rather a collection of various religious and secular individuals and groups with diverging if not actually contradictory tendencies. Ayatollah Beheshti's party has consequently been unable to come up with a political, economic and social program, which is indispensable for any government party. More important, it cannot count on the support of a sufficient number of cadres to administer the state. The Iranian intelligentsia, refractory or hostile to the Islamic regime from its very outset, is split between the liberal Right and the Left, Marxist and otherwise.

VIII

Given Imam Khomeini's apparent determination not to share power with either of these two currents, the situation seems to be heading toward an impasse. A coup d'état in the immediate future seems doomed to failure. The high echelons of the army are divided, and the rank and file remains—until proof to the contrary—loyal to the Imam.

The question of how Imam Khomeini's eventual disappearance from politics would affect the Iranian situation has often been raised. At the very least, the response to this question is risky. The plethora of forces on the political scene, each one obeying its own internal and evolving dynamic, makes it impossible to hazard any valid prognosis for the medium or long term. The most that can be done is to evaluate the balance of forces such as it exists today and could remain for the near future.

As of mid-August 1980, three phenomena could be noted: the decline in the prestige of the clergy as an instrument of government; the strengthening position of the leftist parties as a whole, but nowhere near the point of offering a credible alternative to Islamic rule; and the persistent influence of the moderate nationalists and the Right—liberal or otherwise—in the state institutions and economic centers of decision. This last category would thus appear the only one with a serious chance of taking control in the event of Khomeini's disappearance, especially since in all probability it would be supported by a good part of the army and a not inconsiderable part of the clergy. It should not be forgotten that the other five "grand ayatollahs" besides Khomeini all either disapprove of or are hostile to the Imam's religious and political concepts.

Yet for all that, a victory of the Right would by no means be assured. A great deal would depend on its own actions. If it follows Khomeini's "anti-imperialist line," satisfies the demands of the minorities, and guarantees the continued existence of the leftist parties, its chances of success would be increased. Otherwise, the obstacles in its path would probably be insurmountable. The Muslim and Marxist Left, supported by segments of the ethnic minorities and part of the clergy, is strong enough to keep the country in a state of anarchy comparable to that prevailing at present.

The role of the U.S.S.R. could weigh heavily in this regard. Up to the present, Moscow has treated Khomeini's republic with kid gloves, generally restraining itself from responding to the criticisms, attacks and threats showered upon it, particularly concerning Afghanistan. The Kremlin's tactic seems to have been to wait for the end of the power struggle in Iran while hoping to reap the fruits of America's errors. If a definitely unfriendly—i.e., pro-American—government took over in Tehran, undoubtedly the Soviet Union would no longer hesitate to support all the forces hostile to the new rulers. The major leftist movements—the Mujahidin, Fedayeen and Tudeh—already favor a tactical or strategic alliance with the socialist camp. Moreover, pro-Soviet sympathies are far from negligible in the provinces of Kurdistan, Gilan, Mazanderan and even Azerbaijan. In the light of all this, the civil war which would be triggered by the installation of any Pinochet-type regime would undoubtedly be long and, perhaps, even dangerous to world peace.

Thus, the prospects with or without Khomeini seem bleak indeed. The wisest course for the international community is surely patience, and most especially nonintervention in a country still in the throes of revolution. All things considered, the lesser of all evils for the Western powers is surely to leave to the Iranians themselves the freedom to determine their own future.

The Politics of Extremism in Iran

JAMES A. BILL

Three years after the revolutionary overthrow of the regime of Muhammad Reza Shah Pahlavi, Iran found itself fighting to survive as a coherent nation-state. In 1981, the country had three different Presidents and four different Prime Ministers when the post-revolutionary fissures in the body politic deepened and an undeclared civil war broke out in the form of a contest between regime repression and opposition by terrorism. With the destruction of the government of President Abol Hassan Bani Sadr in June, much of the legitimacy was stripped from the political process in formation. Ayatollah Ruhollah Khomeini began to lose control and indeed contributed to the violence and bloodshed in the wake of Bani Sadr's downfall. The shocking bomb incident of June 28, 1981, represented the end of the first stage of the Iranian revolution and introduced a time of terror. In the bombing, 74 members of the political elite were assassinated, including the founder of the Islamic Republican party (IRP), Ayatollah Muhammad Hussein Beheshti, four Cabinet ministers, six deputy ministers, and 27 majlis (parliamentary) deputies.

In retaliation for the massacre, the IRP initiated a campaign of repression that resulted in over 2,000 executions and many thousands of arrests. This, in turn, provoked the opposition groups to a policy of terror that had as its targets the most powerful leaders of the IRP itself. The crippling campaign of violence resulted in the unprecedented assassination of both the Head of Government and the Head of State on August 30, when Prime Minister Muhammad Javad Bahonar and President Muhammad Ali Rajai died together in a bombing of the heavily guarded prime ministry. Many other extremist leaders died violent deaths in 1981, including influential majlis deputy Hassan Ayat (August 5), prosecutor-general Ayatollah Ali

Reprinted by permission of the publisher, Current History, Inc., from *Current History,* Vol. 81, No. 471, January 1982, pp. 9-13, 36.

Qodusi (September 5), and Khomeini's personal representative in Tabriz, Ayatollah Asadollah Madani (September 11). The IRP, through its coercive arm, the Revolutionary Guards (*Pasdaran*), reacted by intensifying its own campaign of repression and execution. The last half of 1981 found Iran trapped in a vicious and spiraling cycle of violence.

The social and political fabric of post-Pahlavi Iran has continued to unravel in the face of government factionalism, personal rivalry, ethnic cleavages, religious fanaticism, ideological confrontation, and economic malaise. Externally, Iran's debilitating war with Iraq moved into its second year, sapping Iran of badly needed human and economic resources. (This war helped deter the Iranian military from intervening directly in the chaotic domestic political situation.) The release of the American diplomatic hostages on January 20 did little to temper internal hostilities as new debates developed about the terms of the agreement. The United States was continually held responsible for failures and upheaval in Iran while the Soviet Union was also deeply distrusted and often condemned. In both its internal and external policies, the Iranian political elite pursued programs that can best be described as incoherent and counterproductive.

To students of the history of revolution and to analysts of social forces and class conflict, the contemporary situation in Iran is not unexpected. The failure of the Pahlavi system to build political institutions and its promotion of social cynicism and factionalism provided a legacy that ensured chaos and violence. The character and ethnicity of the Iranian people, the mind-set of certain leaders of Shi'i Islam, the violence of the revolution itself and the untimely actions of the superpowers contributed their share to today's morass in Iran. The revolution itself was a multiclass movement that has inexorably come under the control of the lower and lower middle classes. In the process, it has devoured many of its own supporters; the upper middle and middle classes have been attacked and peeled away from the post-Pahlavi power structure. While 1979 and 1980 saw the demise of the old aristocracy and the upper middle classes, 1981 was the year in which the professional middle class and liberal intelligentsia became political casualties. At the same time, the merchants and the bourgeois middle class of the bazaar found themselves increasingly under attack and began to organize quietly against the revolution that they helped to finance.

The competing political forces in Iran can be placed along a continuum from the radical left to the extremist right. It is useful to attempt to freeze the alignments temporarily in order to compare and contrast the forces involved. There are four major groupings of the radical left, two centrist coalitions, and one major extremist force on the far right.

The Fedayan-e Khalq is a nationalist, Marxist group on the far left whose support is drawn from young students and the radical wing of the intelligentsia. Although the theoreticians of the Fedayan differ sharply over tactics, they all condemn what they consider capitalist and imperialist exploitation and seek to build a radical socialist state in Iran. The members of the Fedayan are intensely ideological and have a 15-year history of guerrilla resistance. In 1980, the Fedayan splintered into three main factions including the Fedayan Guerrillas (*Cherikha*), the Minority (*Aqaliyyat*), and the Majority (*Aksariyyat*).[1] While the Guerrilla and Minority splinters have sought to pursue their radical goals independently, the Majority group has revealed a willingness to compromise and to form a front with the powerful extremist right-wing IRP in order to protect their existence and to promote their interests. The Paykar group is a small radical Marxist force with organizational strength among both students and workers. Its position approximates that of the minority wing of the Fedayan.

The Mujahedin-e Khalq is an impressive constellation of Islamic radicals who profess many of the same goals as those of the Fedayan and Paykar groupings. The major principle of the Mujahedin is the concept of *towhid* which refers to "a divinely integrated classless society, a society with total equity."[2] In this ideal society, there will purportedly be an end to the exploitation of man by man. The Mujahedin's campaign against capitalism, imperialism and ethnic exploitation is carefully articulated within an Islamic context and as such represents a major challenge and potential alternative to rule by the extremist right.

Unlike the Fedayan, whose credibility has been severely damaged by its policy of compromise, the Mujahedin forces have refused to waver from their professed goals. Although they initially expressed support for Ayatollah Khomeini, they consistently attacked the rule of the religious leaders on the right, whom they regard as repressive, reactionary and revolutionary dilettantes.[3] In the last months of 1981, the Mujahedin became the major armed force fighting the extremist regime of IRP clerics. Well over 60 percent of those arrested or executed by the government were Mujahedin members or sympathizers.

The other group of the left is the old Tudeh party, which has from the beginning of the revolution sought to protect itself by entering into an accommodation with Khomeini and the IRP. Unlike the Mujahedin who have stood by their ideological commitments, the Tudeh party has pursued a policy of political pragmatism. This has worked as a double-edged sword, protecting the existence of the party in the violent early years of the revo-

1. The dramatic split between the Majority and Minority factions can be traced to June 10 and 11, 1980, when the two groups each published their own version of the party's official newspaper, *Kar*. Both editions were presented as number 62. See *Kar,* 20 and 21 Khordad 1359.

2. "The History of the PMOI, 1965–1971," *Mojahed—The Organ of the People's Mojahedin Organization of Iran,* vol. 1 (May, 1980), p. 20.

3. See "The Content of the Islamic Republic," *ibid.*, p. 24.

lution while damaging its long-term credibility and recruitment potential. The Tudeh party has one further weakness: it has always maintained very close associations with the Soviet Union, an external great power considered a dangerous imperialist threat by the other nationalist groups on the left.[4]

In the center of the political spectrum of Iran today stand two important but amorphous groupings: the secular professional-intelligentsia and the bourgeois middle class. Unlike the more radical and disciplined parties of the left, the moderate groups of the center are large, loose aggregations of individuals who more closely approximate objective social classes than political parties or movements. As such, they have been highly ineffective in the political conflict that has dominated Iran since the overthrow of the Shah. This has been especially true of the professional-intelligentsia, which proclaimed liberal goals and preached progressive reform. Trapped between the ideological radicalism of the left and the religious extremism of the right, the members of the professional middle class have watched their influence destroyed and their revolutionary role disparaged. From Shahpur Bakhtiar to Mehdi Bazargan to Bani Sadr, the demise of the center has been complete.

Unlike the secular professionals, the bazaar middle class is somewhat more difficult to write out of the Iranian revolution. It was the merchant class that formed a tight alliance with the Shi'i religious leaders who were central to the success of the victory over the Shah. The bazaaris provide the economic backbone of Iran, and the period of radical-extremist politics has slowly but steadily alienated them. In late 1980, Ayatollah Khomeini himself publicly and frontally attacked the bazaaris, accusing them of preferring profits to the revolution. The forced resignation of President Bani Sadr in June convinced many merchants that the revolution as interpreted by the extremist right had gone astray. Bani Sadr had become popular in merchant circles and his downfall increased their doubts. In mid-July, two prominent bazaar merchants were executed by the regime for allegedly engaging in anti-state activities. Throughout, the bazaaris had become more and more disturbed by the deteriorating economic status of Iran. During 1981, the bazaar began to detach itself from the hardline IRP programs and policies.

The final bloc of players in the contemporary political drama of Iran is the huge and influential religious right, which stresses fundamentalist Islam and which is dominated by the Islamic Republican party. Supported by the masses of Shi'i believers who compose the lower and lower middle class of society, the extremists of the religious right have come to control Iran. This control, won at the expense of both radicals and moderates, is backed by

4. For an excellent introduction to the various radical political forces in Iran and an informed discussion of their organizational antecedents, see Ervand Abrahamian, "The Guerrilla Movement in Iran, 1963–1977," *MERIP Reports,* no. 86 (March–April, 1980), pp. 3–15.

organized Revolutionary Guards and street mobs known as the Partisans of God (*Hezbollahis*). Although many of the most eminent religious leaders of Iran are moderate progressives, not hardline extremists, the role of these clerics has become submerged by the torrential influence of the Hashemi-Rafsanjanis, Mahdavi-Kanis, and Khamene'is. These are individuals who, like their leader Ayatollah Khomeini, were aptly described by Crane Brinton when he wrote that "... only a sincere extremist in a revolution can kill men because he loves man, [can] attain peace through violence, and free men by enslaving them."[5] History teaches that extremism inevitably comes to prevail as revolutions run their violent courses.

THE DIALECTICS OF DISASTER

By mid-1981, the turbulent revolutionary process in Iran had reached a crossover point. The struggle between the left and right for control of the revolution had been filtered and buffered by the existence of a moderate center consisting of technocrats, professionals, writers, intellectuals, social scientists, and a wide variety of other liberal thinkers. In the early months of the revolution, these professionals, represented by individuals like Mehdi Bazargan, Ibrahim Yazdi, Abbas Amir-Entezam, Sadeq Qotbzadeh, Ali Akbar Moinfar, Hassan Nazih, Ali Asghar Haj-Seyyed-Javadi, Ali Reza Nobari, and Abol Hassan Bani Sadr, were a critical and visible force in the revolutionary government. Although radical and immoderate in many of their ideas and actions, these leaders were neither Marxist believers nor religious zealots. One by one, they were defeated and their influence was destroyed by the inexorable attacks of the extremists; they found themselves caught in the withering crossfire between the true believers of left and right.

In 1981, the remnants of the moderate center felt increasing pressure from the religious right in general and the IRP in particular. The moderates' last hope resided in the person of Bani Sadr, the first popularly elected President in the history of Iran. Partly out of desperation and self-defense, Bani Sadr and his group of supporting technocrats began to move leftward on the political continuum and to form an alliance with the Mujahedin-e Khalq. At the same time, Bani Sadr worked hard to establish close relations with the military leaders and was in constant consultation with them at the Iranian-Iraqi war front. This alarmed the religious rulers, who stepped up their campaign against Bani Sadr whom they regarded as part and parcel of the radical Marxist left.

As the IRP accelerated its attacks on moderate and radical groupings, the middle classes were driven further left as they sought refuge from the

5. Crane Brinton, *The Anatomy of Revolution,* rev. ed. (New York: Vintage Books, 1965), pp. 159–160.

repression of the right. This convinced the religious extremists that their own predictions concerning the liberals had been correct and that the moderates were little more than Westernized Marxists who sought to turn the revolution away from Shi'ism and toward a form of atheistic imperialism. Thus they intensified their campaign against the moderate intelligentsia in general and against Bani Sadr in particular. Bani Sadr managed to stave off defeat by utilizing his special relationship with Ayatollah Khomeini who at first stood somewhat above the conflicting factions while acting as supreme political arbiter.

In late May, 1981, the leaders of the IRP convinced Khomeini that Bani Sadr was plotting a coup against them and provided the Imam with enough evidence to convince him that Bani Sadr was a threat to the revolution. Khomeini was particularly upset about Bani Sadr's growing relationship with the Mujahedin and asked the President to sever that relationship immediately.[6] Bani Sadr's refusal to do so convinced Khomeini that the President had become a serious liability to the Islamic revolution. On June 10, Khomeini dismissed Bani Sadr from his important post as commander-in-chief of the armed forces. The Majlis voted to declare the President politically incompetent on June 21; a few hours later, the prosecutor-general ordered his arrest and on the following day he was dismissed as President. His term of office had lasted less than 17 months.

On July 29, 1981, Abol Hassan Bani Sadr, accompanied by Mujahedin leader Massoud Rajavi, fled from Iran and was granted political asylum in Paris. The extremist victory over the moderate center was completed on July 24 when Muhammad Ali Raja'i, with substantial IRP backing, won an overwhelming electoral victory for the presidency.[7]

With the fall and flight of Bani Sadr, Khomeini turned his attention to the organized threat from the radical groups on the left. His scorn for liberal elements in exile was summarized in an address he gave on August 10, 1981.

> Hypocrites and infidels can never form a united front. All they do is interview and talk and curse each other, talk and curse us and curse themselves. A group of bankrupt politicians who escaped the country disguised in women's clothes are now claiming that they are going to lead the country! They claim that the whole of Iran supports them. Well, if the whole of Iran supports you, why have you gone? You were in Iran and you had all your supporters with you.[8]

During the vicious struggle for power between the President and the IRP, the other major political groupings were also undergoing significant shifts

6. Mansur Farhang in an interview on "The MacNeil-Lehrer Report," September 29, 1981. Farhang was a close adviser to Bani Sadr and was forced into hiding before fleeing from Iran.
7. Officially, Raja'i received 12,900,619 votes out of a total 14,642,552 votes cast. *Jomhuri-ye Islami,* 6 Mordad 1360/28 July, 1981, p. 1.
8. *Teheran Times,* August 11, 1980, p. 1.

in policy and popularity. The major loser during this period was the Fedayan-e Khalq, which splintered into three pieces, with the largest section forming an alliance of convenience with the extremist right. In the process, the Fedayan lost much of its appeal, and its supporters became visibly fewer in number. The Mujahedin-e Khalq, on the other hand, increased its strength enormously and found itself drawing large numbers of supporters both from the ranks of the Fedayan to its left and from the moderate intelligentsia to its right. In the first few months after the overthrow of the Shah, the Fedayan's supporters outnumbered those of the Mujahedin by an approximate ratio of five to one. By April, 1981, this ratio remained essentially the same except that the positions of the two parties were reversed: now the Mujahedin clearly outnumbered the Fedayan. With its commitment to Islam, its unwillingness to compromise with the repressive right, its history of opposition to Pahlavi despotism and Western intervention, its appeal to the growing numbers of Iranians disaffected with the state of the revolution, the Mujahedin will undoubtedly be a critical force in the political future of Iran. It is important to note that as the Mujahedin have increasingly challenged the ruling religious extremists, they have slowly but clearly moderated their own social and political positions.

AFTER BANI SADR

The fall of Bani Sadr had four major consequences. First, the middle classes are without genuine political representation for the first time in revolutionary Iran. This can only result in their continued flight from the country, their cynical withdrawal from serious technical and professional responsibility, and their increased radicalization born of desperation. Second, with the fall of Bani Sadr and his skilled team of administrators, subsequent governments have lacked the professional talents necessary to run a country in today's world. Already, the revolution's cannibalistic tendencies have stripped Iran of its top three or four layers of technocrats. Piety and ideological purity seem poor substitutes for the skills of talented professionals, administrators, technicians, doctors and scientists. Third, Bani Sadr's fall cracked the legitimacy of the post-Pahlavi revolutionary government of Iran. Elected in a popular election in which he received over 70 percent of the vote, the President was a primary creature of the new constitution. His ignominious collapse helped undercut the entire edifice of legitimacy erected so painstakingly after the Pahlavi overthrow. If the Republic's first President is subject to such dismissal, where does political legitimacy reside?

The major consequence of the disappearance of President Bani Sadr was the direct conflict between radical left and extremist right. Positions immediately hardened and confrontation could no longer be filtered through intermediary forces. Khomeini himself lost important leverage. The result was a war for political survival. Very shortly after Bani Sadr's disappearance, an assassination attempt on President-to-be Seyyed Ali Hussein Khamene'i nearly succeeded and seriously injured the powerful mullah. It is especially significant that it was exactly one week after Bani Sadr was impeached that the bombing at IRP headquarters brutally and instantly eliminated much of the leadership of that party. This incident only strengthened the resolve and determination of the extremists in power, who counterattacked the left through a massive wave of arrests, imprisonments and executions. The killing of Ayatollah Muhammad Beheshti transformed an increasingly unpopular political figure into an instant martyr. Despite the deaths of Beheshti and six dozen other members of the IRP political elite in June and the subsequent assassinations of leaders like Raja'i and Bahonar, several important extremists have survived. These included mullahs like Seyyed Ali Hussein Khamene'i and his brother Muhammad Hassan Khamene'i, Ali Akbar Hashemi-Rafsanjani, Muhammad Reza Mahdavi-Kani, Hadi Ghaffuri, and Muhammad Yazdi.

The cleavages that divided yet balanced the contending forces in the two years after the fall of the Shah were sharply reduced in the third year. The balanced tension exploited by Ayatollah Khomeini gave way to direct confrontation between two dedicated forces of fanatical commitment. By the latter half of 1981, the clash had become so intense that the hardened but outnumbered groups on the left reverted to their old tactics of urban guerrilla warfare and terrorism. Both the extremist mullahs of the IRP and the fighting brothers of the Mujahedin have repeatedly demonstrated their willingness to die for their beliefs. This mind-set of martyrdom has seasoned the internal conflict in Iran with a flavor of fanaticism, the effects of which will remain in the society for many years.

AYATOLLAH KHOMEINI AND THE FUTURE OF IRAN

The major hero of the Iranian revolution, 81-year-old Ayatollah Ruhollah Khomeini, found the revolution's third year to be difficult. On several occasions, the wily old leader expressed his exasperation at the continuing conflict and political turmoil that dominated Iranian politics in 1981. His manipulative tactics proved to be increasingly inadequate to the tasks at hand. For example, his policy of playing the two extreme forces off against one another backfired when the Bani Sadr balancing wheel disintegrated and when, by mid-1981, Khomeini threw all his support to the side of the religious extremists. This compromised his leverage and brought him for the first time directly into the political melee.

During 1981, the Shi'i religious leaders played an unprecedented direct role in the day-to-day business of government. The clerics penetrated the Cabinet, where Muhammad Reza Mahdavi-Kani and Muhammad Javad Bahonar held the influential posts of Minister of Interior and Minister of

Education respectively. With the bombings of June and August, the places of fallen mullahs like Beheshti and Bahonar were taken by other clerics, and an even higher percentage of extremist religious figures found themselves in government positions. Bahonar had moved from the ministry of education to the prime ministership and head of the Islamic Republican party. In the new government established on September 1, Muhammad Reza Mahdavi-Kani was the provisional Prime Minister and Ali Hussein Khamene'i was the new head of the IRP. Another mullah, Abol Majid Mo'adikhah, remained on as Minister of National Guidance. On October 2, the militant cleric, Hojjat ol-Islam Ali Hussein Khamene'i, was elected President of the Islamic Republic of Iran.

An examination of the mullahs in key political positions indicates that the most extremist among them were precisely those who wielded the most political influence. Moderates like Ali Golzadeh-Ghaffuri, Muhammad Javad Hojjati-Kermani, and Ayatollah Hassan Lahuti were noticeably and abruptly silenced. The triumph of the extremist clerics was clearly seen in a revealing Majlis incident in June, when the parliamentary body was debating the fate of President Bani Sadr. One of the few individuals who courageously stood to speak in defense of Bani Sadr was a mullah, Muhammad Javad Hojjati-Kermani. Hojjati-Kermani's comments were suddenly smothered when the Majlis Speaker, Ali Akbar Hashemi-Rafsanjani, simply cut off the power to Hojjati's microphone. In the heyday of extremism, the moderate mullahs found themselves in a position of suffocating impotence.

Subsequently, Ayatollah Khomeini threw his support to the religious extremists. Early in 1981, he had attempted to curb the expanding and arbitrary power of the religious right. On March 16, for example, he made an effort to contain the influence of the extremists who were already threatening to destroy Bani Sadr. On this occasion, he called a private conference of eight feuding political leaders and told them to terminate their quarrels. Khomeini's grandson also worked to keep the extremist right under control. In an April interview, the grandson, Seyyed Hussein Khomeini, made the following statement:

> I think generally it is not necessary to have a religious government. The people should decide and they themselves should run the country. It is not impossible for religious leaders to rule or reign, but it is not necessary. In our country, religious leaders have neither the merit nor the ability to run the country, therefore we need statesmen.[9]

Despite these incidents, by mid-1981 events had begun moving too quickly, and Khomeini abandoned his balancing policy. Doubtless, the Ayatollah's sympathies had rested primarily with the religious extremists all

9. *Iran Times,* April 24, 1981, pp. 5, 11.

along, but he joined them openly and decisively for two reasons. First, he was worried about Bani Sadr's flirtation with the military and especially with the Mujahedin-e Khalq. Most important, Khomeini maintained a special antipathy against the radical left. He carefully noted the huge demonstration of April 27 in Teheran, where the Mujahedin rallied over 100,000 supporters to protest regime repression. The more the Mujahedin gained in popularity and strength, the more concerned and involved Khomeini became. By June, he decided that it was time to strike a crippling blow against them.

Once Ayatollah Khomeini had cast his lot with the extremists, Iran became the scene of a bitter if undeclared civil war between radicals and extremists. During the last six months of 1981, violence and bloodshed became the order of the day, and the Ayatollah found himself losing both popularity and influence.

By the end of 1981, the internecine political conflict in Iran had hardened into internal war in which true believers faced off against one another. In this setting, the future of the country seemed to rest more and more in the hands of some kind of military rule. The disappearance of the moderate center, the increasing anarchy and violence, the worsening economic situation, and the inexperience of the latest round of government leaders brought closer the time when a new authoritarianism would appear in order to prevent the disastrous disintegration of a critically important and venerable member of the community of nations.

PART THREE

THE PRODUCING STATES AND INTERNATIONAL POLITICS

The Producing States, The Middle East and the Third World

It is difficult, if not impossible, to separate the political significance of the international oil revolution from the economic. Yet it is obvious that the oil-producing states have distinctly managed to enhance their capabilities and leverage in both spheres. It is the political aspect that is of interest in this and the following chapters. The extent and permanence of the producers' newly acquired power, while highly visible, is nevertheless extremely limited. However constrained, whether real or merely perceived as enhanced, the new power of these states has pushed them into new relationships with countries around the world. For the Middle Eastern oil producers, the growth of influence can be traced through a series of circles radiating outward from the Middle East and proceeding through the Islamic world, the Third World, the industrialized countries and the superpowers.

In comparison with the situation of twenty or even just ten years ago, the movement of the oil states to center stage in the Middle East has been particularly noticeable. For most of these states, the first step in this process was necessarily one of re-integration into the Arab world. While this process undoubtedly would have occurred in any event, through the emergence of independent nation-states out of the web of European domination, the events of 1973–1974 did result in the crossing of a distinct threshhold. In the decade since then, there has been a steady stream of visitors to Riyadh, Kuwait, Algiers, Abu Dhabi and other capitals, in search of advice, approval and assistance. A new tradition has become firmly established, one of financial grants to friends and enemies alike in hopes of ending squabbles or creating consensus. More than one leader of these states has been successfully involved in the mediation of regional or neighborhood disputes.

Gradually, a shift in the balance of power within the Arab world has been discernible, working to the advantage of the oil states and at the expense of the traditional core of the Arab world. While the oil revolution has been the principal cause, this shift has also been prompted by such factors as the explosion of Lebanon and Egypt's isolation as a result of the Sadat initiative. Credit for the October 1976 ceasefire in Lebanon's civil war must go in large part to Saudi Arabia, which gathered the leaders of the forces involved together in Riyadh and prodded them to reach a *modus vivendi.*

From another angle, the oil states have found it necessary to play increasingly more active roles in perhaps the most pressing Arab concern: the Arab-Israeli conflict. As early as the Khartoum conference following the 1967 war, the oil states accepted as their responsibility the financial underpinning of the frontline states which bore the military burden in this conflict. Further involvement by the time of the 1973 war was indicated by the institution of the oil embargo. Later, Egypt's unilateral participation in the Camp David process and signing of a separate peace treaty with Israel relegated that country at least temporarily to the sidelines. This development resulted in an apparent obligation for Saudi Arabia to take the reins in creating Arab consensus. The 1981 unveiling by then-Crown-Prince Fahd of a Saudi proposal for an Arab-Israeli peace was a significant departure from normal Saudi policy of seeking agreement behind the scenes before making any public commitment.

The apparent rejection of the Saudi proposal by the hardline states at the November 1981 Arab summit conference in Fez points out the limitations of this balance-of-power shift. Saudi Arabia and its fellow Arab oil producers can only persuade and not force; their only weapon lies in *dinars, dirhams* and *riyals.* Their weakness was later demonstrated by the Arab world's complete inaction during the Israeli invasion of Lebanon in the summer of 1982. Furthermore, the shift has not always seemed to produce constructive results or even intentions. Libya's capital surpluses have not only been used to restructure the domestic scene but also to pursue aggressively a hardline anti-Israel stance and to engage in a series of foreign adventures, as Lisa Anderson outlines below. Likewise, a contributing factor in the Iranian revolution was the Shah's insistence on transforming Iran into the major power of the Gulf, with the only indigenous deep-water naval force in southwest Asia, to the dismay of the Arab littoral.

The Islamic world comprises the next circle of relationships. The Middle Eastern oil producers regard themselves as bearing particular responsibility for the welfare of fellow Muslims throughout Asia and Africa. Bilateral aid has been substantial to such states as Ghana, Mauritania, Somalia, Bangladesh, India and Pakistan, in addition to the utilization of institutions such as the Islamic Development Fund. The emerging community of interests between the Middle East oil producers and other Islamic states is best seen in the example of Pakistan. Strengthened ties are observable both in the economic realm—with thousands of Pakistanis working in the Gulf and the receipt of considerable development aid in Islamabad—and political—

additional support for each side's foreign policy concerns follows financial support for Pakistan's defense plans and its secondment of military and naval personnel to the Gulf states.

The bond between OPEC and other Third World states remains strong, despite the adverse impact of the oil price spiral on a number of developing countries. As pointed out in the introduction to Chapter 4, OPEC was the first successful blow by Third World countries to improve their terms of trade with the industrialized bloc. Consequently, the developing countries tend to regard the oil producers' action in much the same light as Japan was viewed after its 1905 victory over Russia. The potential this suggests for stronger Afro-Arab relations is stressed by Ali A. Mazrui:

> Enormous wealth in the hands of a big country creates a temptation to play the role of an imperialist Power. But great wealth in the hands of a small country is an opportunity for generosity. . . .
>
> In this situation the Gulf States of Eastern Arabia have a historic role to play. The low density of population in Eastern Arabia is itself an opportunity. When oil is discovered in a densely populated country like Indonesia or Nigeria, that is good news for Indonesians and Nigerians. But when oil is discovered in a country with a small population that could be good news for the world. Enormous wealth in the hands of a small country could lead to the distribution of benefits far from home. A greater variety of human beings—of different nationalities—could stand to gain.[1]

Over the last decade, however, other Third World attempts to emulate OPEC's success in other spheres and commodities have not had similar results, and strains have appeared between OPEC and Third World oil-importing nations, as shown in the Hallwood and Sinclair article below.

Common agreement on North/South issues also goes beyond economic matters, with often similar viewpoints on neo-imperialism and superpower activities. The 1973–1974 Arab oil boycott was extended to South Africa, Rhodesia and then-colonialist Portugal; in return, black Africa generally has supported the Arab states on Arab-Israeli matters. Further strengthening of relations has been helped by oil states' conscious efforts to do business with Third-World firms in areas such as oil exploration, construction, hotel management and industrial concerns. A number of projects are handled from start-to-finish by Third-World firms, supplying management, materials and labor, and the savings sent home by thousands of Pakistani, Indian, Korean, Philippine and other workers is a major contribution to those countries' balance of payments.

For Europe and Japan, the decade of the 1970s also saw growing economic ties, as increasingly expensive but still vital oil imports were offset by

1. "Afro-Arab Relations and the Role of the Gulf States of Eastern Arabia," *India Quarterly,* Vol. 31, No. 3 (July-September 1975), p. 304.

mushrooming exports to the Middle East. Seemingly, the low points of Middle Eastern/European relations—including the Anglo-French attack on Suez (1956) and the bitter Algerian revolution (1954–1962)—have been superseded as Europe's position in world affairs has shrunk and attention redirected inwards. A million Algerians at a time have found employment in France and French presidents have recently been welcomed in Algiers. British expatriates, consumer items and construction firms are ubiquitous in the Gulf; meanwhile, London has replaced Beirut as a summer haven for many Arabs. Japan's dependence on imported oil, even greater than that of the poor cousins of Europe who have no share in North Sea oil, has resulted in the operation of numerous Middle Eastern oil concessions by Japanese companies. It goes almost without saying that Japanese car, truck, watch and electronics manufacturers find one of their most important markets in the oil states.

Finally, both superpowers have demonstrated a remarkable upsurge of concern over events affecting the oil-bearing regions of the Middle East. Since discussion of Gulf security has become an almost daily occurrence in both Washington and Moscow, and because the development of strategies to protect superpower interests there—not to mention the divining of the other side's intentions—has become a growth industry in those same capitals, this subject rightfully claims its own chapter.

The selections for this chapter focus on three important aspects of the opportunities and problems arising from the Middle Eastern oil producers' changing status in the world. Hermann Frederick Eilts begins by discussing the implications that the continuing Arab-Israeli conflict has for security in the Gulf. Rather than being more able to exert influence over this thorny issue, the Gulf states seem to find themselves just as impotent while being placed more on the spot. Although this article was written prior to the unfolding of several relevant events—the assassination of Anwar Sadat in October 1981, the Israeli invasion of Lebanon in the summer of 1982 and the Reagan proposals for an Arab-Israeli peace in September 1982—the conclusion remains valid that this still-volatile conflict constitutes a potentially destabilizing force for the Gulf.

Paul Hallwood and Stuart Sinclair next discuss the sometimes-rocky relationship between OPEC and NOPEC—the non-oil exporting countries. They point out that the OPEC states remain full-fledged members of the Third World but that their relationship with their fellow developing countries has been troubled by the economic dislocations suffered by spiralling foreign-exchange costs of oil imports. In the authors' view, the future of this relationship will rest to a large degree on OPEC's commitment to the forging of a new international economic order, as well as the extent of their concern over the effects of oil policies on the NOPEC states.

In the third selection, Lisa Anderson summarizes the vast changes in Libya, transformed from one of the poorest and most obscure states in the world to a highly visible actor in the international arena, thanks to the gift of oil and the quixotic adventures of Mu'ammar al-Qadhdhafi. But the attention of the outside world has been focused far less on the near-total reconstruction of the domestic economy and society than on Qadhdhafi's controversial support of world revolutionary movements in the name of anti-imperialism and mercurial pursuit of pan-Arab and/or Islamic unity schemes. One result of growing American apprehensions, particularly obvious during the Reagan administration, was the highly publicized dogfight between American and Libyan jets in August 1981. Professor Anderson places this confrontation in the broader context of Libyan foreign-policy goals and problems.

Bibliographic Sources for this Chapter:
11, 17E, 17H, 17I, 20, 23, 32, 49, 56, 68, 77, 78, 80, 85C, 88, 91, 99, 108, 112, 119, 124, 125, 126, 135, 136M, 136O, 139, 150, 156, 157, 171, 177, 204, 210, 216, 220H, 220L, 234F, 256, 257, 269, 296C, 297, 298, 301, 304, 309B, 309I, 309J, 314, 315, 316, 328, 340, 353D, 353I, 354E, 366, 371, 374O, 375, 377, 378, 379, 380, 381, 388E, 390, 392, 393, 398, 400, 405, 406.

Arab-Israeli Spillover

HERMANN FREDERICK EILTS

The Arab-Israeli dispute as a potentially destabilizing factor in the Gulf merits serious consideration. The Gulf is indeed a part of the Middle East region. A distinguished former high-ranking Department of State official has affirmed that United States policy successfully sought to separate the Arab-Israeli problems from those of the Gulf. To the extent that such a separation existed in the 60s and 70s, it was more by chance than by design. The Arab states of the Gulf (and now apparently Iran) are more than interested spectators in the Arab-Israeli problem; they are directly involved in it—emotionally, politically, and financially. The drain on their financial resources may be becoming increasingly onerous, but they dare not cease to continue that involvement and therefore make the best of it. Nor do the sizeable expatriate Palestinian communities in the Gulf states allow them to forget their responsibilities toward the Arab cause, whatever that implies at any particular time.

If any reminder of Gulf state involvement in the Arab-Israeli issue is needed, it is worth recalling that in both 1967 and 1973, the Arab Gulf states embargoed the sale of petroleum to the United States and to certain European states because of real or imagined American support for Israel. In June 1967, even the late King Faisal of Saudi Arabia, though he did not want to do so, felt he had no alternative in the face of the Saudi public and Arab pressure but to embargo the sale of oil to the United States. On that occasion, the cutoff lasted only a few weeks and passed almost unnoticed. In October 1973, as a result of Egyptian urging and because of his own disquiet over United States policy, Faisal was far less reluctant to take such a step. This second embargo lasted almost six months after the war had

Reprinted from Hermann F. Eilts, "Security Considerations in the Persian Gulf," *International Security,* Vol. 5, No. 2, Fall 1980 by permission of The MIT Press, Cambridge, Massachusetts.

ended, notwithstanding efforts of Secretary Kissinger and President Sadat to have it lifted. An Arab oil embargo against the United States, despite disavowals of some Saudi leaders to want to use oil as a political weapon, is likely to be a feature of any new Arab-Israeli hostilities. Arab public opinion will overwhelmingly demand it and Egypt's isolation in the Arab world, following the Camp David accords, precludes the Egyptian leadership from helping, even if it wished to do so. (Iraq, though it proclaimed devotion to the Arab embargo, quietly continued in 1973 to sell oil to regular customers. Whether it would do so again will doubtless depend upon the then prevailing circumstances.) The unanswered question is whether, in a non-wartime situation, but with no progress on the Palestinian problem, an Arab oil embargo against the United States might be applied as a form of pressure tactic. Nothing can be excluded, but an overall peacetime oil embargo seems unnecessary except as a last resort. Instead, such pressure devices as reduction of production levels and oil price increases could as first steps be equally effective. The sizeable financial investments which the Gulf states have in the United States and Western Europe, and their resultant stake in Western economies, may to some extent limit Gulf leaders' actions, but should not be expected to preclude them from using oil as a pressure device, if they conclude that the United States is deliberately working against Arab interests.

The Gulf states, Saudi Arabia among them, are disappointed with what they perceive as a lack of progress in the American-sponsored Middle East peace effort. The Camp David accords, in their view, failed adequately to focus on the core issue of the Palestinians. To them, the skeletal language of the "Framework for Peace in the Middle East" accord looked like a sham. Prime Minister Begin's statements in New York immediately after Camp David, indicating that he had in effect accepted no more than the limited West Bank/Gaza autonomy plan which he had tabled ten months earlier, belied American explanations of what was intended. The collapse of what was construed by Presidents Carter and Sadat to be Israeli agreement to a protracted freeze on settlement activities in the West Bank and Gaza, within less than a day after signing of the Camp David accords, added to the general Arab disillusionment. The peace treaty between Egypt and Israel, despite American and Egyptian avowals that it is the cornerstone of a comprehensive Middle East settlement, was branded a separate peace. Subsequent Israeli statements about sharply limited autonomy, coupled with the establishment of new Israeli settlements in the West Bank and Gaza, seemed to confirm Arab suspicions. Regrettably, Camp David and the Egyptian-Israeli peace treaty, despite their seminal success in bringing some measure of peace to the Middle East area, also had the corollary effect of tarnishing the American image in the eyes of most Gulf leaders. It is not peace that the Gulf leaders oppose; on the contrary, most of them fervently want a just and durable peace. It is their perception of continued shadow boxing with the critical Palestinian issue that evokes their censure.

The Gulf leaders' impatience is understandable. They have the Palestinians on their backs. The want the Palestinian problem resolved, yet have the nagging misgiving that the United States, for domestic reasons, will be unwilling to grasp the nettle. Somewhat naively perhaps, but with all earnestness, they believe that it has the capability to do so, if it wishes. So, it should be noted, did President Sadat. As a result, the Arab states of the Gulf—even Saudi Arabia, a long-time close friend of the United States—have, since Camp David and the peace treaty between Egypt and Israel, drifted toward the rejectionist orbit of Iraq and Syria. Considerable political discomfort attends such a drift, but they have concluded that for the time being at least they have no alternative.

The troubled West Bank/Gaza autonomy negotiations are being watched by most Gulf leaders with no confidence that anything even passably acceptable to the Palestinians—read the PLO—is likely to emerge. Only a few elements in the Saudi Arabian leadership retain a "smidgin of hope" that something meaningful is still possible, if the U.S. presses for the implementation of pronouncements of a "homeland" or "entity" for the Palestinians. The Jerusalem issue also looms large for the Saudis and other Gulf leaders.

In connection with the West Bank/Gaza autonomy negotiations, the Gulf states, as hosts to large numbers of Palestinian refugees, are conscious of one salient fact that tends to be overlooked in most media reporting on the talks. Even if the stalled autonomy negotiations succeed, in terms of an agreement on a self-governing body with broad powers and responsibilities, this will *not* solve the heart of the Palestinian problem, namely the 1.5 million diaspora Palestinians. They, unlike the smaller number of displaced persons among the Palestinian community who fled during the 1967 war, are not effectively covered by the Camp David accords. This critical issue, in accordance with Israeli wishes, was indefinitely deferred for future international discussion. Israel, for reasons which are understandable from its vantage point, but which are rejected by the Arabs, does not want any large numbers of Palestinian refugees to return to the West Bank or Gaza, let alone to Israel proper. Thus, to the Gulf states, the autonomy negotiations deal at best with peripheral aspects of the Palestinian issue and ignore that element of the problem which is of primary concern to them. Moreover, implicit in the Israeli position is the view that the Palestinian refugees be resettled in the Arab world. Notwithstanding their political and humanitarian sympathies for the Palestinians, few Gulf leaders—or peoples—favor any such solution. As guests, the Palestinians are welcome; as permanent residents or citizens, they are not. They would create unwanted demographic, political, and social changes. For its part, the United States itself might ponder whether its interests are likely to be furthered through the permanent resettlement and enfranchisement in the Gulf states of a totally embittered, anti-American Palestinian community.

One might wish otherwise, but the Palestinian problem, unless steps are taken to grapple with its core element, i.e., the refugees, carries the seeds of further deterioration of the Middle East area, including in the Gulf. It could adversely affect continued Western access to Gulf oil in the quantities desired. Former French President Giscard d'Estaing recognized this on a recent trip to the Gulf, when he publicly endorsed Palestinian self-determination. Other European states are moving in that direction. These gestures offer little promise of changing anything, but symbolize our European allies' uneasiness that United States-sponsored negotiations may be faltering. An equitable solution to the Palestinian problem, one satisfactory not only to Israel but also to the Palestinians, is essential if a political climate facilitating continued Western access to Gulf oil and optimum production levels is to be fostered for the years ahead. This is not mission impossible. It will require compromise by both sides, as was the case in the negotiations between Egypt and Israel. Above all, it will require active and impartial American participation and ideas if any meaningful progress is to be made. The scope of the West Bank/Gaza autonomy talks is clearly inadequate. If they are to offer any promise, they need to be considerably expanded in objectives.

To talk, as some eminent senators have after the Egyptian-Israeli peace treaty, of some sort of United States-Israeli-Egyptian military protection of the Gulf oil fields is illusory. Israeli participation in any such operation would create a wave of Arab and Islamic revulsion. Whatever its short term success might be, and even this is doubtful, it would be disastrous to American interests in the Middle East and the Gulf. Widespread sabotage and international terrorism would assuredly ensue. We should not underestimate the Arab capacity for self-destruction, when confronted by what are perceived to be "honor" violations. As the late President Sadat emphatically told an eminent American senator who raised the point, "The Arabs are fully capable of protecting the oil fields. Israeli participation would be altogether harmful." A unilateral American military action to protect the oil fields, unless on invitation of Gulf leaders, would have almost equally negative area and Third World repercussions. It would be viewed as blatant imperialism and provide grist for the mill of anti-American elements, including the Soviets. Only in a demonstrably *in extremis* situation, where the very survival of this nation is at stake, does it deserve consideration.

Without belaboring the point, one further factor needs be noted about the current United States position in the Gulf. While there was no Arab love for the Shah, he was viewed throughout the Gulf—often to his annoyance—as the bastion of the American power position in the Southwest Asian area. Inevitably, therefore, the Shah's demise and the failure of the

United States to try to save him (and no Gulf leader is convinced of U.S. inability to do so) hurt American credibility in the Gulf region. It was made worse by what was perceived to be Washington's inept handling of the Shah's departure and subsequent search for domicile. To friendly Gulf leaders, the question of United States dependability in times of crisis inevitably arose.

OPEC's Developing Relationships With the Third World

PAUL HALLWOOD AND
STUART SINCLAIR

The publication, in 1980, of the report of the Independent Commission of Inquiry into Development Issues—better known as the Brandt Report—firmly identifies the members of OPEC, the Organisation of Petroleum Exporting Countries, as part of the 'Southern' or 'developing' world. In so doing, the report implicitly accepts the judgment on this question of OPEC itself. For no theme emerges so consistently from the public pronouncements of OPEC (the secretariat and the various aid agencies) as that which stresses the consistency of interest and unity of purpose between OPEC and the non-oil exporting developing countries. References to OPEC being 'full members of the Third World'[1] and 'a pioneering group of developing countries'[2] pepper OPEC press statements. But what justification is there for this? To what extent are there in practice conflicts of interest between OPEC members and non-oil less developed countries (LDCs)? And over what issues? Furthermore, what weight would the considerations of the non-oil LDCs' problems assume in OPEC's policy deliberations?

In clarifying and going some way towards answering these questions, this article attempts to outline the evolution of OPEC's relations with the rest of the Third World. It begins with a brief review of the state of relations between OPEC members and non-oil LDCs immediately prior to the announcement of the oil price increases of 1973. It then traces the strains which developed in that relationship as the non-oil LDCs began to experience certain types of economic dislocation in the mid-1970s, and then goes on to analyse the responses which these problems prompted from OPEC.

1. A. A. Al-Anbari, President of the Iraqi Fund for External Development, at OPEC Headquarters, Aug. 1980. Quoted in *OPEC Bulletin,* Sept. 1980, p. 14.
2. Address of René Ortiz to UNCTAD V, May 1979, *OPEC Press Release,* 1979.

Reprinted by permission of the publisher, The Royal Institute of International Affairs, from *International Affairs,* Vol. 58, No. 2, Spring 1982, pp. 271-286.

This leads finally to an appraisal of the extent of the consistency of interests between the two groups of countries and a discussion of the likely course of developments between the two in the 1980s.

EARLY ILLUSIONS OF OPEC–THIRD WORLD SOLIDARITY

The announcement of the Shah of Iran, on December 23, 1973, that henceforth a barrel of light crude oil would sell for $11.65 rather than $5.04—the price which had prevailed since October—came as a considerable and unpleasant shock to oil consumers. Immediate reaction was intemperate, with references to an 'explosion in north–south economic relations'[3] and the assertion that 'a critical turning-point in history'[4] had been reached with the prospects for the Third World irremediably worsened. Rather than have such reaction stand unchallenged, however, certain OPEC member states swiftly took up the diplomatic initiative and attempted to turn the perceived disaster into an opportunity. In particular, the Algerian government, which had lately instituted a new and busy foreign policy, was keen to have OPEC's action seen as part of the effort to establish some form of 'new international economic order.' At the sixth special session of the United Nations, hastily convened at the behest of such nations as Algeria, no opportunity was lost to pose the issue in this light. With United Nations secretary general Kurt Waldheim calling the session 'a timely and opportune initiative' by President Houari Boumedienne of Algeria, it appeared that any problems anticipated to arise from the oil price rise could be submerged in talk of a new world order. Similarly, at the first Conference of Sovereigns and Heads of State of OPEC member countries, held in Algiers in March 1975, the Algerian memorandum spoke of the oil price rises as advantageous to the entire Third World, and as 'a practical illustration of what the developing world can do.'

Although most observers appeared unimpressed by these claims, there was already some substance behind the rhetoric. The Venezuelan government, with the enhanced scope allowed it by domestic stability after years of upheaval, felt a shared interest with those countries which had as yet not succeeded in wresting control over their own mineral wealth. The Caribbean Development Bank received extra Venezuelan financing very shortly after the oil price rises, while elsewhere among OPEC members, Nigeria was the first to implement—albeit diffidently—an oil-price discount scheme to neighbouring oil-importing countries. Up to 3 per cent of Nigeria's crude oil exports were soon being received by other West African States; the formation of the Economic Organisation of West African States (ECOWAS) was hastened by this gesture of goodwill.

3. R. N. Gardner *et al.*, 'A Turning Point in North–South Economic Relations,' in *Trilateral Commission, Task Force Report* (New York: New York University Press, 1977).

4. Trilateral Commission, *ibid.*

While, as will shortly be seen, to some extent these measures could be seen as early and hasty efforts at marshalling some defence against the industrial countries' criticisms of OPEC, they also reflected the prolonged effort to foster a greater degree of unity among Third World countries. One of the oldest of such efforts was the Afro-Arab impulse. For several years prior to 1974, the Middle Eastern OPEC states had been attempting to consolidate the alliance they had forged with a number of black African states, having induced them away from earlier commitments to Israeli aid and technical assistance. This manoeuvre went some way to fulfilling the long-standing aspirations of the pan-Africanists and pan-Arabists such as Gamal Nasser, Julius Nyerere and Sekou Touré. Having had little success in achieving their rather vaguely formulated desires for closer ties between African and Arab states at the Accra Conference of 1958 or the Addis Ababa conference five years later, these statesmen found everything coming together rapidly. As Ali Mazrui observed, 'never in modern history did there seem to be as much solidarity between Arab Africans and black Africans as there emerged . . . in the course of 1972 and 1973.'[5]

These, then, were the political relationships put in question in late 1973 when the oil price was raised in a sequence of three steps in June, October and December.

GROWING DISILLUSION AMONG NON-OIL LDCs

Strains were not long delayed. This 'political shrinking of the Sahara' as Mazrui so judiciously put it,[6] first encountered difficulty at the Organisation of African Unity meeting of late 1973 at which ministers from Kenya, Ghana, and Tanzania requested concessional oil prices. This request, the first of several as it was to turn out, was rejected, but notwithstanding that rejection the OPEC aid effort began, with a $200 million funding, immediately afterwards. At the Arab summit in Algeria two aid agencies, the Arab Fund and the Arab Bank for Economic Development in Africa, were instituted to begin disbursing this aid. Similar tensions were visible at an African Conference in September 1975, with the Tanzanian Minister of Commerce accusing OPEC of having 'turned their backs on the developing countries, particularly the least developed.'[7] Elsewhere on the continent, President Mobutu of Zaire was reportedly even ready to restore relations with Israel. An effort was made to put the OPEC-African relationship on a new footing at the Afro-Arab summit of March 1977. Held in Cairo and attended by twenty-one Arab representatives and thirty-nine Africans, this meeting heard many appeals to unity. But still, the intentions of goodwill

5. A. A. Mazrui, *Africa's International Relations* (London: Heinemann, 1977), p. 145.
6. Mazrui, *op. cit.*, p. 138.
7. *New African Development,* Feb. 1977, p. 117.

and intended munificence so frequently voiced failed to find a suitable vehicle.

By 1979 tensions had become still more marked and more coherent in form. At the fifth session of UNCTAD in Manila in May, Costa Rica was instrumental in organising a lobby calling itself OPIC, the Organization of Petroleum Importing Countries. At the same meeting a Colombian delegate called for oil to be put on to the agenda after it had been excluded by OPEC members. OPIC members argued that Venezuela, as the OPEC state closest to its members, should sell crude oil at concessional prices to its poorer Central American and Caribbean neighbours. Venezuelan ministers had in fact long been calling for more OPEC aid, suggesting that its scope be extended so that others than Bolivia, El Salvador, Honduras and Haiti in that part of the world would be recipients. Indeed, the Venezuelan Energy Minister had made a two-week tour of North Africa and the Middle East seeking support, or at least attention, for his ideas earlier in 1979.

In the event, the Iraqi deputy trade minister proposed a new fund to compensate the poorest oil importing nations, but the idea was felt to lack conviction. A characteristically robust retort came some five weeks later from the OPEC meeting in Geneva on June 26–28 when it was recommended that its member governments supplement the OPEC Special Fund by $800 million. Apart from this they proposed that a permanent OPEC and OECD fund to compensate LDCs for the rising prices of both oil and manufactured goods to be instigated. A communiqué then reiterated the fundamental OPEC position—by now familiar—that no dialogue on energy issues could be entertained unless it also embraced a more comprehensive discussion of all North–South development matters. Nor did the Special Fund's announcement of a few days earlier—that a selection of non-Arab African countries (including Cameroon, Lesotho, Mali, and Zambia) were to share in a further $38.5 million of project assistance—assuage disappointment in Third World capitals.

By the end of the 1970s then, many Third World governments were increasingly asking whether the burgeoning—if arbitrary—aid they had begun receiving was sufficient compensation for the oil price rises they supported—implicitly or explicitly—at so many international conferences.

The obvious question which arises here is: to what extent was this disillusion justified? The 1970s had opened to witness growing alignment between Arab countries hitherto on the periphery of Third World institutions and councils; they were, seemingly, finishing with widespread ill-feeling, suspicion and not a little jealousy. How far, indeed, had the talk of new economic orders, of Third World solidarity, and of OPEC having blazed a trail for others to follow, been vacuous? Were there still reasons for believing that something could be accomplished in the next decade? The following section begins to analyse these questions.

OIL PRICES, DEBT AND ECONOMIC PERFORMANCE

In order to understand the legitimacy or otherwise of the objections from Third World governments to OPEC policy it is necessary to review briefly the nature and extent of the economic impact of OPEC oil price decisions upon the non-oil LDCs. This is itself no small question: determining the criteria by which this should be judged is difficult, as is judging the relative impact of the industrialised countries' policy actions as against OPEC's. However, the main economic consequences of OPEC price policy can be considered with reference to five subjects: the terms of trade between oil prices and the non-oil LDCs' export prices; the non-oil LDCs' growing foreign payments deficits and debt growth; the growth of OPEC trade with the rest of the Third World; the evolution of the OPEC foreign aid effort; and, finally, through an examination of the rates of economic growth experienced in the Third World after 1973.

The terms of trade (any one of several possible ratios between export and import prices) has for long featured as a central issue in the North–South debate because it largely determines the distribution of benefits between trading partners. Such writers as Prebisch and Singer[8] have pointed to the dependency of Southern economies upon Northern industrial countries and particularly to the alleged long-run deterioration of the South's terms of trade. The latter, essentially an empirical question, has been subjected to close scrutiny and found to be wanting. To the extent that there is a systematic relationship in the South's terms of trade with the North, the evidence suggests it is cyclical rather than secular in its rhythm.[9] In any case, comparing international price relationships between just two trading blocs is far too aggregative to allow useful conclusions to be drawn as it obscures important details. As far as OPEC is concerned, when the organisation took control of oil prices from the oil companies in the early 1970s (the 1971 Tehran agreement between OPEC and the companies marking the watershed), it not only changed the nature of the members' economic relationships with the North, but clearly with the non-oil LDCs also.

The statistical facts can be stated quite simply. In the third quarter of 1973 the price of primary commodities (as measured by *The Economist's* dollar index) relative to crude oil (Arabian 34° light) stood at the decade's peak of 136 (1970 = 100). By the following quarter this term of trade index had fallen to 83, and in 1974 (I), it had tumbled much further—to 37. In all, it was down 73 per cent in just three months. This remarkable shift in the relative prices of the two groups of traded commodities has persisted,

8. R. Prebisch, 'Commercial Policy in Underdeveloped Countries,' *American Economic Review,* Papers and Proceedings, 1967.

9. For example, Paul Hallwood, *Stabilization of International Commodity Markets* (Greenwich, CT: JAI Press, 1979), pp. 19–28.

and has even deepened with the second oil 'shock' of 1979 serving to compress the price of primary commodities relative to oil even further.

Clearly, OPEC members now have a degree of control over their export prices that is denied to the non-oil exporting LDCs. Furthermore, the success of OPEC in controlling oil prices and production was not only a shock for industrial countries, it also created new relationships within the developing world. Indeed, a new 'centre-periphery' relationship (to use Prebisch's terminology) has, to some extent, now arisen. The OPEC members manage and try to optimise oil prices within the confines of existing economic and, especially, political constraints, while the non-oil LDCs are left with little countervailing power as far as their export prices are concerned. Moreover, to the extent that OPEC's price policies have weakened the economic performance of the industrial countries (a matter upon which opinion is still divided) the non-oil LDCs have indirectly suffered from the loss of dynamism in their Northern markets.

Fortunately, however, some non-oil LDCs have been able rapidly to increase their exports of goods and services to OPEC members and so partly mitigate the effects of higher oil prices. Traditional primary commodities as well as manufactures have been exported in greater volume as have exports of services such as those provided by civil engineering contractors, consultants and migrant workers. Six years after the 1973 oil shock the LDCs as a group were exporting close to $10 billion worth more goods to the OPEC countries than they were in 1973. Saudi Arabia and the United Arab Emirates were the two OPEC members which increased their imports from non-oil LDCs the most rapidly—at over 50 per cent, per annum, during the 1973–79 period—while all the other OPEC members had high import growth rates of over 20 per cent, per annum. However, it is apparent that only a few non-oil LDCs have succeeded in penetrating to any significant extent the rapidly expanding OPEC members' markets. Thus only thirteen non-oil LDCs (most prominent being South Korea, India, Taiwan, Brazil, Hong Kong and Singapore) accounted for about 90 per cent of OPEC's growing imports from the Third World during this period. Indeed, such was their exporting success that nine or ten of OPEC's biggest LDC trade partners had by 1978 actually improved their export/import ratio with the OPEC members. But then again, few of even the dynamic non-oil LDC trade partners have ever had surpluses on their trade accounts with OPEC countries, and the situation is much worse for the mass of non-oil LDCs.

There is a degree of concentration similar to that in exports amongst LDC suppliers of labour to OPEC members (mainly those labour-scarce members in the Middle East). In fact, other Middle Eastern countries such as Egypt, South and North Yemen, Lebanon, and Syria, have sent the largest numbers of workers and have benefited most substantially from workers' remittances of wages. Outside the Arab world other large labour-sending countries are Pakistan, India and the Philippines. It remains an open question, however, whether remittances are always sufficiently large to offset those costs associated with the loss of the labourers' services in their home countries—especially when skilled labour is involved. Thus not too much should be read into recorded international remittance flows without fuller examination of the offsetting costs involved in some cases.

While some non-oil LDCs were in these ways eventually able to derive some benefits from OPEC's growing wealth and so partially mitigate the effects of higher oil prices, for the non-oil LDCs as a whole, higher oil prices immediately worsened their foreign accounts. For in 1974 their aggregate current account deficit was multiplied by a factor of four, compared with the previous year, to $24 billion. It was never again to fall below $20 billion and reached $97 billion in 1981. The importance of high oil prices to the non-oil LDCs' payments deficits can be more easily judged from the fact that of an aggregate merchandise account deficit with the world of $280 billion over the 1974–80 period, about two-thirds was with the thirteen members of OPEC. Indeed, in three of the last four years of the 1970s, three-quarters of the non-oil LDCs' deficits were with OPEC members. Alternatively, looked at another way, in 1972 only Brazil amongst non-oil LDCs allotted more than 10 per cent of its total import bill to oil. Just two years later, twenty-one did so—a figure which was to rise further over the next seven years.

Thus, most of the deterioration in the non-oil LDCs' foreign payments balances after 1973 can be explained by developments in their foreign trade with the oil exporting countries. Further examination of the matter by the authors rules out other possible systematic influences.[10] Since 1973 there has been no permanent deterioration in the price of primary commodities relative to manufactures; export performance in the non-oil LDCs taken as a group held up well in the late 1970s as the real value of exports (nominal export values deflated by the IMF's industrial countries' export price index) recovered after 1975 returning to the 1961–68 trend levels; nor can explosive real import growth be considered as a general contributory factor, for, following the 1973 oil shock, the real value of the non-oil LDCs' imports was at or below the long-term trend levels.

Foreign aid has been given by OPEC states on a large scale in recent years and it might be thought to have gone a long way in helping to finance the non-oil LDCs' payments deficits. However, this is not the case. A fuller analysis of the political dimensions of OPEC aid and its importance to OPEC's relations with the non-oil LDCs appears in the fifth section of this analysis. However, since the dimension of this aid is relevant in the context of the financing of the non-oil LDCs' payments deficits, some observations on the matter ought to be made here. A comparison of the non-oil LDCs'

10. Paul Hallwood and Stuart Sinclair, *Oil, Debt and Development: OPEC in the Third World* (London: Allen and Unwin, 1981), pp. 82–83.

trade deficits with receipts of foreign aid from OPEC shows that compensation is far from complete—with, in the 1974–77 period, only about 28 per cent of the aggregate deficit being financed by OPEC aid. Since then, the level of OPEC aid has not increased significantly. Useful insights into the relationship between OPEC aid and the non-oil LDCs' oil imports costs can also be made if oil prices are 'index linked' to OPEC import price inflation so as to take account of OPEC's claim that oil prices were adjusted in 1973 only to account for inflation over the previous ten years or so. A reasonably objective assessment can be made by constructing a series of notional oil prices. These are prices which would have been obtained during 1973–79 had the crude oil price (Arabian 34° light) been linked to primary product prices (*The Economist*'s dollar index) ever since 1960. On this basis, during 1973–79 Arabian 34° light cost some $7 to $8 per barrel more than its 'index linked' level. (This is also true if crude oil prices had been linked to industrial countries' export or consumer goods prices.) At an extra $8 per barrel in the four years after the 1973 oil shock, then, the non-oil LDCs spent some $38 billion on imported crude oil more than if prices had kept to the indexed value. At a notional $6 per barrel extra cost, they spent $28 billion more. OPEC aid disbursements during this period amounted to a little over $20 billion. However, this aid was very heavily concentrated upon just a few recipients so that only a few countries could use it to offset their higher oil import bills. In fact, just six countries (Egypt, Syria, Pakistan, Jordan, India, and North Yemen) received 70 per cent of aid disbursements from the OPEC members during the 1973–77 period; ten received 80 per cent of the total; and of the largest twenty recipients of OPEC aid disbursements (87 per cent of the total) nineteen were Muslim countries.

Accordingly, the foreign debts of most non-oil LDCs (of which ten account for about two-thirds of the total) have had to be largely financed from other aid sources and, more particularly, by the international banking system. For between 1974 and 1980 the outstanding foreign debts of the non-oil LDCs owed to private creditors increased from 29 per cent to over 40 per cent of the total. According to World Bank statistics the non-oil LDCs' foreign debts are now increasing at a rate of over 20 per cent per annum and, by the end of 1981 had reached some $400 billion. Even by 1978 the external debt outstanding of the largest twenty non-oil LDCs[11] amounted to 25 per cent of their combined GDPs when it had been only 19 per cent in 1973.[12] These debts are proving to be worrisome not only to the borrowers themselves but also to their creditors, and it has been questioned whether financing on such a scale can continue through the existing institutional framework.

The worsening foreign exchange constraint and growing burden of foreign debts upon the non-oil LDCs has contributed to an apparent break in

11. Ranked by the size of their GNPs.
12. Morgan Guaranty Trust, *World Financial Markets* (May 1979), p. 8.

their aggregate average rate of economic growth. Thus, in the six years prior to 1973 the aggregate growth rate of real GDP averaged 6 per cent, per annum. Subsequently, it has fallen to average some one-and-a-half percentage points less than this (even this is a performance somewhat better than that achieved by the industrial countries). There is also some evidence to show that the jolt of the first oil shock increased the degree of economic instability in the non-oil LDCs. 'Economic instability' can be measured as the average proportionate deviation of GNP from a long-term constant growth rate trend. Using this measure, then for a sample of twenty-four non-OPEC LDCs, the average annual coefficient of variation increased from 0.91 during 1967–72 to 1.01 during 1974–78.[13]

However, while the foregoing account of the economic impact of oil price rises upon the non-oil LDCs shows that deleterious trends were indeed instituted and old problems exacerbated, it is inappropriate to talk of an 'unmitigated disaster for many of the LDCs' as was predicted in 1974 in certain quarters.[14] Oil imports and economic growth were to a large extent financed by an increase in foreign debt.

OPEC INITIATIVES IN THE 1970s

It is instructive to note that immediately after 1973, the OPEC members' responses to complaints from oil-importing LDCs were to a large extent *ad hoc*. Few coherent arguments had been developed: the sophisticated public affairs and information systems which would later characterise the secretariat's operations in Vienna had yet to be organised. A common theme in debate was nonetheless beginning to emerge, and this was the argument that OPEC's actions would shortly serve as a prototype for other groups of primary commodity exporting countries. There was much discussion along these lines at the CIEC, or, North–South dialogue, in Paris. Indeed, this idea has still not completely disappeared, and as late as 1980 OAPEC ex-assistant secretary-general, Fadhil al-Chalabi, wrote that, in addition to OPEC's price action having 'effectively strengthened inter-third world solidarity' it had created 'a model for correcting the prices of other raw materials.'[15] Similarly, speaking in 1980 at OPEC in Vienna, H. E. Abdul Amir al-Anbari of Iraq argued that OPEC remains 'a model and . . . pioneering group of developing countries,' whose success 'is not necessarily because of the nature of the commodity.'[16]

13. Hallwood and Sinclair, *op. cit.*, Table 9.4. The post-1974 sample is less than twenty-four countries owing to lack of data availability.

14. R. M. Dunn, 'The Less Developed Countries,' in Z. A. Yager and E. B. Steinberger (eds), *Energy and US Foreign Policy* (Cambridge, Mass.: Ballinger, 1974), p. 193.

15. F. al-Chalabi, *OPEC and the International Oil Industry* (Oxford: Oxford University Press, 1980).

16. *OPEC Bulletin*, Sept. 1980, pp. 14, 17.

The lack of substance in these propositions has, however, become evident during the 1970s. For in no other raw material has a comparable cartel been established. Many primary commodities are still priced on free international markets which are, from time to time, subjected to severe fluctuations in demand from both speculative and final user sources. Sometimes primary commodity export prices are agreed on long-term contracts between host countries and multinational corporations and very often it is the latter who have the upper hand in the negotiations over prices. No group of primary producers has been as successful as the OPEC members in acquiring control of their primary product exports from the foreign corporations, yet this is a necessary step in the creation of a successful cartel. As is clear by now, economic fundamentals are also unfavourable for producer country cartelisation, in that industrial country importers have open to them many substitution possibilities which amount to a form of countervailing power against producer country cartels should they be established.

It is significant, however, that no OPEC member has been involved with any scheme to invest surplus funds in other Third World commodity cartels. A minor exception to this was Venezuela's participation in the Bogota Fund group of Latin and Central American coffee producers, who for a time tried to influence prices by intervening in the coffee futures markets. However, even this was more a defensive scheme to support a weak market than anything else. The closest OPEC members have come to assisting other raw material producers' aspirations has been their support for the various commodity buffer stocks that are planned in the UNCTAD's Integrated Programme for Commodities. But this is now becalmed for lack of finance, although the OPEC Special Fund did offer $100 million—which was somewhat greater than the members' pro rata share.[17]

There are two main reasons why OPEC members have not wished to further other Third World commodity cartels by providing financial support, although sometimes moral support has been forthcoming. The first reason is the wish to avoid further heightening tensions between OPEC and industrial countries—a factor which remains of importance even in the determination of oil prices, especially on the part of the Western-leaning Gulf states. Secondly, as is argued later, the OPEC members have alternative means of progressing in the major political, economic and military spheres of importance to them other than through the rather blunt instrument of 'Third World solidarity' purchased through this channel. Besides, since successful Third World commodity cartels are likely to be possible in relatively few markets, even with considerable OPEC financial support, the uneven geographical distribution of the benefits consequent upon that support is likely to be, at most, of only minor diplomatic advantage to OPEC members.

17. OPEC, *Annual Report: 1979,* Vienna, p. 14.

Support of the nature actually given illustrates how the grounds of the debate into which OPEC has felt itself being drawn have altered since 1973. For, rather than arguing, as was initially the case, that OPEC's actions could be replicated for other primary commodities, and could thus be seen as to the benefit of the un-industrialised countries generally, OPEC tends now to argue that its own wealth, while created in a manner not replicable among other groups of raw material exporters, can nonetheless still be to the benefit of the LDCs. This benefit could be realised both through the aid effort and by using the threat of oil supply interruptions to widen and deepen the set of negotiable issues between the industrialised world and the rest. This second proposition is both more and less ambitious than the first. It is less ambitious in so far as it replaces fairly clear and measurable objectives—helping foster successful and operational cartels in other raw material markets—with more ethereal objectives whose degree of success is intrinsically hard to measure. This is clear. But the new proposal, to use whatever power OPEC can muster to assist the rest of the world, is, by dint of its less parochial sweep, also more ambitious. Thus, writing of the problems caused by the stockpiling policies of industrialised countries in upsetting oil prices during 1979 and 1980, Fadhil al-Chalabi forewarned that 'OPEC is prepared to use its ability to counter such moves.'[18] Similarly, on a different plane, René Ortiz, OPEC Secretary General, has proposed that 'OPEC can act as a moderating political force and as a catalyst in achieving an 'equitable interdependence' between developing countries, including OPEC countries, and the industrialised nations.'[19]

This idea of a new OPEC initiative on aid, rather than commodity market intervention, is an attractive one. From initially modest beginnings OPEC aid blossomed in the mid-1970s only to stagnate for lack of clear political direction later in the decade. The proposals of the Committee on Long Term Strategy (the Yamani Report) have tried to remedy this but it has itself been shelved for various reasons. OPEC's aid has been mainly channelled bilaterally, but with a trend towards multilateralism as a plethora of OPEC and Arab aid agencies have sprung up. Despite the heavy geographic concentration of Arab aid pointed to above, it has been largely given on an equitable basis, mainly because many Muslim country recipients are among the poorest in the world. Poor Muslim countries, especially in North Africa, have received more OPEC aid as a proportion of either per capita income or increased oil import bills than have the better off LDCs such as the so-called newly industrialising countries. Even so only eight countries (Afghanistan, Bangladesh, Pakistan, Egypt, Jordan, Somalia, Sudan, and Syria) have received more OPEC aid than they have paid out for higher priced oil.[20]

18. Editorial, *OPEC Review,* Vol. 4, No. 2, Summer 1980, p. iv.
19. *Ibid.,* p. i.
20. Hallwood and Sinclair, *op. cit.,* pp. 110–28.

The marked geographic concentration of OPEC aid, together with policy statements, suggests that political motivation has been a major factor behind most of the Arab countries' aid programmes. All the Arab foreign aid donors, Saudi Arabia, Kuwait, Qatar, UAE, Iraq, and Libya, have used their financial resources to aid the so-called 'front line' states against Israel. Other Arab states have benefitted from acts of Arab solidarity. Regional security has been another important motivating factor behind Arab aid. For example, Saudi Arabia has given aid to Somalia, North Yemen, and Oman in an attempt to counter Soviet influence. And, of course, Saudi Arabia in particular and other Arab aid donors have used aid in support of ideological and religious causes, especially the strengthening of Islam, both in Arab and in 'fringe' Muslim countries.

Despite the new thrust of the OPEC effort being partly dependent upon improved aid flows, however, OPEC—or, more precisely, the OPEC Special Fund for International Development—has marshalled two arguments which point to certain constraints on the capacity and willingness of OPEC to provide further aid. The first is that OPEC aid does not come from 'renewable annual income,' as Director-General Ibrahim Shihata put it recently[21] and that, given this, OPEC's ability to offer aid is necessarily less than that of industrialised countries.

The second argument used now to defend OPEC against charges of an inadequate aid record (such as was suggested by Chancellor Schmidt at the opening of the eleventh world energy conference in Munich in September 1980) is that, irrespective of the weight of the first argument, OPEC need feel in no respect duty bound to offer aid to crude oil consumers. This is since no other raw material producers do: as Shihata put it 'the principle of compensation [has not] been invoked in respect of any other commodity.'[22]

In essence, the position of OPEC spokesmen is that the present imperfect world economic system is one largely fashioned by the agencies of the industrialised north. All OPEC members can hope to do is to try, probably at the margin, to redress the balance of advantage. This echoes arguments put forward about the fundamental inequity of the system at the Islamic Council of Europe conference of 1978. There, 'the oil weapon was not forged by the Arabs: it was forged by Gulf, Shell and B.P.'[23] was a leitmotif in the speeches.

This sentiment well illustrates the thinking of the major OPEC donors throughout the 1970s. On the one hand they were anxious to be seen mitigating the worst consequences of the oil price rises for certain LDCs—with their priorities between those LDCs depending on their strategic value and a shared religion as much as anything else. On the other hand the aid

21. *Financial Times,* July 17, 1980.
22. *Financial Times, loc. cit.*
23. Islamic Council of Europe, *The Muslim World and Future Economic Order* (London: Islamic Council of Europe, 1979).

donors resisted having this aid seen, by recipient country governments or by industrial country governments, as 'conscience money.' Instead it was persistently described by the donors as payment *ex gratia,* and, moreover, the type of payment long overdue from the industrial countries as well.

THE NEXT FEW YEARS: BARRIERS TO UNITY

The preceding sections have sketched out the ways in which relations between the thirteen OPEC member states and the non-oil LDCs unfolded in the 1970s. Above all it has been apparent that while certain interests were shared by both sets of countries, strong elements of antagonism also persisted in their relationship. It remains to pull together these observations to generate some insights into the likely development of these relations in the 1980s.

Two sets of issues which arise from the foregoing discussion will predominantly determine the future course of these relations. The first is the degree to which the non-oil LDCs' interests will come to assume more significant weight in decisions arrived at by OPEC members and thus affect oil and aid policy in the future. The second is the question of what sort of OPEC it is which claims to have a serious interest in the evolution of some form of new international economic order.

Given the efforts expended by OPEC member states, both individually and acting together through their various new agencies, at enhancing relations with non-oil LDCs, might one expect the 1980s to be characterised by greater harmony between the two groups? There are three answers to the question.

First, one must consider what interests OPEC truly has in securing approval or ratification elsewhere in the Third World. In the mid-1970s most observers believed these interests were considerable, chiefly for reasons of the Gulf rulers' internal insecurity. Thus, it was commonly argued that in order to protect their own position as privileged elites in a sea of resentful poverty, Gulf rulers in OPEC would have to be seen providing largesse for the poor in other countries. Only in this way could the anachronistic rulers hope to retain power. Since then this view has lost adherents—not least because the demise of the Peacock Throne in Iran occurred despite a large programme of foreign aid and treaty links with some non-oil Muslim countries.

A variant of the argument that 'Third World solidarity' would flourish other than at the level of mere rhetoric was that OPEC states, again particularly those in the Middle East, could not afford to jeopardise their new-found allies among the rest of the Third World because as large an anti-Israel vote as possible was needed in the United Nations. Moreover, support for OPEC actions in the conferences of the United Nations specialised agencies such as UNCTAD was valuable in so far as it tended to force the industrial countries into taking the initiative in offering enhanced development proposals. Thus, it could appear that it was up to the industrial countries' representatives to propose new forms of relief for the Third World generally rather than being up to OPEC states to offer recompense for their Third World compatriots.

The evidence against this hypothesis of evolving OPEC-Third World relations is, however, mounting. Above all, it is increasingly likely that it is not votes in the United Nations which will eventually yield some form of Palestinian homeland so much as the momentum of ideas and heightened fears of interruptions in oil supplies arising from the tensions set up by shifting allegiances around the 'front line' states. The evident interest of the EEC foreign ministers—as well as the Israeli Labour Party—in the principle of a Palestinian state has become vastly more important to Arab OPEC members than the fractured and uncertain commitment to that objective they could hope to obtain from elsewhere in the Third World.

A second factor which tends to confirm the erosion of the non-oil LDCs' governments' influence over OPEC members is the changing nature of international trade in crude oil. After the nationalisation of the multinational oil majors by certain oil-exporting countries, and the reduction of the companies' autonomy in others (processes intrinsic to the development of OPEC, of course) it was inevitable that crude oil trade would be to an increasing degree politicised. That politicisation is apparent in such decisions as that of Saudi Arabia to suspend crude shipments to Italy in late 1980 pending the clarification of government scandals. Similarly, the Nigerian authorities have threatened more than once to curtail crude supplies to companies whose affiliates are suspected of on-trading to South Africa. Another reflection of the intensified use of oil as a lever is the increasing frequency with which 'incentive crude' is made available to companies (or, more typically, state trading companies closely tied to their governments) investing in oil-related infrastructure. Japanese and British firms and consortia from the Far East are now guaranteed a certain volume of crude oil per day commensurate with the size of their commitment to a new refinery, petrochemicals plant or similar projects.

An impression of the extent of oil-producer country governments' stake in crude oil sales can be obtained from estimates made by the International Energy Agency. In 1970 it is estimated that OPEC producer entitlement was 2 per cent of total OPEC output. By 1973 that share had grown to 20 per cent and by 1979 to 80 per cent. As for importing countries, it is estimated that for non-oil LDCs in aggregate, 32 per cent and 45 per cent of crude imports in 1978 and 1979 respectively arrived through government bilateral agreements. (In Europe, by comparison, the percentages were 12 per cent and 18 per cent in the same two years.)[24]

24. J. M. Mohnfeld, 'Changing Patterns of Trade,' *Petroleum Economist*, Aug. 1980, pp. 329–32.

To the extent that bilateral state trading grows, OPEC members' oil agencies involve themselves more deeply in the distribution of crude (and refined products) and incentive crude contracts spread, then OPEC members will clearly be expanding the policy instruments at their disposal. M. Ait-Laossine, previously of Sonatrach, put it thus at the 1980 Oxford energy seminar:

> The price of oil will less and less be expressed in terms of cash and more and more in terms of resource transfers, of technology as well as hardware. This non-commercial aspect of oil trade will inevitably lead to a further restructuring of the traditional supply channels than we have seen so far.[25]

What these trends point to for developing country statesmen, then, is an increasingly ambitious and imaginative use of crude oil in international affairs—an extension of the diplomat's armoury not foreseen twenty years ago. This enlarged authority of OPEC member governments over oil-importing governments' actions (in some areas at least) points again to a greater degree of tension in OPEC relations with the Third World.

Although security of crude oil supply is coming to acquire a position of some importance—almost a primary objective of foreign policy in certain countries—there are, however, also instances of closer co-operation between OPEC members and non-oil LDCs being brought about. A recent example is the decision of the United Arab Emirates (UAE) and India to construct in the UAE a 12 million tonnes oil refinery, half of whose output of refined products will be assigned to the Indian government.[26] This plan, whose implementation has been enhanced by the UAE offer of 1.5 million tonnes of crude in 1981 to help India over the shortages caused late in 1980 by the Iran-Iraq war, is one of the first large-scale Indian investment projects in the Gulf. It is also an example of countries jointly planning to process more intensively their own primary products while at the same time trying to avoid the excess capacity that frequently accompanies such efforts at industrialisation. Many Third World countries have suffered from refinery excess capacity. Indeed, there are presently around three times more oil refineries in the non-oil LDCs than in the OPEC member states.

There is a further reason for OPEC as an entity to resist the creation of an enlarged oil-aid facility, and still more, a concerted and thorough diplomatic effort to secure Third World support for its position. This reason stems from the fact that each OPEC member has regional as well as global interests, and those regional interests will generally take precedence over global interests whenever the returns appear to be higher from a given outlay of effort. Venezuela, for instance, has for decades harboured a

25. Speech at Second Oxford Energy Seminar, Sept. 1980. Reported in the *Financial Times,* Nov. 3, 1980.
26. *Financial Times,* Dec. 22, 1980.

desire to speak with authority in its region. Now that Brazil and Mexico are so frequently seen as being in the vanguard of the Third World, as newly industrialising countries, there is greater incentive for Venezuela to try to recapture some of the diplomatic initiative. The regional oil concession scheme announced in conjunction with non-OPEC member Mexico in mid-1980 reflects this regional imperative. Under this scheme, the two countries will between them sell some 160,000 barrels per day of crude, available through a five-year 4 per cent credit, to nine nearby countries (Barbados, Costa Rica, El Salvador, Guatemala, Honduras, Jamaica, Nicaragua, Panama, and the Dominican Republic). This scheme should, however, be understood as a concession on credit terms rather than on oil prices. More widespread concessions by OPEC members on prices charged to Third World countries are unlikely, partly because automatic price concessions would jeopardise whatever political leverage OPEC members obtain by negotiations over foreign aid, and partly because of difficulties in arranging and policing such schemes with private oil companies who still control tanker fleets. Similarly with Nigeria, whose various governments since independence have felt that the most populous black African nation should play more of a part in continental affairs. With the eclipse of Ghana and the arrival of multi-party parliamentary democracy in Lagos, the Nigerians can also be expected to take a more prominent part in regional affairs.

As well as different regional priorities between OPEC members, different ideological priorities will also persist as barriers to any greater interest in the idea of Third World solidarity. In contrast to the heady brew of socialism espoused by Algerian spokesmen in referring to future developments in the Third World, the Nigerians (fairly clearly committed to capitalist development) and the Venezuelans (in possession of a relatively sophisticated industrial economy) would prefer not to enter the debate at that level of aggregation.

Internal dissent over collective OPEC policies *vis-à-vis* both oil prices and the non-oil LDCs illustrates these differences in objective. Of importance here is the paralysis of Iraqi–Iranian relations which has in turn adversely affected the unity of OPEC. Thus, instead of a comprehensive OPEC approach on oil prices and foreign aid, as proposed by the Committee on Long Term Strategy, the members have, for the time being at least, been thrown back upon their own devices. Oil price unity has partly collapsed and the function of OPEC as a price-setting organ has been brought into question. But more important in the context of this article, each OPEC member, or a sub-group like the Gulf States, will continue to arrange its relations with individual non-oil LDCs with its own national interests the main priority. Wider OPEC interests, to the extent that they exist, will meanwhile be given lesser weight, although the trend towards channelling the members' foreign aid through OPEC's multilateral institutions is likely to continue.

CONCLUSIONS

In conclusion it is clear that much about OPEC states' interests and priorities remains only hazily comprehended. As was pointed out in *Foreign Affairs* not long ago, 'the forces involved are far from completely understood'.[27] Not least among the barriers to understanding, of course, is the heterogeneity of the Third World itself. Embracing countries so diverse in economic circumstances, in political priorities and in integration in the world economy as Mali, Nicaragua and Singapore, for example, it is little surprise that assessment of OPEC's impact upon this group of states should give rise to confusion and uncertainty. But a few principles can probably be outlined.

In instances where Third World support is forthcoming, and can be acquired at relatively little cost, OPEC as an agency will in all probability continue to seek it.[28] Furthermore, even in instances where substantial aid outlays are involved, OPEC is likely to continue assisting countries as cases of clear need arise. But the OPEC aid agencies will certainly continue to resist the institutionalisation of that aid in the form of a placebo for LDCs' having to put up with rising oil prices. Indeed, neither set of countries is happy to have the OPEC aid effort viewed in that light. For their part leaders of the oil-importing LDCs rarely feel that they receive 'sufficient' aid to mitigate their higher oil outlays. And for their part OPEC members will continue to deny the proposition, implicit in a formal oil-aid scheme, that their oil price policy necessarily harms the rest of the Third World.

27. A. Q. Corradi, 'Energy and the Exercise of Power', *Foreign Affairs,* Vol. 57, No. 5, Summer 1979, pp. 1,144–66.

28. See R. D. Hansen, 'The Political Economy of North-South Relations: How Much Change?' *International Organization,* Winter 1975, Vol. 29, No. 1, pp. 921–47, where southern solidarity is referred to as 'a free good'.

Libya and American Foreign Policy

LISA ANDERSON

The nature of Libyan politics and the appropriate American responses to Libyan foreign policy became subjects of heated debate in the United States during the Reagan Administration's first 18 months in office. Concern that the Libyan leader Mu'ammar al-Qadhdhafi represented—as he claimed—a genuine threat to American interests in North Africa and the eastern Mediterranean, perhaps even further afield, fueled discussion of appropriate American responses. Much of the debate about Libya and American policy took place, however, with little appreciation of the historical context and domestic circumstances of the Libyan regime or of Libyan views of US-Libyan relations.

Libya's experience of the 20th century has provided its people and its leaders with rather dismal expectations about Western behavior in the international arena. Whether these perceptions are entirely fair is less important than the fact that Libyans can be expected to act upon them. General agreement in Libya on the fundamental inequity of the international system and on the necessity of its revision to the benefit of non-industrialized countries did not, however, extend to support of many of the Qadhdhafi government's specific aims, or of most of its means. By the time the Reagan Administration took office in the United States, Qadhdhafi faced widespread opposition at home and no small measure of ridicule abroad. Failure to understand both these characteristics of the contemporary Libyan political scene—general sympathy with revolutionary movements and symbols, and simultaneously, specific local opposition to the ostensibly revolutionary Libyan government—led many Americans to misconstrue the consequences of American policy towards Libya.

Reprinted by permission of the publisher, The Middle East Institute, from *The Middle East Journal,* Vol. 36, No. 4, Autumn 1982.

LIBYA IN THE FIRST HALF OF THE 20TH CENTURY

Libya became independent in 1951. At that time it was the poorest country in the world; its population of approximately one million enjoyed, if that is the word, a per capita income of about 50 dollars a year. The country's major export was scrap metal from the debris of the North African military campaigns of World War II, and the literacy rate was about 20 percent among adult males. The country was, for all intents and purposes, a ward of the United States and Britain, who subsidized the operating budget of the monarchy, and provided foreign exchange in rents for military bases on Libyan territory.

Despite appearances to the contrary, however, Libya was not a sleepy desert backwater. Its career during the first half of the 20th century had been marked by almost constant upheaval, and the accumulated residue of these successive turns of fortune continue to weigh heavily on the contemporary political scene today.

Libya entered the 20th century as the last North African province of the Ottoman Empire. It had been exposed to the reform efforts of the imperial government during the 19th century and, like the rest of the Empire, was shaken by the Young Turk Revolution of 1908. The province was consumed with the ensuing internal upheavals when the Italians declared war on the Ottomans and invaded their North African province in 1911. Although the Europeans were to acknowledge Italian sovereignty in Libya, the legitimacy of Italian control was recognized by neither the Ottoman Empire nor most local Libyan leaders. The Italo-Turkish War lasted only a year, but the Ottoman political and military influence in Libya did not end until the close of World War I, by which time the battles for control of the province among the Italians, the Ottomans, and the provincial leaders had severely undermined the local administration and economy.

The aftermath of World War I saw a number of local efforts to create independent states in the territory, including the short-lived Tripoli Republic, which sent delegates to the Paris Peace Conference, and the Sanusi government. This last, organized around a religious brotherhood which had cooperated with the Ottoman provincial authorities, found a protector in the British in Egypt, who briefly lent their good offices to negotiations between the Italians and the Sanusi leader, Idris.

By 1922, however, the new Fascist government in Rome decided to forego negotiation and undertake the military conquest of the territory. In one of the most brutal colonial wars of the 20th century, the Italians captured control of their North African territory kilometer by kilometer, facing fierce resistance for well over a decade. In the course of the fighting, large portions of the Libyan population were interned in concentration camps to prevent civilian cooperation with the resistance fighters. Wells were poisoned, civilian settlements bombed, and captured resistance fighters hanged on the spot.

By the middle of the 1930s, when Italian control of their North African colony was finally uncontested, the Libyan population had been halved by famine, war casualties, and emigration. Particularly significant was the loss of almost the entire educated elite and much of the middle class. Settled coastal agriculture and domestic trade had been completely disrupted. Even with that, however, the Italian victory was to prove short-lived; the North African campaigns of the Second World War left Libya in British and French hands after a long series of difficult and destructive battles.

During the war the British in Egypt promised Idris, then in exile in Cairo, that eastern Libya—Cyrenaica—would not be returned to Italian control if he would provide troops from among his followers to fight with the Allies. After the war, this promise proved decisive in the debates over Libya's future. The newly established United Nations was unable to agree on a suitable mandatory power—among the various aspirants were both Italy and the Soviet Union—and so Libya was granted independence in 1951 under Idris as King.

LIBYAN INDEPENDENCE: FROM RAGS TO RICHES

The monarchy's administrative structure was simple in the extreme and its decentralized federalism—each of the three major regions, Tripolitania, Cyrenaica, and the Fezzan, constituted an almost autonomous province—reflected the weakness of the central government, the tenuous character of national identity, and the political consequences of poverty. At that time Libya could not afford a large or elaborate bureaucracy, nor would it have been able to staff one with its small population, boasting as it did seven college graduates at independence. Most genuine political and administrative power was delegated to locally prominent families or members of the King's immediate circle, an economical and efficient arrangement in the face of the country's overwhelming poverty.

As if the upheavals of the previous half-century had not been enough, however, oil was struck in 1959. Within less than a decade, per capita income rose to more than 1,500 dollars a year, and despite efforts to revamp and strengthen the administration—the federal system was abolished in favor of a centralized state in 1963—the bureaucracy proved too fragile to withstand the pressures of the new wealth. It proved virtually impossible to separate the public administration from the private patrimonies permitted in the poverty-ridden 1950s. Development planning and even everyday administration soon faltered in the face of system-wide corruption, as the King's Court and entourage failed to distinguish the public treasury from the privy purse.

The monarchy, designed as it had been for a small, poor, and ill-educated populace, was equally poorly equipped to accommodate the new demands of politically conscious students or a new and economically savvy middle class. Political parties had been banned at independence and politics had taken the form of conflict over patronage within the aging King's circle. By the late 1960s, as Nasir's Arab nationalism was beamed over the radio waves from Cairo, the monarchy appeared to many observers to be an anachronism on its way out.

The military coup which brought Colonel (then Captain) Mu'ammar al-Qadhdhafi to power on September 1, 1969, was therefore greeted with more relief than surprise; many plots had been afoot and the only question on the morrow of the coup was the identity of the successful plotters. The King, who was out of the country, abdicated; the militia thought to be his loyal guard stayed in its barracks and within several months of the coup it was disbanded.

QADHDHAFI: THE MAN IN HIS CONTEXT

In many respects, Qadhdhafi presented an uncanny reflection of the "average Libyan" he claimed to represent. Said to be about 27 years old at the time of the coup—he was probably several years older—he symbolized the youth of a country in which well over half the population is under 15 and a mere ten per cent over 50 years old. (With one of the highest population growth rates in the world, this age profile will likely continue to characterize Libya well into the 21st century.) A graduate of the military academy and of a six months' signal course in England in the late 1960s, Qadhdhafi and his fellow Free Officers could claim to be reasonably well-educated by Libyan, if not international, standards.

Qadhdhafi himself was born, as he says, "in a tent" in humble circumstances in Sirte, the border region between Tripolitania and Cyrenaica. His very accent in Arabic, being characteristic of neither of Libya's major regions but of this border area, suggested his commitment to transcending regionalism and developing wider loyalties. Like most Libyans, he combined a rural background with urban residence—fully a third of the population of two and a half to three million lives in Tripoli alone—and like most Libyans he remained ambivalent about his loyalties to his kin. Disdainful of tribalism as a principle of political organization, he was nonetheless proud of his heritage as a member of a *marabut,* or saintly, tribe, and his own relatives became increasingly prominent in the upper echelons of the military and intelligence services during the late 1970s.

Having grown up in poverty and come to political maturity with the influx of oil wealth, Qadhdhafi and his generation of Libyans remained profoundly ambivalent about the country's riches and what they might buy. Their respect for Western technology, particularly military hardware, was born of an intimate acquaintance with its destructive capacity, and was accompanied by a profound distrust of the constructive intentions of its developers. Qadhdhafi was also well aware that the Libyans toiled very little to earn the 9,000-dollar annual per capita income they enjoyed in the early 1980s, and this awareness translated into a domestic policy which oscillated between wasteful spending and artificial austerity and arbitrary traditionalism.

THE REVOLUTIONARY REGIME IN THE FIRST SIX YEARS: BUILDING POPULAR SUPPORT

Although the military coup—or, as it was known in Libya, the revolution—of September 1, 1969, was relatively popular, its leaders had had little contact with civilian opposition to the monarchy and they enjoyed little organized support among the general populace when they came to power. They therefore inaugurated the new regime with declarations of their new, and still somewhat vague, philosophy, and began what would be a series of experiments with administration reorganizations designed to mobilize popular support for the new order.

Qadhdhafi and his comrades declared themselves to be followers of Egyptian President Gamal Abd al-Nasir, devoted to his goals of Arab unity, anti-imperialism, and Arab socialism, to which they added an explicitly political attachment to Islam. The concern with Islam, which prompted the closing of churches and nightclubs, the banning of alcohol, and the reinstitution of Islamic criminal penalties soon after the coup, was largely a response to the character of the old regime in Libya. The King had been the head of the Sanusi religious order, and his overthrow required that the new rules establish their own religious credentials as an element of their legitimacy. Qadhdhafi himself was widely reputed to be personally pious, although the character of his Islam, and his relations with the orthodox *'ulamā'* and the fundamentalists in Libya, was to change quite dramatically in the late 1970s. Nonetheless, the austerity of strict Islam continued to ensure an atmosphere of popular sacrifice appropriate to a revolution but difficult to instill otherwise amid vast wealth.

The first several years of the new regime were consumed with ridding the country of corruption and the trappings of imperialism. The upper echelons of the King's civil administration and military apparatus were purged of corrupt politicians and officers, and public trials were held as the wealthiest entrepreneurs found many of their properties expropriated. The British and Americans were asked to evacuate their military bases, which they did; the remaining Italian residents were expelled; new agreements were negotiated with a number of the oil companies operating in Libya to provide much greater Libyan participation—some were eventually nationalized—and Qadhdhafi took a quite successful hardline activist stance in international negotiations over oil pricing policies; Arabic was restored as the language of all government business; and a number of other substantive or symbolic moves were undertaken to impress upon the world the commitment of the new regime to Libya's honesty and independence.

Simultaneously, the model of the Egyptian single political party, the Arab Socialist Union, was adopted as a vehicle for popular participation, as the regime began a serious and reasonably successful effort to redistribute the oil income more equitably. Public housing, agriculture, education, and industrial development were all targeted for investment in ambitious, if often poorly planned and executed, projects designed to benefit the previously disadvantaged sectors of society. By 1977, investment in education particularly was beginning to show results: 75 per cent of the adult male population was literate and university enrollment went from 3,000 in 1969 to 20,000 in 1979. By this time, no Libyan wanted for housing, medical care, or transportation.

THE REVOLUTIONARY REGIME IN THE SECOND SEVEN YEARS: IDEOLOGY AND ERODING SUPPORT

Nonetheless, as early as 1973, Qadhdhafi and his colleagues felt that for all their efforts, the Libyan population did not exhibit the fervor for revolutionary change they had anticipated. That year the "popular" or "cultural" revolution was launched, as Qadhdhafi announced a campaign against, among others, the feebleminded, the bureaucrats, and the bourgeoisie. As the age of government officials began dropping—through forced retirements and recruitment of youthful enthusiasts—popular committees were established to supervise the country's administration. The Arab Socialist Union was soon moribund, as the committee system was expanded, eventually to be codified in Volume I, "The Solution to the Problem of Democracy," of *The Green Book,* Qadhdhafi's philosophy of the revolution.

In September 1975, the General People's Congress was established as the umbrella organization for the popular committees. At about the same time, a plot against Col. Qadhdhafi was discovered in the army. Although neither the first nor the last of such attempts to overthrow the regime, this aborted coup proved to be particularly significant, for it marked a major split within the ruling Revolutionary Command Council, and with its failure, the defeat of the regime's pragmatic wing. Qadhdhafi, who had resigned all his government posts to become the Secretary General of the General People's Congress, was opposed by his Ministers of Planning and of Foreign Affairs. They objected to expenditures on military and foreign involvements to the detriment of domestic development investment and to Qadhdhafi's continued involvement in the daily affairs of government after his supposed resignation to take a political position. After the discovery of the plot, they fled into exile, a number of their supporters in the military were arrested, and soon the ruling circles of Libya were composed entirely of Qadhdhafi loyalists: ideologues at best, sycophants at worst, but none willing to stand up to the Leader of the Revolution.

Within a year, serious disturbances at the University in Benghazi led to numerous arrests and, for the first time, civilian executions, as seven students were publicly hanged. The implementation of the dictates of *The Green Book*'s Third International Theory (as Qadhdhafi calls his philosophy, to distinguish it from both capitalism and communism) proceeded apace, however, in the face of ample and growing evidence of its less than universal popularity. In 1977 Qadhdhafi resigned all his political positions to become Philosopher of the Revolution, declared "the authority of the people," and Libya was transformed into the Socialist People's Libyan Arab Jamahiriyyah, a "state of the masses" or "peopledom." By the time the second volume of *The Green Book,* on economics, came out in 1978, the reshaping of the economy had already begun. There was to be no private retail trade, no wages, no rent—all these had been declared forms of exploitation—but only profit participation partnerships and worker self-management committees. By 1981, individual bank accounts had been seized and the currency demonetarized, policies designed to guarantee that all Libyans had equal assets.

The implementation of *The Green Book*'s precepts unmodified by pragmatic or practical considerations sparked considerable opposition throughout the late 1970s and early 1980s. The defeat of the pragmatists within the ruling elite prompted an increasingly accelerated brain drain, as technical qualifications became a less important criterion for political influence than ideological loyalty. Corruption became increasingly evident, as a sizable black market developed and military rank became an important qualification for desirable housing. In 1981, it was estimated that as many as 100,000 Libyans lived outside the country.

The Libyan private sector was obviously beleaguered, a development which represented a significant turnabout, since it was the Libyan petite bourgeoisie which was originally favored by the regime. The "non-exploitative capitalists" were encouraged by government economic policies during the early years of the revolution and it was this class which particularly profited from government projects in health care, housing and education. By 1978 the policies of worker self-management had been expanded and the petite bourgeoisie had been abandoned by the regime.

The only economic stratum which continued to support the Qadhdhafi regime was comprised of those who had owned little or nothing before 1975. This does not constitute a particularly numerous group, since there is no Libyan industrial proletariat and most wage earners had been employed in small family-owned enterprises. Insofar as this stratum did exist, however, by the early 1980s it made up—with a segment of the youth—the foot soldiers of the regime, staffing the revolutionary committees and providing the "masses" for demonstrations.

The students, and more broadly, the intelligentsia, were deeply divided over the regime. Many students had received educational opportunities for which they could not have hoped under the monarchy, and among half the population which had grown up under the Qadhdhafi regime, there were a goodly number of pro-government zealots. On the other hand, arbitrary and unpredictable military conscription policies combined with a revolution of rising expectations and a desire for genuine influence in policy-making on the part of the young intellectuals and technocrats to create significant numbers of disaffected and disappointed young people.

Also disappointed was the religious establishment, which was courted in the early years of the regime in the well-publicized Islamic references of the government. The initial support of the orthodox *'ulamā'* in Libya appears to have been a significant aid to Qadhdhafi in his challenge to the Sanusi King's religious supporters. By the late 1970s, however, Qadhdhafi found the religious authorities less than enthusiastic supporters of the Third International Theory, particularly after *waqf* (religious endowment) lands were expropriated as part of the new economic order. Qadhdhafi reacted to religious opposition to his economic policies by lashing out at the *'ulamā'*, instructing the masses to "seize the mosques," and proclaiming that the Islamic calendar should be revised, leaving Libya with a calendar ten years out of step with the rest of the Muslim world.

The widespread opposition among students, the religious establishment, as well as the erstwhile private sector, was met with increasing repression. In March 1979, Qadhdhafi had felt constrained to establish "revolutionary committees" to ensure the attachment of the masses to the Third International Theory and its author. It was these committees who were responsible for the seizure of Libyan Embassies and for the liquidation of a number of self-exiled "enemies of the revolution" in Europe.

By the beginning of the 1980s it was evident that Qadhdhafi's position in Libya was not based on broad popular support. The students were divided, some his most loyal supporters, others, his most bitter opponents. The local artisans, retailers, and other small entrepreneurs, who had been the backbone of his support as beneficiaries of his early social and economic policies, had seen their property nationalized and their businesses closed, and they deserted the regime. The *'ulamā'* stayed home, in sullen but symbolic protest. The increasing number of mutinies and attempted coups reported during the early 1980s suggested eroding support within the military as well.

THE DETERMINANTS OF LIBYAN FOREIGN POLICY: ECONOMIC DEPENDENCE AND POLITICAL IDEOLOGY

Qadhdhafi's international position was also troubled by the beginning of the 1980s, although for somewhat different reasons. An adroit manipulator of international media attention, speaking reasonably good English when the occasion demanded, and concerned with his public image to the point of vanity, Qadhdhafi had managed to convince his foreign supporters and detractors alike that he was a significant regional if not international power. Few outside Libya were aware of the weakness of his domestic position and numerous analysts and policy-makers in the West took him at his word when he claimed to lead world-wide opposition to Western influence at the helm of the vanguard of the Libyan revolution.

In fact, he not only faced widespread if disorganized opposition among Libyans at home and abroad, but much of the structural support of his regime was foreign, dependent upon the global interest in recycling petro-dollars. In 1980, more than half the managerial and professional workforce was non-Libyan, as was one-third of skilled and semi-skilled, and almost 60 per cent of unskilled labor. All told, foreigners comprised 40 per cent of the active workforce in Libya, including 5,000 British, 20,000 Italians, tens of thousands of Tunisians, Egyptians, Chadians and Sudanese. Maltese telephone repairmen, Pakistani English instructors, Palestinian lawyers, Bulgarian road crews made up much of the human infrastructure of the Libyan economy, and few of these foreign laborers can be said to have been devoted to "popular authority" and its leader.

The economy in general reflected the limitations of a capital-surplus, labor-short structure. Fully 70 per cent of the GDP was accounted for by a single export, petroleum, while agriculture, which employed 30 per cent of the workforce, accounted for two per cent of GDP, and provided less than 40 per cent of the country's food requirements. Like its sister oil-exporting countries in the Persian/Arabian Gulf, Libya would find its absorptive capacity very severely restricted if it did not rely heavily on foreign labor and imports. Unlike his fellow oil-exporting rulers in the Gulf, however, Qadhdhafi felt no compunction about declaring his opposition to the international system upon which his country was so dependent.

Qadhdhafi came to power declaring his Nasirist persuasion, and in foreign policy he wavered little thereafter. He remained devoted to Arab nationalism and Arab unity, which he interpreted as entailing unremitting opposition to Israel as an outpost of Western imperialism and an interloper in the Arab world, as well as continuous efforts toward political unity among the Arab states. Over the course of his first 13 years in power, however, Qadhdhafi had little to show for his efforts, and his ideological inflexibility and occasionally erratic behavior alienated many of his counterparts in the Arab world. Indeed, for all his ardent and lavish courtships, he had not a single example of permanent union to show for his efforts and he was widely considered in the Arab world to be a difficult and annoying member of the political family, tolerated but disliked.

Like Nasir, Qadhdhafi came to power believing that unification of the Arab nation is a necessary prior condition to economic and social development and the recapture of dignity in the Arab world. His failure to achieve unity led in the mid-1970s to greater attention to consolidating the Libyan revolution at home and preparing for its export abroad. Qadhdhafi apparently came to feel that Arab unity might be successfully achieved in certain cases only after the overthrow of the present—reactionary—Arab governments. His willingness to support opposition groups in Arab countries whose regimes he had decided are hopelessly anti-revolutionary—like Morocco, Tunisia, Egypt and Sudan—did little to endear him to his Arab colleagues and much to raise concern among Western countries.

The concern voiced in Western capitals over Qadhdhafi's alleged subversive activities in countries thought to be friendly to Western interests, particularly those of the United States—in Africa as well as the Arab world—reflected an accurate assessment of Qadhdhafi's views of the West, if an exaggerated sense of his capabilities. He began his tenure in office as an exponent of anti-imperialism and non-alignment in the Third World and his government became increasingly opposed to specifically Western influence—in his terms, more concerned with anti-imperialism and less with non-alignment—over the course of the 1970s.

Qadhdhafi's views of the inevitability of Arab unity and of the malevolence of Western influence are exaggerated versions of perceptions widely shared in Libya. Few Libyans are as convinced of the feasibility of a single Arab state or of the likelihood of Third World revolution against imperialism. Many, however, share a strong sense of brotherhood with Arabs elsewhere and a mistrust of the West which both grow out of their experience of Arab and Muslim support against a particularly brutal form of European imperialism.

Qadhdhafi's means—particularly his lavish use of Libyan oil revenues to fund international movements, and his rigidity on questions of revolutionary purity and of uncompromising opposition to Israel—are widely opposed within Libya. His ends—a diminution of Western influence in the world and closer ties among Arab, Muslim, and African countries—are by contrast widely considered legitimate.

THE LIBYAN MILITARY'S "INTERNATIONAL VOCATION"

Out of both ideological conviction and extrapolation from what is in fact Libya's anomalous experience, Qadhdhafi has held little brief for international frontiers and formal citizenship. Libya's reliance on foreign labor, particularly from the Arab world and Africa, confirmed Qadhdhafi's perception of the arbitrariness of national boundaries and the interdependence of the Arab and African states. As he frequently declared, the Libyan revolution was based on an international ideology, not on a national movement. The foreign workers in Libya, whether they appreciated it or not, were to form the vanguard of this international revolution, providing the

Libyan regime with a recruiting ground for the international wing of the Popular Militia, or People's Army.

The Libyan military establishment included two major elements: the regular military, including the three conventional branches, and the People's Army. The regular military was a reorganized and expanded version of the monarchy's services; the King's militias were disbanded and incorporated into the regular army, and within a year of the 1969 coup, the army included about 22,000 troops. By 1979, 37,000 troops were on active duty and in late 1981, this figure was estimated to have risen to between 55,000 and 65,000. They were, by general consensus, poorly trained, and they have not notably distinguished themselves in what combat they have seen—against Egypt in the spring of 1977, in Uganda in 1978, or in Chad in 1981.

This regular army, and the small naval and air forces, were extremely well-equipped. It was indeed one of the major sources of concern in the West that since the mid-1970s, particularly after the abortive coup of 1975, the Eastern bloc and especially the Soviet Union had been lavish in its supply of modern, sophisticated equipment to Libya. An estimated 12 billion dollars' worth of equipment had been transferred to Libya by the USSR since 1976, and there were estimated to be about 2,000 Soviet military advisors, and an equal number of East German and Cuban technicians, in Libya.

Apart from prestige, which was no small element in the calculations of the Libyan leadership, it was not altogether clear what Libya intended to do with its sophisticated equipment. Its purchases from a variety of suppliers including Britain, France, and Italy, and the use of American trucks for military transport created standardization problems which were reported to have seriously hampered logistics while the army was in Chad. The importance of prestige and of disorganized procurement policies should not be underestimated in assessing Libyan arms purchases. Qadhdhafi had shown himself to be enamored of high-technology expensive solutions to agricultural as well as military problems, and his willingness to pay cash provided him with many eager suppliers. The Libyan regime would have liked to boast of being able to match Israel's military strength—something well beyond the price the Libyans were likely to be willing to pay in work if not in money—and this seemed to underpin Libyan efforts to aid development of an "Islamic bomb." Libya had had a Ministry of Nuclear Energy since the mid-1970s and was thought to have been quite lavish in underwriting Pakistani efforts to develop a nuclear capability, presumably in hopes of sharing in the proceeds.

The Soviet intentions in providing this equipment were also somewhat obscure, permitting many Western analysts to suggest that they were stockpiling, or prepositioning, equipment for their own or their allies' eventual use. Since Libya had pointedly refused to sign the standard Treaty of

Friendship Soviet clients ordinarily agreed to, however, the relationship appeared to be primarily a business arrangement. The Soviet leaders were cordial to Qadhdhafi but rather open about their mistrust of him; they evidently found him very unreliable, and the terms they used to describe him—an "idealist," a "Muslim fanatic"—suggested their skepticism. It was probably an element in Soviet calculations, however, that Qadhdhafi had paid for his military purchases in hard currency—much of it dollars—and this has always been an added incentive for the Soviets and their allies. Indeed, for some it is quite significant: Cuba's second-largest earner of foreign exchange after sugar in 1979–80 was its commercial, military, and technical ties with Libya and Iraq, undistinguished in the Cuban accounts.

In contrast to the Soviets, who appeared to advise the regular army, and the East Germans, who provided Qadhdhafi's bodyguards and advisors to the intelligence services, the Cubans, along with Palestinians and other Arab advisors, helped train the Popular Militia, or People's Army. This force was created after several attempts in the mid-1970s to supplement the regular army with a nation of reservists, incongruously said to have been modeled on the Israeli military reserve system. By the end of the decade it was estimated that some 45,000 Libyans had received some training in the People's Army. There appeared to have been little or no coordination between this force and the regular army. It also appeared that it was in the People's Army that most of the non-Libyans who had been trained in Libya got their military education.

LIBYA IN AFRICA AND THE ARAB AND ISLAMIC WORLDS

Apart from the isolated volunteers in, for example, southern Lebanon with the Palestinians operating there, Libyan nationals had seen service outside the country in only two instances: Uganda and Chad. The airlift of Libyan troops (and, it was said, some non-Libyans resident in Libya) to aid the failing regime of Idi Amin in 1979 cost the Libyans perhaps as many as 500 casualties and provoked widespread disaffection at home. It appeared to have been a serious miscalculation on the part of Qadhdhafi himself, since he was said to have ordered the mission against the recommendations of his military advisors. Thereafter, and with the exception of Chad, Qadhdhafi avoided sending Libyans into action and instead contented himself with training, equipping or funding dissidents in Arab or African countries whose governments he found objectionable.

In doing so, Qadhdhafi seemed to feel he was justified in supporting national liberation movements, a category that not only included Tunisian opponents of the Bourguiba regime, like those who launched the abortive raid on Gafsa in January 1980, but also the Palestine Liberation Organization and the Irish Republican Army. In many of these cases, efforts to impose the strictures of the Third International Theory, as well as less than completely reliable funding, have created serious strains in the relationship between Libya and the beneficiaries of its aid. Several times in the late 1970s and early 1980s, the PLO offices in Tripoli were closed, for example, while PLO leader Yasir Arafat and Qadhdhafi feuded, only to reopen when they patched up their differences.

Among the very few successful movements Qadhdhafi's government aided was the Khomeini opposition to the Shah of Iran. Qadhdhafi apparently expected that Khomeini, upon taking the reins of power in Iran, would immediately transform the country into a second outpost of Libyan-style "popular authority." Qadhdhafi's second in command, 'Abd al-Salam Jallud, was among the first to arrive in Tehran to congratulate Khomeini, and the Libyans were surprised by the chilly reception they were given. Despite his revolutionary rhetoric, Qadhdhafi was not notable in his support of Islamic movements unless their interests coincided with another of his concerns—anti-imperialism, for example—and he apparently forgot, in welcoming the Iranian revolution, the unhappiness within the Shi'i Muslim community over the disappearance of the Lebanese Shi'i leader Musa Sadr in Libya in August 1978. Despite his disappointment in the failure of the Iranian revolutionaries to acknowledge his aid, he still seemed to feel that the most recent deposers of Kings should stand together, and he supported Iran in its war with Iraq.

Qadhdhafi's fortunes in Africa were mixed over the decade of the 1970s. His early successes in replacing Israeli technical aid with Libyan financial aid in Black Africa had lessened Israeli influence in the continent and weakened African support for Israel in international organizations, although parallel Saudi efforts must also be counted in this outcome. Qadhdhafi's support for Idi Amin—which had grown out of his "conversion" of the Israeli-trained Ugandan ruler—well after most of Africa had abandoned him tarnished Libya's image in Africa, however, and in the early 1980s, Israel appeared to be regaining some of its lost ground. The Libyan support of Jerry Rawling's second coup in Ghana in early 1982 may have been partly a consequence of the civilian government's revived relations with Israel.

In the Maghrib, Qadhdhafi's support of opponents of the Bourguiba regime did little to endear him to the Tunisian government, once the object of Libyan unity efforts. Qadhdhafi's support of one of the factions of Polisario, the liberation movement of the Western Sahara, damaged his relations with Morocco as well as with Algeria, which supported another faction. Since his aid to Polisario had apparently been, as with other such movements, inconsistent, occasionally manipulative, and certainly unreliable, he won few permanent supporters even among his erstwhile allies. Moreover, he had publicly said he did not even support an independent Sahraoui state, preferring union of the Western Sahara with Mauritania.

LIBYA AND CHAD: AN EXCEPTIONAL CASE?

The major exception to Qadhdhafi's policy of providing financial aid, training and equipment to "national liberation movements," but not troops, was his action in Chad. Interpreted by many in the West, and in sub-Saharan Africa, as the first step toward a Qadhdhafi-led "Saharan Islamic Republic," the Libyan move into Chad in December 1980 raised the spectre of Libyan-backed destabilization efforts throughout northern and western Africa. Certainly Qadhdhafi himself did little to reassure his neighbors of his limited intentions, calling on the Tuareg nomads who inhabit much of the Sahara to advantage themselves of military training in Libya before rising up against the governments of the states in which they live.

Within three months of the Libyan move into Chad, six West African countries—Senegal, Gambia, Mauritania, Niger, Nigeria and Ghana—announced that they had found evidence of Libyan interference in their internal affairs, and Gabon and Cameroons implied as much. That the Qadhdhafi regime had suddenly begun a concerted campaign of subversion in West Africa was possible—there was certainly speculation at the time that his failures to realize unity in the Arab world had led Qadhdhafi to shift his gaze southward—but it was equally possible that the African governments had found a plausible foreign scapegoat for domestic unrest. Indeed, the truth of the matter very likely lies somewhere in between: as in Tunisia, Qadhdhafi was willing to finance and train dissidents, but he did not provide them with the rationale for their dissent; that developed at home.

Libyan involvement in Chad was, however, of a different order. What is now northern Chad was considered by the Ottomans an integral part of their North African province and the border between the two countries was not established until the late 1920s, in French and Italian negotiations. The seizure of the Aouzou strip, a 60-mile section of desert along the border, occupied by Libya in 1973 and formally annexed two years later, was considered by the Libyans a rectification of a border disputed by even the Italians in their era.

Apart from the border itself are the long-standing tribal and ethnic ties of southern Libya and northern Chad. This is a very permeable frontier, and Libyan resistance fighters found refuge from the Italians in Chad well into the 1930s. The 16-year civil war which followed Chad's formal independence from France had claimed 5,000 lives and left at least 20,000 refugees by the time the French pulled out their last paratroopers in May 1980. The French withdrawal prompted a defensive agreement between Libya and the nominal Chadian government—in fact, only one of a number of the factions in the war—and by the end of the year, Libya was asked to intervene to support the government.

Although the Libyan intervention was greeted with dismay in African capitals, it was not until the fall of 1981 that, under pressure from the

United States and France, the Organization of African Unity agreed to establish a peacekeeping force to replace the Libyans. The Chadian government, in response to a request from the OAU, asked Libya to withdraw its forces, their place to be taken by those of the OAU. Qadhdhafi agreed, the OAU proved unable to replace all the Libyan troops, and by the beginning of 1982, the situation in Chad had once again begun to deteriorate. It appeared that Qadhdhafi, designated as host of the OAU summit conference in 1982 and the Organization's next chairman, had bowed to African pressure rather than risk losing this prestigious and potentially influential post. It also seemed unlikely, however, that Chad had seen the last of the Libyans.

US-LIBYAN RELATIONS AND AMERICAN POLICY TOWARD QADHDHAFI

Libyan-American relations were difficult throughout the four American administrations which span Qadhdhafi's tenure in Libya. Qadhdhafi came to power espousing non-alignment and although his opposition to communism was a cardinal element of his international orientation, he was no less opposed to Western influence in the world, which he regards as imperialism. Indeed, over the course of his first 13 years in power he made it clear that he viewed Western imperialism—which he interprets as including the existence of Israel—the greater threat to international equity and national independence.

The Nixon, Ford and Carter Administrations responded to Qadhdhafi's opposition to American influence with combinations of carrots and sticks. His initial request that the US Wheelus Air Force Base outside Tripoli be evacuated was met with alacrity and the United States made efforts to cooperate with the new regime. It is in fact often alleged both within Libya and in the United States that Qadhdhafi received warnings of at least one anti-government plot in the Libyan army from the CIA. Certainly military transport aircraft ordered during the monarchy were delivered after the military regime came to power.

Qadhdhafi proved unmoved from his anti-imperialist position by apparent American flexibility and he continued to put pressure on American firms operating in Libya, particularly the oil companies. By 1972 American diplomatic representation in Libya had been reduced to the level of a *chargé d'affaires*. The most important of the differences between the two governments came, however, over the issue of Israel. Qadhdhafi participated enthusiastically in the oil boycott of the United States which followed the 1973 Arab-Israeli war, nationalized several US oil companies when Washington attempted to create a consumers' union in 1974, and broke with Egyptian President Sadat as he began to move toward closer ties with the United States after the 1973 war.

By 1975 the United States refused to permit delivery of military transport planes ordered—and paid for—by the Libyan government. Through the second half of the 1970s relations were quite strained as the United States accused Libya of supporting international terrorist groups, and the Libyan government accused the United States of supporting imperialism and terrorism in its backing of Israel. In December 1979, soon after the taking of American hostages in Iran, a government-backed mob sacked and burned the US Embassy in Tripoli. By May of 1980 the US pulled out its last diplomats from Libya and a year later, accusing them of coordinating threats against Libyans in the United States, expelled the Libyan People's Bureau, or Embassy, staff from the United States.

Throughout the 1970s the apparent strain in relations was partially belied by ambiguous signals from both sides which suggested that each was interested in better relations. Libya continued to enjoy cordial relations in the American private sector, awarding lucrative contracts to American business and providing funds to support Middle Eastern studies programs at several American universities. Until quite recently 40 per cent of Libyan oil exports went to the United States, constituting as much as 15 per cent of American oil imports. (High Libyan prices and a glut on the world oil market reduced this trade in early 1982 to less than 20 percent of Libyan production and one per cent of American consumption.) Conversely, the United States was represented in Libya not only by private businessmen of impeccable credentials, but also private entrepreneurs who made no secret of the good, if informal, contacts with their former employer, the CIA; a number of other well-connected political and social figures, from Muhammad Ali and J. William Fulbright to President Carter's brother Billy, visited Tripoli during the late 1970s. Indeed, during the crisis in Iran, it was reported that Qadhdhafi was approached by American government officials to explore his acting as an intermediary with the Iranian government.

Although from both sides of the US-Libyan relationship the "carrots" were viewed as no more than incentives to modify what each deemed as the other's reprehensible behavior in international relations, they were received—again on both sides—as indications of more flexibility than probably existed. The lack of congruence between the hard-line hostility of the rhetoric and the continuing contacts of reality proved to be a source of ambiguity and misunderstanding. Both the Libyan and the American policy-makers interpreted the mixed signals as suggesting each government was more amenable to pressure than was the case.

The Reagan Administration came into office determined to take a new and apparently more consistent position toward Libya. Qadhdhafi was selected for special attention by the United States as the symbol of all the United States finds repugnant in international affairs—support for international terrorism, opposition to a peaceful solution of the Arab-Israeli conflict, and support for a diminished US role in the world. The expulsion of the Libyan diplomats was accompanied by much fanfare and the long-standing tacit US agreement not to challenge Qadhdhafi's declaration of 200-mile territorial waters was abandoned. By August 1981, US naval maneuvers in the Gulf of Sirte—waters to which Libya's claim was unrecognized in international law—had precipitated a dogfight in which two Libyan aircraft were shot down. The publicizing of Libyan transgressions against "international norms of behavior" continued throughout the fall to culminate in announcement of a "hit squad" allegedly sent by Libya to assassinate President Reagan. The United States then cancelled all passports for travel to Libya, requested all Americans leave the country and, in March 1982, announced the embargo of oil imports from Libya and the banning of high-technology exports.

The Administration argued that this new policy demonstrated that the United States would not tolerate lawlessness in the international arena. The relatively high profile the Libyan issue was accorded by American officials was designed to suggest the increased resolve of the Reagan Administration not to tolerate what it viewed as repugnant behavior, particularly from Third World countries.

Beyond the symbolic utility of acting upon "moral outrage" in relations with governments like Qadhdhafi's, the Administration hoped to modify Qadhdhafi's behavior. This would have presumably served two purposes—to correct the behavior itself, and to demonstrate the power of the United States to accomplish such a change in another country's policies. The seriousness with which this goal was taken may be indicated by President Reagan's own suggestion that the US would have liked to see Qadhdhafi out of power if he was not willing to accommodate American demands.

Libya appeared to be an appropriate focus for the Administration's attention. It is not a particularly strong or important country—its high visibility being entirely a function of its strategic position and its oil reserves—and hence it was an easy target for American opposition. It had close ties with the Soviet Union and therefore any change in Libyan policies could have been interpreted as a loss for the Soviets. Insofar as Qadhdhafi's policies did constitute a threat to pro-Western governments in Africa and the Arab world, and to Western interests in the Middle East and Mediterranean, American opposition to his government was thought to illustrate the Administration's interest in being a reliable friend and ally.

Although there were a number of grounds upon which the Reagan Administration policies were questioned, serious criticism took two major lines. Observers asked both whether the Administration's aims were desirable and whether the policies were likely to produce those desired goals.

In the first instance there was considerable skepticism, particularly in the Third World but also in the United States and Europe, that it is in fact appropriate for the United States to take upon itself the determination and policing of international political norms above and beyond its already preponderant role in formal international order and alliances and *de facto* leadership of the Western world. Qadhdhafi posed this problem most starkly, since he denied the legitimacy of the international order. Few Third World leaders went so far in absolutely denying historical legitimacy to the international system, but many do view it as inequitable and in need of significant reform. The imposition of American values and interests upon other countries, particularly through threats, is widely considered inappropriate to a country which already benefits from a major role in world affairs.

Even many who shared the Reagan Administration's abhorrence of communism and concern about Soviet and Eastern bloc influence in the world, who acknowledged the legitimacy of American leadership of the Western world—that is, who broadly agreed with American aims—nonetheless found the approaches of the US government dismaying. Both the Europeans and their ex-colonies in the Third World found the belligerent American rhetoric and declarations of moral outrage all too reminiscent of an age when the imperial powers made all the rules and claimed all the moral rectitude.

Finally, the likelihood that the US government policies would produce the desired ends was also questioned. Although Qadhdhafi had demonstrated considerable flexibility on tactics over the previous 13 years, he had shown little interest in altering his basic ends, that is, a general lessening of outside interference in the Middle East and Africa, and particularly a diminution of Western influence and a reordering of the international balance of power to the benefit of the Third World. It therefore seemed likely that he could be persuaded to lower Libya's profile as a refuge for opponents of the international *status quo* or of governments supporting that *status quo,* but very unlikely that he would abandon the role altogether. Pressuring him appeared unlikely to produce a significant change in his position, but it did serve to illustrate his longstanding contention that the United States will stop at nothing to maintain its control over Third World countries.

There was no small irony in this, since by the time the Administration's request that all Americans leave Libya was carried out in early 1982, the potential for US influence in the country through personal or diplomatic channels was nil. The US oil companies operating in Libya had replaced their US personnel with nationals of other countries, and the Libyans themselves found they could purchase technical expertise from countries like Iran, which sent at least twenty technical petroleum experts to Libya by late 1982. Insofar as the United States denied Libya access to its goods and services, be they commercial exports, technical expertise, political or educational knowledge, the Americans made genuine non-alignment a virtual impossibility and increasing Libyan reliance on Eastern bloc support became almost a self-fulfilling prophecy.

The considerable publicity given to American opposition to Qadhdhafi also, and perhaps most importantly, undermined the position of his domestic opposition. Although there was widespread disaffection at home and dismay abroad with Qadhdhafi's government, there was also equally widespread suspicion of the West and particularly the United States in the Third World—as had recently been demonstrated in Iran. Few politically knowledgeable Libyans would have been willing or able to establish the legitimacy of a new government in a context suggesting covert American support. Thus, even if there were no such support, the appearance would have been enough to create serious questions about the *bona fides* of a successor regime.

Moreover, the United States could not expect that Qadhdhafi's successor would be any more amenable to American interests. Although Qadhdhafi's tactics, his ideological fervor, and his lack of pragmatism were very unpopular within as well as outside Libya, many of his political perceptions were widely shared. As a rule, the Libyans view high oil revenues as reparations for their devastation by Western imperialism and, unlike the Saudis, for example, they see no reason why these payments should be limited by the economic or political needs of the West. Cynicism about the potentially benign role of the West, or of industrialized countries in general, is deeply rooted in Libya, and an enthusiastically pro-Western foreign policy would not have been popular.

Libyans often remark on the vertigo with which their recent history has left them. From Ottoman province to Italian colony, from the devastation of a colonial war to the destruction of World War II, from abject poverty to splendid riches, from conservative monarchy to radical revolution, the Libyans have had barely a decade of peace, stability and continuity in the last 80 years. Many educated Libyans acknowledge that this had confused their sense of what their country should be, has undermined efforts at calculation and investment for the future, and has created a cynicism and alienation well beyond what is already typical of the Third World.

American policy has yet to reflect a sensitivity to the historical legacies and domestic politics of most countries of the Third World, and in this Libya is but a case in point. The Libyan case may illustrate more starkly than most, however, the futility—indeed, the negative consequences for both the United States and the countries of the Third World—of policies based on preconceptions and misunderstandings. To expect the Libyans to view their national interests as consequent with those of the United States is to expect them to forget their own history, and although most Americans may not know that history, every Libyan schoolchild does.

Domestic opposition to Qadhdhafi weakened his claim to leadership of national, to say nothing of worldwide, revolution, by the early 1980s and suggested that although his government may have merited watching it did not constitute a serious threat to the international order. The Libyan opponents to the regime were as well-versed in Libyan history as its supporters, however, and among the lessons they learned from that history were those drawn from the frequency with which foreign countries had stationed troops on Libyan soil, imposing and deposing governments. They felt the need to maintain their country's independence as strongly as they did their opposition to its government. International pressure to act against the regime merely served to reinforce their commitment to Libya's freedom and independence and to postpone a genuinely Libyan resolution of their differences with their government.

The Question of Gulf Security

The Middle Eastern oil producers discussed in these pages, with the exception of Algeria and Libya, are all located in the Gulf. Because of the preponderance of oil in this region, the Gulf has been the recipient of increasingly high visibility in recent years. At the same time, however, it is also a region of great vulnerability, both because of its central importance in the world's economy and by the prevalence of regional disputes and internal schisms.

By and large, the Gulf states are small countries, relatively underdeveloped even by Third World standards and of recent creation as nation-states. Their fragile political systems must regularly contend with a multitude of domestic divisions, whether social, ethnic or ideological. The revolution in Iran is the most obvious example of a government unable to cope with the combination of strains caused by these existing factors and the additional pressures brought about by the influx of oil wealth. It may be true that Iran was a special case; that the mixture there of national fragmentation, unfulfilled expectations and—above all—the singular nature of the Shah's tight-fisted rule, really does not obtain elsewhere in the Gulf.

Still, dangerous pitfalls are painfully obvious in Iran's neighbors. Ethnic cleavages can lead to civil war, as has plagued Iraq and its Kurdish population for decades. Ideological opposition may move from underground into open rebellion, as seen in the Marxist challenge in Oman's Dhufar province. Events in one country provoke ripples in their neighbors: Kurds are numerous in Iran (and other nearby countries) as well as Iraq; Dhufar raised tremors throughout the monarchical regimes of the Gulf. The problems raised by these and other divisions are treated in the first selection. Hermann Frederick Eilts places particular emphasis on the basic dichotomies between Arabs and Iranians, Sunnis and Shi'is, as well as on the various national rivalries around the Gulf littoral. Disputes and hostilities between neighbors are unfortunately endemic to the Gulf, with territorial claims adding another dimension to the above-mentioned factors. The outbreak of war in 1980 between Iraq and Iran was the most recent and serious manifestation of this source of instability. (A brief summary of the principal causes of the war and a chronology of its progress are contained in Appendix F.)

At the same time, there are many external issues which may adversely affect the Gulf. Two prominent examples of the intrusion of Arab-Israeli issues into the Gulf in 1981 were Israel's airstrike on the Iraqi nuclear reactor and the controversy surrounding the sale of five American AWACS aircraft to Saudi Arabia. Upheaval in the Horn of Africa and Soviet moves in Afghanistan have also contributed to nervousness in Gulf capitals. Adeed Dawisha provides a succinct view of the concern held in one of these states, Saudi Arabia, on the matter of the various threats surrounding the Gulf. He points out that potential threats to the stability of the Saudi regime arise from both internal and external factors, the pressures generated by rapid change inside the kingdom and ideological challenges posed by Riyadh's neighbors as well as the Arab-Israeli spillover and the specters posed by Iran and the Soviet Union.

The third article deals with one central area of concern which has already erupted: the Iran-Iraq war, which began in September 1980 and was still very much alive into the beginning of 1983. John Duke Anthony's contribution is not concerned so much with the course of the fighting itself but with the ramifications of the war on both regional and extra-regional actors. While the expense and counter-productive shocks of the fighting have weighed most heavily on the two combatants, the war has had a strong ripple effect on the conservative Arab monarchies of the Gulf, as well as the entire Arab world which sees itself caught between the Iranian threat on the east and the Israeli threat on the west. For both the superpowers, the war has been an embarrassment: both Moscow and Washington have little to gain from the victory of either Iran or Iraq and have generally attempted to remain neutral, insofar as that has been possible.

The Gulf states are, of course, fully aware of the worldwide attention their oil has generated and of the discussion of strategies for "protection" or control of their fields and production facilities. Their governments naturally object to outside interference in what they see as properly a regional concern. Their announced intention of providing for their own security resulted in the 1981 formation of the Gulf Cooperation Council (GCC), envisioned as a joint economic/political/military venture. But as Valerie Yorke points out below, serious disagreements remain present among the members over the direction of the GCC. Its effectiveness is further sapped by the nonmembership of the two "great powers" of the Gulf, Iran and Iraq. Apart from their war, serious difficulties complicate their admission: Iraq is not trusted because of its radical ideology and Soviet ties while the second is an outsider by virtue of being the only non-Arab state in the Gulf.

But the small size and military weakness of these countries raises the serious question of whether they can really be masters of their own destiny. The West's vital dependence on oil and perceptions of a growing potential for its denial has meant the undertaking of more direct—and unilateral—commitments for the defense of the Gulf. The assertion that Soviet moves in Southwest Asia and the Horn are directed at the Gulf has added to American determination to secure continued access to Gulf oil by whatever means necessary. Reaction from the six conservative members of the GCC has been ambivalent. Economic, political and even military ties to the West have been and will continue to be extensive, but there is considerable concern that American interests in the Gulf may not always coincide with those of the people of the region.

Bibliographic Sources for this Chapter:
2, 8, 13, 16, 17, 45, 51, 52, 60, 62, 69, 71, 72G, 72H, 73, 76, 80, 85C, 85L, 86, 91, 92, 99, 108, 115, 117, 118, 124, 137, 143, 144, 148, 164, 174, 175, 182, 188, 192, 193, 219K, 219L, 226, 227, 228, 229, 237, 240, 241, 242, 247, 255, 259, 262, 263, 264, 270, 271, 296N, 296O, 303, 306, 308, 310, 311, 312, 327, 345, 348, 353H, 361, 362, 373, 382, 387, 388C, 389, 396, 397, 399, 404, 407, 412; see also entries for Iran/Iraq War (Appendix F).

Security Conditions

HERMANN FREDERICK EILTS

Conceptually, potentially destabilizing factors endemic to the Gulf exist on both vertical and horizontal planes.

On a *vertical* plane, they are essentially two in number: the Arab-Iranian jousting for political pre-eminence in the Gulf and the related, but differently motivated, Sunni-Shi'i sectarian conflict. The latter, of course, has political overtones. Both are vintage conflicts in the sense that they have afflicted the Gulf peoples for centuries; both are rooted in age old religious-cultural disparities; both are hegemonial in objective; and both have in the past been characterized by a high level of mutual intolerance and distrust.

Iranians, ethnically and linguistically dissimilar from their Arab neighbors and legatees of a 2,500-year-old civilization, look with a certain contempt upon Arabs and often act accordingly. Arabs are aware of this Iranian condescension, and frequent past heavy handedness, and resent it. During the Shah's regime, while Iranian relations with the Gulf littoral states were superficially improved, the Arabs were disquieted by Iranian efforts to impose a *Pax Irani* on the Gulf. The Iranian handling of the median line controversy with Saudi Arabia in 1965–67, involving renunciation of an already signed but not yet ratified treaty, delineating a median line between the two countries in the Gulf, and the subsequent Iranian naval seizure of an ARAMCO rig, with a Saudi official aboard, aroused deep Saudi bitterness, which even the Shah's visit to Saudi Arabia in 1969 did not entirely overcome. "International piracy," the late King Faisal called it. Similarly, the Iranian military seizure of the Greater and Lesser Tunb Islands and Abu Musa in 1971 was humiliating to Arabs up and down the Gulf. Nor has Iraq forgotten that it was virtually compelled in 1975,

Reprinted from Hermann F. Eilts, "Security Considerations in the Persian Gulf," *International Security,* Vol. 5, No. 2, Fall 1980 by permission of The MIT Press, Cambridge, Massachusetts.

admittedly through Algerian mediation, to agree to a revision of the Shatt al-Arab boundary in favor of Iran.[1] The Iranian claim for the *thalweg* as boundary in the Shatt al-Arab made some sense, but the Shah's heavy handed pressure to assert it disturbed Iraqis and other Gulf Arabs. The Shah's abortive idea of proclaiming the Gulf a regulated semi-enclosed sea likewise found little support among Gulf leaders. They saw it as an Iranian effort to control ingress and egress to the Gulf, which would adversely affect them.

Seizing upon the internal problems confronting the Khomeini regime, the Iraqi Baath government of President Saddam Hussain has taken the initiative to press Iran. Posing as champion of Gulf Arab rights, Baghdad has stridently demanded the return of the Tunbs and Abu Musa to the Gulf principalities from which they were seized, Ras al-Khaima and Sharja respectively. It has likewise publicly called for a reversion of the Shatt al-Arab region to pre-1975 arrangements under which the northern bank of the river, save for two five-kilometer stretches fronting Abadan and Khorramshahr, was the international boundary between the two states.

Similarly, Iraq engaged in promoting ethnic strife in Iranian Khuzistan-Arabistan, as the Arabs dub it. Iraq's "Voice of the Masses" radio exhorted Iran's Arab population to oppose Tehran's rule and Iraqi-trained subversives were infiltrated into Khuzistan. Iran charged Iraq with culpability in recurrent sabotage of oil and natural gas pipelines in that province, while Iraq accused its neighbor of complicity in terrorist attacks, in one of which a Deputy Prime Minister was wounded. A rash of border clashes erupted between the two countries and, amidst mutual recriminations, Iraq and Iran virtually suspended all diplomatic relations. Ever the whipping boy, the United States is charged by each with complicity in the other's nefarious designs—regrettably, a convenient device in recent years to tarnish images in the public eye. Iraq's objective is an old one: the severance of Khuzistan from Iran or at least broad autonomy for Iran's Arab populace. Though less militant on the subject, other Arab states share the conviction that Iran's Arab minority is mistreated.

Conversely, the Iranian Islamic revolution has had its fallout on the Arab side of the Gulf. Khomeini has publicly proclaimed that his revolution is for export. In the Gulf area, his success has evoked two kinds of reactions, partly contradictory. At the grassroots level, it has instilled a new sense of pride and even militancy among Muslims everywhere, which governments will have to take into account. At the governmental level, and especially among Sunni leadership elements, its provenance, Ja'fari Shi'i Iran, makes it suspect. Some contend that the Sunni-Shi'i split has lost its significance. This writer's observations do not bear this out. To be sure, the World

1. The agreement also called for stoppage of Iranian support to Iraq's dissident Kurdish minority.

Islamic Congress has sought to instill a spirit of ecumenicalism among sectarian *'ulama,* or religious leaders. Sunni leaders have been more amenable to this appeal than have their Shi'i counterparts. To the latter, Khomeini's success is a high water mark not only of Islam *per se,* but also a vindication of Shi'ism's millenia-old claim to lead the Muslim community. This inevitably casts a pall over interstate relations in the Gulf. As a factor adversely affecting intercommunal relations, the divisiveness of the Sunni-Shi'i schism also remains strong and especially so in parts of the Gulf and the Arabian Peninsula.

The Iraqi Baath government's policy toward its sizeable Shi'i minority has of late been two pronged: on the one hand, the community is kept under tight control by police methods if necessary; on the other, efforts have been launched to show a greater understanding for Shi'i religious, as opposed to political, concerns. President Saddam Hussain has publicly called upon Iraqi Sunnis and Shi'is to subordinate their sectarian differences in the interests of Iraqi and broader Arab nationalism. The Shi'i community's sense of discrimination precedes the Baath regime. It is unlikely to be changed by rhetoric or superficial governmental actions, and Khomeini's message adds to the potential volatility of Iraqi Shi'i discontent. Reacting to this, Baghdad executed the spiritual leader of Iraq's Shi'i community, the eminent Ayatollah Sayyid Muhammed Baq'r al-Sadr, allegedly for conspiring with Tehran to fan Shi'i dissidence in Iraq. Some 4,000 persons of Iranian descent, many of them long-time residents of Iraq, have been forcibly expelled. (Ed. note. War broke out between Iraq and Iran in September 1980 and was still continuing in early 1983. See Appendix F for more information on the background of the war and a chronology.)

In Bahrain, to which Iran only gave up in 1971 its long-standing claim that the island was the fourteenth province of Iran, the resurgence of Shi'i Islam is equally worrisome. With a Shi'i majority, governed by a Sunni princely house, the attraction of Iran persists in many Bahraini quarters. There is Shi'i-inspired restiveness in Bahrain. After the revolution, two eminent Iranian Shi'i *ayatollahs* publicly reasserted that Iranian claim to Bahrain, though their statements were later somewhat equivocally toned down by Bani Sadr. Were they airing personal views or will the Shi'i regime in Tehran, once it has put its own house in order, revive the Iranian claim to Bahrain? Bahrain has little oil, but its refinery is important. It has also provided, on an informal basis, facilities for United States Middle East Force naval units. Saudi concern for Bahrain's independence has been of long standing. This is not surprising, since Bahrain is only 20 miles off the coast of al-Khobar in the Eastern Province. There is again talk of constructing the long mooted, often postponed causeway between Saudi Arabia and Bahrain, partly no doubt as a gesture of reassurance to the Bahraini authorities.

Nor has Saudi Arabia been immune to the Iranian fallout. Its 250,000 strong Shi'i community in the Qatif area of the Eastern Province has, following Khomeini's success, become increasingly restive. In 1980 two incidents, involving sufficiently serious demonstrations to require National Guard intervention, have marred provincial tranquility. One occurred on *'ashura,* the 10th of Muharram, when Shi'is throughout the world passionately mourn the martyrdom of Hussein, their spiritual progenitor; the other in March on the first anniversary of Khomeini's return. Both were embarrassing to the Saudi government, but both were localized. The first may have been spontaneous, but the second was clearly instigated by Tehran. Later, Radio Tehran's Arabic language service repeatedly beamed anti-Saudi regime diatribes at the Saudi Shi'i community, exhorting its members to protest by painting inflammatory slogans on walls, to arm themselves and, generally, to engage in anti-government militant acts. Nominally, Iran alleges Saudi discrimination against the Shi'i community as justification for its unfriendly actions. Traditional Shi'i distaste for Sunni Unitarianism (*Wahhabism*) is yet another motivating force. But an equally compelling factor is doubtless the Khomeini regime's concern over Saudi Arabia's close relations with the United States. By thus seeking to foment problems for Saudi Arabia, the Iranian regime presumably hopes to weaken the residual American position in the Gulf. Since at least a third of the Saudi labor force employed at the ARAMCO-managed oil installation in the Eastern Province is Shi'i, a seriously disaffected Shi'i community poses a potential threat to these installations. To date, however, the Shi'i labor force has not sought to translate its grievances into acts of sabotage.

The Saudi government has sought to deal with the problem on two levels. Despite Tehran's provocations, it has endeavored to maintain correct relations with the new regime in Iran. Concurrently, it has discussed with the Saudi Shi'i community a series of reform measures designed to allay existing discontent. A new *modus vivendi* with the small Shi'i community is clearly needed if Iran's meddling is to be countered. This may be difficult for the guardians of Unitarian orthodoxy, the Saudi *'ulama,* to swallow, but the Saudi leadership has in the past shown that it can manage intransigence from that quarter. The Mecca incident notwithstanding, it should again be able to do so.

NATIONAL ACTORS, REGIONAL ASPIRATIONS

On a *horizontal* plane, the Gulf may be viewed in four separate categories: Iran, Iraq, Saudi Arabia and the smaller Gulf principalities, notably Kuwait, Qatar, Bahrain, the United Arab Emirates and Oman. Oman is geographically outside of the Gulf, although by virtue of a detached sliver of national territory, Ras Musandam, it shares with Iran sovereign control over the Straits of Hormuz.

The Shah, with American encouragement following the British departure from the Gulf in 1971, envisaged Iran as the linchpin of Gulf security and beyond this, had begun to aspire to a broader Indian Ocean security role. The Shah's objectives fitted neatly into the concept of the post-Vietnam Nixon doctrine. Arrogant and opinionated the Shah might have been, but he was strong, well disposed toward the United States, and ready and able to assume area security responsibilities against a putative communist threat. His imperial ambitions may not have been consonant with the wishes of the Iranian people, who have historically preferred neutralism to alignments, but public opinion was not the determining element.

The Shah did provide surrogate security in the region. His dispatch of troops and supplies to the hard-pressed government of Oman, coupled with British military support, enabled it to contain and eventually to crush the South Yemeni (and Iraqi) supported insurgency in Dhufar, Oman, of the Popular Front for the Liberation of Oman and the Persian Gulf. It won the grudging approbation of the Arab states of the Gulf, Iraq excepted. Since the Shah was a valuable security asset in the Gulf, Washington preferred to ignore or to downplay the domestic shortcomings of his misrule and oppression.

Post-Pahlavi Iran has been largely, though as we have seen not entirely, introspective. It has been preoccupied with firmly rooting its Islamic revolution and, concurrently, with controlling centrifugal forces omnipresent in the polyglot Persian empire. Some of these, like the Kurds, have seized upon the prevailing chaos and uncertainty to press for autonomy. Others, like the Baluch, Azerbaijanis, and Turkmans, would like to do so. Keeping its mosaic of minorities quiet has consumed much of Tehran's time and energies and may do so for some time to come. Nevertheless, as Radio Tehran broadcasts already suggest, a resurgent, nationalist Iran will at some point seek to reassert itself politically and militarily in the Gulf, in the pattern of its predecessor. Will it act responsibly or not? This will, of course, depend upon Iran's emerging power hierarchy, but experience to date gives little cause for optimism.

With the collapse of Iranian imperial and military power, Iraq stands as the strongest state of the Gulf region. Like Khomeini's Iran, it is antagonistic to the United States, at least politically so, though some of its attitudes are showing signs of change. Seeking American technology and know-how, it welcomes apolitical American business enterprise, while consistently rebuffing any resumption of diplomatic relations with the United States. Iraq's strength derives from three main sources: a) its strong ideological compulsion, based upon Baath Party concepts of socialism, pan-Arabism, and anti-imperialism (in which it includes the United States and Israel); b) its growing military capability (Iraq's military establishment, some 250,000 strong organized into twelve, largely Soviet-equipped divisions, is next to Egypt's, the strongest Arab military force in the Middle East), and c) the economic power of the country, derived from oil revenues, and the Baath government's willingness to allocate resources to the achievement of political objectives. Saddam Hussain, Iraq's President, is energetic and able and, if he survives attempts on his life, will be a Gulf leader to be reckoned with in the future.

Paradoxically, while Iraq purchases modern military equipment from the Soviets, Soviet political influence on Iraqi policies is sharply limited. Despite its "progressive front" policy, nominally extolling Baathi-Communist political cooperation, the Baath leadership has not hesitated to jail or even to execute Iraqi communists. Soviet protests have been unavailing and Iraqi relations with the Soviets are becoming increasingly strained. In terms of the superpower competition in the Gulf, any such development is helpful to American interests, but the long-established mutuality of Iraqi-Soviet dependence is likely to constrain total estrangement between the two. In any case, the Iraqi leadership, rightly or wrongly, is convinced it can "manage" the Soviet tie.

The Iraqi power potential suffers from certain weaknesses. First, its political contest must be conducted on three fronts: Iran to the east, Syria to the west (the most recent effort at Iraqi-Syrian collaboration dissolved without result) and the Gulf states in the south. Second, the Baath government is a minority government. Despite its avowals of internal harmony, it must cope with serious internal problems—the Kurds in the north (not fully pacified) and the Shi'is in the south. Third, Iraq has only a ten mile coastline on the Gulf, though it still entertains a latent irridentist claim to Kuwait, based upon suzerainty arrangements extant in the Gulf area during Ottoman times.

Of the Gulf leaders, Saddam Hussain—after first consolidating his regime at home by the ouster of potential rivals—has been the most outspoken in opposing any American effort to establish a security umbrella for the Gulf. Speaking to a popular rally held in Baghdad on February 8, 1980, he adumbrated a new Arab National Charter, an obvious effort to counter the Carter doctrine. The proposed charter, *inter alia,* rejects all "foreign" military forces or bases in the area, and threatens to ostracize and to boycott any Arab regime which fails to adhere to this principle. Eschewing Arab involvement in international conflicts (except that against Zionism), it calls for total neutrality and nonalignment toward any foreign party in the event of war, so long as Arab territorial integrity is not violated. Immediately afterwards, Iraq's official newspaper called for Arab endorsement of the new charter. Most of the smaller Gulf states (as well as many non-Gulf Arab states and the PLO) were quick to give approval in principle. Saddam Hussain's initiative scored a point in enhancing Iraq's political prestige in the Gulf region. Despite correct relations between Saudi Arabia and Iraq, it was an indirect challenge to Saudi influence in the Gulf. Though maintaining a discreet silence, the Saudi leadership, far more concerned over a possible Soviet threat to the Gulf than is Iraq, could not but resent this Iraqi one-upmanship. The charter, it recognized, was also an oblique slap on the wrist for its association with the United States.

Saudi Arabia, the largest and wealthiest state of the Gulf littoral and a leader of the Arab world, continues to enjoy considerable influence in the Gulf region. This is especially true with the lower Gulf principalities and Oman. Its influence on Kuwait is somewhat less, mainly because of Kuwait's own wealth and concomitant desire for a more independent role in the Gulf and the Arab world. The Saudi leadership, due to growing Arab criticisms that the Saudi association with the United States has not brought expected political benefits, is currently somewhat defensive. The late King Faisal once bitterly complained to an American ambassador, "You Americans do not make things easier for your friends. You disregard their concerns and thereby weaken their (political) influence in the area." That thought will on occasion have crossed the minds of the present Saudi leadership. The King's lament may not have been fully justified, since Washington, neither now nor in the past, has deliberately sought to ignore Saudi views. Nevertheless, through miscalculation or otherwise, it has sometimes projected the impression of taking the Saudis for granted. The fault is not one-sided. In part, the Saudis are themselves to blame by too often projecting an impression of inflexibility on the Arab-Israeli issue. In some American eyes, therefore, they are perceived as unhelpful to United States efforts to seek solutions to this complex problem. In terms of future Gulf security, the special relationship that has long existed between the United States and Saudi Arabia needs refurbishing. Saudi Arabia has in the past played a constructive role in Middle East and Gulf affairs and can, if properly supported, continue to do so. Disagreement on some issues should not obscure the two countries' mutuality of interest on broader global issues.

Saudi Arabia is in the throes of a gigantic economic boom which some contend has seriously eroded its moral fiber. Any such boom is bound to be attended by some waste and charges of corruption. The management of Saudi Arabia's oil reserves and its massive development program have posed major problems for King Khalid and Crown Prince Fahd. They are seeking to grapple with these problems, but the task is awesome. It has become faddish to criticize the Saudis, yet few appreciate the magnitude of the development problems that the Saudi leadership faces. Regrettably, Saudi Arabia suffers from a public relations problem in this country. This is partly of its own making, but the Saudi leadership deserves better from this nation. The media has too often cast the House of Saud as an effete body of idle princes. The image is a distortion. The family has its full share of able, energetic men, both in the older and younger generations. It has far more resiliency than some critics choose to believe.

Despite past strains in relations between the United States and Saudi Arabia, occasioned by such diverse things as improper conduct by a former

American diplomat stationed in Jidda, disappointment over the Camp David accords and what is viewed by the Saudis as unfair Congressional and media criticisms, Saudi Arabia has been a long-time friend of the United States. It is no American client state, as some would believe, but is proud of its independence. It has consistently sought to be helpful to the United States in matters of OPEC pricing policy and oil production levels, often incurring the ire of its fellow OPEC members for doing so. The Saudi leadership, especially King Khalid and later King Fahd, has kept production levels high. It has done so in the face of economically justified criticism from Saudi conservationists that doing so hurts the future of the country. King Fahd had full confidence in President Carter's determination to strive for an equitable, comprehensive Middle East peace settlement, one which will give the Palestinians their legitimate national rights. But he wants to see results. Other Saudi leaders, especially among the younger generation, many of them trained in the United States, are less enamored of their American connection.

The Saudi leadership remains intensely concerned about South Yemen and Ethiopia, viewing them as Soviet surrogates, bent upon fostering instability in the Arabian Peninsula and the Gulf region. Equally troublesome to the Saudis has been the tendency of the leadership of the Yemen Arab Republic to flirt with the Soviets, including acceptance of Soviet arms and perhaps advisors. They have put strong pressure on the YAR to desist from doing so and and may have had some success on that score. A slight thaw may also be developing between Saudi Arabia and the People's Democratic Republic of Yemen following governmental changes in the latter.

The November, 1979, incident in Mecca, involving seizure by fundamentalist religious zealots, of the Masjid al-Haram, the holiest shrine in Islam, aroused widespread worry in this country and abroad about Saudi internal security. This is natural, but in this writer's judgment, the threat was overdrawn by the media. The political legitimacy of the House of Saud has for the past two centuries been rooted in its strong adherence to and espousal of Unitarian (*Wahhabi*) Islam. Despite massive modernization, the family prides itself on that association and the *Qur'an* remains the constitution of the land. The Madhist zealots, under two disaffected 'Utaiba tribesmen, Juhayman bin Muhammed bin Saif al-Utaibi, and the self-styled Mahdi, Muhammed bin Abdullah al-Qahtani, seemed to be challenging that claim. Did the Mecca incident represent a neo-*ikhwan* movement, which has seriously eroded the religious basis of the regime or could do so? Admittedly, the evidence is not entirely in and the Saudis have not helped themselves by their handling of information on the incident, but no hard facts have yet been adduced convincingly to support such a contention.

The Mecca incident was indeed a shock to the Saudi authorities. It hurt the Kingdom's image in some Western and Muslim circles, and it exposed a disturbing degree of inefficiency on the part of the Saudi security apparatus. Nevertheless, despite rumors and innuendoes coming out of Kuwait and elsewhere, there is no good reason to believe that the Mahdists represented any broadly based national constituency. They were largely Unitarian religious zealots, mainly from the 'Utaiba, one of the *ikhwan* tribes forcibly suppressed by the late King Abdul Aziz in 1928–29, with a scattering of Kuwaiti, Egyptian, Yemeni, Pakistani and other religious fundamentalists. South Yemeni complicity, at Soviet behest, is suspected by the Saudi leadership. The insurgents were well armed and the operation showed evidence of intensive planning. How so heterogenous a group could have organized itself without arousing Saudi security suspicions remains a mystery. One explanation is that established Saudi religious leaders, incapable of believing that their erstwhile students could go so far, befriended the leaders when the latter were earlier harassed by the police. While clearly only a part of the story, this is credible, since no self-respecting Muslim could believe that any group of professed Muslims, however zealous they might be, would desecrate the Masjid al-Haram. One must assume that the astonishing security lapse has been or is being corrected by the Saudi authorities. Lest an Iranian connection be assumed, the Mahdist leaders, as fundamentalist Unitarians, sharply attacked what they charged was excessive official tolerance of the small Saudi Shi'i community.

The writer, on a recent visit to Saudi Arabia after the Mecca incident, sensed no nationwide groundswell for the return of pristine Islamic values, as demanded by the Mahdists. Even those who criticize the Saudi leadership for allegedly tolerating corruption do not favor any such radical notion. Over the years the conventional wisdom has been that a potential future opposition to the Saud family might come from the growing number of Western-trained Saudi technocrats, many of whom aspire to more responsibility in the power structure, and/or from the military, whose training and equipment makes it a significant sector of society. Since the arrests of some officers in 1969, for Nasserite connections, there have been no significant signs of military restiveness on the part of the Saudi officer corps. Nasser, as a symbol of Arab unity, is gone; Nasserism, as a military force, was discredited by the 1967 defeat. Baathism, while it may appeal to some military officers, is not likely to have much attraction in Saudi Arabia. It is too regimented an ideology and regarded as too tolerant of Soviet activities. Nor is there any evidence that Qaddafi has any appeal. The Saudi leadership is well aware of the spate of military coups which has marked the history of the area. Under the vigorous and active Prince Sultan bin 'Abdul 'Aziz, a full brother of King Fahd, a concerted effort has over the years been made to win the loyalty of the military through the provision of modern equipment and perquisites. The National Guard exists to balance the army. This may not always be enough, but available evidence suggests that for the foreseeable future it is.

The Saudi technocrats and bureaucrats are well paid, which may not stifle their desire for more responsibility, but defers crisis. Moreover, and this should be emphasized, neither discontented military officers—and there are bound to be some in any Middle East society—or restive technocrats would welcome a return to the fundamentalist values advocated by the Mahdist zealots. To the extent that these sectors of Saudi society have an objective for the country, it is to go forward, not backwards. Thus, the Mecca incident, regrettable though it was, may have forged a new and stronger consensus between the Saud family and the educated, non-royal national elite. There is widespread hope that the Mecca incident will catalyze a measure of broader political liberalization in Saudi Arabia, and as Crown Prince Fahd has spoken of a new organic law, a consultive assembly and provincial assemblies that may shortly be established. This would be the beginning of wider participation in the decision-making process. In the interests of the future stability of the regime, it is clearly desirable that it expand commoner participation in the decision-making process. Equally essential to the present image and future well being of the regime is an active campaign to curb corruption, including excessive commission payments by American and Western firms.

The United States attaches considerable importance to its relations with Saudi Arabia. To be sure, the United States misjudged likely Saudi reaction to Camp David. It has since sought to redress its error. While the Saudi drift toward the rejectionist camp after Camp David has disturbed Washington, relations remain good. One element in this relationship has been the substantial U.S. military sales program to Saudi Arabia. Because of Congressional criticism and slow deliveries, the Saudis may no longer rely entirely on American military equipment. In most instances, however, they still prefer it and want the supply relationship to continue. Related to this, while the Saudi leadership desires no formal defense association with the United States, it sees the sizeable United States Military Training Mission in Saudi Arabia as a kind of "plateglass" assurance of Washington's concern for the country's territorial integrity and political independence.

The smaller polities of the Gulf have little political power. With the exception of Kuwait, they are largely engrossed in their own mini-problems. These are not without interest, but except in a few areas where they impinge upon broader Gulf security considerations, they need not detain us here. In the latter context, four general observations applicable to all are germane. First, their indigenous populations are startlingly small and totally incommensurate with the vast wealth that some of the Gulf principalities command. Second, they are heavily dependent upon expatriate labor of all nationalities, most of it disenfranchised, in order to implement development programs. (The most glaring example is Kuwait, where only 47% of the population is native born). Third, despite facades of modernity and individual cosmopolitanism, their societies remain essentially tribally oriented, and act accordingly. Finally, the durability of their princely houses in the years ahead depends in large measure, especially after the fall of the Shah, upon a strong Saudi Arabia under its present form of government.

Kuwait, because its population is concentrated within a small area and because of its longer experience with economic development, is a partial—but only a partial—exception to the above rules. Now fourth among Gulf Arab oil producers after having cut production for reasons of conservation, Kuwait has created an effective welfare state and, equally impressively, a remarkably sound and economically grounded regional development program. On Arab issues, especially the Palestinian issue, it is holier than the Pope, partly as a self-defense mechanism. For the past several years, it has been engaged in a minor border controversy with Iraq, initiated by the latter, which it seeks to handle by negotiation. Much palaver has thus far brought no resolution. It professes not to be concerned about its security, although the disappearance of Pahlavi Iran, previously the principal restraining force on Iraq, leaves Kuwait vulnerable to Iraqi pressure. Whether or not they like to admit it, few Kuwaitis have forgotten 1961, when a large British force had to be deployed to Kuwait, at the request of the Kuwaiti government, to counter what appeared at the time to be an Iraqi intention to enforce militarily its broader irridentist claim to all of that principality. It hopes any such claim has been permanently renounced, but cannot be sure that this is the case.

The presence of large numbers of Palestinians among the already mentioned expatriate communities in the Gulf states also represents a potentially destabilizing factor. They cover the spectrum of the PLO organization from moderates to extremists. In Kuwait, they number over 20 percent of a population of just over a half a million; in Abu Dhabi and Dubai, they may also be as high as 20 to 25 percent of a population of 450,000; in Qatar, perhaps 10,000 out of 100,000; in Bahrain, perhaps 10 percent of 230,000. The Palestinians, it should be emphasized, have contributed significantly to the administrative and economic development of the Gulf countries. But they represent unassimilated communities, both because the host governments wish it so, and because most Palestinians consider themselves to be on a higher intellectual level than that of their Gulf brothers and have little desire for naturalization, even if they could more easily obtain it. The indigenous residents of the Gulf states and the expatriate Palestinians are in effect separate and parallel societies. Host governments have generally managed to keep the political activities of Palestinians confined to narrow Arab-Israeli issues, but it has not always been easy. A Minister of State for Foreign Affairs of the United Arab Amirates was killed by a Palestinian three years ago at the Abu Dhabi airport while greeting visiting Syrian Foreign Minister Khaddam. The shot was intended for Khaddam, but the incident showed the difficulty that smaller Gulf host countries have in keeping out of internecine Palestinian feuding and out of Palestinian differences with other Arab states. Palestinian distrust of the United States prompts Gulf leaders to move warily in dealing with Washington.

Oman is not in, but is certainly of the Gulf. Its security role in the Straits of Hormuz has already been mentioned. Through the waterway pass almost 60 percent of the Western world's petroleum supplies and 80 percent of Japan's. The feasibility of blocking the roughly 52-mile-wide Straits of Hormuz has provoked much debate. Terrorists, the Saudi Minister of Petroleum has warned, could with Katusha rockets, fired from the barren coast of Ras Musandam, blow up a tanker in the Straits and thereby interdict passage of other vessels. This may have been conscious hyperbole, intended to spur on greater Western efforts to resolve the Arab-Israeli problem. More recently, former Iranian Foreign Minister Ghotbzadeh threatened Iranian action to block the Straits, should American punitive military action against Iran be attempted. He hoped thereby not only to urge the Europeans and Japanese to dissuade the United States from any such action, but also to warn the Gulf states that they too would suffer if they condoned or cooperated with any such American design. Iranian naval maneuvers took place in May 1980 in the Straits of Hormuz area under American and Soviet surveillance. They may have been more posturing than real, given the decrepit state of the Iranian Navy, but were clearly intended to underscore Iran's territorial waters claim to this strategic waterway and to remind Gulf neighbors of Tehran's determination to assert it. While total closure of the Straits of Hormuz chokepoint would be difficult to achieve, partial blockage and resultant transit slowdowns are plausible.

The Government of Oman, under its Sandhurst-trained ruler, Sultan Qabus, has taken seriously this putative threat. It has sought to strengthen its capability to safeguard the Straits by asking military and naval assistance from the United States and others. This is a reasonable position, even if Oman's capability to protect the Straits—whatever that may mean in practical terms—is likely to remain limited. Qabus, like his father before him, has opted to keep Palestinians out of Oman, relying instead on Baluchis, Indians and Pakistanis, all of whom can more readily be deported if necessary. He is receiving considerable British assistance, in the form of seconded British military officers and mercenary officers, to train the fledgling Oman defense forces. Egypt has also offered assistance to Oman, such as pilot instructors and ground personnel, and discussions are under way to look into the feasibility of Oman's requiring and accepting such help. Unlike his father, Qabus has encouraged an active economic development program, which is gradually changing the face of the country. But the Sultan is authoritarian, latent internal political problems do exist, and Oman has a long way to go before it becomes modern.

Qabus was one of the few Arab or Gulf rulers to endorse the Camp David accords and the peace treaty between Egypt and Israel. This earned for him the opprobrium of some of his neighboring Gulf rulers, especially the Iraqi leadership. Worse still, in their eyes, he accepted American overtures for military facilities on Masirah Island, prompting sharp Iraqi charges that Oman's motives were suspect and not in accord with the newly proposed Arab National Charter. Omani leaders have denied that any arrangements for foreign—read American—bases have been concluded, but the Iraqi warning is troublesome to them. They have emphasized that only limited military facilities have been agreed upon. In proceeding with negotiations for such facilities for the United States, they counted upon Saudi understanding of the rationale for their decision. Since the termination in 1960 of the Dhahran airfield agreement, Saudi Arabia has opposed American military facilities in the Arabian Peninsula. In the wake of the Soviet invasion of Afghanistan, the Saudi leadership appears willing to take a more tolerant, if still unenthusiastic, view of possible American military facilities on Masirah Island and also in Somalia. (During World War II, the United States Air Force used RAF facilities near Salalah, in the Omani province of Dhufar.) U.S. use of the Masirah facilities in staging the abortive Tabes rescue effort brought Omani charges of failure to consult in advance and further embittered Iranian-Omani relations.

Prospects for Saudi Arabia's Security

ADEED I. DAWISHA

The stability of Saudi Arabia, and indeed of the peninsula as a whole, has been viewed with increasing concern in the wake of the Iranian revolution, and particularly after the attack on the Grand Mosque in Mecca. As a *status-quo* power pursuing moderate and pragmatic policies, Saudi Arabia constitutes a crucial Western interest in the area, and as such, the maintenance of the Saudi regime has been a primary objective of the foreign policies of the United States and the Western world in general. There can be no doubt that a mass uprising in the oil-rich, pro-Western kingdom would be a major blow to the West, and it is therefore hardly surprising that the Western circles have become apprehensive about the possibility of the Iranian revolution repeating itself in the more exposed and seemingly less secure Saudi kingdom.

It could be argued, however, that these fears are somewhat exaggerated, and that the danger of Saudi Arabia undergoing the kind of upheavals which befell Iran is, at present, minimal. The conditions that contributed to the Iranian eruption are not found in Saudi Arabia. In the first place, the Saudi population is one-eighth of Iran's so that it was possible for all levels of society to benefit from the dramatic post-1973 increase in the country's wealth. The Saudi government, with its massive surpluses, has been able to undertake wide-ranging social projects (such as housing, free education and medical services, irrigation systems, transport and road improvement) and, unlike its counterpart in Teheran, could extend the benefits of these projects right down to the lowest levels of Saudi society. It has, moreover, been extremely successful in impressing upon the people that the considerable progress in their social and economic conditions is due entirely to the monarchy's beneficence, and therefore that any threat to the Royal Family

Extracted by permission of the publisher, The International Institute for Strategic Studies, from Adeed Dawisha, "Saudi Arabia's Search for Security," *Adelphi Papers,* No. 158, Winter 1979–80, pp. 31-35.

jeopardizes the interests and well-being of the Saudi population. Thus, although a wide disparity between the upper and lower levels of society does persist, the standard of living of the poorer Saudis has improved to such an extent that they now seem to identify their interests fully with the preservation of the Saudi monarchy.

Another important difference between the Iranian and Saudi cases can be found in the role played by the religious establishment in the political system. A major contributing factor to the Iranian revolution was the mounting resentment of the Shi'a *mullahs* against a secularist monarch intent on curbing their political role and influence. This does not apply in the Saudi case, where the political system is consciously steeped in Koranic values, and where the religious establishment is constantly consulted and made to participate fully in the decision-making process.

Finally, although the Saudi political system is undoubtedly authoritarian, even archaic, it is nevertheless based on tribal tradition and has its own peculiar brand of grass-roots democracy. The Royal Family, through the institution of the *Majlis,* have constantly made themselves available to the public. They have thus ensured that the population has access to them, and that communication between the rulers and the ruled is maintained. This is in stark contrast to the situation that existed in Iran, where the Shah increasingly isolated himself from his public and their problems and aspirations.

There are, however, clear dangers resulting from the rapid modernization process that the country is currently undergoing. It is difficult to predict the eventual effect of this societal transformation on the population's traditional allegiances and society's existing system of values; it could very well affect the country's stability by undermining the very same social patterns and cultural values that underpin the legitimacy of the monarchy. The Saudi rulers, however, seem well satisfied that the chances of serious societal dislocation resulting from rapid modernization are at present slim. Oil Minister Yamani has observed that rapid change was bound to produce tension, but he has said that he believes it would take at least another generation for the present material revolution to bring about substantial social change. Ghazi al-Gosaibi, Minister of Industry, insisted that the strong influence of family and religion would see the people through the transformation, and these sentiments were echoed by Minister of Planning Hisham Nazer, who described the institution of the family as 'the arsenal of Islam which will defend the Saudis against the corruptions of the modern world'.[1]

It remains to be seen whether these unifying forces, which stem from the very roots of Saudi society, can combat the undoubtedly disruptive effects of this rapid modernization, but there are no indications at present of an

1. *New York Times,* 15 March 1978.

impending explosion. Thus Prince Naif, Interior Minister, could boast in 1977, 'we never had more than 250 political prisoners at any time, and we released the last 63 two years ago'.[2] This might sound as if it were a standard princely claim, but few would strenuously contest it. Indeed, a US State Department report released by the Senate Foreign Relations Committee in February 1978 concluded that arbitrary political arrest in Saudi Arabia was neither practised nor condoned, and that there was no evidence of individuals being imprisoned for their political beliefs. The report continued that torture was not sanctioned or practised, and that closed trials had rarely been held in recent years.[3] It could thus be safely argued that at least in the foreseeable future the Saudi rulers have very little to fear from their own population.

However, the same might not be true of the immigrant labourers who have been working in Saudi Arabia for many years. The 1975–80 Five-Year Plan called for nearly a three-fold increase in their number, and one in every three workers employed in Saudi Arabia was thought to be a foreigner by the end of 1980. On the whole these workers, many of whom are radical Shi'i Yemenis, have been excluded from the considerable state benefits available to Saudi nationals, and, with their numbers constantly increasing, and their social and economic position deteriorating in relation to the indigenous population, the immigrants have come to represent a growing source of concern for the Riyadh authorities. As a frustrated community with mounting grievances, they could become a potent destabilizing agent in Saudi domestic politics. It was hardly surprising, therefore, that when a Saudi Communist Party was clandestinely formed in November 1975, its leadership declared that it would concentrate much of its activity 'on the workers from abroad'.[4] The Saudi authorities consequently endeavoured to tighten their control over the immigrants and the Saudi security forces, with reports of thirty people killed and 120 injured in one riot, and a newspaper editorial denounced 'recent attacks on security forces by foreign elements who were being checked for proper identification'.[5] It thus seems clear that, as long as immigrant workers continue to be excluded from the full social and economic benefits and opportunities the state extends to Saudi nationals, such friction is bound to manifest itself in a growing confrontation that may eventually undermine the stability of Saudi Arabia's social and political systems.

Strictly speaking, however, the immigrant workers do not constitute an internal threat to the regime, for foreign nationals can always be deported if they are perceived as such. The bulk of this community consists of unskilled labourers who can be easily and quickly replaced by workers from

2. *The Guardian,* 21 June 1977.
3. *Arab Report and Record,* 1978, p. 137.
4. *Arab Report and Record,* 1975, p. 614.
5. *Okaz* (Jeddah), 16 October 1978.

other countries, with few adverse effects on the development plans. Thus, for example, placid and more conformist workers from Pakistan, Baluchistan, the Philippines or South Korea could, if the need arose, take the place of radical Yemenis. In all probability, such a move would receive the overwhelming support of the indigenous Saudi population.

If these conclusions make it appear that all is well internally, there remain a number of problems on the horizon which, if ignored, may shake the complacency of the Royal Family. There is widespread corruption; there is a high degree of materialism which rides uneasily with the spiritual commitment to Islamic ideals; there are strains within the Royal Family and a degree of factionalism; and there may soon come a time when the pursuit of wealth alone no longer satisfies the aspirations of the Saudi populace. In Kuwait, the merchants are beginning to challenge the political power of the sheikhs, and it would be surprising if the commoners in Saudi Arabia did not also seek a measure of political power in the years to come. Much will depend upon the willingness of the Royal Family to allow some levers of power to leave their hands. Permanent division of the country into first- and second-class citizens by birth could incite disaffection among the lower strata of society, exerting pressure that could ultimately lead to a civil explosion.

In addition to these possible sources of disruption, attention is focused on external threats aimed at destabilizing the Riyadh government. Given Saudi Arabia's experience of the Nasser-backed republican revolution in North Yemen in the 1960s and her intense antipathy towards the Marxist regime of South Yemen, it is only natural that she considers the southern border to be her most vulnerable strategic point. In explaining the decision in early 1979 to withdraw the Saudi contingent from the Arab Peace-keeping Force in Lebanon, Foreign Minister Prince Saud told the Lebanese Prime Minister 'Just as you are burdened with your problems in the south, we also have a southern problem at the Saudi-Yemeni border'.[6] The great unspoken fear is that the two Yemens might merge and, with a combined population of nine million, present a formidable challenge to Saudi Arabia. Such a merger would be all the more ominous because the more dynamic and better organized Marxist leadership of South Yemen would be bound to emerge as the dominant political authority. A major strategic objective of the Saudi regime has therefore been to prevent the unity of the two Yemens at all costs. It appeared more than coincidental that North Yemen's President Hamdi was assassinated the day before he was due to travel to Aden to discuss possible unity measures. Saudi Arabia will no doubt continue to exploit her influence with the influential tribes of North Yemen to try to frustrate any efforts to create a united Yemen. However, as she has

6. *Al-Hawadith* (Beirut), 13 April 1979.

discovered elsewhere, there is a limit to her ability to manipulate events in other countries, and the Royal Family must consider the possibility of confronting a Soviet-backed, Marxist-led united Yemen in the future.

A further possible threat, in the context of any future Arab-Israeli confrontation, is that of an Israeli air attack against Saudi oil fields. With her increasing regional and international influence, her considerable financial support for the Palestinians and the Arab confrontation states, and her publicly-declared hostility towards the Egyptian-Israeli accords, Saudi Arabia is no longer considered by Israel as a peripheral actor in the Arab-Israeli conflict. Consequently, to strike against the very source of the Arabs' growing regional and international power must be a very tempting proposition. Even with the recently purchased (but as yet undelivered) F-15 fighters, Saudi Arabia can hardly expect to stand up to the efficient Israeli Air Force.

Another threat could arguably come from her radical northern neighbour. Iraq, led by the socialist, left-wing Ba'ath Party, and having oft-declared hegemonial ambitions towards Kuwait and ideological antipathies towards Saudi Arabia, has long constituted a major destabilizing threat to the Saudi regime. Paradoxically, however, this threat has receded noticeably in the wake of the Iranian revolution. The majority of the Iraqi population belongs to the Shi'a sect of Islam, and might therefore come to look upon the Iranian *mullahs,* rather than the minority Sunni government in Baghdad, as its moral and political leaders. This potentially explosive domestic situation in Iraq has been exacerbated recently by the re-emergence of Kurdish secessionist activities after the fall of the Shah. As a result, the Iraqi regime has become very reluctant to indulge in disruptive regional operations. Instead it has increasingly endeavoured to pursue moderate and co-operative policies towards its southern neighbours. Indeed, Baghdad has even been willing to consider the hitherto much-derided notion of a Gulf security system as a possible policy option. It might in the past have considered such concepts as 'regional alliances' and 'filling the power vacuum' to be reactionary efforts aimed at maintaining American influence in the area, yet it was interesting to note that, in the wake of the Shah's deposition, Iraq was beginning to consider such concepts quite seriously. In April 1979, Saddam Hussein, the most influential figure in the Soviet-backed Iraqi regime, made this startling statement:

> We must take up arms against any foreigner, regardless of his colour, who may violate Arab sovereignty. In this context, we do not differentiate between a progressive, or a Zionist, or a Frenchman. Nor do we differentiate between American and Soviet. Irrespective of the formal friendly ties it may have with some of us, the Soviet Union, and it is a friend of Iraq, cannot be allowed to occupy Saudi territory. This is

because Saudi land is not outside the Arab map, and what applies to it applies to the rest of the Arab countries.[7]

Present indications, therefore, suggest that the hitherto potent Iraqi threat to Saudi stability has considerably decreased.

A further reason for Iraq's moderation towards Saudi Arabia relates to the obvious need of the radical Arab regimes to close ranks with their conservative counterparts in order to confront the Camp David Agreements and the Egyptian-Israeli Treaty. To a large extent, therefore, the pursuit of co-operative policies towards Saudi Arabia was implicitly contingent on her adoption of a militant stand against the policies of President Sadat. And it is here that Saudi Arabia might ultimately face the greatest challenge to her survival, for an adamant opposition to the Camp David Accords could potentially erode the hitherto strategically overriding relationship between Saudi Arabia and the United States.

In February 1979, then Senator Frank Church, Chairman of the Senate Foreign Relations Committee and a well-known supporter of Israel, admonished Saudi Arabia for allegedly being more of a detriment to peace than the United States was willing to admit. He said that the projected sale of the F-15 fighters should be reassessed and suggested that Saudi Arabia should be told that a 'special relationship could not be a one-way street'.[8] The White House immediately disassociated itself from this highly partisan speech, but the Senator's remarks nonetheless served to highlight the painful dilemma in which Saudi Arabia now finds herself. A perceptible coolness in the 'special relationship' inevitably followed, and this was accompanied by rumours of schisms within the Royal Family and tension inside the armed forces. Whatever the truth of these reports, there can be no doubt that in the summer of 1979, an intense debate regarding Saudi-American and Saudi-Arab relations took place within the ranks of Saudi Arabia's ruling elite.

The choice facing Saudi rulers is both difficult and very consequential. On the one hand, they could not condone the Egyptian-Israeli Accord for fear of alienating the Arab world and undermining their country's legitimacy, which is based on its status as the guardian of Islam and Islam's holy places. On the other hand, they were clearly unwilling to adopt a militant stance and thereby run the risk of weakening the moderate regime of President Sadat[9] and, more importantly, of losing crucial American politi-

7. *The Guardian,* 11 April 1979; see also *Al-Destour* (London), 2–6 April 1979, p. 10. A further startling development occurred in June 1979, when the Baghdad regime announced a $300 million aid agreement with the conservative, pro-Western and Saudi-backed government of North Yemen (see *The Middle East,* July 1979, p. 10).

8. *MEED Arab Reports,* no. 3, 1979, p. 29.

9. In the wake of the Saudi-Egyptian break that occurred as a result of Camp David, a high Saudi official was reported to have likened Saudi attitudes towards Egypt to 'someone who was forced to amputate a limb in order to keep alive' (*Al-Hawadith,* Beirut, 13 April 1979, p. 14). (President Sadat was assassinated in Cairo in October 1981. Ed.)

cal and strategic support. Obviously, Saudi Arabia would prefer not to have to choose between these two alternatives, for both are intrinsically linked to the survival of the Saudi regime itself. Yet only if there were a dramatic breakthrough in negotiations on the West Bank, Jerusalem and the status of the Palestinians, would Saudi Arabia be extricated from her dilemma. Otherwise, a clear-cut choice will have to be made in the end, and in making that choice the House of Saud will face the gravest crisis it has yet had to meet.

Regional and Worldwide Implications of the Gulf War

JOHN DUKE ANTHONY

INTRODUCTION

Speaking about the regional and worldwide implications of the Iraq-Iran war is reminiscent of John Dewey's observation that "a sculptor may see many different figures in a block of stone." So, too, may an analyst view the present Gulf war from many diferent perspectives. Defense and security specialists who perceive the importance of a balance of forces in the region, for example, have been unable to determine the exact military nature of the conflict inasmuch as neither combatant to date has demonstrated the ability to inflict a decisive defeat upon the other. Other observers, who minimize the significance of the confrontation between the two countries' armed forces, insist that the larger implications of the struggle stem from the Sunni-Shi'a religious differences between Iraq and Iran—and among Muslims elsewhere in the region. Still others have been acutely apprehensive all along about the possibility of increased superpower involvement in the conflict.

All of this is but to say that the regional and worldwide implications of the conflict to date have been and remain characterized by extreme complexity. Such phenomena are all the more confusing in the midst of the disorder accompanying any war in progress. Hence, an analyst of its implications must contend with a myriad of sometimes not-so-obviously related forces and factors. While these phenomena may at times appear quite vague and inseparable, a tentative assessment of the war may be reached if one but gauges the respective responses to the war by outside parties.

An earlier version of this article was presented at the "Symposium on Iraq: The Contemporary State," University of Exeter, 23-25 July 1982.

EVOLUTION OF THE WAR

Perceptions of the war to date have passed through at least six stages. In the first, there was a widespread perception among the Arab Gulf states that the war might sooner or later involve everyone in the region. This was the inspiration behind the early *de facto* move of these states to align themselves with Iraq. Kuwait, Saudi Arabia, the Emirates and Oman, for example, indicated their willingness to extend important logistical and financial support to Iraq only days after the fighting erupted.

Perceptions later entered a second stage whereby these same states reasoned that it was indeed possible to remain detached from the actual military dimensions of the conflict. A major factor was Western diplomatic intervention during the early stages of the war which resulted in Iraq's curtailing its earlier intentions to wrest control from Iran of three disputed islands near the mouth of the Gulf. For the next fourteen months, maritime traffic through the strategic Strait of Hormuz remained relatively unimpeded; US-dispatched AWACS airplanes to Saudi Arabia served to bolster the air surveillance and defense capabilities of a considerable portion of the Gulf's southern littoral; and foreign access to and local production and export of the non-combatants' petroleum resources continued apace.

A third phase came with the December 1981 *coup* attempt in Bahrain, which illuminated much more clearly than previously the broad implications of the war. The conflict was still perceived as confined to Iraq and Iran, but began to carry with it a potential for Sunni-Shi'a confrontation on a scale much broader than envisioned earlier. A fourth phase evolved when a number of Arab states with close ties to Moscow—Syria, Libya, South Yemen—plus Israel and North Korea, extended various forms of political and military support to Iran. This demonstrated to some analysts that the aforementioned third phase, which had seemed to raise the specter of sectarian conflict, had affected as well a coalition of "radical" Arab states and such unlikely bedfellows as Israel and North Korea. Arrayed in opposition to this coalition, in the eyes of these analysts, was a "moderate"-to-"conservative" grouping in alignment with the heretofore-considered "radical" Iraq. To still other observers, these developments were less relevant or significant than the irony of such countries as Soviet-supported Syria, Libya, and South Yemen, together with Israel, aligning themselves with Iran not so much in pursuit of radicalism, but rather as a manifestation of anti-Iraqi sentiments. This placed the regional role and involvement of the USSR in a position of potential influence on matters affecting Gulf security quite unlike any series of developments in recent memory.

A fifth phase began in July 1982 with the Iranian invasion of Iraq in what appeared to many as a possible means for Tehran obtaining a better bargaining position in an eventual settlement. But the Iraqi resistance succeeded. Finally, a sixth phase evolved after the failure of the Iranian invasion. The war reached a military and diplomatic stalemate with no end in sight to what, all analysts were forced to concede, had degenerated to an inconclusive situation.

SOME STRATEGIC CONSIDERATIONS

Among the many regional and global implications associated with the conflict to date, the war has served to highlight the view that Iraq, in regional geo-strategic terms, is in some ways less significant strategically to the Gulf states, the Soviet Union, and the West than is Iran. Some analysts, for example, consider that Iraqi ambitions to play a regional security role remain seriously constrained by its having the shortest littoral of any of the eight Gulf states. The length of its coastline is less than fifty miles, with most of that lying in shallow water and of uncertain access and sovereignty (the latter as a result of Iraq's territorial disputes with both Kuwait and Iran).

Iran is seen by contrast as sharing not only contested land and water frontiers with Iraq, but also offshore boundaries with the entire north Arabian Peninsula littoral across its more than six-hundred-mile coast from the Shatt al-'Arab all the way to Pakistan. Iraq, moreover, lacks Iran's strategic significance due to the latter's position astride the northern shores of the Strait of Hormuz. Through the latter waterway passes the bulk of the oil bound for sale on the international market. Thus, on the matter of applying laws of the sea to international waterways, it is Iran's policies, not those of Iraq, that have come to matter the most in regional and international councils. If Iraq were to prove unable to secure its own border with Iran, let alone make good its claim for undisputed sovereignty over the strategic Shatt al-'Arab waterway—the country's only outlet to the sea—how much less credible, in the eyes of many analysts, would be its pretensions to a leadership role in matters of regional security.

Iraq's *global* geo-strategic significance has likewise paled in comparison with Iran's. The Soviet Union, for example, neighbors not Iraq but Iran. And no small neighbor it is. The two share 1,500 miles of common frontier. Though little noticed by the rest of the world, the day-to-day interaction between Iranian and Soviet citizens exceeds that of the Soviet Union and any other Middle Eastern people.

Iraq to date may have had a longer and more comprehensive military relationship with Moscow than has had Iran. Libya, Ethiopia and Syria may have served Soviet ideological and military interests more effectively in North Africa, the Red Sea, and the eastern Mediterranean. Both Ethiopia and South Yemen, in addition to ideological compatibility and geographic complementarity, may have better met important Soviet needs for access to and a physical presence in the Horn of Africa. But these roles notwithstanding, none of them has had quite the potential global and regional geo-strategic significance that has characterized Iran.

MILITARY UNCERTAINTIES

It has not been conclusively determined thus far whether one of the combatants is militarily stronger than the other. Although Iran is clearly by far the greater in terms of size and overall population, it remains to be seen whether this can or will be translated into an unmitigated defeat of Iraq. In any case, whether the war's outcome would be determined on the battlefield remained open to question. Matters of a logistical nature were also important. Certainly, when Syria closed its frontier with Iraq in 1982, that left open only the routes to and from Iraq by the way of Jordan, Turkey, Kuwait or Saudi Arabia. In theory then, it remained plausible that Iran might resort to such economic measures as blockades or aeriel and maritime interdiction of key supply routes rather than a deep military incursion to achieve its aims in Iraq.

If in the long run clear military victory of either party alone seems improbable, an external factor contributing to such an inconclusive outcome may be Turkish and Egyptian interests. Turkey, the most militarily powerful Islamic Middle Eastern state, a member of NATO and a neighbor of both countries, has stated its opposition to an Iranian invasion and occupation of Iraq and has indicated that it would not hesitate to intervene if its own national security and related interests were to become endangered. Egypt, still regarded by many as the most militarily powerful Arab country, has also stood opposed to Iran in this context and, as indicated below, has enhanced Iraq's strategic and military prospects for resisting Iran.

FOREIGN ASSISTANCE

Throughout the war, Iran has had more limited outside assistance than Iraq. To date, Tehran's main supporters have been Syria, Libya, South Yemen, North Korea and Israel. Helping to facilitate the assistance of the first four have been Kremlin officials who, for the reasons indicated, regard Iran as the greater prize to be sought and who have retained the hope that, the growth in Iranian anti-Soviet propaganda and other setbacks notwithstanding, some way might be found to further Soviet interests there. Yet Moscow has not had a free hand in choosing whether or not to directly support—or oppose—Iran. A potentially enormous cost of heavy Soviet assistance is that most of the Arab states would view such action as hostile toward their own interests.

In addition, the Soviet Union has domestic restraints. It has a vested interest, for example, in not becoming over-extended or precipitously involved in events beyond its borders, given the recent leadership succession at home plus uncertainties along its European and Asian frontiers. Moscow's behavior in the 1980s to date has indicated a reluctance to allow events within client states to exceed the point where it might be obligated to intervene on their behalf.

A further constraint to Soviet intervention thus far has been the nature and orientation of the Iranian regime. Many have questioned the extent of a Soviet inclination to intervene in support of a regime in Tehran which is neither communist nor socialist nor likely, in its makeup, to extend a significant degree of recognition or tolerance to either ideology. Beyond the insecure political climate in Iran, staying the hand of the would-be Soviet interventionist further has been the lack of a sufficiently mass-based, pro-USSR political constituency.

Still another inhibiting consideration has been the fact that just as Iran's many geo-strategic attributes have warranted ongoing Soviet attention in connection with Moscow's hopes to enhance Soviet global and regional interests, so has the reverse argument also been valid. Iran, far more than Iraq, has a potential for destabilizing an important region inside the Soviet Union—the Central Asian Soviets—because its Islamic perspective is closer than Iraq's Ba'thist outlook to the sentiments of the Soviet Union's Muslim citizens. In the absence of the above-mentioned grass-roots support, it cannot be ruled out that a Soviet intervention against the Khomeini regime in Iran, following the Soviet invasion and costly occupation of Afghanistan, could provoke negative consequences inside the USSR that would outweigh whatever international strategic benefits might be gained.

The prospects for Soviet intervention in Iraq have been similarly bleak. Moscow has been without a secure political base there as well, not only as a result of the Ba'th Party's entrenched position in the government. As in Iran, it has also been due to the absence of a mass-based, well-organized constituency favorably inclined toward the Soviet Union.

If the threat of Soviet intervention has been shown to be considerably less credible than some had imagined, there nonetheless remain other reasons for concern. It is highly questionable, for example, whether the Gulf states will be able to continue to finance Iraq in the period ahead at the level which they did during the first two years of the war. Israel's invasion of Lebanon resulted in the needs of Syria, Lebanon, and the Palestinians for billions of dollars in economic assistance. Yet these needs appeared at a time when, due to dwindling financial surpluses and depressed market conditions in the petroleum industry, the Gulf states are without the kinds of resources that were at their disposal previously.

Quite apart from these considerations, were Syria to succeed in its request for large-scale financing from Saudi Arabia and other Arab Gulf states, this quite possibly could affect the regional balance between Syria and Iraq in favor of the former which, throughout, has supported Iran. Saudi Arabia, Kuwait, the UAE, and Qatar have, therefore, had little option but to proceed with great caution and no small amount of discretion in assessing the weight of various variables affecting their interests. Not the least of these has been avoidance of Khomeini-inspired disturbances by their own Shi'ite Muslims. The regimes of those non-combatant countries in immediate proximity to the conflict have already pumped into Iraq an average of a billion dollars a month since the war began. Many analysts, in assessing the implications of such contributions, have concluded that the GCC states might go to considerable lengths to find ways to be even more generous if it would facilitate a settlement which would accomplish two objectives: an end to the daily possibility of an Iranian air strike on one of their oil installations and an end to Tehran-inspired attempts to spread the fundamentalist Shi'ite movement to the south side of the Gulf.

Failing a settlement, this much is certain: Iraq will be in continuous need of external assistance to defend itself against Iran. Toward this end, some analysts, indulging perhaps in wishful thinking, have opined that Baghdad, in an effort to forge a fundamental strategic realignment in the region, might go so far as to cancel the 1971 Soviet-Iraqi Treaty of Friendship. Whatever stake one may place in the day-to-day substance of such treaties, there is no denying that such documents carry with them a symbolic significance that is ordinarily not treated lightly by either of the parties. Yet in this instance, the treaty has remained dormant since both parties disregarded one of its most important clauses—the obligation of one signatory to inform and consult with the other in advance of any military action—when the Soviet Union invaded Afghanistan in 1979 and when Iraq, a year later, invaded Iran.

There is, moreover, ample regional precedent for the renunciation by an Arab state of a close identification with one or the other superpower. Syria's rupture with the United States some years ago, for example, was viewed both regionally and further afield as both the necessary and expedient thing to do in view of Damascus' shift in orientation toward the Soviet Union. A more dramatic, recent and memorable case, however, was Egypt's severance of the Cairo-Moscow Treaty of Friendship in 1972 as a prelude to turning toward the United States. Reference to these previous instances of abrupt regional-global realignment suggests that if Iraq were in fact to find itself in need of more far-ranging assistance from the West than has been forthcoming to date, it may turn to a European state, such as France, with Saudi Arabia and the other Gulf states helping to pay the bill.

Certainly, this would be a more convenient and expedient option for Baghdad to pursue, for it is quite unlikely that Iraq would be able to receive direct assistance from the United States. The reason has less to do with strategic realities than the widespread public image in the US of the Baghdad regime as one which continues to harbor and sponsor terrorists. Compounding the difficulty is that, in contrast to the regime in Tehran, Iraq has had a far longer and more multifaceted relationship with the Soviet Union, with which the Reagan Administration has been considerably more preoccupied than any US Presidency in a quarter of a century. However much Iraq might wish it were otherwise and hope that an attentive American public as well as US policy might recognize and reward the change in Iraq's international posture in recent years, the legacy of earlier days when most Americans, and particularly Congressmen, perceived Iraqis as villains, lives on in Washington.

IDEOLOGY

Many of the implications for foreign involvement in the war have turned on the ideological dimension of the conflict. In this context, many have seen Iraq as competing at a distinct disadvantage. In this regard, a substantial component of the Ba'th Party's ideology is viewed as secular and imported—from sources which were themselves inspired by Western ideas—and, as a consequence, without enduring impact in depth on the majority of the country's population.

By contrast, Iranian ideology has generally been viewed as more indigenous in its origins and more pervasive, both inspirationally and sociologically, in its extent. In addition to these attributes, the combination of the Ayatollah Khomeini's religious zeal and political power has produced an inner dynamic that has rendered the Iranian "model," on balance, a far greater challenge than Ba'thism to the regional *status quo*. Given the relatively low literacy rates in both countries, few have been surprised to find that a homegrown variety of Islamic ideology has had greater appeal to large masses of Iranians and Iraqis.

On the Iranian side, how much of the motive has been ideological and how much, at the highest levels, has been personal, has been an open question from the day the war began. Khomeini himself, when he has addressed topics of an ideological nature concerning the Baghdad regime, has often confused the issue by making little distinction between Iraqi President Saddam Husayn and the Ba'th Party as a whole. Although some Iranian demands have called for his ouster, others have indicated that the entire Party would have to step down before Iran would agree to a cessation of armed hostilities. The reason for the ambiguous distinction made between Saddam Husayn as Head of State on one hand, and the Ba'th Party as the basis of the government, on the other, may be, as many have claimed, that Khomeini has all along believed fervently that Iraq is destined to become the next Islamic Republic. By all accounts, it has so far been a source of major dismay and disappointment on his part that Iraq, with its majority Shi'a population and the strong cultural ties with Iran of many of it inhabitants, has not yet produced such a republic.

REGIONAL SECURITY IMPLICATIONS

Among the more immediate concerns of the Arab Gulf states since the outbreak of the war was their awareness that both their national security as well as the jugular of their economic well-being could be dealt a devastating blow literally within minutes by actions involving one or the other of the two combatants. The need to find a more credible and effective means to deal with the pressing problem of security was, indeed, one of the most compelling reasons for establishing the Gulf Cooperation Council (GCC).

The reaction of Saudi Arabia to the Iraq-Iran war to date has been quite different than that of the other Gulf states. The Riyadh regime has been and remains profoundly disturbed by the sectarian character of the war. Saudi Arabia has traditionally preferred to align itself with the conservative side of the Sunni Muslim camp. Moreover, the fundamentalist foundation of its own regime has served as the repository of a very different interpretation of Islam. Thus, on a sectarian level, Saudi Arabia has been especially concerned about the potential of the Tehran government to undermine an important dimension of the Kingdom's regional role. Qatar's reaction has been similar to that of Riyadh.

Kuwait has also had numerous reasons to worry about the war but for different reasons than Saudi Arabia. Being a closer neighbor to both combatants than Saudi Arabia, having been bombed several times by Iranian pilots during the first two years of the war, and lying far more exposed militarily than any other noncombatant due to the crucial logistical role it has played in channeling supplies to Iraq from abroad, Kuwait lived as dangerously as any GCC state as the war continued. Kuwait's position in the Baghdad-versus-Tehran ideological competition was also different from the outset; it had been developing for some time as a secular rather than sectarian state, overtly sympathetic to neither the Saudi Arabian brand of Islam nor the Iranian interpretation; neither had it been anti-Sunni or anti-Shi'a in its policies and actions.

In terms of the UAE, there has been, in effect, a mirror image of Kuwaiti policy, *i.e.* the successful management of relations with a far more influential neighbor despite the difficulties emerging out of a highly asymmetrical power equation. Just as Kuwait for most of its existence to date has managed to co-exist with Iraq, so has the UAE managed to live alongside Iran despite disputed claims over islands. In addition, both the UAE and Kuwait have built up a reservoir of considerable international goodwill in return for the vast number of financial favors they have provided others through the generosity of their foreign economic assistance programs—assistance which, with an eye to strategic and security interests, they astutely intertwined with ongoing support for their national independence and territorial integrity.

Bahrain, on the other hand, has been a special case. As the Arab world's only island state and one of the few Gulf countries lacking the financial wherewithal to ingratiate others through aid programs, Bahrain had ample reason to be acutely apprehensive about the implications of the war for its unique population. As in Iraq, a clear majority of Bahrain's Muslim inhabitants have for quite some time been Shi'a. The government itself, such as in the incumbent regime in Iraq, has long been dominated by Sunnis. This factor alone has sufficed to sustain ongoing Iranian interest—and, as already indicated, indirect involvement—in Bahrain's internal affairs.

Oman, which occupies a position on the Strait of Hormuz that puts it in a different situation than that of the other GCC members, has been the Gulf state least worried about the sectarian dimension of the war. The reason is that the Shi'a population of Oman, unlike the situation in the other Gulf states, is not indigenous but consists rather of longstanding emigrant communities from southern Iraq, Bahrain, Pakistan and India. In further contrast to most of the other Gulf states, the implications of the war for Omani national security and related interests have centered mainly on matters of a strategic nature. Muscat's abiding concern has been the extent to which, if at all, the course of the conflict might give Tehran cause to rely to a greater extent than before on the Soviet Union.

In addition to the Arab states of the Gulf, Pakistan—as an Islamic country bordering Iran—has come to figure more and more in the regional security equation. It is important to underscore that the Pakistani government has indicated with increasing frequency that it will not involve its military forces abroad in combat against Muslim people. Some Pakistani defense planners are still smarting from the help that Pakistani soldiers provided King Husayn in September of 1970 when the Jordanian army was fighting Palestinian guerillas. Although actual Pakistani units did not take part in the fighting, there is little doubt that individual Pakistani soldiers, whether as trainers or instructors, got caught up in the fighting with particular units and found themselves in the midst of a conflict with fellow Muslims, in this instance Palestinians. It is unlikely that Pakistan would commit its armed forces to a conflict with Iran due to the long-standing neighborly relations and underlying affinity between the two peoples; historically, they have been much closer to one another than either has been to neighboring Arab nations. In response to a lesser level of challenge, however, it remains possible that the Pakistanis might be willing to extend direct military assistance, if invited, to help protect one or more of the GCC states, should the need arise.

Returning to Iraq, with its Sunni minority regime as compared with Iran's Shi'a majority rule, it withstands underscoring that Baghdad's policies in the past few years have posed far less of a challenge to security in the Gulf than have the policies of Iran. This marks a significant shift from the not very distant past, when the predominantly Sunni Arab Gulf states had reason to be more concerned by Iraq's numerous *coups* and revolutionary rhetoric than they were assured by the Sunni composition of its regime.

In addition, the constraints on intervention elsewhere in the Gulf by either of the two combatants are considerably stronger in Iraq than in Iran. Baghdad remains vulnerable to the unpredictable actions of Khomeini, on one hand, and to the financial largesse of the Arab Gulf states, on the other. Both factors have had the effect of shortening Iraq's political leash. The Arab Gulf states, for their part, have been well aware that should the government of Saddam Husayn happen to fall, its replacement might be much more threatening to their security.

Despite the foregoing, Gulf Arab security concerns *vis-à-vis* Iraq, on balance, have been alleviated in the eyes of some since Baghdad, under Saddam Husayn, has leaned increasingly toward the West and the dynastic Arab states, even while Iran, under the Khomeini regime, more often than not has seemed to be doing its utmost to wrench away from both. From the global perspective of various Western countries, as well as in the context of the other Gulf states' concerns of a regional nature, the consequence, again, has been that the more worrisome of the two countries has clearly been Iran, not Iraq.

The regional implications of Iraq's predicament *vis-à-vis* its Arab Gulf neighbors can withstand still further comment. Certainly, evidence of Iran having moved away from many of the other Gulf states, and of Iraq having moved towards them, has been reflected in Iraq's steadily improved relations with Saudi Arabia. Of additional importance in this regard have been reduced tensions between Iraq and Kuwait, and the non-recurrence, since 1975, of Iraq's previous support for the leftist Popular Front for the Liberation of Oman.

Indeed, since Khomeini came to power, the record indicates that Iraq has forged a broad range of cooperative ties with other Gulf states aimed at enhancing their respective capabilities against Iranian-sponsored infiltration activities. In many other ways as well, Iraq, though not a GCC member, has adopted policies and on occasion taken actions on issues of regional importance that have been complementary to GCC needs. In short, the anxieties occasioned by the Iranian revolution and civil war, as well as by Iraq's failure to win a quick or decisive victory against Iran, has kept Iraq committed to improved relations with Jordan and the other Arab Gulf states. It is these states upon which Baghdad has had to rely continuously for economic, logistical, political, and diplomatic support, and upon which it will likely remain dependent in the near term whether the war continues or reaches a settlement.

The above considerations notwithstanding, it is important to stress that not all Iraqis have discarded a point of view articulated with some force to this writer by a group of Ba'thists a year before the war began. That view advanced the thesis that an Iraqi victory over Khomeini-led Iran would enhance, rather than endanger, Western interests in the Arab world. What these Ba'thists had in mind was mainly the six GCC states, but it was clear that they envisioned Jordan and Egypt as benefitting also. However debatable such a proposition may have been or could yet become, those who proclaimed its validity at the time emphasized that the setback for Western and Gulf states' interests which would occur, should Iran defeat Iraq, was—and would remain—beyond question.

To what extent such Ba'thist reasoning was indicative of mainstream governmental thought in Baghdad was unclear at the time. The fluidity of circumstances occasioned by the war, requiring tactical ingenuity, pragmatism, and at times unbridled opportunism, hereby served to reassure the Arab non-combatants that the recipient of their support was a fundamentally and permanently re-oriented Iraqi Government that was favorably disposed to their long-term interests. Certainly those upon whom Iraq remained the most dependent for supplies, loans, and mediational efforts to bring the war to a rapid end evinced an abiding interest in not seeing Iraq emerge from the conflict with a renewed capacity or intent to export Ba'thism to lands beyond its shores.

A markedly different view could be heard outside the region among the countries which had helped Iran in the conflict. Various officials within the Syrian, Libyan, and South Yemeni governments, for example, indicated that they would not necessarily view with disfavor a scenario whereby Iran, after the war, might serve as a greater inspiration than it had previously for radical and revolutionary forces operating in and adjacent to the Arab Gulf states. Again, the regional implications of the conflict became clearer when one considers that Iraq has had neither the stated intent nor the individual conduits at hand for such actions but Iran, in addition to its considerable advantage over Iraq in terms of geography, demography, and military forces available for these purposes, has had both.

For those who had any doubt, Iranian intentions towards the south side of the Gulf were sufficiently demonstrated by the 1981 *coup* attempt in Bahrain. Not knowing whether other subversive groups might also be training inside Iran for future strikes at a Gulf regime, and, if so, how many and/or which GCC state might be the next target, the GCC states, with Saudi Arabia at the forefront, reacted swiftly. Within a week, Riyadh and Bahrain had signed a security agreement designed for consultation, exchange of intelligence and mutual military assistance in the event of any similar incident occurring in the future. Qatar signed a nearly identical agreement with Saudi Arabia shortly thereafter and within two months the remaining GCC states had entered into similar arrangements.

MINORITIES

An important consideration for security analysts on both sides of the Gulf has been the position and role of minorities in Iran and Iraq and the potential involvement at some point in the conflict of their ethnic and sectarian compatriots elsewhere in the region. Here again, the greatest focus has been on Iran. To the surprise of many Western analysts, the activities of the several minority groups in Iran who had previously sought autonomy from the Tehran government have all been contained since the war erupted. One reason has been that the overwhelming majority of Iranians, without regard to ethnic identity, class, or sectarian orientation, have coalesced in what, in essence, has been a national effort to defend Iranian territory in the face of the original Iraqi invasion.

An equally telling factor has been that this war, like all wars, has created its own conditions of severity. No less than many another government at war, the regimes in Baghdad and Tehran have sought to suppress any domestic challenge of the conflict. In time of peace, most governments have responded to rebellious citizen activity with force, to be sure; in time of war, however, the tendency has been to use maximum brutality to defeat such rebels and brand them as traitors. Given this reality and the almost certain harsh response of either of the two governments in this matter, minority groups in both countries have been exceptionally cautious in their behavior since the war broke out.

It is, of course, important to underscore the fact that Iraq is not immune from such phenomena and that, with particular respect to its Kurdish population, it has had greater experience than Iran in dealing with this kind of problem. Nonetheless, should Iran lose the war, Saddam Husayn may anticipate renewed heavy support by Iran for the Kurds in Iraq and, also, for disaffected Shi'ites who, although outnumbering Iraq's Sunni Arabs, remain very much a political minority.

Iran's minority problems, by contrast, have been more complex to date than Iraq's because: (1) the groups in question have been larger in number, both in category and in terms of overall size (including Kurds, Turkish-speaking Iranians, Arabs, and Baluchis, to name the most prominent); (2) they have been based in areas some distance—and in different directions—from the capital; (3) they have been simultaneous, on more than one occasion in recent history, in their breakaway activities; and (4) there is the consideration that the uncertainty of whether the central government in Tehran will be able to come to terms with these groups, and *vice versa,* is both greater and more recent, being held in abeyance since the Iraq-Iran war began.

Despite the foregoing, one of the implications for Iran's minorities may be the option of resuming active pursuit of their aspirations once the war is ended. In this event, an important question would be whether one or more of the minorities in question might turn to outside groups for assistance, for example, to the Soviet Union or to kindred groups in Pakistan, Afghanistan and the GCC states. If so, it would be important to consider what might be the response. Would the Soviets be able and/or inclined to extend the kind of aid to any of these groups that it provided, from its presence in South Yemen, the insurgents in Oman's southern province of Dhofar? And were it to do so, what might be the expected result? Might this be likely to set off a conflict different from any which has involved the Soviet Union and/or other non-Gulf powers in the region to date?

Conversely, were the minorities to receive aid from non-Soviet sources, what kinds of problems might this entail? More specifically, if minorities thusly aided were to rebel against the Iranian regime, might that regime, or its successor, be inclined to accept foreign assistance for the purpose of quelling the insurrection? In such a scenario, might a country such as the Soviet Union, citing insecurity on its southern flank, be inclined to intervene without regard to the niceties associated with whether or not a formal invitation was extended?

With the war coming to an end, there will therefore be reason to query the near and longer term fate of such groups. Will they renew pressure for greater autonomy and/or a greater voice in the national government? If so, would the policy of the central government be accomodational or confrontational? If at all the latter, to what extent might this compel one or more of the groups in question to seek external support?

Of the various options and scenarios at hand, one similar to that in which Baluchistan were to achieve autonomy or to successfully pursue secession from Pakistan could increase in significance. Analysts who have studied this scenario point to the larger number of Baluchis in Soviet-dominated Afghanistan; their potential ability to provide the Soviet Union with a corridor to the sea; the fact that they are widely dispersed not only in Pakistan and Iran but throughout the Arab states of the Gulf; their reputation as reliable fighters; and their position and role as one of the most economically resourceful minorities in the region.

LINKAGE AND DIPLOMACY

The situation of any war in process defies the certainty of precise analysis. This has especially been the case in the Iraq-Iran war due to the constantly changing events in the eastern Mediterranean, most particularly in Lebanon, but also in the West Bank and Gaza. Looking down the road, if peace were to be established in Lebanon and the Palestinians were to become engaged in a direct dialogue with the United States, the degree of antagonistic and alienated feelings amongst regimes and people alike toward one of the superpowers with a stake in the conflict's outcome—the United States—would diminish. This could have a comprehensive and altogether salutary impact on the region, substantially weakening the export appeal of Iran's fundamentalist revolution in the process, especially if the Palestinians were to be seen as part of a process in which the United States was playing a positive role.

On the other hand, if the reverse were to occur, the potential for a backlash against the more moderate regimes supporting Iraq could be anticipated, and this could only work to the benefit of Tehran. The mutuality and reciprocity of certain Israeli and Iranian strategic interests must, therefore, not be ignored. When Iran crossed into Iraq it was a net gain for Israel in that it diverted the attention of the Arab Gulf states away from Lebanon. To compound matters, Iran advanced its own cause and embarrassed the Arab regimes by being the first—and by most accounts the only—regional state to provide actual military assistance to the Syrians, Lebanese, and Palestinians during the Israeli onslaught in the summer of 1982.

This was all the more ironic with regard to certain other aspects of the Iran-Israel relationship. Evidence of collusion between the two countries has not exactly been elusive. On numerous occasions, each has seen an advantage in operating in the cover of the other's military moves. Certainly Israel has been central to Tehran's ability to prolong the war by sending weapons and spare parts. In May 1982, a senior US Department of Defense spokesman conceded to this writer that the extent of such Israeli assistance to Iran up to that time had been "massive." And one month later, following the Israeli invasion of Lebanon, there appeared fresh evidence of the nature and scope of collaboration between Tel Aviv and Tehran. In one incident where Israeli forces had captured a cache of 50,000 completely unused, Soviet-manufactured Kalashnikov automatic weapons, eyewitness accounts of American military personnel serving with the UN peacekeeping forces indicated that, within 48 hours of their capture, the entire consignment of these weapons, plus two and a half million rounds of ammunition, had been sold and delivered by Israel to the Iranian Government.

For their part, the Arab states of the Gulf, although acutely conscious of the broad range of regional security problems that these events have occasioned, have been dismayed at the inability of international diplomacy to bring the war to a close. To be sure, there have been numerous unsuccessful attempts by third parties—the UN, the Islamic Conference Organization, non-aligned groups, and Turkey, Pakistan and Algeria—to find a peaceful settlement to the war. In regional eyes, the superpowers have been suspected, and in several instances actually accused, of having an interest in seeing the conflict prolonged. Such cynicism, although consistently deemed unwarranted and unfair by US government spokesmen, has stemmed as much as anything else from official US policy statements issued during the course of the war. Aside from emphasizing that the US "has remained from the beginning, and will remain, neutral in the war", official US policy has been "supportive of the independence and territorial integrity of both Iran and Iraq", opposed to seizure of territory by force, and has reiterated the need for "an immediate end to hostilities, and a negotiated settlement."

A formal White House statement of July 14, 1982, declared US support for the security of friendly states in the region which might feel threatened by the conflict, and announced that the United States was prepared to consult with those states in the region which might feel threatened by the conflict on appropriate steps to ensure their security. Statements of intent aside, however, the US had no formal diplomatic relations with either Iraq or Iran, although some American diplomats were in Iraq. Although the outcome could be so potentially serious for Western and Arab interests in the Gulf, the US continued to claim that it held little sway over the course of hostilities, despite its disposal for constraining Israeli support for a crucial aspect of the Iranian war effort.

Europe, meanwhile has been distancing itself from the political quagmire of the Middle East and devoting itself more and more to European problems. To the extent that Europe has remained involved in the Middle East, it has been overwhelmingly in the context of economic interests. European statesmen have continued to apply the proper and locally much-appreciated rhetoric on political questions. However, there is little evidence that such actions have done much to influence US policies and actions on the question of Palestine, Jerusalem, the West Bank, Gaza, or the Golan Heights. Similarly, with regard to bringing the Iraq-Iran war to a close, Europeans have had little impact to date. What might have been an exception, of course, was the mediation effort by the Swedish Prime Minister Olaf Palme. The Palme mission, however, was premature and thus of only marginal significance.

The history of Iraqi-Iranian relations—as well as those of other Middle Eastern countries—demonstrates that only when both parties to a particular dispute are themselves ready for agreement can other parties play a role, not *vice versa.* On no occasion of this or earlier conflicts between Iraq and Iran has a unilateral mediation effort by outside parties produced results of any lasting benefit. An example in this context was the Algiers Agreement of March 1975, when the Shah of Iran and Iraq's Saddam Husayn met in Algiers and signed an accord to end their longstanding disputes over the Shatt al-'Arab, the Kurdish question, and Iraqi support for radical movements in the Gulf. Only then, when both states were ready to conclude an agreement, was mediation effective, even if Iraq's Husayn was beleaguered at the time and the Shah was at the height of his power, thereby lending credence to the view of many that the agreement was adhered to by Baghdad under circumstances akin to duress.

Until recently, neither country has indicated a readiness to resolve the current dispute. As a result, international initiatives to end the war have provided both countries with opportunities to exploit such initiatives for their own purposes. On the Iraqi side, the war in Lebanon provided Baghdad with an exceptional opportunity to orchestrate a diplomatic end to the war. Had these efforts been successful, the Baghdad regime might thereby have avoided the far higher domestic costs which it could expect to face in the event of its defeat on the battlefield. Certainly, many Iraqis were hopeful that a diplomatic settlement could be achieved before the non-aligned meeting, which was scheduled to take place in Baghdad on September 6, 1982. But fighting continued and the conference was transferred to New Delhi.

On the Iranian side, Tehran's confidence during the same period was demonstrated first by its refusal to comply with an Arab-led call in OPEC for ceilings on individual oil production by the main producers, and second by its decision to ignore a unanimous UN Security Council resolution on July 12 for a ceasefire in the Gulf war. Staying the hand of every UN mediation effort to date has been Iran's flat rejection of any actions by that body to end the dispute, with its reason being that the UN made no effort to intervene when Iraq had the upper hand.

POLICY CONSIDERATIONS

If the foregoing attests only in part to the numerous regional and worldwide implications of the conflict to date, it nonetheless indicates several directions that policy formulation might be expected to take in the near future. In terms of Western concerns, a major theme had clearly come to be the degree to which the intensification of the war, on the one hand, and mounting evidence of aggressive Iranian intentions against virtually every Arab Gulf government, on the other, posed direct and immediate challenges to Western strategic, economic and political interests.

Secondly, while a neutral Western response to Iraq's 1980 invasion of an Iran which was still holding the US hostages was seen as only appropriate in the eyes of most Western analysts, the continuation of such a policy in light of the reversal of the war and Iran's increasingly anti-Western vehemence was viewed by many of the same observers as counterproductive to the interests of Western countries, the Arab Gulf states, and most other Arab countries.

Thirdly, all Western countries with interests at stake in the outcome of the conflict have indicated that they see little choice but to ensure that Iraq has the capability to withstand a sustained Iranian invasion. Such steps as these powers have considered undertaking towards this goal have included: (1) providing assurance that Iraq would receive adequate military equipment and supplies, including additional shipments through Arab states friendly to Iraq (an example was a British-facilitated $400-million arms transaction through Egypt in the spring of 1982); (2) recommending that the United States normalize diplomatic relations with Iraq and support an intensification of diplomatic efforts to end the conflict; and (3) urging the US to call Israel unequivocally to account for violating US laws in the process of providing Iran—even while Iran held Americans as hostages—with military equipment and spare parts for its predominantly US-manufactured weaponry, most particularly the crucial aid which Tel Aviv supplied the Iranian air force.

A number of subsidiary themes have commanded attention as well. Prominent among these has been acknowledgement by many that so long as the Khomeini regime remains virulently anti-West, there may be little the West can do in the short run aside from not radically altering its relationship with Iraq to such an extent that a later improvement in Western-Iranian ties would be precluded.

In the interim, the more constrained of the Western powers—the US—has limited itself to lending assistance where possible in the construction of a GCC-centered air shield over the Gulf's southern oil fields as a means of guarding against an Iranian threat, and strengthening American "over-the-horizon" capabilities against the worst case possibility, however remote, that a future Iranian government might seek US help against Soviet intervention.

Influencing the political interests of numerous Western countries was the fact that Iraq, in comparison with Iran, has in recent years become increasingly open to relations with the West in general. Western diplomats were well aware that almost simultaneous with Saddam Husayn's assumption of office, Baghdad initiated a shift in policy away from Moscow. Thousands of Soviet advisors subsequently left Iraq and normal diplomatic relations were being conducted between Baghdad and all of the major Western countries except the US. Further testimony to the reorientation of Iraq's foreign policy in recent years was its severance of diplomatic relations with North Korea and its establishment of consular relations with South Korea in their place.

Western foreign policy officials also seemed to give increasing weight to the fact that, in comparison with Iran, Iraq's economy was clearly the more dynamic of the two. Baghdad had increasingly, in almost ideological aversion to Tehran, intensified its economic ties with the Western world. Politically as well, it was Arab Iraq, not non-Arab Iran, which held a position of leadership among the non-aligned countries, many of which had had longstanding plans to attend the Non-Aligned Conference which was to have been hosted by Iraq.

Finally, and again in terms of any American involvement in settlement of the conflict, it remained a source of widespread concern, one fraught with unpredictable regional and global implications, that the US might send troops to protect what it declared at the time of the Carter Doctrine in January 1980 to be its vital interests in the region. At this writing, it was in no one's interest—neither that of such global actors as Europe, the US, or the Soviet Union, and, among regional actors, least of all the Gulf states, whether Iraq or Iran or the non-combatants—that such a scenario come to pass. Weighing all costs, regional and global actors alike had every reason to reject further passivity toward this conflict and to ensure that neither the Iranian invasion of Iraq, nor the opposite, would be allowed to succeed.

Bid for Gulf Unity

VALERIE YORKE

The first summit in Abu Dhabi on 25–26 May 1981 of the Gulf Co-operation Council—the body through which Saudi Arabia, Kuwait, Qatar, the United Arab Emirates (UAE), Bahrain and Oman intend to co-ordinate all government activities—is a landmark in the political development of these states and a bold experiment in Arab co-operation. The Council, which was founded at the Islamic summit in Taif in February on the basis of a Kuwaiti proposal, formalizes the already extensive bilateral links among the Arabian Peninsula states. It also demonstrates a readiness on the part of Gulf rulers to put aside long-standing rivalries in favour of stronger co-ordination on political, economic, oil and security matters. While the rulers of six Gulf oil-producing states have made this first tentative step towards regionalism in the interest of self-preservation, their initiative is likely to have important implications for Middle East politics as well as for the West.

In the past, the conservative Gulf states have been unable to agree with Iran and Iraq on how to make the Gulf secure. Historical Arab–Persian rivalries were inevitably an inhibiting factor. While Riyadh and Tehran shared a healthy distrust of Baghdad's Ba'athist regime with its pro-Soviet ties, the Saudis suspected the Shah of regional leadership ambitions. They could neither forgive his seizure in 1971 of the three Arab islands in the Strait of Hormuz nor tolerate his ties with Israel. The religious cleavages between Saudi Arabia and the Gulf, which are predominantly Sunni, and Iran whose dominant faith is Shia sharpened disagreements between the Arab Gulf and non-Arab Iran.

Against this background, an attempt to move towards a Gulf Security Pact foundered at a meeting in Muscat in 1979, mainly due to differences between Iran and Iraq. The Shah's Iran favoured a Gulf alliance while Iraq

This article has been reprinted by permission from *The World Today* (Vol. 37, July-August 1981), monthly journal of the Royal Institute of International Affairs, London.

wanted a more restricted arrangement to protect navigation. Yet, the danger of polarizing the Arab Gulf and Iran meant that the exclusion of Iran was never a practical option either. In any case, suspicions about Iraq's aspirations for regional leadership and differences within the conservative Gulf camp worked against a narrowly based Arab–Gulf organization. All Gulf states share an interest in keeping the Strait of Hormuz open, but Oman holds different views from the others, mainly Kuwait and Iraq, over how this should be done and on the role of 'outsiders.'

Two years of turbulence in the Gulf and growing super-power competition since the fall of the Shah appear to have altered the calculations of Gulf rulers regarding the costs and benefits of working together. They came to realize that they will either stand together or fall individually. Indeed, the formation of the new Council was arguably a logical step in the Gulf rulers' post-Shah diplomacy. It is the centrepiece of an elaborate insurance policy of adjustments to foreign and domestic policies. The aim is to forestall escalation of regional and indigenous tensions, which might upset their fragile regimes or spread from one Gulf state to another, and to keep the super-powers out of the area.

This process of policy adjustment began in 1979 with the Egypt–Israel Treaty and Iran's Revolution. Gulf rulers found it prudent to distance themselves from the Treaty which had dangerously divided the Arab world. With the exception of Oman, they felt they could not risk the disruptive effects that support for the Treaty could have on their populations. This happened precisely when Iran's Revolution had left them feeling more exposed. Deprived of Iran as their regional policeman and fearful of the Revolution's demonstration effect, the Gulf states needed reassurances of U.S. support. Yet, the American response to the Soviet invasion of Afghanistan in the form of the Rapid Deployment Force and the recent U.S. determination to forge a 'strategic consensus' have sharpened the fears that the House of Saud and the Gulf Emirs have always harboured of being seen to rely on the United States for protection. In their view, an American presence on Arab soil would fuel opposition to their regimes and encourage Soviet involvement. They also know that American forces are not necessarily intended to save their regimes but to protect Western oil supplies. Local hostilities and domestic vulnerabilities thus encouraged Gulf rulers to strengthen co-operation on security—a process that was reinforced with the danger of the Gulf War spilling over to the Arab Gulf. With Iraq and Iran preoccupied with war, an important constraint on formal Gulf co-operation was also removed.

To ensure a good start for their bid for unity, the Gulf Six wisely intend to consolidate such functional co-operation as already exists among them rather than break new and more controversial ground. Their common history, religion, language and socio-economic make-up combined with traditional political practices place them in a favourable position to tackle together the related problems of planning for a future without oil, political development and internal security. They recognize the need for industrial consultation to limit wasteful project duplication, the growth of immigrant workforces and unsettling competition. At the same time, moves towards resolving outstanding disputes will be important for harmonious relations. The recent agreement between Oman and the UAE to redraw their border is a helpful precedent; the UAE and Kuwait are also mediating between Oman and Marxist South Yemen, though without success so far.

The Council's infrastructure was officially established at the summit in Abu Dhabi. It consists of a Supreme Council of Heads of State, a Council of Foreign Ministers and a Secretariat based in Riyadh over which Kuwait's former Ambassador to the United Nations, the able Abdullah Bishara, presides as Secretary-General. A series of Gulf-wide ministerial committees to unify policies on oil, economic planning, industrial development, finance and trade and social affairs were set up and arbitration procedures for the settlement of disputes were finalized. The Gulf rulers have thus taken a historic step in supplementing traditional decision-making procedures with permanent institutions to cope with the problems of their rapidly modernizing societies. Yet, they have deliberately minimized the scope for disagreement. Importantly, the Six will not surrender sovereignty, though co-operation may eventually lead to some form of political union; they will contribute equally to the Secretariat budget and decision-making will be by consensus.

The Gulf states have a collective interest in pooling their resources to promote stability, but the fact that each has its own incentive for joining the Council further enhances its credibility and strength. Oman, which feels threatened by the Soviet presence in South Yemen, wants to end its isolation from the Arab world. Bahrain needs to rally support against Iran's expansionist ambitions and also welcomes the chance of economic help without political strings. Kuwait does not believe Iraq has abandoned claims to its territory and worries about the growing demands of Palestinian and Shia communities. Qatar and the UAE, whose populations are 75 per cent expatriate, sense that there is little advantage in developing industries that must be run by foreigners. Badly shaken by the siege of the Grand Mosque at Mecca, Saudi Arabia wants to tighten up on internal security throughout the Peninsula. For the Saudis, however, the Council, which excludes the rival Iraqis, will be a useful vehicle for advancing their own bid for regional influence through a more assertive diplomacy.

Ironically, while fears for security brought them together, Gulf rulers are aware that it is precisely defence and foreign policy co-ordination that could ultimately prove divisive. They have, therefore, sensibly postponed the hammering out of a joint security policy. In the summit's final communiqué, which reflected the official thinking of the Saudis, the Heads of State firmly rejected foreign 'intervention' and the presence of foreign bases and navies, insisting that Gulf security was the responsibility of the local states. The rubber stamp of Sultan Qaboos, who alone provides the Americans with 'facilities,' demonstrated the importance that Council members attach to an appearance of unity.

Behind this façade, sharp differences exist. For Oman, the chief threat to the region comes from the Soviet Union from the north and through Soviet-backed South Yemen. The five-day visit of two Soviet warships in South Yemeni waters provocatively coinciding with the summit strengthened its argument. The Omani answer to this threat is contained in a security proposal which was prepared at the request of the Council's Foreign Ministers at their Muscat meeting in March 1981 and discussed in Abu Dhabi. This also revives Oman's 1979 plan for an Arab minesweeping force financed in part by Western oil consumers and underpinned by a Western military presence.

Other Gulf states consider regional dangers like the Gulf War and the unresolved Palestinian issue to be much more immediate. They are concerned lest the Council jeopardize relations with Iran or Iraq, thereby producing difficulties for Gulf states within the Arab League and the Organization of the Islamic Conference. Similarly, they are reluctant to co-operate with America in countering the Soviet Union on account of U.S. support for Israel. The current crisis in Lebanon and Israel's bombing of Iraq's nuclear installations will have reinforced this position and will ensure that the debate on the merits or otherwise of a Western presence in the region will be kept strictly private. Kuwait, the only Gulf Council state which has relations with the Soviet Union and whose Foreign Minister recently visited Moscow, goes further. It wants the Council to move towards a more non-aligned posture.

These inhibitions preventing public discussion of the security issue illustrate the dilemma facing the new Council. While its members are pro-Western by inclination, they cannot afford to alienate the Arab and Islamic worlds. The presence at the inaugural summit of Mr. Chedli Klibi, Secretary-General of the Arab League, and Mr. Habib Chatti, Secretary-General of the Islamic Conference, was a clever device aimed at providing the official endorsement that Gulf rulers felt their new Council required. In the future, the Council's freedom of manœuvre will be circumscribed by its members' desire to hold together an Arab and Islamic consensus even if this means adopting a more hard-line position themselves. Indeed, the message for the West is this. The Gulf Co-operation Council may try to exercise a moderating influence in Arab and Islamic fora. On the other hand, its members also know that the danger which the Arab–Israeli dispute poses to their regimes may compel them to use the power at their disposal to persuade the West to tackle this issue. The fact that the Gulf Six supply over 40 per cent of the West's oil, control the strategic Strait of Hormuz and represent the third most important region for the West's exports give them significant leverage over Western governments.

Competing Superpower Interests in the Gulf

The superpower struggle for influence in the Gulf is an echo of earlier rivalries there, when France, Germany and Russia all attempted to displace British supremacy. More recently, the goal of the United States is to retain access to oil and to prevent it from falling under Soviet control; while Soviet goals are obscure, they presumably would like to replace the United States as the most influential foreign power in the region. Their current rivalry is still largely a one-sided affair: the U.S. begins with considerable advantage over its opponent. One obvious example comes from the observation that Washington maintains diplomatic representatives in seven of the eight Gulf states, Moscow only in three. The U.S.S.R. is still seen as an interloper by most of the Gulf whereas the U.S. is more often made welcome.

Partially, the American position derives from the longstanding ties with Saudi Arabia, a relationship which began with the formative role of Aramco in social and economic change there and which continues with the education of many Saudis in American universities and extensive American official and private sector involvement in Saudi Arabia's ambitious development program. In the smaller states of the Arab littoral, the U.S. has been heir to Britain's century of involvement there. Furthermore, the United States benefits both directly and indirectly through its European allies by the continued strength of their economic and cultural ties with these countries. Oman is a prominent example of how the torch of political and military cooperation has been passed on from Britain to the U.S.

Iran, however, stands out as a warning of the dangers of too close a relationship with a "client" state. Does the superpower control the action in such a relationship or is it often the other way around? While the Shah appeared to be faithfully carrying out American security interests in the Gulf, it is painfully obvious in retrospect that the major consequence of extensive American involvement in Iran was complete identification with the Shah's repressive policies and a bitter estrangement leading to the hostage situation.

The Soviet Union has not fared much better. The development of close relationships with certain Middle Eastern oil states has not come about because of any natural affinity; rather, Soviet intrusion has been possible only after the West's position has been eroded through such factors as the rise of nationalism and the Arab-Israeli conflict. Despite decades of military supplies to and diplomatic support of Iraq, Moscow has gained little leverage in Baghdad. Iraq's Ba'thist rulers have periodically purged the army and government of Communists and severely criticized Soviet policy in the Horn of Africa. Algeria stands as another case of a radical regime accepting Soviet assistance without falling into the Soviet orbit. The emerging sense of self-confidence and pragmatism in Algiers has led to the growth of trade ties with the U.S., without sacrificing its socialist ideology, as well as the ability to serve as intermediary between Washington and Tehran in the resolution of the hostage situation. The cash received for its military hardware may well be the only substantive benefit accruing to Moscow from its supply of Libya: Colonel Qadhdhafi is not likely to embrace Marxism at home or propagate it abroad.

The first three selections below offer a comprehensive survey of superpower interests in the Gulf region and analyze the prospects of their advancement in the foreseeable future. Despite the obvious nature of the ultimate goal—control of oil—there is deep disagreement over the choice of strategies to be employed for its attainment and, indeed, over measurement of progress towards that goal. The picture is further complicated, of course, by the problem of ethnocentricity. Not only is it difficult to understand—let alone come to terms with—the desires and fears of the peoples of the Gulf, but the science of accurately and precisely determining Soviet intentions is far from perfect.

While all of the following selections offer some policy advice, whether explicit or just implicitly, the final article of this book, Christopher Van Hollen's admonition to not engulf the Gulf, deals directly with the continuing and so far inconclusive debate over the formation of an effective American policy for the Gulf. It is preceded by a short note introducing some of the major issues and participants in this debate.

Bibliographic Sources for this Chapter:
17C, 17D, 17G, 23, 26, 31, 34, 40, 44, 47, 53, 59, 63, 70, 79, 80, 81, 82, 83, 85L, 86, 87, 92, 94, 109, 120, 129, 173, 176, 194, 206, 213, 220G, 223, 241, 245, 252, 254, 257, 265, 273, 274, 277, 278, 279, 284, 287, 296O, 296P, 300, 301, 302, 307, 311, 313, 317, 333, 335, 336, 337, 339, 341, 342, 343, 344, 346, 349, 350, 353A, 353G, 360, 362, 364, 368, 379, 380, 383, 385, 386, 388F, 388G, 388H, 394, 395, 397, 407.

Superpower Rivalry in The Gulf

CHARLES W. YOST

Superpower rivalry in the Gulf is not something new. It has been going on for about 150 years, during the first hundred or so of which the United States was not involved. It was rivalry between Britain and Russia which culminated in the effective partition of Iran into spheres of influence in 1907 and in Britain moving into the Gulf in a very substantial way both economically and politically and subsequently, after World War I, acquiring a dominant position in Iran.

What one might call the modern period of superpower rivalry in the Gulf began with World War II, even though at its outset the superpowers in question cooperated rather than competed. It turned out that the main channel of supply for the Soviet Union after they came into the war was through Iran and therefore the country was occupied for a number of years by the three Allied superpowers. In addition, one of the big wartime summit conferences was held in Tehran. All of these served to concentrate U.S. political attention on the Gulf area for the first time. In fact, after the Yalta Conference in 1945 President Roosevelt stopped on his way home and had a meeting on an American warship in the Red Sea with King Ibn Saud where they discussed some of the problems of the area.

It was as the Cold War began after World War II that the Gulf became primarily an area of U.S.–Soviet rivalry. British influence, of course, continued for 25 years or more and still exists to a small degree, but with the British withdrawal first from India and Pakistan and then from Suez, Britain's influence was more and more overshadowed by that of the United States and eventually Britain bowed out almost completely except for its still substantial oil interest. So the United States was really left as the major Western actor in the region. This country had long-established cultural

Reprinted by permission of the publisher from *The United States, Arabia, and the Gulf,* Ronald G. Wolfe (ed.), pp. 37-43, copyright 1980 by the Center for Contemporary Arab Studies, Georgetown University, Washington, D.C.

connections with the Middle East through its missionary endeavors and its schools. It also had a limited commercial interest due to its involvement in the oil business. But to these were added after World War II two major new interests: a much enhanced concern with the strategic security of this region which sat astride the main lines of communication between Europe, Africa, and Asia, and a concern for the security of a newly established state in the area, Israel. The Soviet Union was perceived as the main threat to the first of these interests. While the Soviets initially supported the establishment of Israel, after they commenced supplying arms to a number of Arab states they soon came to be seen as the main threat to Israeli security as well. Hence they were rivals of the United States in the area in two ways.

For convenience, let us divide the history of postwar superpower rivalry in the Gulf and in the Middle East more generally, into five partially overlapping periods. The first lasted from the withdrawal of Soviet troops from Iran early in 1946, which took place only under very heavy U.S. and United Nations pressure, until the active reentry of the Soviet Union into the area by the supply of arms first through Czechoslovakia to Egypt in 1955 and eventually to other Arab states, including Iraq after the revolution in Baghdad in 1958. During this first period the strongest external influences in the region were those of the United States and Britain. These influences were sufficiently prevalent that in 1953 in Iran a government that was perceived by the West as leftist but whose primary sin was probably that it nationalized oil was overthrown with U.S. encouragement, and the Shah restored. While the claims of some Americans that this country did most of the job are probably greatly exaggerated, and it is more likely that the restoration of the Shah was carried out primarily by the Iranian military, this country nevertheless encouraged the overthrow and played a role.

With Soviet reentry into the area in 1955 through the supply of arms, a second period of more acute competition commenced. The United States and Britain felt the region sufficiently threatened at that point to require a conspicuous consolidation of its northern tier and this was carried out through the establishment of what was at first called the Baghdad Pact and subsequently, after the violent withdrawal of Iraq from that constellation, the Central Treaty Organization. At the same time, the United States proclaimed the now almost forgotten Eisenhower Doctrine, under which assistance would be offered to states that believed themselves threatened by Communism. The doctrine attracted very few takers in the region, primarily because of intense Arab resentment of U.S. support of Israel, and it soon faded away.

During the third period of superpower rivalry, embracing the 1960s and early 1970s, attention was focused more on the Mediterranean littoral than on the Gulf. While the period saw the 1967 and 1973 Arab–Israeli wars, the latter preceded by the dramatic expulsion of the Soviets from Egypt in 1972, the Gulf proper was relatively quiet in those years. British withdrawal was completed. Both Iran and Saudi Arabia proceeded to develop and modernize, although still at a relatively modest pace. All states in the area except Iraq were under conservative pro-Western governments. The Organization of Petroleum Exporting Countries was formed but was not yet very active or influential.

The fourth period (1973–1979) began not so much with the Arab–Israeli war of 1973 as with the oil embargo and the subsequent drastic increase in oil prices which the U.S. supply of arms to Israel during the war provoked. The price rise would no doubt have occurred in any case, but probably somewhat more slowly and less dramatically. But the consequence, at any rate, was the sudden entry into the Gulf region of an enormous influx of wealth which had many positive consequences but which nevertheless if handled badly may ultimately come to be regarded as having been a dubious blessing, similar to the vast flow of silver and gold into Spain in the sixteenth century.

Strategically speaking, the influx of wealth has served more to destabilize than to stabilize the region. The Shah's enormous purchases of arms proved irrelevant to the internal political and religious struggle that arose between the forces of modernization and those of traditionalism. Whether or not the large arsenals of modern weapons in other countries around the Gulf will prove more useful in promoting stability remains to be seen. So far they have been primarily an instrument, though not a decisive one, of superpower influence—Soviet influence *vis-à-vis* Iraq, American influence in the countries farther south. Meanwhile, destabilizing consequences of the sudden wealth include conspicuously extravagant elite lifestyles, an increase in corruption which shocks both traditionalists and reformers, and a massive influx of immigrants and workers who are treated as second-class citizens and hence potential candidates for subversion and unrest.

The 1973–79 period included a number of other developments that sharpened the superpower rivalry in the region. First, the Arab–Israeli war of 1973 almost produced a direct clash between the United States and the Soviet Union. Afterward, during the time of Henry Kissinger's shuttle diplomacy, events tended more and more to exclude the Soviet Union from a significant role in an area it considered of utmost importance to its security and prestige; this naturally generated feelings of wounded pride and resentment.

When the Carter administration came to office it first tried to revive U.S.–Soviet cooperation in Arab–Israeli peacemaking along the lines of recommendations made a short while before by a Brookings Institution study group. But after Sadat's dramatic visit to Jerusalem at the end of 1977, the United States was more or less obliged, or at least it chose to revert to a unilateral approach by fostering and participating in the Camp David agreements and their implementation.

As far as the traditional area of rivalry was concerned, the Soviets were left with close associations only with Syria, the PLO, and, to an uncertain degree, Iraq. However, just as in 1955 they had leapfrogged over the pro-Western northern tier of regional states to acquire a foothold with the Arabs through their arms sales, in the late 1970s they leapfrogged over the whole region of Arab–Israeli confrontation to acquire footholds in Somalia (until 1977), Ethiopia, and South Yemen. Thus they established themselves, at least temporarily, on both sides of the Red Sea and not far from the critical exit from the Gulf.

The final significant development of the 1970s was that the United States, its Western European allies, and Japan, which had, during the previous 20 years, carelessly allowed their economies to become dependent on Gulf oil, suddenly woke up to this harsh and dangerous fact of life. A shattering side effect of this dependence was the enormous economic toll levied upon them and the rest of the world by the steady escalation of oil prices. Furthermore, the Soviet Union, which had been both itself self-sufficient in oil and the sole supplier of its Eastern European allies, began to confront potential shortages and to oblige its allies to make substantial and costly purchases from OPEC. By the beginning of 1979, therefore, superpower rivalry in the region had already begun to move into a new phase. This evolution, which in so many respects was ominous from the Western point of view, was accelerated in 1979 by two dramatic events marking the opening of the fifth (and present) period of rivalry, namely the revolution in Iran and the Soviet invasion of Afghanistan.

The first eliminated almost overnight the principal pro-Western bastion in the region and transformed it into an area that appears destined to be at best one of prickly nonalignment and at worst one of outright hostility to the West. The affair of the American hostages obviously enhances the latter possibility. These developments in Iran, which the United States could not have prevented, certainly not during the final critical phase, clearly upset the strategic balance, making the Gulf area more vulnerable both to internal disturbance and external intervention as well as deeply troubling the pro-Western states on the other side of the Gulf. A further side effect of the revolution was a reduction in the supply of oil to the West and a sharp price increase.

Afghanistan, though Muslim, had usually been thought of as peripheral to the Middle East region. Moreover, its proximity to the Soviet Union had always imposed on it, particularly after the British withdrawal from India, something of a tilt toward the Soviets. Nevertheless, the coup in April 1978 which installed an unequivocally Communist government was a decisive shift, pregnant with consequences for Southwest Asia, which was perhaps insufficiently remarked at the time. It took the Soviet invasion, however, to really capture American attention. The invasion had two special features. For one thing, it represented the first time Soviet military forces in substantial numbers had been used outside the Warsaw Pact area and hence created an alarming precedent. Second, it caused concern that if the conquest of Afghanistan were to be successfully completed down to the borders of Iran and Pakistan, opportunities would exist for exerting at least political pressure on both of these countries in order to incite restless minorities in each and to promote possible further movement, either overt or covert, toward the Gulf and the Indian Ocean.

There is a great deal of debate about the motives for the Soviet invasion. It is often spoken of as defensive and in a sense, if one ignores the circumstances of the 1978 coup, one could say that the Soviets were confronted at the end of 1979 with a choice of this Communist government being overthrown and a pro-Western government or at least a thoroughly nonaligned government being installed or of moving in and suppressing the widespread revolt which was taking place against that government. A successful revolt in Afghanistan could have faced the Soviets with the prospect of that country becoming more and more infected with Islamic militancy in an area immediately adjacent to the Soviet Union's millions of Muslims. On the whole, explanations along these lines are more plausible than the argument that the Soviets consciously planned this as a first step in moving in a military way down to the Gulf and the Indian Ocean. They may have been genuinely surprised at the vigor of the Western and Islamic reactions to their move into Afghanistan but they certainly would not find surprising the vigor of the reaction that would occur if they move farther south.

Nevertheless, whatever the Soviets' motives may have been, either in 1978 or 1979, the fact that if and when they are able to digest Afghanistan, they will be adjacent to Iran along much of its border and to Pakistan on one of its most unstable frontiers necessarily must create grave apprehension throughout the region and among all countries dependent on the Gulf for oil and energy. This apprehension could be expected to grow if there came a time when the Soviet bloc was itself becoming seriously short both of oil and of acceptable means of payment for it, and hence would be increasingly tempted to acquire Gulf oil supplies by pressure or even by force.

The two destabilizing events of 1979 have obliged the United States to reassess and readjust its policies toward the critical Gulf area, policies which, though certainly not uniformly successful, yet proved sufficient during the preceding three decades to keep the Middle East in fairly stable equilibrium and hold Soviet influence within reasonable bounds.

Now the keystone of the northern tier, Iran, has been shattered and both its neighbors, Turkey and Pakistan, have become unstable politically. Iran itself, at such time as it acquires a government capable both of governing and of sober reflection, will no doubt perceive that it has more to fear from the Soviet Union than from the United States. At that point it will presumably seek a partial rapprochement with the United States, but this is likely to be delayed both by the still unhealed scars of the past and by the controversy of the hostages.

The initial strong reaction of the United States to the invasion of Afghanistan was both sound and reassuring. The embargo on grain and high technology, the boycott of the Olympics, the solemn warning that the United States will not tolerate further aggression in the area, the deployment nearby of naval elements, and commencement of the organization of a so-called rapid deployment force to resist further incursions, all these steps must show the Soviets that they have miscalculated and that further moves southward would be fraught with very grave dangers to themselves.

On the other hand, one must frankly face three reservations about the effectiveness of these measures. The first relates to whether our European and Japanese allies will participate sufficiently in helping present an impressive united front. They manifest little zeal for anti-Soviet embargoes and boycotts which would be commercially costly and politically embarrassing. The dependence of most of them on oil from the Gulf, including Iran, is so great that they may lean over backward to avoid offending any of these states and possibly throwing them into Soviet hands.

Second, it remains unclear at this early stage precisely what a U.S. rapid deployment force could effectively do in the area to enforce what is loosely called the "Carter Doctrine." The deterrent value of having an enhanced naval element there is clear and useful but beyond that the concept of such a force becomes somewhat more dubious. The force as planned could not hold back a full-scale Soviet invasion, though such an invasion would so clearly risk provoking World War III that it appears extremely unlikely.

Third, it seems apparent that most of the pro-Western states of the area would not welcome a conspicuous U.S. military presence on their territories which would open them to internal and external charges of being American stooges. It is far from clear that they would even welcome American military intervention to help them suppress internal revolts if those should occur, which are the chief threat that most of these countries might have to fear. Such help, even if heeded, might be deemed too politically embarrassing to be acceptable. These attitudes of reserve among even the most pro-Western Arab states are, of course, greatly intensified by massive and conspicuous American support of Israel and by this country's failure to persuade Israel to make significant concessions on those elements of the Arab–Israeli conflict of deepest concern to the Arabs of the Gulf, namely Jerusalem and the Palestinians.

If the autonomy talks and the Camp David framework collapse or prove wholly abortive, as unhappily seems likely, Arab resentment of the U.S. role is almost certain to increase with a resultant further reduction in the American ability to play an effective role in and around the Gulf. This factor among others means that far from objecting to Britain and France playing a semi-independent role in the Middle East, the United States should encourage them to do so. There are a number of areas in which Arab states might feel it politically safe to cooperate with them at a time when cooperation with the United States might seem risky. That does not mean that this country should bow out in their favor but there are opportunities for all parties to play useful roles. After all, Japan, France, and some other European nations are much more dependent on Gulf oil than is the United States and they do have, the French at least, some military capabilities in that area.

Some comment should perhaps be made about what seems to be the most immediately dangerous spot, not in the Gulf itself but in the adjacent region. And that is Yemen, both Yemens. South Yemen is very close to being a colony of the Soviet bloc. It is in effect a Soviet military base which has the capability of threatening, though only at considerable risk to the Soviets, access both to the Gulf and to the Red Sea. Moreover, land offensives could be launched from South Yemen either against Oman or against North Yemen. The latter has for years existed in a state of acute instability, among other reasons because Saudi Arabia, which clandestinely exercises a strong influence there, has never been able to make up its mind whether it prefers North Yemen to be an unreliable and indigestible Saudi satellite or to be self-reliant, independent, and unmanageable. In any case, it might be that if South Yemen should undertake overt aggression, either northward or eastward, this would be one situation in which a U.S. rapid deployment force could effectively be used with at least the tacit approval of most of the states of the region.

During the coming decade, the Gulf region is almost certain to be both a locus of rapid sociopolitical change and a focus of external, including superpower, rivalries. The former phenomenon, social and political change, seems more likely to produce instability and violence in the near future than the latter. Though the other states of the region lack some of the characteristics that brought revolution to Iran and have that upheaval as a warning, they share a number of others with pre-revolutionary Iran. The ability of their governments to master the disruptive aspects of rapid social change seems firm at the moment, but whether this will remain true over a period of years is more problematical.

Superpower rivalry in the region will continue and probably intensify. This rivalry will probably not lead to critical confrontation in the near future, simply because neither superpower wishes either to risk nuclear war or to become involved in costly Vietnam-style hostilities in a region so distant and so difficult for military operations. Both, one must hope, will realize that oil supplies cannot be seized and exploited by external military force. This moderately sanguine forecast, however, calls for two reservations.

The first is the possibility that serious internal disorders might spontaneously or otherwise break out in some part of the region, that the superpowers would gradually be drawn in through their support of opposing states or factions and that they might, as a consequence, quite unintentionally suddenly find themselves in a direct confrontation from which neither would wish to retreat. This hypothetical situation, of course, is similar to that immediately preceding World War I and it is always a most worrisome possibility.

The second risk, a more remote one but not completely negligible, is that through failure to reduce oil consumption sufficiently or to develop alternative sources of energy rapidly enough, either West or East or both might at some point feel themselves in desperate need of more Gulf oil to keep their economies alive, and so possibly undertake actions which would now be considered quite irrational.

The Soviet Union and the Gulf

THE ECONOMIST

HONEY FOR THE BEAR

A western diplomat in the Gulf sums it up briskly:

> There are two interpretations of Soviet actions in the area. One is apocalyptic, that the Soviets are encircling the Gulf with a view to taking it over; the other is that they are opportunists, exploiting openings as they arise. One cannot dismiss either theory. So the west must set itself two tasks: to deter them from invasion, and to ensure that such openings do not arise.

Russia's record in the Middle East has on the whole been so dismal that it is easy to dismiss the Soviet potential there. Russia has failed to find any client state bigger than tiny South Yemen, and has succeeded in selling arms in quantity only to Iraq and the Yemens. The Russians have diplomatic relations with only two other Gulf countries, Kuwait and Iran. Valerie Yorke[1] concludes that the Russian approach has for the most part been "low-key and low-risk, opportunistic but with some ideological content, and reactive rather than aggressive".

Since the late 1960s, the Russians have gradually built up a fleet of some 20 naval ships in the Indian Ocean, which is fully 8,000 kilometres from the nearest Soviet port.

To accommodate the ships, a smattering of Russian anchorages and bases has been set up. Aden is now being extended with the addition of eight to 12 submarine pens. The island of Socotra, off South Yemen, has become a major naval base with a large floating dock towed from Berbera, from

1. The Gulf in the 1980s. Valerie Yorke, Chatham House, 1980.

Excerpted by permission of the publisher, The Economist Newspaper Ltd., London, from "Defending the Gulf: A Survey," *The Economist,* June 6, 1981.

which the Russians were recently kicked out. They are seeking to use a former French base in Madagascar, Diego Suarez, and a former British base on Gan, in the Maldive islands. The huge Soviet interest in Ethiopia seems geared to controlling the oil routes as much as to securing a foothold in Africa. In the Red Sea off Massawa, Russian installations have been set up on the Ethiopian islands of Dahlak and Perim. The Dahlak base reportedly includes submarine pens and storage tanks.

The Russians have never made much secret of their aim of securing a stake at least equal to that of the west in the countries on their southern flank. Russia has alternately backed a proposal by the UN Indian Ocean conference to set up a "zone of peace" and suggested that Russia and America draw a line between their interests in the area.

The Saudi oil minister, Sheikh Ahmed Zaki Yamani, believes that by 1987 the Russians will have to import large quantities of oil from OPEC. The latest CIA estimates suggest that the Russians will still be producing 10m-11m barrels of oil a day by 1985, maybe just enough to cover consumption, but a gap could develop after that. And the impact on their economy of buying even 10% of Soviet requirements abroad would be dramatic. At the moment half of Russia's hard-currency earnings (some $6 billion) come from oil exports; these would disappear and Russia would certainly have to pay hard currency for OPEC oil.

At this stage, as Sheikh Yamani says, "it remains to be seen whether the Soviet bloc will try to acquire the oil sources and pay for its oil in roubles instead of dollars." Or whether Russia will lose its interest in high oil prices (which it has favoured as long as it is a net exporter of the stuff) and propose to the west that it join in intimidating Gulf producers to lower their prices; this Mr. Brezhnev has already leeringly half-suggested in his plan for sharing influence in the Gulf. Russia's presence in the area also owes much to the larger strategic consideration that, in the event of a world war, the Soviet Union would want to deny Japan and most of western Europe its primary source of fuel.

Russia has a formidable force on its southern border. There are at present 12 divisions stationed along the border with Iran in the Trans-Caucasus Military District and 20 more nearby, in the military districts of North Caucasus, Turkmenistan and Central Asia. Three of these are tank divisions, two of them are airborne and the rest are motorised rifle divisions. Amnon Sella[2] points out that Russian roads and railways for switching troops rapidly between their borders with China and with Europe link up with a north-south communications system leading down eventually through both Iran and Iraq. So maybe 30 more divisions could be despatched fairly quickly to the southern front if the need arose.

The crack airborne divisions (each containing 7,500 men and a cluster of

2. Amnon Sella: Soviet Political and Military Conduct in the Middle East. Macmillan, 1981.

anti-tank guns) could be airlifted to take over an oilfield (although only one at a time). The Russians have seven such divisions altogether and another one is being assembled; they are trained in desert warfare. In manoeuvres, the Russians have shown that they can further airlift as many as 100,000 men in less than 10 days.

The Russians have recently switched their approach to the two countries, Iran and Iraq, which offer them a highway to the oilfields. During the Gulf war, they have shown more concern for Iran than for Iraq, a long-standing socialist ally with which they signed a treaty of friendship in 1972. In September, Russia concluded a similar treaty with Syria, the one Arab country that has aligned itself with Iran during the war; a joint Soviet-Syrian declaration was issued backing Iran's "inalienable right to determine its own destiny independently and without any foreign interference"—a clear slap at invading Iraq.

The Iraqi deputy prime minister, Mr. Tariq Aziz, went to Russia to beg for supplies at the beginning of the war, only to be met by junior officials. The Russians did not even bother to dignify his talks with the adjective "cordial". The Iraqis have received no significant shipments of Soviet arms and ammunition. In contrast, after a request for arms from the Iranian ambassador in Moscow, the Russians seemed keen to sell; but the Iranians promptly recalled their ambassador for making the request without authority. Russia is now Iran's chief trading partner.

The Russians were understandably furious with Iraq for not giving them advance warning of its attack on Iran. Mr. Brezhnev may have been angered by the unsettling effect of an act of aggression so close to Russia's borders. The Iraqis were probably being punished, too, for straying out from under Russia's wing and towards the west in recent years. All the same, the Russians have risked provoking Iraq to make a clean break with them when the war ends. Ominously, the Russians seem to consider that the opportunity of friendship with Iran is more important than the loss of Iraq, a country only a third as big. For Mr. Brezhnev, three birds in the bush are worth one leaving his hand.

STEPPING STONE

Of all the countries that Russia could have picked for its first full-scale invasion outside its Yalta-defined empire, Afghanistan might seem the least enticing. A mountain fastness in the north-east, a barren plateau in the centre, flat and empty in the west, a desert in the south, it is fertile only in the centre-east. It offers the Russians little more than poppies, wild tulips and picture-postcard views. There are no natural resources to speak of except rugs—although, one day, there may be a little oil.

The people are poor, although not oppressively so. Except in Kabul and a handful of cities they live in a state of tribal backwardness. And they curl up into spiky balls when handled by outsiders. The British, being blooded in the nineteenth century, decided Afghanistan just wasn't worth the effort. Czarist Russia, after watching the British experience, decided not to bother.

So what makes a modern superpower well versed in the strength of anti-colonialism in third-world countries go and grab the unappealing one next door? One probable reason is that Afghanistan is on the way to places. It offers, in the words of Pakistan's foreign minister, Mr. Agha Shahi, "strategic options". To go east, enlisting the help of the Indians to put a squeeze on Pakistan and to deal a blow at China's influence in the area. Or to go south towards the Gulf. Or both.

It has been argued that, if the Russians were really interested in the Gulf's oil, they would have made for Iran. But Russia is cautious. An attack on Iran might have led to military confrontation with the west. Iran's oil is anyway at the bottom of the country, an arduous land campaign away. Afghanistan's annexation carried a low risk, improved Russia's strategic position, and provided a useful test of western reaction.

Russia's conclusion, after several years in Afghanistan, must be that the bite of local resistance is far worse than the west's bark. Nothing, but nothing, has gone right for the 85,000-man Soviet occupation force. It quickly became obvious that the inoffensive President Karmal, because he was backed by a Russian occupying army, was less acceptable to most Afghans than his sadistic predecessor, President Amin. The guerrilla army swelled from about 40,000 men to about 80,000 men in a year; the Afghan army which the Russians had fondly hoped would do most of their fighting for them, shrank from around 80,000 men to around 20,000, as soldiers deserted in droves to join the guerrillas.

The guerrillas could roam at will through the countryside; villages were prepared to offer them shelter and support; the Russians could only move about in strength, protected by army columns. Otherwise they stuck to their garrisons in the main cities; large parts of even these were effectively controlled by the guerrillas.

There are some signs that the Russians are preparing for a long stay and that they intend to use Afghanistan as a stepping stone. The runways at Kabul airport are being lengthened; a new airport is being built near Kandahar at Askargh; oil and ammunition depots are being built south of Herat, near the Iranian border.

Permanent barracks are being built for Russian soldiers, although a recent visitor to Kabul told your correspondent that many of them are still billeted in tents. New bridges are being built, and existing ones strengthened, across the Oxus river between Afghanistan and Russia. The main road from Russia to Kabul is being widened and a pipeline is being installed along it.

The war is unwinnable on present Russian tactics with their present strength; yet the Russians have meekly kept their heads down, making little effort to reinforce themselves. There are a number of possible explanations for this. If Afghanistan is merely a corridor for the Russians they would be less concerned with subduing the whole country, and more with keeping open vital highways. Or they may have planned to send in reinforcements but got sidetracked by the Polish crisis. Or they may be wary of the west's reaction to an escalation of the war. Or they may be looking for a way out.

Some mistake surely? Yet the governments of Britain, Saudi Arabia and Pakistan all believe there is enough reason to hope that the Russians are tired of their Afghan bed of nails to offer them the prospect of a way out. Pakistan's Mr. Shahi fell for an old Russian ploy in January 1981, when the Soviet Union suddenly seemed willing to talk about the future of Afghanistan. The Russians dropped their previous objections to allowing a representative from the United Nations to convene talks between Afghanistan and its neighbours, Pakistan and Iran; they also accepted that Mr. Karmal's government could go to the talks unrecognised by either country; he could be invited as an interested party, not as a head of government.

Mr. Shahi accordingly persuaded his fellows to tone down their condemnation of the invasion of Afghanistan at a meeting of Islamic leaders at Taif in February; no sooner was this safely past than the Russians decreed that Afghanistan should only meet Iran and Pakistan separately, and that the UN representative must keep his mouth shut at the meetings. For the Pakistanis, this would have looked too much like straight government-to-government talks, conferring recognition upon the Karmal regime. They backed off.

Still, Mr. Shahi believes there is something to talk about. The objective of the talks would be to set up an Afghan government which its neighbours could recognise—namely one which could stand without the support of Soviet troops. As in Mr. Shahi's view the present regime would not be acceptable to the Afghans, a government of national reconciliation would be set up which could include some members of the ruling Communist party.

The settlement would have to be guaranteed by the five permanent members of the UN security council, then approved by the UN general assembly as a whole. Afghanistan's permanent non-aligned status would be guaranteed and the Russians would be forbidden to go back in should the government of national reconciliation fall apart—as it surely would.

The guerrillas say they will never join a government which includes the Communists. "Anyone who says he is prepared to do so will instantly be discredited", one mujahedin leader told your correspondent. The guerrillas are realistic enough, though, to say that they want to form a government which would "not be hostile" to the Soviet Union.

The much bigger objection is that the Russians show no interest. The irrepressible Mr. Shahi, though, has an idea that might appeal to them: the

Americans, he says, could reach a direct deal with the Russians in which they would agree not to set up their Rapid Deployment Force for use in the region, in exchange for a Russian withdrawal from Afghanistan. The Russians might indeed be glad to withdraw from a place they had no business to be in anyway in exchange for an American decision to give up protecting the Gulf after the fall of Iran.

If the Russians stay in Afghanistan they will have to choose between continuing to endure the present level of violence, or launching a massively reinforced drive against the guerrillas. In the second event, the west has a response: to provide the weapons and ammunition the mujahedin badly need. For the present scale of fighting, the mujahedin are adequately equipped with the .303 Lee Enfield rifles they love to use, and which can be supplied by local gunsmiths.

But they are short of ammunition, which is expensive. They need more anti-tank guns, particularly the excellent Russian-made RPG-7. They badly want ground-to-air missile launchers to defend themselves against the Russians' lethal Mi-24 helicopter gunships. And they could do with more mines. If the west cannot help its friends in a case where right is considered by most of the world to be on their side, it will never convince the Russians that it is serious about fighting for the Gulf.

FIVE DOORS

After Afghanistan, where? There are five soft spots in south-west Asia that Russia, if it takes a risk or two, might be tempted to make for. The aim of western policy must be to deliver Russia from temptation by hardening up those spots. They are:

- Pakistan, which faces a serious danger that the Russians may try a dash into its North West Frontier Province in order to clean out the Afghan camps there, and a lesser one of dismemberment by Russia and India between them. There is also a temptation for Russia to make a grab for grumbling Baluchistan, in order to get a slice of Indian Ocean coastline to build a warm water port.
- Iran, still in danger of disintegration, civil war, or both. The Russians might be presented with an opportunity they could not let pass if bits started to fall off, or if their friends called on them for help in grabbing or holding power in Teheran.
- Iraq, which offers the Russians a geographical highway to Gulf oil (Iran, by contrast, is a difficult country to cross). The Iraqi army, though, would be a tough one to beat.
- Saudi Arabia, where the danger is less one of outside attack than of an internal coup or revolution by oil-stopping nationalists. Saudi Arabia is so vital to the industrial west that its loss, unlike that of Iraq in 1958 or Iran in 1979, simply could not be allowed to occur.

- The Strait of Hormuz, teeming with sharks. Four tenths of western oil supplies go through the strait, which is just 24 miles across. This heart valve could be blocked by (a) terrorist action; (b) a Soviet tactical air strike from Afghanistan; (c) a hostile, or Russian-occupied, Iran on the north shore; (d) the fall of Oman's pro-western government on the south shore either to revolution or to invasion by South Yemen.

To the dangers from Russia and from internal upset must be added a sixth: that of internecine war in the Gulf. Iraq and Iran are by no means the only neighbours who cordially detest each other. Just about everybody has a claim on everybody else (the Iraqis on the Saudis, the Saudis on most of their smaller neighbours, the Iraqis on the Kuwaitis and vice versa). Only if these local upheavals threaten to stop the flow of Saudi oil or to result in a blockage of tanker traffic through the Gulf do they seriously affect western interests. Then they should be planned for.

U.S. Security Policy in Southwest Asia: A Case Study in Complexity

HAROLD BROWN

The interests of the industrialized democracies, and specifically of the United States, in Southwest Asia have been of major importance ever since and indeed even during World War II. The region, for the purposes of my discussion, extends from Turkey and Egypt on the west—although Turkey has part of its territory in Europe and Egypt lies almost entirely in Africa—to Pakistan and Afghanistan in the northeast, and can be considered to include in geostrategic terms the Horn of Africa on the south. There are further fringes that might also be considered part of the region. So we're dealing with a complex and extensive portion of the earth's surface.

U.S. interests were expressed in the region, so defined, during the late forties, for example, through the application of the Truman Doctrine to Turkey and in U.S. support for the creation of the state of Israel. Shortly after the end of World War II, antedating both these events by a year or two, the United States and Britain had pressured the Soviets to evacuate northern Iran, which the Soviets had occupied during World War II in a spheres-of-influence kind of arrangement, and where they had subsequently attempted a continued domination through support of a puppet Azerbaijani regime in northwest Iran.

All of these early U.S. actions recognized the importance of the region as a geographical area of major strategic significance in terms of communications among Europe, Africa, and Asia. But in the 1940s and through the 1950s, the oil-producing capacity of the region, while significant, was by no means the principal source of the oil in international trade. It was only during the 1960s, and even more during the seventies, that the West Europeans, the Japanese, and to a considerable extent the U.S., became so dependent upon the oil from the Persian Gulf region, which has corre-

Reprinted by permission of the publisher, The Johns Hopkins Foreign Policy Institute, School of Advanced International Studies, Washington, D.C., from Harold Brown, *U.S. Security Policy in Southwest Asia: A Case Study in Complexity*, 1981.

spondingly become the strategic focus of the Middle East and of Southwest Asia.

The full nature of the dangers that this dependence poses became clear only through a sequential and convergent series of events. These included: the 1973 Arab-Israeli war and the Arab oil boycott and major oil price rise that followed it; the Iranian revolution, which brought with it the recognition that the dependence through much of the 1970s on Iran as a pillar of political and military stability in the region, and of resistance to Soviet expansionism, was a mirage—and also brought with it another major oil price rise; the Soviet invasion of Afghanistan and its implications with respect to Soviet willingness to use their own forces outside the borders of the Warsaw Pact; and the Iraq-Iran war.

U.S. policy for the region is influenced by as complex a set of interacting needs, tensions, conflicts, power centers and histories as has ever beset U.S. policymakers—and that's just talking about these factors in the region itself. It doesn't even take account of the corresponding factors within the U.S. that influence policy. The complexities include the political and diplomatic, the economic and the military. Synthesizing a U.S. security policy for the region has, not surprisingly, proven to be a major undertaking. Significantly, the expression of that policy by the Carter administration in 1980 and by the Reagan administration so far in 1981 shows more resemblance than difference. But even if this represents a policy consensus within the United States—which I considerably doubt, at least so far—we will have a truly enormous task in terms of carrying out that policy.

If it is to be managed at all, we have to provide an international diplomatic and political framework for it, both among the nations of the region and among our allies. The U.S., in my judgment, must lead in creating a strengthened military instrumentality built on that framework in order to support our policy objectives. However necessary, it is not clear how feasible these actions will prove to be. Moreover, we require an energy program—both short-term and medium-term—and an encompassing economic program to ease dependence on the energy resources of the Persian Gulf region during the period between now and the end of the century, beyond which those resources will have been substantially depleted in any event at the current rate of utilization.

Such an energy program is required for the industrialized democracies in general. With more specific reference to the United States, continuing major efforts in conservation and a new effort in synthetic liquid fuels both seem to have, unfortunately, dropped from the national agenda in recent months. Both of them are vital to a U.S. policy along the lines I've mentioned. Otherwise, the dependence on Persian Gulf oil—which darkens the 1980s and which, as we will see, will be so difficult to manage—will persist through the 1990s and beyond. This doesn't have to be the case, in my judgment.

The degree and nature of that energy dependence are, at present, a major fact of geopolitical life. It's notable that Western Europe and Japan rely considerably more than the U.S. on the oil of the Persian Gulf. We get about 10 or 11 percent of our total oil consumption from that region. The figures for European countries vary among them, but Western Europe as a whole gets about 50 percent—perhaps more—of its oil from the area and Japan, 75 percent. More than half of the net oil exports in international trade come from that region. If you think of it in U.S. terms, a reduction of, say, 10 or 11 percent of our consumption might be considered manageable, if painful. But such an approach completely misstates the case, because the rest of the 50 percent that we import would be very considerably disrupted as to reliability and price. It would be enormously increased in price if there were to be a cut-off of exports from the Gulf. Since we produce about half of our oil consumption, we might, with draconian government action, limit the economic consequences to the U.S. of such a cut-off to a major depression. But Europe and Japan would be economically devastated to a substantially greater degree than that.

Consequently, U.S. policy with respect to the Persian Gulf faces two nightmares. The first is a cut-off of exports as a result of disruption within the region; and the second is that of Soviet political or military control of the resources of the region, which would put Europe and Japan in the role of Soviet vassals. If the Soviets clearly control the exports of the Gulf states, or even just control those exports above the level that the Gulf states need for their current foreign exchange needs, Soviet political influence over Western Europe and Japan would clearly be immense. In fact, it would not have to be made explicit in order to loosen or even destroy the ties of alliance that those countries have with the United States. We've talked, from time to time, of sanctions and their effect on international relations, their effectiveness as a tool of international relations. The threat of that particular Soviet sanction, in my judgment, would really work and work very well.

It has been argued that Persian Gulf oil is so obviously vital to the industrialized democracies that the USSR will be deterred by that fact from an attempt at military conflicts through the region. Presumably, the argument goes, the Soviets know that Western responses would, in one way or another, lead to thermonuclear war. If that chain of reasoning is correct, this should be a pretty substantial deterrent.

Such considerations certainly are one element of deterrence of certain extreme Soviet military action. But how it would work would depend very much on circumstances. A rapid enough military success would change what the Soviets call the correlation of forces suddenly and shockingly enough so that the coherence of U.S. alliances might dissolve very quickly, both in the region and throughout the world. What steps the U.S. would then be able to take is not an easy problem.

In any event, Soviet political/military encroachment, country by country or "autonomous region" by "autonomous region," is even less likely to be deterred by the threat of massive retaliation. It is more likely to be deterred by conventional military strength, but conventional military strength can be only one, and perhaps not the most important although a necessary, element of a strategy to prevent such a Soviet encroachment. The degree to which the Soviet encroachment approach or the military attack approach is deterred will depend very strongly on the military as well as on the political balance in the region. Moreover, those two factors are also related to each other. In particular, the political balance in the region will be the more favorable to us the more the West is, and is seen to be, able to offset Soviet military capabilities.

One danger that exists, and will continue to, is that of internal conflicts within a number of the major oil-exporting states. A serious enough change in political complexion in any of them, as we've seen in the case of Iran, can cause a major reduction in its oil exports. If this happened in a number of the smaller or one of the bigger producers, or some combination thereof, the deprivation to the international supply of oil could run anywhere from a few to 10 million barrels a day.

Even more severe could be the result of major wars within the region. An internal change could well reduce production to a level sufficient to provide what the new regime would see as the need for foreign exchange. Some of the Gulf states' exports are far larger than this, at least in part out of a recognition that too low a production rate or too high a price will cause economic catastrophe in the industrialized world with very negative repercussions on the Gulf states themselves. Radical regimes are unlikely to recognize that line of reasoning. Hence their tendency to reduce output to the level of foreign exchange needs or, if they're disorganized enough, perhaps below that level. Some radical regimes—Libya, for example—produce at a higher rate than what we think of as their foreign exchange needs. However, they may well have additional uses in mind for that foreign exchange. But regimes at war are likely to have their oil facilities destroyed or disrupted in a way that leads to a production rate reduced below that corresponding to foreign exchange requirements. Certainly, that happened in the case of the Iran-Iraq war. Wider-spread regional conflicts, even without superpower participation but involving a number of major oil-producing states, could cut oil exports by 10 or even 15 million barrels a day, depending on which and how many were involved.

This consideration reminds us that the Arab-Israeli conflict is, by no means, the only one in the region. Egypt fought in Yemen during the 1960s, the two Yemens have had military collisions, Iraq and Iran are at war even now, and other wars could well occur. But the Arab-Israeli conflict has an intensifying effect on most of the others, though perhaps not all. Not, for example, on the Iran-Iraq war except as another charge (of being friends of the Israelis) for both sides to hurl at each other. One can argue about just how great an intensifying effect it has.

A third, more severe, scenario involves a gradual or creeping Soviet political dominance over various elements of the region; more severe because it's less likely to be reversed than the effects of the other developments.

Radical indigenous regimes find that they do have to get foreign exchange. Even in a war whose effects may be lasting, those effects are unlikely to be irrevocable in terms of resuming production of oil. But Soviet control, as many countries have found, tends to last a long time when it is of a country or a region contiguous to the Soviet Union. Such a creeping Soviet political dominance would involve increasing, and eventually major, Soviet control over decisions about oil exports and about pricing.

The chain of events could involve modest Soviet incursions in border areas in the guise of supporting indigenous ethnic or political movements, accompanied by major arms transfers and by military advisors.

The fourth scenario presents the most formidable military challenge in terms of what would be needed to defeat it. Yet I consider it a rather unlikely contingency. This is a massive and direct Soviet military attack aimed at military conquest of the Persian Gulf oil fields, destroying or gaining control of them.

A Soviet military takeover would, for reasons already indicated, clearly be a mortal threat to Western access; and a legitimate interest of the industrialized democracies is that of access to the oil in the region, not control over it. The oil belongs to the people who live there.

This raises the question of the right of access to Persian Gulf oil. We don't now have a situation of economic warfare between the West and the Soviet Union. It could develop in certain circumstances, even in a state of formal peace, but it would not be a happy omen for the future. And absent such a development we should not oppose the Soviets' legitimate interest in being able to purchase energy from the region providing that they have the necessary foreign exchange. Clearly, there are some people in the West who think that OPEC prices at some level could be, or maybe are now, such as to be a denial of access to oil—denial by OPEC, not the Soviets—and a threat to the existence of the industrial democracies. Some offer the argument that such a situation would or does justify military action against the authorities in the region itself. But it wasn't the countries of the Persian Gulf that made the West depend on their oil. No Persian Gulf fleet appeared in New York Harbor demanding that we open the United States to foreign trade. So I find such an argument repellent in ethical terms, reminiscent of 19th century and more recent (specifically non-Western) imperialism. I also judge that an adventure based on it would be unsustainable in practical terms, partly because of the existence of the Soviet Union as a superpower. Such a move would certainly give the Soviets the opportunity to become the welcomed protectors of the Persian Gulf nations from "Western imperialism."

How then can a security structure be built in the region that will sustain U.S. security interests? It should be noted that with U.S. leadership, such security structures have been built both in Western Europe and in East Asia with considerable, albeit imperfect, success.

Such success can never be said to be permanent either. Changing circumstances, changing relationships, and mistakes by one or another U.S. administration certainly can undermine such a security structure. Those security structures, both in Europe and East Asia, have had their vicissitudes but they still remain. They comprise both military and non-military elements, the latter including political, economic and cultural sub-elements. In Southwest Asia more than in the other regions, these two major categories—military and non-military—display a degree of conflict with each other in any attempt to build a security structure.

Some factors in the regional situation are a real aid in resisting Soviet political and military penetration. The strongest supranational factor in the region is its Islamic nature, which in general produces a strong resistance to communism as well as a considerable resistance to Westernization or modernization in some cases. The sense of individual nationhood is growing in many of the countries of the region and this nationalism is usually expressed as anti-Soviet, though also often as against American influence. Both the Islamic and the nationalist elements have been experienced, to their sorrow, by the Soviets in Afghanistan.

Moreover, if we look for bright spots, the region as a whole is a reasonably well off one economically—in fact, rich. But this possible element of strength is undercut by the fact that the richest countries are the least populous and the least able to defend their own independence, although there are some marginal exceptions to this. Finally, I would note the presence in the region of Israel as a democratic nation, with the greatest single national military strength in the region, or very nearly so. That contributes elements of strength to the Western position, though Israel's conflicts with its neighbors are, at the same time, a major element of regional instability.

Now, let's look at some of the difficulties in building a security framework in the region. On the non-military side, these include the various ethnic, tribal, religious and national splits within Islam as well as the Arab-Israeli conflict. They include the intense poverty in some countries of the region and the lack of political development in many. Some commentators have questioned whether the necessary political substructure for a security framework can be put together in a region with so many internal conflicts. They've also contrasted the situation unfavorably with those in Europe and in the Pacific after World War II. Indeed, both Europe and Japan had been much more industrialized before World War II than has ever been the case in Southwest Asia, and there was in those regions a strong heritage of literacy and technical training as well as of existing governmental institutions, although the war had devastated both the institutions and the economy. But if one looks back realistically to that time, the rivalries that had existed between Germany and France or between Korea and Japan were long-standing. In the early forties, Britain was looking largely away from what was a disunited Europe and there was also, not incidentally, a 30-percent communist vote in France and in Italy. The European framework did not seem so easy to build then. It is more difficult in Southwest Asia, but that does not make the task hopeless.

Still, it's hard to see us sponsoring a regional alliance of the NATO type or signing, in the near future, as explicit a security treaty as we have with Japan with most of the nations of Southwest Asia.

At least in part, this is a consequence of the political instabilities in the region that would be exacerbated by such moves. The countries in the region are, for the most part, very different from us—more different than the Europeans or the Japanese. But we have no choice, in my view, but to try to create a security framework, loose and shifting though it may be, with the nations of the region. We cannot expect it to have the characteristics we have come to associate with our existing alliances in Europe or in East Asia.

An overall program will have to include, along with allied military capabilities and the various political commitments between the industrialized democracies—particularly the United States—and the states of the region, at least two other elements. The first is economic development assistance to those more populous and militarily important nations of the region that do not have economic resources that go with oil production. Examples are Egypt, Turkey and Pakistan. The help will have to come from the U.S., from Western Europe and Japan, and from the oil-rich states of the region.

The second additional element is movement toward the resolution of the Arab-Israeli disputes, including the prospect of some settlement of the Palestinian issue. Only the U.S. can provide the assurances to Israel that are necessary, but perhaps even then not sufficient, to gain Israeli acquiescence to any such development.

Moreover, the U.S. will have to provide the great bulk of the military strength to assure the moderate Islamic states—and even the radical ones need some reassurance about this, except perhaps for South Yemen, the only Marxist state in the region—that there is a deterrent and a counter to Soviet military intervention in the region. Both of these latter factors cast the U.S. in a key role in the region among the industrialized democratic states.

U.S. arms sales both to Israel and to the moderate Arab states are a major tool in building such a framework. But both the history and the current state of sales to Saudi Arabia illustrate that it can be a very sharp-handled as well as sharp-edged tool. Without getting into a detailed discussion about F-15s, added equipment for them, refueling capability, and AWACS, I'll make two observations.

The first is that arms sales packages for Saudi Arabia and Israel can be tailored to meet any reasonable set of self-consistent American policies for each. The problem, and it's a real one, arises when those policies conflict whether because of Israeli and Saudi attitudes or because of U.S. failure to reconcile its own objectives as well as it might.

Second, in the end, U.S. arms sales policy must be determined, where there are conflicts, by what the U.S. thinks is in its own security interests, broadly construed to include all the factors: military, diplomatic, economic and moral. U.S. and Israeli security interests are inextricably entwined. So are U.S. and Saudi interests. We can't decide unilaterally for either of them their own security policies, but we have to lead in fashioning a security framework in the region, if one can be fashioned.

Turning to some of the other military aspects of such a security framework, it should be noted again that the military balance in the region is important even in peacetime. A dangerous political contest between the Soviets and the West is sure to occur in the region. Indeed, it's already under way. Even in the probable absence of a Soviet military attack, the likely military prospects in case of such a Soviet invasion will influence peacetime political attitudes in the region.

The West faces a number of obvious disadvantages in this military balance. The Soviets are much closer geographically, or at least the Soviet borders are closer geographically, to the region. It's easier to shut down or to destroy the oil production and export facilities than it is to preserve them; and the Soviets could do almost as much damage by destroying them as they could by controlling them. The Soviets have very extensive bases both in the PDRY (South Yemen) and in Ethiopia, both relatively close to the Persian Gulf. And Soviet military equipment, with the influence and potential leverage that it implies, is present in a number of the major non-aligned states in the region, including Iraq and Syria, of which the former has been moving away from the Soviets politically during the past couple of years while the latter has drawn somewhat closer.

But the military advantages are not all on the Soviet side. The U.S. can dominate the seas adjacent to the region. Indeed, U.S. naval power in the Arabian Sea outclasses anything that Soviets have there at sea or could realistically bring to bear. Of course, our naval power, even when aided by the French, British and Australian units currently there, has its limitations, too. The Persian Gulf is too confined to make it prudent to bring aircraft carriers into it in the face of the threat from land-based air around its periphery. This rather undercuts the advantage claimed for a so-called naval strategy in Southwest Asia—for example, of being able to reach the major oil fields of the Gulf from the surface of that body of water with the 16-inch guns of battleships taken out of mothballs.

Two possible strategies for the defense of the oil fields of the Gulf are holding a line on the Zagros Mountains or holding an enclave around the oil fields themselves. Without going into the political or military feasibility of either one of these in detail, neither is feasible through the use of guns from surface vessels or solely with sea-based air power or even land-based air all by itself. Some ground forces would clearly need to be able to be moved into the region quickly.

A second military factor that potentially could favor the West is that Turkey lies athwart Soviet invasion routes reaching from the Soviet Caucasus to the Persian Gulf. Turkish bases and military strength, as a threat in being, must give the Soviets pause. However, the Turks, understandably, are not inclined to participate in a regional security framework that commits them to action in the absence of a NATO-wide commitment to the defense of the region upon which the well-being of the NATO countries so strongly depends. The military capabilities of some of the other countries in the region, should they be used in self-defense, offer some military advantages, though many of them have done poorly even in conflicts within the region.

Finally, there is some distance and some difficult terrain between the Soviet borders and the Gulf itself. From the Caucasus to the head of the Gulf is 600 miles. It's true the distance from the southwest border of Afghanistan to the Straits of Hormuz is less, but the oil lies around the head of the Gulf. If opposing forces could be interposed and interdiction accomplished early—both of those are very big ifs—the same sort of prospect which operates to help deter the Soviets from a military adventure in Europe could be introduced in Southwest Asia: the prospect of extended conflict on the ground with U.S. forces and of escalation of intensity and of geographic spread of the conflict. The longer such a conflict lasts, the more uncertain the Soviets or anybody else can be about what may happen, both at the scene of the conflict itself, elsewhere in the world, and in the Soviet Union itself. Thus, the anticipation of a more protracted conflict operates as a better deterrent.

What's the overall military balance in Southwest Asia in the light of some of these factors? Not very favorable. We've done rather a lot fairly quickly since the late 1970s, but not nearly enough to produce the same degree of stability in terms of military balance that exists in Europe. This means that at least for a time the Soviets could well see less military risk in an adventure in Southwest Asia than they would in Europe. The prize is nearly as great as would be offered by a successful military move in Europe because control of Persian Gulf oil would carry the fruits of dominating Europe and Japan. That combination makes Southwest Asia the most dangerous area of the world in terms of the prospects of a superpower confrontation.

The chances of early Soviet military action along these lines are nevertheless rather low. The risks, though not as high as in Europe, are still considerable. Moreover, the Soviets probably see less risky political moves in Southwest Asia which also offer the prospect of very significant advantages to them.

What has the U.S. done on the military side? We've set up a new planning command which would assume operational control in wartime. We've deployed maritime prepositioning ships based at Diego Garcia which contain the equipment and supplies for a reinforced Marine brigade and several Air Force fighter and attack squadrons. These could marry up with ground force personnel and with aircraft, all of which could be flown into the region for the purpose within a week or 10 days to a total of about 15,000 or 20,000 personnel.

We've had, for over a year, two carrier battle groups and other naval forces in the Arabian Sea, dominating the naval situation. We've flown airborne warning and control aircraft into Saudi Arabia and we operate them there now. These would be able to monitor and control air operations over the Persian Gulf, which in turn could be dominated by U.S. F-15 aircraft that could be flown into the region within a matter of days.

We've carried out joint exercises in the region with Egypt and with other nations at a modest level of size, up to a battalion. Congressional funding has been sought and obtained to begin to design and execute construction and expansion of facilities around the region, in Egypt, Kenya, Somalia, Oman, as well as at Diego Garcia. And, for the longer term, major expansions of airlift and sealift capabilities have been projected and, in some cases, even begun.

In fact, all of these activities need, in terms of the military balance, to be very substantially expanded. We need to preposition equipment in the region for air and ground forces. We need to overbuild the facilities in terms of what the indigenous military forces require. We need to rotate temporarily into the region and exercise there a great deal of air and sea power. And we need more frequent and more substantial ground exercises, jointly, with the friendly countries of the region. In the meantime, deterrence of Soviet military adventurism will have to depend largely upon Soviet concerns that such an action on their part would lead to direct U.S.-Soviet conflict or confrontation, in turn leading to military conflict of unknowable intensity and location. Because the degree to which that deterrent operates depends on the balance of the military forces and capabilities in the region, and because that balance is now an uncomfortable one for us, we need to do more in the long run.

The key factors will be: Can we build up our ability to get there quickly and in large numbers? That involves a need for a very substantial additional airlift and sealift. How much? A reasonable near-term, mid-term goal is to be able to get two or three divisions there in a period of a couple of weeks and sustain them indefinitely. A longer-term goal would be the ability to perhaps double that. Corresponding tactical Air Force capabilities would be, say, three or four wings within a few days as a mid-term goal—actually we can do pretty close to that now within a week or 10 days—and, over a longer period of time, perhaps double or triple that amount. In other words, if these figures are right—and they are merely one estimate to go with one military strategy—we need to be able to reinforce Southwest Asia with about the same magnitude of ground forces, although rather less in air forces, as we now program for reinforcement of Europe in the mid-1980s. Southwest Asia is more than twice as far away and staging is much more difficult. So the task is substantially more than twice as great, even in purely logistic terms, leaving out the political problems.

Turning to that more difficult task, the other key factor is: Can we strengthen the political infrastructure of the region? That will include economic development among the poor and less stable countries and progress on settlement of the Arab-Israeli dispute, which does affect the stability of the Islamic countries in the region, rich or poor, moderate and more radical. Not the least of the reasons for strengthening the political structure is that the countries that are threatened must be willing and must be seen to be willing to fight in their own defense, as have the people of Afghanistan. The political framework needs to project the perception, and reflect the fact, that the U.S. is not trying to dominate the area itself, only to offset the threat of Soviet domination.

Can we handle the potential adverse effects of an increased U.S. presence? That's, perhaps, a third key question. There's a strong current in the region of opposition to both of the superpowers. Too visible a U.S. presence will increase the risk of internal unrest in many of the nations of the Gulf. This is why some of our friends have asked that we not be present at all in peacetime, but somehow be able to appear suddenly with massive military forces immediately on indications of an active or imminent Soviet threat. Well, unfortunately, that's not possible. We don't have that kind of magic carpet. In order to be able to move rapidly into the region on the basis of such signs, we need major prepositioning of stocks and modest handling capability to receive U.S. forces. Very little of this need be in the form of an on-the-ground U.S. military presence. How much permanent U.S. military presence it would take would depend on how much contact or indigenous support could be made available instead.

I've mentioned the value of Israel as a democratic, moderately industrialized, militarily strong element in the region. I've also mentioned the need to promote an Arab-Israeli settlement including movement toward a resolution of the Palestinian issue. Failure to progress perceptively toward such a settlement will be both a reason and an excuse for unrest and for denial of facilities needed for access by the U.S. military in an emergency.

Finally, we need creation of an adequate political infrastructure in the region and maintenance of domestic U.S. support for the U.S. commitment to deny control of the resources of the region to the Soviet Union, a denial even more in order to protect our European and Japanese allies than to protect ourselves. That commitment as embodied in the Carter 1980 State of the Union Message and also as embodied in Reagan administration foreign policy statements will therefore require some sort of European participation in the security framework. The French have significant forces in the region and the UK has contemplated commitment of modest forces. That's a substantial step in the right direction. Others among our allies should participate if possible. That will be difficult for many of them. The Germans, for example, are apparently in the process of deciding that they can't really even decide to participate in substantial military sales in the region—a step related to, and easier than, participation in a direct military role or potential wartime military role.

I recognize that a wish to include in the military aspects our European allies in turn introduces more complexity because it will require increased consultation by the U.S. on political as well as military matters connected with the region. Various of our allies may see or may have different interests or different estimates of the situation. All of our allies will have to cooperate in terms of providing en route staging and airlift and sealift contributions to increase the capability of responding militarily in the region if possible. And because much or most of the U.S. reinforcement capability that would have gone to Europe will, in the case of simultaneous threats to Europe and Southwest Asia, have to go to Southwest Asia instead, our allies, both in Europe and in Japan, will have to increase their own military efforts to substitute for those reinforcements that may, in the event, be going elsewhere.

It will be necessary to balance these overlapping, often conflicting factors to create the political and economic structures and military capabilities that, together, constitute a security framework, a security policy, and a security capability for the Persian Gulf region and Southwest Asia. Doing so in order to deter Soviet military adventurism, contain Soviet political penetration, and help the people who live in the region to stabilize it is probably the greatest current challenge to the United States' geostrategic and military plan, and to our international political skill.

Towards an Effective American Policy

It is obvious that the degree of American concern over the Gulf has been steadily heightening in recent years. Among the number of disturbing developments prompting this process have been:

1. continuing dependence on Gulf oil by the United States and its allies;
2. the turmoil in the Horn of Africa and Soviet entrenchment in Ethiopia;
3. the collapse of the Shah's regime in Iran and the emergence of fears in the West that other states of the Gulf were in imminent danger of similar eruptions;
4. The 1979 war between the Yemens and suspicions of South Yemen's designs on the Gulf through subversion in North Yemen and Oman; and
5. the Soviet occupation of Afghanistan in the face of considerable opposition of the Afghani people.

Perceptions of steadily intensifying threats to the maintenance of American interests in the Gulf have resulted in the parallel development of increasingly more direct American commitments to preserve those interests. In 1971, there was the Nixon Doctrine, formulated in the aftermath of the American experience in Vietnam. Iran and, to a lesser extent, Saudi Arabia as regional powers would serve as surrogate policemen for the U.S. in the Gulf area through the so-called "twin-pillar" system. The concept of reliable surrogates, however, was severely jolted by the Iranian revolution and suspicions of Soviet intentions were increased by their actions in Afghanistan. Resultant American anxiety produced the Carter Doctrine in 1980: the U.S. would use military force if necessary to repel any attack on the Gulf.

The disappearance of America's most suitable "pillar" has resulted in two essentially independent but interrelated strategies. On the one hand, there is a growing tendency to rely on increased economic and military coordination with the remaining pillar, Saudi Arabia. Much is made of a "special relationship" between the two countries and the U.S. government has become heavily involved in the enhancement of that kingdom's military and security capabilities, in addition to its considerable role in Saudi socio-economic development. American perceptions of Saudi Arabia's central importance were underscored by the 1981 Reagan "codicil" to the Carter Doctrine: "Saudi Arabia we will not permit to be an Iran." Nevertheless, controversy over the $8.5 billion sale of five Airborne Warning and Control System (AWACS) aircraft, together with various other items of arms and equipment, to Riyadh in 1981 reflected continuing domestic concern over American involvement with that country.

Essentially, opposition within Congress, which had the right to veto the sale, centered on two arguments. One was that the provision of such sophisticated airborne radar planes to Saudi Arabia would constitute a direct threat to Israel. The possibility that the Saudis would be able to use their AWACS planes to coordinate an Arab attack on Israel was, however, discounted by the Reagan administration. The other concern focused on the possibility of internal strife in Saudi Arabia leading to the loss or compromising of the highly secret computers contained within the AWACS planes. The parallel to Iran was frequently, if inaccurately, drawn. In the final event, the assassination of Egyptian President Anwar Sadat in early October seemed to swing the necessary votes in the Senate to allow the sale to go through.

The other post-Iran strategy for the protection of American interests places emphasis on the necessity of "going it alone". The focal point of this approach is the Rapid Deployment Force (RDF), a joint-service task force created to send limited American forces quickly to the Gulf in the event of crisis there (see Appendix G). The RDF was redesignated the U.S. Central Command at the beginning of 1983 and given full responsibility for all U.S. military activities in the Southwest Asia and Northern Africa theater. Despite this change, numerous operational and conceptual problems are still thought to exist.

Both the strategy and design of the RDF have been severely criticized. One observer, Kenneth N. Waltz, argues that "An RDF should serve vital interests only and in serving them should be guardedly used."[1] Other means to further American interests should be utilized first.

> Even in situations where non-military means do not suffice, we should act militarily only if a strategy promising success can be devised. We should avoid the temptation of resorting to force because nothing else will avail. We should use force only if we can see a way of doing so that will enable us to get our way.[2]

1. "A Strategy for the Rapid Deployment Force," *International Security,* Vol. 5, No. 4 (1981), p. 57.
2. *Ibid.*

He goes on to suggest that the goal should be "an asset-seizing, deterrent force that can handle both internal and external threats" and not a "war-fighting, defensive force."[3]

The difficulty of moving U.S.-based forces to the Gulf in time and in large enough numbers is often cited as well, as is the question of the RDF's ability, once in place, to serve as an effective deterrent to larger and heavier-equipped opponents. Current planning for the RDF assumes that it will arrive on target unopposed. Even though the RDF was not involved in the attempt to rescue the American hostages in Iran, similar doubts about the readiness of the force are raised. Since the RDF so far exists only in times of emergencies and then draws upon forces from other units of the armed forces, American readiness to respond to other, simultaneous crises would presumably be negatively affected.

Despite the verbal commitments for the defense of the Gulf and the establishment of the RDF, a detailed, optimal policy has never been fully articulated nor a consensus of opinion reached. There remains a basic ambivalence: is American defense of the Gulf directed principally against external or internal threats? If the answer is the latter, or even both, can it not be justifiably construed by the states of the Gulf that the U.S. itself is an external threat through insistence on its absolute right to Gulf oil regardless of circumstances? Debate is not just concentrated on the ends of American policy but also embraces the means. Is the proper American role a relatively modest one of cooperation in strengthening Gulf allies, with assurances of more direct support if necessary and when invited? Or should this country seek to control access to oil directly by establishing a permanent presence in the Gulf, thus permitting the accomplishment of its central goal by intimidation of the Gulf region as well as the Soviet Union?

The debate features proponents of radically opposing views. A prominent advocate of the "hawkish" position has been Professor Robert W. Tucker who argues that "there is no reliable substitute for Western power in the Gulf."[4] This power entails the "permanent stationing of land-based aircraft and ground forces" in the Gulf, or in the Sinai as a "next-best alternative," in order to defend that region "against a hostile coup, intra-regional aggression, or a Soviet incursion. . . ."[5]

From the other side, Professor I. William Zartman questions how American military power in the Gulf is to ensure access to oil supplies: "The two places where that access is now denied are Iraq and Iran, where the intervention of American troops is no more conceivable than the stationing of American troops as a way of getting the oil to flow."[6] Christopher Van

3. *Ibid.*, p. 63.
4. "The Purposes of American Power," *Foreign Affairs,* Vol. 59, No. 2 (1980–1981), p. 251.
5. "American Power and the Persian Gulf," *Commentary,* Vol. 70, No. 5 (November 1980), p. 39.
6. "The Power of American Purposes," *Middle East Journal,* Vol. 35, No. 2 (1981), p. 165.

Hollen, in his presentation below, ends by suggesting that a " 'laid-back' rather than a forward military posture will better serve the interests of the United States." Similarly, David D. Newsom, a former Under-Secretary of State for Political Affairs, counsels moderation.

> Given existing doubts about the nature of the threat, the reception of an American presence in the area, the willingness of Western Europe to cooperate, and U.S. military capacity, a strategy that places U.S. ground forces in the Persian Gulf should not be undertaken without a thorough national and congressional debate. That debate has yet to begin.[7]

Bibliographic Sources for this Subchapter:
30, 40, 64, 66, 86, 92, 110, 120, 127, 168, 213, 218, 256, 320, 321, 322, 323, 324, 331, 342, 355, 356, 359, 363, 365, 372, 376, 388I, 388J, 403, 415; see also entries for Rapid Deployment Force (Appendix G).

7. "America EnGulfed," *Foreign Policy,* No. 43 (1981), p. 32.

Don't Engulf the Gulf

CHRISTOPHER VAN HOLLEN

Five months after Ronald Reagan's inauguration, his Administration's approach to the vital Persian Gulf region is clear—but the logic behind its policies remains obscure. More disturbing, the policies themselves are contradictory and self-defeating; unless modified, they won't work.

Eager to demonstrate American resolve and assertiveness, the new Administration has embarked on a geostrategic East-West military strategy that compounds the errors of the Carter/Brzezinski White House and revives memories of some of John Foster Dulles' more ill-conceived ventures. By concentrating single-mindedly on the Soviet threat to the Gulf, the Administration thus far has failed to shape the type of comprehensive strategy needed to meet the most likely dangers to Western interests—and it has badly misjudged regional realities.

The Reagan approach is based on several assumptions:

- the Soviet Union is the primary threat to a region which supplies about 35 percent of the oil consumed in the free world economy;
- in order to concentrate on thwarting Moscow's expansionism, Washington should deemphasize efforts to resolve the Arab-Israeli dispute;
- Soviet power can best be contained militarily by putting more money and muscle into the Rapid Deployment Force and encouraging a European contribution, even if indirect, to the RDF;
- a "consensus of strategic concerns" should be developed among the countries stretching from Pakistan westward through Saudi Arabia to Egypt and Turkey, and including Israel;
- if a "strategic entity" can be created among these countries, some of the regional states may be persuaded to accept U.S. ground forces—a necessary requirement if the Soviet threat is to be effectively curbed.

This conceptual framework is deficient on a number of counts. The Soviet invasion of Afghanistan upset the regional geopolitical equation, but this does not mean that the Soviet Union is the primary threat to Western interests in the Persian Gulf area. The Islamabad-to-Ankara "strategic entity" theme evokes memories of the containment policies of the 1950s, but it lacks political coherence or a structural underpinning. Any attempt to shape an effective regional security framework, while downplaying the Arab-Israeli issue, is doomed to failure. Equally unrealistic is the belief that Washington can line up the Gulf states in an anti-Soviet phalanx, possibly including U.S. troops, at a time when the littoral countries want to exclude any superpower presence.

A U.S. military buildup *is* an essential response to the Soviet Union's invasion of Afghanistan—the first full-scale Soviet invasion of a country outside Eastern Europe since World War II. But an exclusively anti-Soviet military approach can be dangerously destabilizing if pressed with excessive zeal in the politically volatile Persian Gulf region. A more sophisticated multilateral approach is essential—one that is better attuned to the most likely threats and to regional sensitivities, and that defines "security" in other than simply military terms. Closer attention ought to be directed to the interrelationship between U.S. military measures and the political, economic, energy and alliance policies which are integral parts of a comprehensive security strategy. While military scenarios must be prepared to meet all threats, longer term political planning is also needed to defuse U.S.-Soviet confrontation in the region.

II

Overt Soviet seizure of the Gulf oil fields would—and should—trigger an American military response. Any such Soviet action must be deterred by the clear prospect of a U.S. response not only on the ground but if necessary through retaliation outside the immediate Gulf region that could draw on the full range of military options, not excluding the use of nuclear weapons. A direct Soviet attack on Pakistan could invoke the 1959 bilateral agreement, which provides for possible use of U.S. military force to protect Pakistan against communist-inspired aggression.

But these contingencies are near the bottom of the threat list. The most likely challenges to Western interests will come from wars between regional states, transborder incursions, civil disturbances, oil embargoes or production cuts, or the overthrow of existing regimes. Whether the Soviet Union abets or exploits these developments depends upon how it assesses its interests, including its ability to project political, military and subversive power beyond its borders—weighed against the Kremlin's concerns about regional and Western retaliation, and the security of its Muslim Central Asian Republics.

Although a direct Soviet attack on the Gulf is improbable, we must continue to strengthen American global military capabilities and the U.S. naval presence in the Indian Ocean. A substantial naval force is required as long as the Soviet Union remains in Afghanistan, Ethiopia and South Yemen, and Soviet ships operate in adjacent waters. A naval presence provides deterrence against Soviet adventurism, and military and psychological reassurance to Saudi Arabia and other regional states. It also signals American determination, as demonstrated at the outset of the Iraq-Iran war when the two carrier battle groups off the Gulf of Oman reminded both belligerents that the United States along with others would act to ensure free passage through the Strait of Hormuz. Diego Garcia, although 2,200 miles from Hormuz, is an indispensable naval support base, and a staging area for P-3 reconnaissance aircraft and, later, for B-52 bombers.

It is also important to create a mobile force which, *in extremis,* can be quickly deployed if truly vital U.S. interests are jeopardized. But as some military analysts and members of Congress have suggested, a small, highly mobile force, based at sea and supplied by sea, might be superior to the ungainly Rapid Deployment Force—which could become "a standing invitation to a military disaster."[1] Aside from its military deficiencies, the RDF is encumbered with political liabilities because of the interventionist rhetoric which has flowed out of Washington since Henry Kissinger's we'll-seize-the-oil-if-we-need-it interview in *Business Week* in 1975. Some Gulf states view the RDF as a threat to their oil resources, rather than as a protector of their national integrity. Moreover, the inevitable escalation in demands for logistical support for the projected 225,000-member RDF carries the risk of imposing intolerable political strains on the four states—Egypt, Kenya, Oman and Somalia—which have granted the United States military facilities.

A naval presence and small mobile force are key instruments of U.S. policy. But the severe limitations on the use of military power to ensure the flow of Gulf oil must be faced realistically. The task of protecting the oil fields from attack by regular armed forces or by terrorists is vastly complicated by the expanse of the major fields, the exposed character of the wells, pipelines and loading faciliites, and the fact that about two-thirds of the oil is exported through only three ports, controlled by eight pump stations that can be easily destroyed. Even conservative states such as Saudi Arabia have declared that they will blow up the oil fields rather than allow them to fall into foreign hands.

III

The inherent limitations on the use of force in the Gulf do not negate the value of American military power as a deterrent against developments inimical to U.S. interests. They do require a subtle orchestration of U.S.

1. Jeffrey Record, *The Rapid Deployment Force and U.S. Military Intervention in the Persian Gulf,* Cambridge, Ma. and Washington, D.C.: Institute for Foreign Policy Analysis, Inc., February 1981, p. 2.

military and political-diplomatic policies. And, as the Carter Administration learned the hard way, and the Reagan Administration is beginning to learn, a successful political-military strategy requires a more sensitive understanding of the new regional realities.

When Secretary of State Cyrus Vance traveled to Saudi Arabia after Camp David, when Defense Secretary Harold Brown visited the Saudis after the Iranian Revolution, and when National Security Adviser Zbigniew Brzezinski flew to Pakistan after the Afghanistan invasion—each was rebuffed in his efforts, respectively, to persuade the Saudis to endorse the Camp David Accords, to induce the Saudis into closer security ties, and to encourage the Pakistanis to accept U.S. arms. These rebuffs to senior Carter officials, and the Reagan Administration's early experiences, underscore the far-reaching but imperfectly understood political and psychological currents that have swept the Gulf and Southwest Asia during the past four years. The geopolitical transformation is attributable in varying degrees to the repercussions from Camp David, the Iranian Revolution, and the Afghanistan invasion—reinforced by the rise of oil power and Islamic fundamentalism, and the surge of nationalistic and nonaligned sentiments. Concerned about the Soviet threat, the local states nonetheless do not want to become pawns of the superpowers or overly beholden either to Washington or Moscow. Yet they continue to hold the United States responsible for the ultimate resolution of the Arab-Israeli conflict.

Secretary of State Alexander Haig's direct exposure to Gulf realities came in early April 1981 when he visited Saudi Arabia as part of a Middle East trip designed to demonstrate that the United States is a "credible, reliable, long-term security partner." Saudi Crown Prince Fahd and his colleagues expressed appreciation for this stronger American resolve, as they voiced their customary concern over the Soviet Union's intentions in the area. But, dispensing with their typically obscurantist diplomatic style, the Saudi leaders also made it clear that the Arab-Israeli conflict was the centerpiece of their concerns. In effect, they rejected the new Administration's thesis that the Arab-Israeli dispute should be handled "in a strategic framework that recognizes and is responsive to the larger threat of Soviet expansion."

Two weeks later, this thesis was dealt another embarrassing public blow. Citing the "Soviet threat" as the primary reason for the decision, the White House on April 21 announced the President's intention to include five Airborne Warning and Control System (AWACS) aircraft with the F-15 enhancement equipment already promised Saudi Arabia. Israeli Prime Minister Menachem Begin protested bitterly, arguing that the AWACS were a direct threat to the security of Israel. The next day, the Saudi Minister of Petroleum, Ahmed Zaki Yamani, told a New York audience that Israel, not the Soviet Union, posed the primary threat to Saudi Arabia. "Israel," Yamani said, "is the opening through which the Russians creep into the Arab world."

In the aftermath of the educative events of April, there is evidence that the Reagan Administration tempered its earlier "Russians first" rhetoric. Some officials, for example, now suggest that the efforts to counter Soviet expansion and the Middle East peace process are "mutually reinforcing"; the Lebanese crisis over Syrian missiles has highlighted the complex, multifaceted nature of Middle East problems. But senior policymakers in the Administration still need to develop a more sophisticated appreciation of the political environment in the Middle East and Persian Gulf, rather than to push ahead doggedly in an attempt to construct an anti-Soviet security framework. They should also face up realistically to the weak political infrastructure upon which any "strategic entity" to protect the Gulf might be built.

IV

Those who talk glibly about projecting U.S. military power into the Gulf area convey the impression that there are strong and stable local regimes receptive to U.S. security overtures. In fact, the United States lacks diplomatic relations with the only two littoral states with significant military forces—Iran and Iraq, whose armies are still locked in indecisive combat. Egypt, Turkey and Pakistan are of strategic importance, but they are far distant from the Gulf—as are Somalia and Kenya. Thus, in the immediate Gulf area, U.S. on-shore security ties hinge essentially on relations with Saudi Arabia, on the facilities agreement with Oman, and arrangements with Bahrain for calls by the small naval Middle East Force.

Saudi Arabia, the world's largest oil reservoir, is an extraordinarily weak state. With a native-born population of only about five million, and a small 35,000-man army which has never been tested, the Kingdom must rely on outside manpower to run its economy and on outside help for its security. Concerned about the Soviet threat, and about dangers from Iran and South Yemen, the Saudis view the United States as their ultimate protector. But they are schizophrenic. The dominant faction within the House of Saud wants to retain a "special relationship" with Washington despite post-Camp David discords. Yet the Kingdom's leaders fear the effect of highly visible U.S. ties on their role in the Muslim world, on the one million Palestinians who inhabit the Gulf, and on leftist political groups inside and outside the region. Moreover, with still recent memories of the November 1979 Grand Mosque occupation, Saudi leaders also worry about internal dissidence from those who believe the regime is sacrificing Wahhabi orthodoxy for forced-draft Western modernization.

U.S. policies must take careful account of this Saudi ambivalence. Washington needs to adopt a lower public profile in its relations with the Kingdom, to limit the number of high-level officials visiting Riyadh, and to desist from pressing the Saudis excessively on issues clearly inconsistent with their perceived interests—such as on military access and bases, oil production and pricing, and sensitive Arab-Israeli questions. We also need to try to monitor the size of the American community in this most Muslim of Muslim countries, where there are now about 45,000 Americans, including 1,000 military advisory and support personnel. This is a small number when compared with more than a million Yemenis and other foreigners but, relative to the size of the Saudi population, it is about seven times as large as the American presence in Iran during the heyday of the Shah.

With the Iranian example also in mind, we ought to keep a closer eye on the level and type of U.S. arms supplied to Saudi Arabia—a country with much less absorptive capacity than the Shah's Iran. Assuming delivery of the equipment now proposed, Washington will have provided between 1975 and 1981 over $10 billion in military hardware to the small Saudi military establishment, which is also buying military equipment from the British, French and Austrians and wants to buy from the West Germans. The AWACS imbroglio highlights the problems inherent in a symbiotic relationship between U.S. professional military personnel and their counterparts in an oil-rich country with almost unlimited financial resources. The U.S. Air Force apparently encouraged the Saudi military leadership to purchase the highly sophisticated airborne radar system despite the Saudis' lack of technical competence to operate it. Senior U.S. civilian officials saw the requirement for long-term U.S. technical support for the AWACS as a way of expanding and extending American military involvement in Saudi Arabia. But the coordinating mechanisms of the National Security Council failed to weigh the importance of advance consultation with the Congress and the Israelis. The result was a potentially "no-win" situation for the Administration.

If we plan primarily against the Soviet threat, when the real threat is internal or regional, our actions may themselves prove destabilizing. By pressing such countries as Saudi Arabia or Oman to accept a large American military involvement, we may unwittingly contribute to our own insecurity. For example, the June 1980 facilities agreement with the Sultanate of Oman is a military asset, at least in the short term. But, on the liability side, Sultan Qaboos bin Said's open association with the United States has already made him more vulnerable to domestic opponents and such enemies as the South Yemenis and Libyans, and the People's Front for the Liberation of Oman. This vulnerability may increase if Washington urges the Omani ruler to expand the American military presence to meet a perceived geostrategic threat from the Soviet Union. If radical local forces overthrow Qaboos, they will be well positioned to seize the strategic Musandam headland, directly adjacent to the Strait of Hormuz chokepoint.

In the case of Pakistan, Reagan Administration officials are negotiating an estimated (but still undefined) $2.5 billion arms and economic assistance package over five years. Pakistan deserves substantially more U.S. support in the aftermath of the Soviet invasion of Afghanistan than President

Carter's $400-million two-year "peanuts" offer. But any new U.S. relationship needs to take account of President Zia ul-Haq's lack of popular support, Pakistan's nuclear program, and important American interests in India. It should also recognize Zia's domestic and international political concerns.

The Pakistani President is anxious to protect his country's recently acquired nonaligned status and its role in the Islamic world; he does not want excessively to provoke India or the Soviet Union or stir up anti-American domestic opponents. Accordingly, Zia refuses to accept U.S. troops in Pakistan or permit its territory to be used overtly as a conduit for weapons for the Afghan nationalists. Having withdrawn Pakistan from the Central Treaty Organization in 1979, he is unwilling to reestablish security ties with the United States. Therefore, Washington should treat the relationship as essentially a credit arms sales arrangement, while confining the hardware to defensive-type equipment designed primarily to protect Pakistan's frontiers from Soviet-occupied Afghanistan. It would be unwise to press the Pakistanis to enter into formal security arrangements as a quid pro quo for U.S. arms.

Instead of an overly close American identification with Pakistan and the ruling Gulf regimes, we ought to fashion more detached and flexible state-to-state relationships in keeping with local political predilections, and as a hedge against uncertain political futures. A looser posture will make it easier for the United States to weather inevitable regime changes; it will also reduce the temptation to intervene on behalf of incumbent political leaders who are challenged in local power plays where there is no evidence of Soviet manipulation. This somewhat less structured style will not appeal to those with a formalistic approach to foreign affairs. It is a far cry from the days of the Baghdad Pact or CENTO or even the Iran-Saudi Arabia "twin pillar" arrangement. Yet it will be the most practical and durable way to advance American interests in the fluid circumstances of the 1980s.

Similarly, it will be more productive to support local security initiatives than to attempt to shape regionwide strategic arrangements in the face of countervailing political trends. If the current Iraq-Jordan tactical alliance continues, with ties to other Arab Gulf states, there may be a tendency to convert it into a more formal security system after the Gulf war ends. The Saudi-sponsored Gulf Cooperation Council, consisting of all Arab Gulf states except Iraq, may assume increasing security responsibilities. And talks between Saudi Arabia and Pakistan may result in a substantial increase in Pakistani military personnel in the Kingdom. Those regional initiatives which serve U.S. interests should be tacitly supported and underwritten by appropriate levels of military supply. Instead of fruitlessly trying to create a "strategic entity" stretching from Pakistan to Turkey, the Reagan Administration will achieve greater success by quietly encouraging indigenous security ventures.

V

U.S. interests in the Persian Gulf and Southwest Asia are integrally linked to developments outside the region. In particular, they are interconnected with U.S. relations with Egypt, Israel and Turkey; with the West European and Japanese allies; with progress toward Middle East peace; and with U.S. energy security policies and those of the 21-nation International Energy Agency.

President Anwar Sadat, who shared most of the Reagan Administration's strategic views, offered to permit the use of Egyptian territory in time of need for defense of the Gulf. Sadat, however, refused to allow U.S. troops in the Sinai after that territory was transferred from Israel to Egypt in April 1982—except as part of a multilateral force to monitor the Egyptian-Israeli peace treaty. He was understandably sensitive to possible charges that he recovered the Sinai from Israel only to turn part of it over to the United States for "bases." Sadat also resisted signing a formal agreement with the United States for military facilities at Ras Banas on the Red Sea. He believed that his countrymen would consider such an agreement an encroachment on their sovereignty, perhaps comparable in their eyes to the Egyptian-Soviet agreement that Sadat scrapped in 1972 when he expelled the Soviet advisers. Because U.S. ties with both Egypt and Israel are critically important, Washington needs to find a formula which will advance the security interests of the three countries without pushing the Egyptian regime toward political self-destruction.

Different sensitivities arise in dealing with Turkey, the strategically located NATO country which shares borders with Iran, Iraq and Syria as well as the Soviet Union. Unlike Sadat, General Kenan Evren and other Turkish leaders are unwilling publicly to offer Turkish territory to facilitate the defense of the Persian Gulf, or even to provide explicit private assurances. They are reluctant to have Turkey be seen as a surrogate for U.S. interests in an area from which Turkey derives more than half its oil. But Turkey could be a major asset for Gulf security, particularly if trouble arises in northern Iran. Without expecting formal advance commitments, we ought to try to develop a dialogue with Ankara on contingencies in the Gulf region, with the hope that the Turks will provide support in time of crisis.

In the case of our West European and Japanese allies, we need to take greater advantage of their better access to Iran and Iraq, while also encouraging additional allied economic and military assistance to the neighboring countries—such as Turkey and Pakistan. Any improvement in U.S. relations with the Khomeini regime will be slow and tedious in the aftermath of the 14-month hostage ordeal. Commercial links with Iran should be gradually reestablished, but it is not in Washington's interest to force the pace of political normalization or depart from the essentially neutral stance it has adopted toward the Iraq-Iran war. Our objective ought to be to work through our allies, particularly the Germans and Japanese, to ensure that this strategically most important Gulf state does not tilt toward the Soviet Union through a sense of isolation from the West. Similarly, although the United States has not had diplomatic relations with Iraq for 14 years, the network of French ties with President Saddam Husayn's regime, as well as American, Italian and Japanese commercial connections, can be used to encourage the Baghdad government to keep its distance from Moscow.

But U.S. cooperation with its allies, and particularly with the regional states, can be jeopardized unless Washington is prepared to address the Arab-Israeli problem. The Reagan Administration's early decision to shelve the Palestinian autonomy talks made sense if it was a tactic to buy time until after the Israeli election. A decision to downgrade the peace process for the longer term is fraught with danger—as attested by the crisis in Lebanon.

Peace between Israel and the Arab states will not be a panacea, ensuring an uninterrupted flow of oil and replacing conflict with tranquillity. The Iranian Revolution and the Iraq-Iran war had nothing to do with the Arab-Israeli dispute; permanent regional stability will remain a chimerical hope. Nonetheless, the Arab-Israeli dispute imposes an enormous burden on U.S. policies.

Specifically, failure of the United States to reengage in the peace process will increasingly strain U.S.-allied relations, limit U.S. political and military access in the Gulf region, enhance the chances of use of the "oil weapon," and play into Moscow's hands. Gulf leaders believe that the Arab-Israeli dispute is a primary source of Soviet influence in the region and is a major contributor to radical political currents throughout the Middle East. By early fall, after the new Israeli government has settled in, Reagan officials must actively resume the quest for a settlement which fully protects the security and integrity of Israel; satisfies the legitimate aspirations of Palestinians; and, ultimately, accommodates Jewish and Muslim interests in Jerusalem.

It is also important to continue to give priority to energy security policies. Although U.S. oil imports have declined 20 percent over the past 18 months, it is essential, indeed vital, to avoid the complacency of the 1975–78 period. Europe and Japan remain extraordinarily vulnerable to Gulf oil disruptions, and their impact not only on the economies of key nations but on allied solidarity has been all too vividly demonstrated both in 1973–74 and 1978–79.

The new Administration's complete decontrol of gasoline and crude oil prices was a positive step, but Reagan Administration officials have retrogressed by dismantling many of the Carter conservation programs—and they have failed to press the Congress for funds to fill the Strategic Petroleum Reserve more rapidly. It does not make sense to spend up to $75 billion over the next decade on Persian Gulf defense, and to exhort our allies to do more, unless we concurrently demonstrate a strong national will to reduce dependency on Gulf oil and to reduce the alliance's vulnerability to supply cutoffs.

VI

Oil may eventually prove to be the catalyst for discussions with the Soviet Union on the Gulf region, but prospects for dialogue are remote in the short term. The Reagan Administration's priorities are to enhance American military strength and to try to develop a consensus with the regional states and U.S. allies. On their part, the Soviets—despite recurrent propaganda gestures—have not shown a serious interest in discussions since they invaded Afghanistan in December 1979. They have limited themselves to unrealistic political proposals for Afghanistan and equally unrealistic proposals for the Persian Gulf area.

Afghan Prime Minister Babrak Karmal's initiative of May 14, 1980, remains the centerpiece of the Moscow-Kabul diplomacy. In effect, it rules out the withdrawal of Soviet troops until after the cessation of the insurgent uprising and after international recognition of Babrak Karmal's puppet regime. Leonid Brezhnev's doctrine of "peace and security," put forward by the late Soviet President in New Delhi in December 1980, bypassed Afghanistan entirely in calling for the removal of foreign military bases from the Persian Gulf and "adjacent islands." Brezhnev said in essence that the Soviets would remain in Afghanistan, but the United States should get out of the Gulf and Diego Garcia. At the 26th Soviet Communist Party Congress in February, he acknowledged that Afghanistan's "international aspects" might be discussed in conjunction with Gulf security, but he rejected any consideration of "internal Afghan affairs."

Although early U.S.-Soviet discussions are unlikely, policy planners in Washington should look ahead to possible future exchanges with Moscow, including practicable international security arrangements. The Soviet Union already has substantial interests in the region, based upon its superpower status, geographic proximity and security concerns. These interests may increase if the U.S.S.R., although now the world's largest oil producer and second largest exporter, faces declining oil production levels by the middle of the decade. Revised CIA estimates now indicate that the U.S.S.R. itself will not become a net oil importer by 1985; nonetheless, the Kremlin's imperative need to supply Eastern Europe might cause it to develop an increasing involvement in Persian Gulf affairs. Moscow might also seek to drive a wedge between Western Europe and the United States by reviving an earlier Soviet proposal for an all-European energy conference to include the question of access to Gulf oil.

In these circumstances, planners in the Administration might want to explore whether American interests would be served in the future through international agreements involving the United States and the Soviet Union, the regional oil-producing states, and major oil-consuming countries, notably Western Europe and Japan. Consisting of both bilateral and multilateral agreements, such arrangements might include guarantees of free passage, equal access to Persian Gulf oil for all nations, and possibly international endorsement of regional nonalignment. If such proposals demonstrated positive U.S. advantages, they could be put forward on a more formal basis after appropriate soundings among the allies and within the region. They would be linked to Soviet withdrawal from Afghanistan but would not envisage any reduction of U.S. naval forces in the Indian Ocean.

Some might contend that such proposals would "invite" the Soviets into the Gulf and legitimize their presence there through unenforceable international agreements. This argument should be examined on its merits. We are deluding ourselves, however, if we assume that the Soviets are now excluded from the Gulf and lack the capability to expand their activities there. The U.S.S.R. has diplomatic relations with Iran, Iraq and Kuwait, as well as treaties with Iraq and Syria, and one should not rule out the possible resumption of formal diplomatic ties with Saudi Arabia. In the recent past, there have been substantial Soviet arms sales and other trade activities with Iran and Iraq and, on a lesser scale, with Kuwait. Soviet naval ships have visited Umm Qasr in Iraq and Soviet hydrographic vessels have called at Kuwait, whose foreign minister visited Moscow in April 1981. Internationally guaranteed arrangements are worth exploring if they can help counteract the Soviet Union's geopolitical and military advantages, while protecting the interests of the United States, the allies, and the regional states.

VII

Some of the foregoing suggestions may have little initial appeal to an Administration that came to power with a Soviet-centered view of the world and the conviction that more American military power would set things aright. But pragmatism and the realities of the Persian Gulf and Southwest Asia should prove better guides to future policies than campaign rhetoric and early public pronouncements, or traditional theorems of international politics.

When the Shah of Iran was overthrown, conventional wisdom held that a dangerous "power vacuum" had been created in the Gulf which the Soviet Union would quickly fill unless the United States created a military "balance of power" in the region. Similar themes are still being echoed by writers who contend that the power vacuum will eventually be filled by one or the other superpower or by both. Members of this school believe that the imbalance of military power in favor of the Soviet Union can be overcome only by stationing enough American troops in the area to offset the Soviet advantage. Otherwise, they warn apocalyptically, the situation is "hopeless." It is only a matter of time before the Soviets seize control of the Gulf as an essential step toward their objective of global predominance.[2]

Such mechanistic and simplistic concepts of "power vacuum" and "balance of power" have only limited relevance today as applied to the United States and the Soviet Union in the Persian Gulf. The Iraq-Iran war has not yet determined what new regional balance of power will prevail, but one thing is clear: the Soviets are as much on the sidelines as the Americans and neither is likely to fill the power vacuum or "pick up the pieces" after the war. The strategic power vacuum thesis tends to overlook the 60 million people living in the eight Gulf states, whose armed forces exceed half-a-million men, or more than double that number if the population and force levels of neighboring Pakistan are included. Motivated by a strong sense of Islamic and national identity, manifested most heroically today in Afghanistan, these peoples are determined to avoid domination by either superpower. This is not to say that certain countries, notably Saudi Arabia, do not look to the United States for their ultimate security. But they prefer that support "over the horizon," fearing that a closer military embrace will increase their political vulnerability.

We tend seriously to downgrade our potential strength in the Persian Gulf, and substantially inflate that of our opponents, when we measure our security solely in terms of the comparative military forces in the region and our relative power projection capabilities. In geographic and military terms, the Soviet Union's local advantages are overwhelming. Moscow's 24 divisions in the Transcaucasus region cannot be offset by U.S. ground forces even in the unlikely event that some of the countries from Egypt to Pakistan allow Gulf-earmarked troops on their territories.

Despite these handicaps, the United States' position is by no means "hopeless" in a region where Marxist-Leninist doctrines have little appeal. The U.S. naval presence in the Indian Ocean is a valuable element in the American security posture, supplemented by British and French naval units and the French military contingent in Djibouti. These forces provide political and military reassurances to the countries in the region, particularly Saudi Arabia, and are flexible deterrents to Soviet adventures. But the ultimate military balance will not be determined exclusively in regional terms. It depends upon our overall conventional and nuclear military power, the fundamental strength of the American economy, and the alliance's collective ability to maximize its political and military assets in conjunction with those of the regional states.

2. For example, see Robert W. Tucker, "The Purposes of American Power," *Foreign Affairs,* Winter 1980–81, pp. 241–274, and "American Power and the Persian Gulf," *Commentary,* November 1980, pp. 25–41.

As we build up our global military strength, if we can reduce our reliance on Persian Gulf oil, restore NATO's confidence in American leadership, and develop subtle regional bilateral relationships, we will be able to strengthen our security in Southwest Asia despite our local military inferiority. Instead of attempting to obtrude external security arrangements, we need to develop more nuanced ways to capitalize on the local forces of independence and national assertiveness—recognizing that our success will be constrained as long as the Arab-Israeli issue remains unresolved. And because the Soviet Union will continue to have substantial interests in the Gulf area, which may grow with its oil needs, we ought to examine privately the possibility of developing longer term international arrangements to meet the interests of the two superpowers, the regional states and our allies.

A short- and long-term strategy along these lines should have more chance of success than the Reagan Administration's one-dimensional military approach. Remembering the Soviet Union's experiences in Egypt, Somalia and Sudan, and the United States' experience in Iran, we should not try to engulf the Gulf with troops or advisers, bases or facilities, or excessive military hardware. A "laid-back" rather than a forward military posture will better serve the interests of the United States.

APPENDICES

APPENDIX A

Glossary

Abadan
Iranian industrial city and site of large oil refinery, located on Shatt al-'Arab River along border with Iraq

absorptive capacity
the ability of a country to absorb or utilize resources within its domestic economy; the limited absorptive capacity of some oil-producing states has resulted in part of their oil revenues remaining as surplus capital and therefore forming petrodollars (q.v.)

Abu Dhabi
town and state on lower Gulf coast; largest member of United Arab Emirates (q.v.) in terms of physical size, population and oil revenues; Ruler of Abu Dhabi, Shaykh Zayid Al Nuhayyan, elected in 1981 to third-term as UAE President

AGIP
Italian oil exploration and production company with operations in Middle East and elsewhere; 100% owned by ENI (q.v.)

Al Bu Sa'id
the ruling family of Oman

Al Khalifa
the ruling family of Bahrain

Al Maktum
the ruling family of Dubai

Al Nuhayyan
the ruling family of Abu Dhabi

Al Sabah
the ruling family of Kuwait

Al Sa'ud
the ruling family of Saudi Arabia

Al Thani
the ruling family of Qatar

Algiers
largest city, port and capital of Algeria

amir
commander; prince; ruler of amirate

amirate (also, emirate)
one of the smaller states along Arab littoral of the Gulf (Kuwait, Bahrain, Qatar, United Arab Emirates) ruled by an amir; also known as shaykhdom

Anglo-Iranian Oil Company (AIOC)
originally formed as Anglo-Persian Oil Co. to exploit newly-discovered Iranian oilfields in early part of 20th century; British government purchased a 55% interest in 1914 to guarantee oil supplies for Royal Navy; nationalized by Iranian government in 1951 but following long boycott, former owners formed consortium to operate concession as previously; original interests in AIOC later known as British Petroleum (BP; q.v.)

Arabian Gulf
see Gulf

Arabian Oil Company
formed in 1958 in Tokyo to operate concession covering former Saudi-Kuwaiti Neutral Zone (q.v.) and still in operation

Aramco
Arabian American Oil Company; major producing company in Saudi Arabia, originally formed as subsidiary of Exxon, Mobil, Standard Oil of California and Texaco, but now wholly owned by Saudi government

AWACS
Airborne Warning and Control System; a modified Boeing 707 aircraft with airborne radar equipment providing a wider range of surveillance than ground-based radar facilities

Baghdad
capital and largest city of Iraq, located in center of country on Tigris River

Bakhtiar, Shahpour
Iranian politician and longtime opponent of the Shah, for which he was imprisoned on several occasions; a leader of the opposition National Front Party, Bakhtiar nevertheless was appointed Prime Minister by the Shah in early 1979 but resigned on Khomeini's return to Iran and went into exile in Paris

Bani Sadr, Abolhasan
anti-Shah activist in exile in Paris from 1963 to 1979; returned to Iran with Khomeini in 1979 and served in various republican posts until election as President in 1980; subsequently dismissed as result of opposition from conservative politicians and went into exile in France

Basra (al-Basra)
major city and principal port of Iraq, located on Shatt al-'Arab River near border with Iran

Bazargan, Mehdi
Iranian businessman and politician; longtime opponent of Shah's rule, for which he was imprisoned in 1962, and leader of opposition National Front Party; appointed Prime Minister of Iranian Republic by Ayatollah Khomeini and held office during February–November 1979

Bedouin
corruption of Arabic word for nomad

Beheshti, Muhammad
Iranian religious leader who played a prominent role in the revolution and early stages of the republican government; served variously as Secretary of the Revolutionary Council, President of the Supreme Court and Chairman of the Islamic Republican Party; killed in July 1981 bombing of Party headquarters

Benjadid, Chadli
elected President of Algeria in early 1979 following death of Boumediene

Berbera
city and seaport on Gulf of Aden coast of Somalia; American use of military facilities at Berbera has been negotiated with Somali government

Boumediene, Houari
President of Algeria from 1965 until his death from illness in December 1978

BP
British Petroleum Company; one of "Seven Sisters" (q.v.); formed in 1909 as Anglo-Persian Oil Company; name changed to Anglo-Iranian in 1938 and to BP in 1954; British government has held controlling interest in BP since 1914; Middle East operations gradually expanded from Iranian concession into Kuwait and Iraq and later in cooperation with Japanese and national oil companies of Middle Eastern states

Bright Star, Operation
American military operation held in and with cooperation of several Middle Eastern countries in late 1981

Camp David
Maryland retreat of U.S. President noted as site of negotiations between Israeli Prime Minister Menachem Begin and Egyptian President Anwar al-Sadat under mediation of U.S. President Jimmy Carter, leading to signing of Egyptian-Israeli Peace Treaty

Compagnie Française des Pétroles (CFP)
major international oil company based in Paris and owned by the Total group; has participated in numerous Middle Eastern oil concessions

Dhahran
city in the Eastern Province of Saudi Arabia; center of Aramco operations

Doha (al-Dawha)
largest city, port and capital of Qatar

Diego Garcia
island of Chagos Archipelago located in center of Indian Ocean; administratively part of British Indian Ocean Territory (BIOT) but leased to United States government which has built airstrip and naval facilities

downstream operations
areas of activity within the petroleum industry lying "downstream" of exploration and production; includes transport, refining and marketing

Dubai
town and state on lower Gulf coast; one of seven members in United Arab Emirates (q.v.) and second in importance in that federation to arch rival Abu Dhabi (q.v.)

Dukhan
name of two hills, one in central Bahrain and the other in south-central Qatar, where oil was first discovered in both countries

Elf-ERAP
Elf—Entreprise de Recherches et d'Activités Pétrolieres; French company involved in all phases of oil industry; subsidiaries and affiliates operate in a number of Middle Eastern countries, including Algeria, Iraq, Iran, Libya, and Oman

ENI
Ente Nazionale Idrocarburé; government-owned holding company for Italian equities in various petroleum, natural gas and related enterprises, including 100% ownership of AGIP (q.v.); under leadership of Enrico Mattei, ENI was aggressive in breaking the Seven Sisters' control of the international petroleum industry

Exxon
one of the "Seven Sisters" (q.v.), Exxon is the largest oil company in the world; originally known as Standard Oil of New Jersey and later as Esso; Middle Eastern activities began in 1928 with participation in Iraq Petroleum Co. and later in Aramco (q.v.); while Aramco provides a major portion of Exxon's foreign oil, the company is also active in Egypt, Qatar, Abu Dhabi, Syria and Lebanon, pulling out of Libya in 1981

Front de Libération Nationale (FLN)
political movement which organized Algerian revolution against French and subsequently took control of independent Algerian republic

Gulbenkian, Calouste
British-educated, Armenian oilman; first became involved in oil deals during Baku (Russia) boom, later instrumental in formation of African and Eastern Concessions, later Turkish Petroleum Company; Anglo-Persian Oil Company's purchase of TPC left Gulbenkian with a 5% interest in APOC; this holding, and similar holdings in later companies in the Gulf earned Gulbenkian the nickname of "Mr. Five Percent"; Gulbenkian family interests later incorporated in Participation and Exploration (Partex)

Gulf
neutral term for Persian Gulf or Arabian Gulf; shallow body of water separating Arabian Peninsula from Asian mainland; approximately 570 miles long and 125–275 miles wide, its only outlet is the Strait of Hormuz (q.v.); it forms part of the boundaries for the states of Iran, Iraq, Kuwait, Saudi Arabia, Bahrain, Qatar, the United Arab Emirates, and Oman

Gulf Cooperation Council (GCC)
founded by Bahrain, Kuwait, Oman, Qatar, Saudi Arabia and the United Arab Emirates (UAE), in February 1981 to provide stronger coordination among members in political, economic, oil and security matters; headquarters and secretariat located in Riyadh

Gulf Oil
major American-based international oil company regarded as one of the "Seven Sisters" (q.v.); founded in 1901 but active in Middle East only from 1928 with participation in Iraq Petroleum Co. and 50% ownership in Kuwait Oil Co.; Gulf Oil continues to purchase much of its oil from Kuwait, as well as other Middle Eastern producers, but has little regional involvement in other ventures

Hormuz
40-mile-wide strait separating Gulf of Oman from Arabian or Persian Gulf; since provides only inlet or outlet from latter Gulf, great majority of Gulf's oil must be transported through this strait; strait is bordered by Iran on the north and Oman on the south, in whose territorial waters shipping channels lie

Husayn, Saddam
Iraqi politician and leader of Ba'th Party; President of Iraq since 1979 but *de facto* ruler for some years previously

International Energy Agency (IEA)
organization of oil-importing countries formed in November 1974 at U.S. prodding; membership limited to some OECD states, accounting for 80% of world's oil imports in 1974; IEA goals are to strengthen long-term cooperation among oil-importers, formulate policies to alleviate potential oil shortages and establish voluntary national oil import ceilings

Jabal 'Ali
major industrial complex just south of Dubai town, site of a giant port as well as a projected steel complex, aluminum smelter and oil refinery

Jubayl and Yanbu'
ports and industrial complexes located on Saudi Arabia's Gulf and Red Sea coasts respectively and joined by giant oil pipeline crossing the Arabian Peninsula; built from scratch in recent years, Jubayl and Yanbu' are planned sites of refineries, petrochemical complexes, fertilizer plants, *etc.*

Kharg
island off Iran's northern Gulf coast, site of major oil terminal

Khomeini, Rouhollah
Iranian religious scholar, of ayatollah rank, and politician; exiled to Iraq after 1963 uprisings against government from where he continued outspoken criticism of Shah's regime; expelled from Iraq in 1978, he relocated in France and became living symbol of revolution, returning to his country after departure of Shah in 1979, to assume supra-governmental role as lifelong "leader" of Iran

LNG
liquified natural gas

Manama (al-Manama)
largest city, port and capital of Bahrain; principal offshore banking center of Gulf

Masira (al-Masira)
low, sandy island off Arabian Sea coast of Oman; British RAF base there was turned over to Omani air force in late 1970s and subsequently used on an occasional basis by U.S. military

mbd
million barrels per day; frequently used measurement of oil production and consumption

Mobil Oil
major international oil corporation and one of the "Seven Sisters" (q.v.); formerly known as Standard Oil of New York (Socony), Socony-Vacuum and then Socony-Mobil; first gained interests in Middle East through part ownership in Iraq Petroleum Company with later acquisitions of minority holdings in Aramco (q.v.) and Anglo-Iranian; despite nationalization, Aramco is still Mobil's major source of Middle Eastern oil and Mobil remains active in Saudi Arabia through various petroleum industry projects

Mombasa
major port of Kenya; occasional use of facilities there by American military negotiated with Kenyan government

Mossadegh, Muhammad
Iranian politician opposed to Shah's regime in early 1950s; became Prime Minister with support of National Front and Tudeh parties during period of Anglo-Iranian Oil Company nationalization and subsequent boycott by Western companies; ousted from power in 1953

Mosul
second-largest city of Iraq, located on Tigris River in north of country in major oil-producing area

mullah
religious leader of Twelver Shi'a sect of Islam

Muscat
capital of Oman, located on Gulf of Oman coast

Neutral Zone
formerly a jointly-administered territory lying between Saudi Arabia and Kuwait but partitioned between the two countries in 1969; oil production from this area, however, remains in the hands of the original concessionaire, Arabian Oil Company (q.v.), and income from this concession is divided equally between Kuwait and Saudi Arabia; another neutral zone lies between Saudi Arabia and Iraq

OAPEC
Organization of Arab Petroleum Exporting Countries; formed in 1968 to safeguard interests of its members and promote economic cooperation in the petroleum industry; joint undertakings under OAPEC auspices have included the Arab Engineering Co.; Arab Maritime Petroleum Transport Co.; Arab Petroleum Investment Corp.; Arab Petroleum Services Co.; and the Arab Shipbuilding and Repair Yard Co.; headquartered in Kuwait with Algeria, Bahrain, Egypt (suspended in 1979), Iraq, Kuwait, Libya, Qatar, Saudi Arabia, Syria, Tunisia and the United Arab Emirates (UAE) as members

OECD
Organization for Economic Cooperation and Development; founded in 1961 to promote economic growth and coordination among its members and to contribute to economic development in the Third World; membership consists of 19 European countries plus the U.S., Canada, Australia, New Zealand and Japan

OPEC
Organization of Petroleum Exporting Countries; formed in 1960 to promote petroleum-related interests of member states and to establish pricing policies; headquarters and secretariat located in Vienna; membership consists of Algeria, Ecuador, Gabon, Indonesia, Iran, Iraq, Kuwait, Libya, Nigeria, Qatar, Saudi Arabia, the United Arab Emirates (UAE) and Venezuela

Persian Gulf
see Gulf

petrodollars
surplus oil revenues of OPEC countries deposited in Western financial institutions; frequently, but not always, in the form of American dollars

posted price
first established as artificial transfer price for crude oil "sold" to international oil companies by their producing subsidiaries; later evolved into tax reference price in calculating host governments' share of oil profits and, in last decade, has served as basis for determining common OPEC price

proven reserves
oil deposits which have been reliably established as to volume, tonnage and quality

al-Qadhdhafi, Mu'ammar
leader of 1969 *coup d'état* in Libya which abolished the monarchy and established a republic; served as Chairman of Revolutionary Command Council and then as President of Libya; later resigned official functions to serve only as "Leader of the Revolution"

Ra's Tanura
port on Saudi Arabia's Gulf coast, just north of Bahrain; bulk of that country's oil exports are shipped from terminal located here

RDF
Rapid Deployment Force; American mobile military force with projected quick-strike capability designed for emergency use particularly in Gulf region; for more information, see Appendix G

recycling
movement of surplus oil revenues from OPEC countries back to oil-importing countries in the form of petrodollars (q.v.)

Riyadh
rapidly-growing capital of Saudi Arabia and ancestral home of Al Sa'ud ruling family; located in Eastern Province, near center of country

Royal Dutch Shell
one of "Seven Sisters" (q.v.) group of international oil companies; formed by 1906 merger of British and Dutch firms; acquired quarter interest in Iraq Petroleum Co. in 1928, which led to subsequent involvement in subsidiary companies in Qatar, Abu Dhabi, Dubai and Oman; also acquired part ownership of Iranian consortium in 1954; in addition, Shell is involved in petroleum and petrochemical projects in Saudi Arabia, Syria and Egypt

"Seven Sisters"
popular term for the seven largest international oil companies which long dominated the world petroleum industry; in order of size, these are Exxon, Royal Dutch Shell, British Petroleum (BP), Gulf, Texaco, Standard of California and Mobil

Shatt al-'Arab
120-mile-long river joining Tigris and Euphrates to Gulf; sovereignty over its waters has long been matter of contention between Iraq and Iran

shaykh (sometimes shaikh or sheik)
honorific term of respect; also title of leader of a tribe

Shi'a (adjective, Shi'i)
largest surviving sect to break away from Sunni (q.v.) or "orthodox" Islam; divided into various subsects, the largest of which (Twelver) predominates in Iran with large communities also found in Iraq, Saudi Arabia and Bahrain

Sonatrach
the government-owned petroleum and hydrocarbons corporation of Algeria

Standard Oil of California (Socal; Chevron)
formed in 1906 out of original Standard Oil and has remained essentially a producing company; involvement in Middle East traditionally in cooperation with Texaco (as Caltex), especially in Bahrain Petroleum Company (BAPCO) and Aramco (q.v.); heavily involved in refinery projects throughout region and particularly in Saudi Arabia

Sunni
refers to major body of, or "orthodox," Islam, from which all other sects have broken off; Sunni Islam is further divided into four schools of jurisprudence (Hanafi, Hanbali, Maliki, Shafi'i); the majority of Muslims around the world are Sunni; of the oil-producing states of the Middle East, only Iran, Iraq and perhaps Oman possess predominantly non-Sunni populations

Texaco
formed in 1901 as the Texas Company; in 1936, joined with Standard of California to form Caltex in operations in Bahrain and Saudi Arabia; Aramco (q.v.) still remains Texaco's most important source of Middle Eastern oil and the company participates in various petroleum industry projects within Saudi Arabia

Tikrit
Iraqi town located on Tigris River, midway between Baghdad and Mosul; a small clique from this town presently dominates Iraqi politics

Tripoli
largest city, port and capital of Libya; not to be confused with city of same name in northern Lebanon

Tudeh
Communist Party of Iran; formed *ca.* 1920 but particularly active in 1940s and early 1950s; outlawed in 1949, it nevertheless played an active role in the nationalization of the Iranian oil industry and challenge to the Shah's rule in the 1951–1953 period

United Arab Emirates (UAE)
federation of seven small Gulf states, formed upon independence in 1971; dominated by oil-rich Abu Dhabi (q.v.) which serves as the capital and whose Ruler is also UAE President; other members are Dubai (q.v.), Sharjah, Ra's al-Khayma, Umm al-Qaywayn, 'Ajman and al-Fujayra

Wahhabi
fundamentalist religious movement predominating in Saudi Arabia as a result of 18th-century alliance between its founder and Al Sa'ud family; not a separate sect but a stricter interpretation of Hanbali school of law within Sunni Islam; adherents prefer to be termed *muwahhidun* (unitarians)

Yanbu'
see Jubayl

APPENDIX B

Oil Production and Reserves in the Middle East

Country	Estimated Oil Production, 1982 (1000 b/d)	% Change in Production from 1981	Estimated Proved Oil Reserves (1 Jan. 1983) (1000 bbl)
Abu Dhabi	883.0	−22.8	30,510,000
Algeria	750.0	−7.6	9,440,000
Bahrain	45.0	+2.3	197,000
Dubai	357.0	−0.6	1,440,000
Egypt	667.0	+11.5	3,325,000
Iran	1,896.0	+42.3	55,308,000
Iraq	914.0	−0.3	41,000,000
Israel	0.3	. . .	893
Kuwait	675.0	−28.1	64,230,000
Libya	1,127.0	+2.1	21,500,000
Morocco	0.2	. . .	290
Neutral Zone	315.0	−15.1	5,840,000
Oman	328.0	+2.8	2,730,000
Qatar	340.0	−16.0	3,425,000
Saudi Arabia	6,484.0	−32.7	162,400,000
Sharjah	7.0	−30.0	404,000
Syria	175.0	+4.8	1,521,000
Tunisia	106.0	−10.2	1,860,000
Turkey	45.0	−4.3	280,000
Total Middle East	15,114.5	−10.5	405,411,000
United States	8,655.0	+1.0	29,785,000
Total World	55,002.0	−5.4	670,189,406
Middle East as Percentage of World	27.8%		60.5%

Source: *Oil and Gas Journal*, 27 December 1982

APPENDIX C

A Chronology of OPEC and Oil-Related Developments, 1973–1982

1973	
September	OPEC members agree in Vienna to press for negotiations with international oil companies on major alterations of basic oil price agreements
October	Fourth Arab-Israeli war begins; talks between OPEC states and oil company representatives in Vienna end inconclusively and are followed by unilateral price rises by Gulf producers; Arab oil producers announce a mandatory cut in exports as consequence of Arab-Israeli war and institute an embargo on exports to the United States and the Netherlands
November	OAPEC states announce a 25% cut in production from September 1973 levels; Arab oil embargo extended to Portugal, Rhodesia and South Africa
December	Six Gulf producers raise posted price of marker crude oil to $11.65 a barrel, effective 1 January 1974; extent of production cutbacks revised to only 15% of September level
1974	
January	Arab-African ministerial meeting in Cairo results in assurances of continued supplies of oil to Africa and the formation of a $200 million fund to finance economic needs of African states
February	Effect of oil embargo and production cutbacks discussed at meeting of Arab heads of state in Algiers and decision made not to lift embargo until more progress on the Arab-Israeli front is visible
March	Decision to end embargo against United States and restore production to October 1973 levels made at meeting of Arab oil ministers in Tripoli and announced at second meeting in Vienna, despite opposition of several states; OPEC meeting in Vienna results in posted price freeze

1974	
June	Cairo meeting of Arab oil ministers results in continuation of embargo against the Netherlands, Portugal, South Africa and Rhodesia; Quito OPEC meeting extends price freeze another three months but provides for increase in royalty payments
July	Arab oil ministers decide in Cairo to raise embargo against the Netherlands
September	Average price of oil to international oil companies is raised 3½%, with exception of Saudi oil, at OPEC meeting in Vienna
November	At special meeting of Gulf oil producers in Abu Dhabi, Qatar, UAE, and Saudi Arabia announce intention to increase oil revenues from international oil companies by raising royalties and tax rate on companies' share of oil
December	OPEC freezes prices for nine months from 1 January 1975 resulting in decrease in real terms due to inflation
1975	
January	Price freeze until September 1975 is reaffirmed at Algiers OPEC meeting
March	Future measures such as a possible commodity stockpile are discussed at OPEC Heads of States meeting in Algiers and some agreement is indicated on a willingness to "negotiate the conditions of a stabilization of oil prices"
April	King Faysal of Saudi Arabia is assassinated in Riyadh
June	Determination is expressed at OPEC meeting in Gabon to raise oil prices after September 1975 amid speculation that new Saudi leaders will not be as adamant on holding line on prices
September	Price of oil is formally increased by 10½% at OPEC meeting in Vienna and price is frozen until July 1976; Saudis ensure moderation in price rise only by threats to flood the market
December	OPEC meeting in Vienna is stormed by terrorists who hold twelve OPEC oil ministers hostage for several days
1976	
May	At OPEC meeting in Bali, Saudis gain acceptance on price freeze until 1 January 1977, despite strong dissent from Iraq, Libya, Venezuela and Nigeria
August	Saudi Oil Minister Yamani indicates inevitability of a price rise in 1977 and raises possibility of a 10% increase; change in Saudi stance thought to be a result of greatly increased demand in the first half of 1976

1976	
August–December	Oil stockpiled in consuming countries at record rates in anticipation of price hikes
December	Absence of agreement on uniform price increase at OPEC meeting in Qatar results in two tier system: Saudi Arabia and the UAE raise prices by 5% while other members announce a 10% increase for January and another 5% in July
1977	
January	Saudi Arabia announces intention to raise production ceilings to 10 mbd, to be reviewed quarterly; measure is designed to put pressure on OPEC nations charging higher prices but bad weather in the Gulf allows only small increase in Saudi production
March	Iran announces increased oil production and exports despite higher prices
June	OPEC compromise results in Saudi and UAE price rise of 5% while other members maintain prices established at Qatar meeting; single price tier restored with effect of overall 10% price increase since December 1976
July	Saudi proposal for an eighteen-month price freeze discussed at OPEC meeting in Stockholm since market conditions considered not conducive to price hikes and glut anticipated because of increased production in Alaska and the North Sea
December	OPEC meeting in Caracas results in agreement that real value of OPEC revenues has decreased due to falling value of US dollar, but that existing oil surplus makes a substantial price hike infeasible
1978	
April	Impossibility of a price rise in the near future because of market conditions and the condition of the dollar acknowledged at "informal" OPEC meeting in Geneva
July–August	Saudis reject any method of oil payments other than in US dollars; Prince Fahd declares that his "definite information is that the dollar will return to its former strength, starting at the end of this year"
December	Eighteen-month oil price freeze ends at OPEC in Doha; 10% increase for 1979 announced, with 5% as a first step effective January 1
1979	
January	Strike by Iranian oil workers is expected to keep 5 mbd off the market; spot prices rise dramatically
February	Abu Dhabi and Qatar raise prices by 7% or $1 per barrel; greater rises anticipated

1979	
March	At OPEC meeting in Geneva, 9% increase is announced with effect from 1 April; individual members expected to add surcharges of $1–2; continuing low level of Iranian exports allows increased prices to stick
June	Prices are increased by 13% at OPEC meeting in Geneva; cumulative rise totals 80% since December 1978, with prices at this time ranging from Saudi Arabia's $18 per barrel to $23.50
December	OPEC meeting in Caracas fails to result in uniform pricing structure; members continue to unilaterally raise prices dramatically
1980	
January	Saudi Arabia raises price to $26 per barrel; followed by several OPEC members' price increases to more than $30 per barrel
May	Saudi Arabia announces a $2 raise to $28 per barrel
June	A floor price of $28 per barrel and a ceiling price of $37 per barrel is established at OPEC meeting in Algiers
September	Iran, Libya and Algeria are reported to have blocked an agreement on long-term strategy for oil pricing, involving an indexation plan for automatic increases, at OPEC meeting in Vienna; Saudis agree to raise their prices by $2 to $30 per barrel
October	Iraq-Iran war causes postponement of OPEC meeting scheduled for Baghdad in November; UAE, Saudi Arabia and Qatar agree to increase their combined production by 1 mbd to cover shortfalls caused by Iraq-Iran war
December	OPEC meeting in Bali sets official price of marker crude at $32 per barrel with a ceiling of $36; maximum price for any OPEC crude oil set at $41 per barrel
1981	
January	Individual price rises announced by various OPEC members
April	Saudi Oil Minister Yamani announces that Saudi Arabia will not raise prices or lower production until other nations reduce their prices significantly
August	OPEC meeting in Geneva ends without agreement on pricing; Saudi offer to lower production if prices are lowered is rejected
October	Consensus is reached at OPEC meeting in Geneva: Saudi Arabia will increase marker price to $34 per barrel, differentials will not exceed $4 per barrel and prices will remain frozen until the end of 1982; after

1981	agreement is reached, Saudi Arabia announces a decrease in Saudi production from 9.5 mbd to 8.5 mbd
November	Saudi Arabia announces increase of productive capacity to 11 mbd, although production remains at 8.5 mbd; Yamani threatens to boost production if other OPEC members do not maintain agreed-upon price structure
December	OPEC meeting in Abu Dhabi discusses price differentials and results in a slight reduction in the prices of most Gulf crudes
1982	
January	Iran announces plans to increase oil exports as Kuwait announces that its oil production has dropped to the lowest level in years
February	Iran reduces price of its light and heavy crude by $1 per barrel twice in same month; Iraqi Oil Minister Tayih 'Abd al-Karim calls for extraordinary OPEC ministerial conference to discuss common threat to OPEC from Iran's price-cutting
March	Qadhdhafi accuses Saudi Arabia of flooding oil market to force prices down and enforce Saudi will on other oil producers; Tunisia joins OAPEC after Libya withdraws objection; Saudi Arabia blames OPEC militants for oil glut and announces a reduction in production to 7.5 mbd, agrees to joint construction with Kuwait of pipeline from Iraq to Red Sea; OPEC emergency session calls for cutbacks in crude production, establishes ceiling of 18 mbd for total OPEC production and retains official price of $34 per barrel
April	Saudi Arabia's Oil Minister Yamani announces 20% increase in Saudi oil revenues for 1981 to $101.2 billion; OPEC President Mani' Sa'id al-'Utayba announces that OPEC oil production declines to 15.85 mbd—2.15 mbd below official ceiling; U.S. agrees to buy 1.8 million barrels of Iranian crude in first purchase since November 1979
May	Kuwait announces oil production in 1981 declined 37%; annual OPEC meeting in Quito, Ecuador, results in agreement to maintain $34 a barrel price and to reduce production ceiling to 17.5 mbd
June	Mobil Oil Corporation announces its withdrawal from exploration and production interests in Libya
July	Special OPEC meeting in Vienna fails to resolve disputes over production quotas and pricing, leaving possibility open for intense price competition among OPEC members; Iranian oil minister rejects proposed Iranian production quota of 1.2 mbd; Libyan oil

1982

	production rises to 1.3 mbd, 550,000 barrels above its quota
August	OPEC's monitoring committee meets in Vienna to discuss ways of maintaining OPEC base prices for oil at $34 per barrel
September	OPEC Chairman Mani' Sa'id al-'Utayba, the UAE's Oil Minister, calls for an oil price freeze through 1985; Saudi Arabia asserts that lowering oil prices would harm world economy; OPEC monitoring committee holds a two-day meeting in Abu Dhabi
November	Iran continues to defy OPEC production and pricing policies and dismisses Kuwait and Saudi Arabia's threats of punitive action as meaningless
December	OPEC meeting in Vienna reaffirms base price of $34 per barrel and rejects any price decrease

APPENDIX D

Middle Eastern OPEC Countries: Foreign Aid Disbursements

In $ million
Figures in parentheses are per cent of GNP

Country	1973	1974	1975	1976	1977	1978	1979	1980	1981
I. Gulf Capital-Surplus States									
Kuwait	356	631	946	532	1,309	991	477	645	685
	(8.62%)	(5.33%)	(7.40%)	(3.64%)	(8.20%)	(5.64%)	(1.79%)	(2.04%)	(1.98%)
Qatar	94	185	338	195	194	109	280	284	175
	(15.62%)	(9.26%)	(15.59%)	(7.95%)	(7.76%)	(3.75%)	(6.03%)	(4.25%)	(2.64%)
Saudi Arabia	1,118	2,153	2,756	3,033	3,138	5,507	4,674	5,944	5,798
	(14.80%)	(9.32%)	(7.76%)	(6.47%)	(5.33%)	(8.45%)	(6.12%)	(5.09%)	(4.77%)
United Arab Emirates	289	510	1,046	1,021	1,060	891	967	906	799
	(12.67%)	(7.04%)	(11.69%)	(8.88%)	(7.27%)	(6.27%)	(5.09%)	(3.38%)	(2.88%)
Total	1,857	3,479	5,086	4,781	5,701	7,498	6,398	7,779	7,457
	(12.76%)	(7.78%)	(8.56%)	(6.33%)	(6.20%)	(7.51%)	(5.05%)	(4.28%)	(3.92%)
II. Other Arab Producers									
Algeria	25	47	41	54	47	44	272	65	65
	(0.28%)	(0.37%)	(0.27%)	(0.33%)	(0.24%)	(0.18%)	(0.89%)	(0.17%)	(0.16%)
Iraq	11	423	215	231	61	173	847	829	143
	(0.21%)	(3.98%)	(1.63%)	(1.44%)	(0.33%)	(0.76%)	(2.53%)	(2.13%)	(0.37%)

MIDDLE EASTERN OPEC COUNTRIES: FOREIGN AID DISBURSEMENTS (continued)

In $ million
Figures in parentheses are per cent of GNP

Country	1973	1974	1975	1976	1977	1978	1979	1980	1981
Libya	215 (3.33%)	147 (1.26%)	259 (2.29%)	94 (0.63%)	113 (0.63%)	146 (0.85%)	105 (0.45%)	282 (0.92%)	105 (0.37%)
Total	251 (1.21%)	617 (1.76%)	515 (1.31%)	379 (0.80%)	221 (0.39%)	363 (0.56%)	1,224 (1.40%)	1,176 (1.08%)	313 (0.29%)
III. Iran	2 (0.01%)	408 (0.87%)	593 (1.13%)	753 (1.16%)	221 (0.29%)	278 (0.37%)	25 (0.03%)	7 (0.01%)	−150 —
Middle Eastern Total	2,110 (3.82%)	4,504 (3.57%)	6,194 (4.09%)	5,913 (3.15%)	6,143 (2.74%)	8,139 (3.40%)	7,647 (1.34%)	8.962 (2.49%)	7,770 (2.61%)*
United States	8,346 (0.64%)	9,931 (0.71%)	17,366 (1.15%)	4,360 (0.26%)	4,682 (0.25%)	5,663 (0.27%)	4,684 (0.20%)	7,138 (0.27%)	5,783 (0.20%)

Source: Organization for Economic Cooperation and Development, Development Assistance Committee, *Development Cooperation Efforts and Policies of the Members of the Development Assistance Committee, 1982 Review.*

*Excluding Iran (for this percentage only)

APPENDIX E

The Iranian Revolution: A Chronology

1941	British and Soviet forces occupy Iran to protect the oil fields; Reza Shah Pahlavi is exiled and succeeded by his son, Muhammad Reza Shah, as ruler of Iran
1942	Tripartite Treaty, signed by Great Britain, the United States and the Soviet Union, calls for the removal of all foreign troops from Iran within six months of the end of the war
1943	The Declaration of Tehran commits the Big Three powers to the preservation of Iran's unity and independence and to promoting its development
1946	British and American troops leave Iran; Soviet troops only leave after condemnation in the United Nations
1947	U.S. and Iran sign agreement for the establishment of an American military mission in Iran
1950	National Front, a broad nationalist coalition of parties led by Muhammad Mossadegh, dominates in Majlis (Parliament) elections
1951	Prime Minister Ali Razmara is assassinated by Islamic fundamentalists; the Anglo-Iranian Oil Company (AIOC) is nationalized by Majlis, international oil companies boycott Iranian oil and consequent lack of revenues disrupts Iranian economy; Mossadegh becomes Prime Minister with support of National Front and Tudeh Party and acts to reduce power of Shah
1953	Mossadegh is ousted from power and the Shah regains control of the country
1954	International oil companies reach agreement with Iranian government to operate former AIOC concession under Iranian ownership

1955	Iran signs Baghdad Pact and joins the Middle East Treaty Organization, later known as the Central Treaty Organization (CENTO)
1962–63	The Shah proclaims his White Revolution, an extensive program of economic, social and administrative reforms
1967	Muhammad Reza Shah and his wife, the Shahbanou, are formally crowned in elaborate ceremony
1971	
March	The 2500th anniversary celebrations of the founding of the Persian Empire at Persepolis draw a glittering international audience
November 30	Iranian forces occupy Abu Musa and the Tunbs islands, previously held by Ra's al-Khayma and Sharjah
1972	
May	President Richard Nixon's visit to Tehran results in promise of unlimited non-nuclear military sales to Iran
June 24	The Shah states that Iran needs "millions of dollars worth of the best military equipment we can find short of atomic weapons"
November 11	The Shah emphasizes requirement for major expansion of Iranian naval power since Iran's defensive needs extend beyond the Gulf into the Indian Ocean
1973	
January 23	The Shah announces that all domestic oil operations will be taken over by the National Iranian Oil Co. by 1979
May 24	Agreement reached with a consortium of Western oil companies giving the NIOC full ownership of its southern operations in exchange for a guarantee of "substantial quantities" of crude oil for 20 years
October	Iran continues to supply oil to the United States despite the Arab oil embargo
October 2	Iranian government announces that twelve people have been arrested for plotting to kill the Shah
October 15	Diplomatic relations resumed with Iraq in midst of Arab-Israeli war
1974	
January 10	US announces that Iran has agreed to purchase 30 F-14A jet fighters for $900 million
August	Iran charges Iraq with periodic shelling of Iranian territory in the Qasr-i Shirin area

1974	
August 30	Iranian sources report that more than 70,000 Iraqi Kurds have crossed the Iranian border following an Iraqi military offensive
December 21	The Shah issues invitation to American businessmen to seek Iranian contracts and encourages skilled Americans to seek employment in Iran to help alleviate the shortage of skilled manpower there
1975	
February 9	Pars News Agency claims Iraqi troops attacked border positions but were driven back
March 5	The Shah of Iran and Iraqi President Ahmad Hasan al-Bakr reached agreement at Algiers on settlement of Shatt al-Arab dispute and cessation of Iranian support for Kurdish rebels in Iraq
May 15	The Shah begins official visit to U.S. and confers with President Ford
May 21	Terrorists kill two US military officers in Tehran
December 7	The Shah pardons 242 political prisoners to celebrate the 28th anniversary of the universal Declaration of Human Rights
1976	
March 14	The Shah warns US that Iran "can hurt you as badly if not more so than you can hurt us" if US Congress imposes an arms sales embargo to the Gulf region
August 27	US agrees to sell Iran 160 F-16 fighters for $3.4 billion
1977	
November 8	Iran confirms the purchase of seven American AWACS planes
November 15	Shah's visit to Washington is marred by riots
1978	
January 1	President Carter visits the Shah in Tehran, describing Iran as an "island of stability" because of the Shah's leadership and "the respect, admiration and love which your people give to you"
Jan.–May	Demonstrations in Tabriz and Tehran end in rioting and deaths; general strike called
August 27	Jamshid Amuzegar resigns as Prime Minister; Shah appoints Jafar Sharif Emami to replace him
October	Ayatollah Khomeini expelled from Iraq; flies to Paris
October 31	Oil workers go on strike; production declines from 5.8 to 1.1 mbd in one week

1978	
November 2–7	National Front calls for a referendum on a new government; violent demonstrations and strikes sweep through major cities; Sharif Emami resigns as Prime Minister and Shah appoints General Gholam Reza Azhari as Prime Minister of a predominantly military government; Ayatollah Khomeini calls for eventual formation of an Islamic republic, by force if necessary
November 30	President Jimmy Carter reaffirms confidence in the Shah and rejects US intervention in Iran's affairs
December 29	Oil production falls below 300,000 barrels a day
December 31	Prime Minister Azhari resigns as demonstrations against Shah intensify
1979	
January 3	Iranian Majlis (Parliament) approves choice of Shahpour Bakhtiar as Prime Minister; Bakhtiar is expelled from the National Front
January 16	The Shah leaves Iran for an "extended vacation"
January 18	In Paris, Khomeini rejects President Carter's appeal to give Bakhtiar's government a chance to succeed
January 25–31	Mass demonstrations lead to riots and deaths as supporters of Bakhtiar, the Shah, and Khomeini clash; Khomeini returns to Iran and is greeted by hundreds of thousands of supporters
February 5	Khomeini announces Mehdi Bazargan as his choice for Prime Minister of a provisional Islamic government
February 11	Bakhtiar and the members of the Majlis resign; General Gharabaghi, chief of staff, declares neutrality of the military
February 14	American Embassy in Tehran is attacked and held hostage for two hours until attackers are dispersed by pro-Khomeini forces
February 21	Iran terminates permission for US monitoring installations in northern Iran
February 28	Prime Minister Bazargan threatens to resign if Revolutionary Council continues to interfere with his government
March 6	Revolutionary Government announces plans to try the Shah in absentia
July 17	Bazargan announces agreement with Khomeini on the sharing of power between his government and Khomeini's Revolutionary Council
August 1	Iran cancels $9 billion arms agreement with US
September 15	1947 law authorizing US military mission in Iran is repealed

1979	
November 1	Khomeini's office encourages students to "expand their attacks" against the US to force the return of the Shah, in the US at the time for medical examinations
November 4	Iranian student sit-in at the US embassy leads to violence and the taking of *ca.* 60 hostages; students call for Shah's extradition
November 6	Bazargan and his government resign; Khomeini directs the Revolutionary Council to manage transition and to prepare for new elections
November 8	US halts the shipment of spare military parts to Iran
November 10	*Ca.* 50,000 Iranian students in the US are directed by US Attorney General to report to the nearest immigration office
November 12	President Carter announces cessation of oil imports from Iran as Iranian officials announce suspension of oil sales to US companies
November 14	President Carter freezes Iranian assets in US banks and their foreign branches
November 19–20	Thirteen hostages, eight black men and five women, are released by student captors
November 23	Finance Minister Abol-Hasan Bani Sadr announces Iran's repudiation of all foreign debts
December 2	Iranian voters approve a draft constitution prepared by Revolutionary Council in light turnout
December 8	Bani Sadr calls for the release of the hostages
December 15	International Court of Justice unanimously rules in favor of the US, ordering Iran to release hostages and restore US property
1980	
January 1–4	UN Secretary General Kurt Waldheim visits Tehran and meets the Revolutionary Council, but is pessimistic about chances for a rapid solution
January 27	Finance Minister Bani Sadr is elected President of Islamic Republic of Iran
February 13	President Carter says US will accept an international commission of inquiry, to investigate the Shah's regime, as a step toward freeing the hostages
February 18	Secretary General Waldheim names five members to a UN Commission on Iran
February 23	Khomeini leaves decision on American hostages to the Islamic Assembly, scheduled to be elected in May
March 11	UN Commission leaves Tehran without resolving hostage situation

1980	
March 27	US Secretary of State Cyrus Vance states that a military option against Iran has not been ruled out
April 7	President Carter severs diplomatic relations with Iran and announces the implementation of an economic embargo against Iran, as well as the inventorying of Iranian assets in the US, the application of US financial claims against Iran to these assets, and the cancellation of all entry visas for Iranians
April 25	White House announces abortive rescue attempt, cancelled because of equipment failure, and the loss of eight lives as result of collision on ground of aircraft at remote desert location in Iran; President Carter accepts full responsibility for the debacle
May 2	Students holding hostages announce that their captives have been dispersed to deter any future rescue attempts
May 10	It is announced that April 30 elections have given the Islamic Republican Party a majority in the new Majlis
May 18	EEC members approve limited economic sanctions against Iran, effective May 22, but exclude all contracts signed prior to the 4 November 1979 takeover of US Embassy
July 27	The deposed Shah dies in Egypt of cancer at age 60
August 16	White House labels reports that the US had formulated a plan for a limited invasion of Iran as "erroneous and totally irresponsible"
September 11	Revolutionary Council is formally dissolved
September 12	Khomeini announces that hostages will be freed if the US returns the property of the Shah, cancels its financial claims against Iran, releases all Iranian assets in the US and guarantees "political and military non-intervention" in Iran
September 13	Presidential candidate Ronald Reagan agrees in principle to all of Khomeini's demands except the matter of the Shah's property, contending that it can not be confiscated without due process of law
September 18	President Carter announces his refusal to apologize to Iran in order to obtain the release of the hostages
October 2	Majlis names seven people to a special commission to determine the course of action on the hostages
October 20	Carter confirms that trade embargo will be ended and Iranian assets "unfrozen" if hostages are released
October 31	Crown Prince Reza, son of former Shah, proclaims himself Shah in Cairo
November 3	Students holding hostages meet with Khomeini and agree to transfer responsibility for the hostages to the

1980

	government; the Algerian Ambassador to the US delivers the official Iranian texts of conditions for their release to the State Department
November 10	US Deputy Secretary of State Warren Christopher flies to Algiers and presents American response to Iranian conditions to the Algerian Foreign Minister
November 26	Algerian diplomats arrive in Washington with Iranian request for "clarification of certain aspects" of the US position
December 2	Christopher delivers a clarification of US position to Algerian officials
December 15	President Bani Sadr says that the return of Iranian assets is the crucial question
December 20	Prime Minister Muhammad Ali Rejai states that main outstanding problem in hostage settlement is the return of the late Shah's wealth
December 21	Most recent Iranian proposals are denounced by Secretary of State Edmund Muskie as unreasonable and requiring actions "beyond the power of the President"
December 28	President-elect Reagan questions any possible payment of "ransom"

1981

January 2	Algerian diplomats arrive in Tehran with new US proposals
January 9	Christopher, in Algiers, announces continued existence of serious problems
January 15	Algerian intermediaries give Christopher a "substantive Iranian response"
January 19	Christopher signs agreement with Iran in Algiers
January 21	Hostages arrive in Algiers
January 27	Iranian Foreign Ministry denies charges of physical abuse of the hostages
January 28	Secretary of State Alexander Haig announces that US will provide no military supplies to Iran and recommends necessity of caution with respect to future commerical transactions with Iran
February 10	Iran deports American citizen Cynthia Dwyer for "espionage and activities against Iran"
February 18	US State Department announces approval of implementation of agreements with Iran "in strict accordance with their terms"
April 7–9	Iranian government takes measures to inhibit expression of opposing viewpoints and places restrictions on press

1981	
May 15	President Bani Sadr claims that Iranian government is carrying out unjust arrests and torture just as the late Shah did and that "there is no more law"
May 27	Khomeini warns feuding Iranian factions to halt attempts to expand power; remark seems to be directed at Bani Sadr
June 8	Khomeini announces he will dismiss any politician who continues to challenge Islamic authorities
June 10	Khomeini dismisses Bani Sadr as Commander-in-Chief of Armed Forces
June 12–13	Bani Sadr denounces policies of Islamic Republican Party and goes into hiding
June 14	Khomeini announces that Bani Sadr can still "carry out his duty" despite the loss of many of his powers
June 20–22	Majlis begins formal impeachment proceedings against Bani Sadr and declares him politically incompetent; Prosecutor-General orders his arrest and Khomeini orders his official removal from office
June 28	Chief Justice Ayatollah Muhammad Beheshti and 71 other leading politicians killed when bomb destroys Tehran headquarters of ruling Islamic Republican Party
July 24	Prime Minister Rejai wins election for a new President
July 29	Bani Sadr receives political asylum in France
August 4	President Rejai names Muhammad Javad Bahonar as Prime Minister
August 10	57 French citizens arrive in Paris after being held up for several days by Iranian government
August 13	Second plane, carrying 50 remaining French citizens, including ambassador, arrives in Paris
August 17	23 leftists are executed in Tehran, bringing total number of executions since Bani Sadr's exile to *ca.* 500
August 26	26 more leftists are executed
August 30	President Rejai and Prime Minister Bahonar are killed when an explosion rips through Prime Minister's offices in downtown Tehran
September 1	Ayatollah Muhammad Mahdavi-Kani is named provisional Prime Minister and Hojatolislam Ali Khamenei is named head of the Islamic Republican Party
September 5	Prosecutor General Hojatolislam Ali Qaddousi is killed in Tehran when a bomb explodes in his office

1981	
September 18	19 more leftists are executed and government announces new rules for quicker execution of dissidents in response to bombings
September 23	41 more dissidents are executed
September 29	43 more dissidents are executed
October 1	Bani Sadr announces formation of a "transitional" government-in-exile in Paris
October 2	Presidential elections are held amidst strict security
October 5	Government announces that Khamenei received 95% of vote for president
October 6	61 more dissidents are executed in Tehran
October 8	Parliament prepares bill to expel former Prime Minister Mehdi Bazargan for his criticism of regime
October 13	Khamenei is sworn in as President
October 29	Parliament confirms nomination of former Foreign Minister Hossein Moussavi as Prime Minister
November 2	Parliament approves Moussavi's 21-member cabinet
December 11	Khomeini's representative in Shiraz is assassinated there by bomb
1982	
January 11	Terrorists wound brother of President Khamenei and kill both of his bodyguards in Beirut attack
January 19	Exiled former Prime Minister Ali Amini announces the formation in Paris of the Movement for the Liberation of Iran
January 28	Government forces restore order in the northern town of Amul after fighting there between anti-government militants and Revolutionary Guards leaves 66 people dead
February 8	Musa Khaybani, the ranking leader in Iran of the Mujahidin-i Khalq, his wife and the wife of exiled Mujahidin leader Masud Rajavi are killed in raid by government security forces; army revolutionary courts announce that 20 individuals involved in escape of former President Bani Sadr will be put on trial
February 15	Explosion in central Tehran kills 15 people and wounds 60; the government blames a monarchist group
February 27	Tehran mulla, Hojatolislam Mustafa Hujjati, is assassinated with the Mujahidin-i Khalq claiming responsibility
March 4	Ayatollah Khomeini grants official pardons to 10,000 prisoners on the third anniversary of the Islamic Republic of Iran

1982

March 6	17 leftists are executed in 3 towns for armed rebellion and membership in an illegal organization
March 15	At least 37 people are thought to have died in fighting between anti-government Kurds and government security forces near Mahabad
March 18	30 government employees are arrested on charges of being members of the Mujahidin-i Khalq
April 10	Former Foreign Minister Qutbzadeh arrested along with several others for plotting to kill Khomeini
April 14	Iranian consulate in Greece is raided by 18 members of Paykar who ransack the building
April 16	In a major offensive against Kurdish dissidents, security forces kill 75 and wound 40; an alleged assassination plot against Khomeini results in arrest of more than 1,000, including religious leaders
April 19	Iran's chargé d'affaires in Rome resigns in protest over executions and political oppression in Iran; former Foreign Minister Qutbzadeh confesses to involvement in Khomeini assassination plot
April 25	30 people killed in fighting between government security forces and the Qashqai tribe of southern Iran
May 2	Iranian government forces kill 50 top leaders of Mujahidin-i Khalq in 10 raids
May 5	Mujahidin-i Khalq attacks office of Imam's Relief Committee and kills 5 employees
May 15	4 wounded in explosion at Imam Khomeini mosque in Tehran; the Movement of the Followers of the Sharia claims responsibility
May 20	14 Mujahidin-i Khalq members arrested in Khuzistan
May 24	50 Mujahidin members executed
May 30	Another 10 Mujahidin members executed in Bandar Abbas
June 22	Revolutionary Guard kills 14 Mujahidin members in Tehran
June 29	64 people killed in Kurdish attack on Revolutionary Guard barracks in Mahabad
July 2	Ayatollah Muhammad Sadduqi is assassinated in Yazd and authorities blame the Mujahidin-i Khalq
July 7	Another 150 Mujahidin members arrested in Khuzistan
July 27	Former navy chief Mahmud Alawi and 95 other political detainees released from prison

1982

July 30	100 "counter-revolutionaries" are killed in attacks on Kurdish insurgents by government forces near Sardasht in west Azerbaijan
August 1	24 Jews, described as "Zionist agents," are arrested for smuggling in Mashhad; Tehran radio reports that 65 members of Mujahidin-i Khalq have been killed or captured in recent raids
August 8	Ahwaz Islamic Prosecutor's Office reports that 300 members of the Mujahidin have been killed, wounded or captured in Khuzistan since 21 April
August 9	Interior Minister Ali Akbar Natiq Nuri replaces Ayatollah Mahdavi Khani as head of revolutionary Komitehs
August 14	Trial of former Foreign Minister Sadiq Qutbzadeh begins in Tehran on charges that he conspired to overthrow the government and assassinate Ayatollah Khomeini; Qutbzadeh denies complicity in any conspiracy against the government or Khomeini but does admit attempts to oust a number of clerics from government; 70 officers who were allegedly involved in Qutbzadeh's plot are executed
September 6	Bomb explodes in central Tehran, killing 20 and wounding 100; the government blames the Mujahadin-i Khalq
September 9	A second bomb in central Tehran explodes, killing 3 people
September 12	Revolutionary Guard reported to have killed 6 members of Mujahidin-i Khalq and wounded four others in Tehran raids; two more bombs explode in Tehran, killing 1 and wounding several
September 20	Bomb explodes in Tehran's central bus station, killing 1 and wounding 20
October 1	Tehran radio reports that Qashqai leader, Khusru Khan Qashqai, was executed in Shiraz; another bomb in Tehran kills at least 60 people and wounds 700 others; from Paris, Mujahidin-i Khalq statement lays blame on Iranian government
October 4	Political dissidents hijack military transport plane and direct it to the United Arab Emirates, where passengers are released and hijackers request political asylum; UAE officials deny asylum request, returning hijackers to Iran where they are arrested by security forces; 100 "counter-revolutionaries" killed in west Azerbaijan during "continuing purging operations"

1982

October 16 Ayatollah Ashrafi Isfahani, an aid to Khomeini, is assassinated in a Bahdaranian mosque; the government blames the Mujahidin-i Khalq

November 13 David Parvizian, head of the Iranian News Agency, seeks political asylum in United Kingdom and declares his support for the Mujahidin-i Khalq

November 14 The sentences of two Iranians convicted in the Qutbzadeh conspiracy to assassinate Khomeini are commuted; they are Ahmad Abasi, the son-in-law of Ayatollah Kazem Shariat-madari, and cleric Abdulreza Hejazi

November 26 Employee of Iran Air defects to Sweden, the fifth key member of Iran Air to flee the country in the last two months

December 3 The list of candidates for the Council of Experts, charged with finding a successor to Ayatollah Khomeini, includes President Ali Khamenei, Majlis Speaker Ali Akbar Hashemi-Rafsanjani and Ayatollahs Khalkhali, Meshkini, Gilani and Musavi Ardabili

December 10 Radical extremists within the Iranian government, many of whom were among students seizing US Embassy in 1979, demand that Ayatollah Montazeri automatically be proclaimed Khomeini's successor; conservatives within the government have so far blocked such an action

December 13 Key Iranian Air Force officer asks for political asylum in Switzerland, claiming that 40% of Iran's army and air force support the Mujahidin-i Khalq

APPENDIX F

Iraq-Iran War: Background and Chronology

The outbreak of full-scale hostilities between Iraq and Iran in September 1980 was as much the product of a number of longstanding disputes between the two countries as it was due to immediate provocations. Brief outlines of some of the more important contributing factors and a short chronology on the war's progress through 1982 follow.

1. RIVALRY AS "GREAT POWERS" OF THE GULF

As the two largest and most powerful states of the Gulf in terms of population, infrastructure and economic potential, Iraq and Iran have long sought to extend competing spheres of influence throughout the Gulf. Increasing ideological divergence, dating from the 1958 Iraqi revolution, and especially the growing ties to opposing superpowers compounded existing areas of friction.

2. ARAB–PERSIAN RIVALRY

A more intangible factor contributing to the war was that of longstanding disputes, hostilities and mutual suspicions between Arabs and Persians in the Gulf region. Arabs led the original Islamic conquest of Sassanid Persia and various Persian states have conquered territories on the Arab side of the Gulf on numerous occasions before and since then. The situation is complicated by the existence of isolated Arab and Persian communities on opposite shores of the Gulf. By the beginning of this century, Khuzistan (the Iranian province located between the Shatt al-'Arab and the mountainous upthrust surrounding the Iranian plateau) had become predominantly Arab and was controlled by the Arab *shaykhs* of Muhammara. The absorption of what had become known as 'Arabistan into the Iranian state was accomplished by Reza Shah only in 1925.

Elsewhere, Arab–Persian relations were troubled by Pahlavi claims to sovereignty over the archipelago of Bahrain, claims resolved only by the Shah's acceptance of a 1970 referendum in Bahrain indicating preference for an independent Bahraini state. In late 1971, Iranian forces took advantage of the impending British withdrawal from the Gulf to seize three small islands (Abu Musa and the two Tunbs) lying just inside the entrance to the Gulf, which had been under the control of Sharja and Ra's al-Khayma. Arab fears of Iranian expansionism were later exacerbated by the Shah's naval expansion and by the dispatch of Iranian troops, fighters and naval vessels to Oman to help put down the Dhufar rebellion there.

3. THE SHATT AL-'ARAB DISPUTE

Formed by the confluence of the Tigris and Euphrates Rivers, and later joined by the Karun River, the 120-mile-long Shatt al-'Arab forms part of the Iraqi–Iranian border before emptying into the Persian Gulf. It has great commercial importance for both countries, as Iraq's chief port of al-Basra and the important Iranian port of Khurramshahr (formerly Muhammara) are both located along the river, as is the large Iranian refinery at Abadan. Major oilfields in both countries lie nearby and the bulk of Iraq's oil is exported along the Shatt.

Sovereignty over the river has been in dispute for some time, and a 1937 Iraqi–Iranian treaty gave Iraq control over all the Shatt al-'Arab except the area near Khurramshahr and Abadan, where the frontier was established along the *thalweg* (the median line of the channel); with the exception of that area, ships using the Shatt were required to employ Iraqi pilots and fly the Iraqi flag. Worsening relations between the conservative Shah and increasingly radical Iraqi regimes led to frequent frontier incidents and the Shah's demand that the *thalweg* determine the border over the entire course of the river. In 1969, Iran, declaring that Iraq had long violated provisions of the 1937 treaty, unilaterally abrogated it and provided Iranian naval escorts for its ships. Hostile incidents between the two countries increased and Iraq broke off relations with Iran, restoring them during the 1973 Arab–Israeli war. Skirmishes continued, however, over the next eighteen months until Iraqi–Iranian differences were settled at the Algiers summit conference.

4. THE KURDISH REBELLION IN IRAQ

Another source of friction between the two countries arose from the Kurdish nationalist struggle. The Kurds are a distinct ethnic group, largely Sunni Muslim and speaking their own language, who occupy the mountainous knot where Turkey, Syria, Iraq and Iran meet. Kurdish nationalism has long existed and became particularly active by the 1930s. In 1946, following the collapse of the short-lived, Soviet-backed Kurdish Republic of Mahabad, the Kurdish Democratic Party was founded and gradually became the largest and most important nationalist movement, under the leadership of

Mulla Mustafa al-Barzani. The 1958 revolution in Iraq seemed to point to cooperation between Kurds and the revolutionary (but Arab nationalist) government in Baghdad, but rebellion broke out in 1961. Despite apparent settlements in 1963 and 1970, the seeming unwillingness of the Iraqi government to implement measures agreed upon and to grant recognition to a separate Kurdish region led to endemic rebellion through the 1960s and in the early 1970s. An already unstable situation deteriorated by 1974 when a major Iraqi military offensive signalled the beginning of all-out war between Baghdad and the Kurdish army. While the Kurds were fragmented politically, their cause benefited from considerable assistance provided by the Shah of Iran, who welcomed the opportunity to put pressure on Baghdad, and from covert help given by the United States and Israel.

5. 1975 ALGIERS ACCORD

Efforts to put an end to a growing climate of Iraqi–Iranian hostility began with discussions in early 1975 between the two countries' foreign ministers. The OPEC Heads of State meeting in Algiers in early March 1975 provided an opportunity, with the benefit of Algerian, Jordanian and Egyptian mediation, for the Shah and Saddam Husayn to sign the Algiers Communiqué on 6 March 1975. For the Shah, this agreement seemed to strengthen Iran's preeminent position in the Gulf and, more specifically, Iran benefited from agreement that the Shatt al-'Arab border would follow the *thalweg* completely. For Iraq, the agreement meant a welcome cessation of Iranian support for Kurdish rebels and the closing of the Iranian border, thus forcing the Kurds to seek peace on Baghdad's terms. The accord reached at Algiers was formalized in the Treaty of Baghdad (13 June 1975), which included settlement of all land and maritime border disputes and an agreement to not harbor *personae non gratae* of either country.

6. THE IRANIAN REVOLUTION

Initially, the growth of opposition to the Shah and then his ouster were favorably viewed by the Iraqi government. Baghdad had never gotten along with the Shah and was not sorry to see him go. Furthermore, the emerging anti-American climate in Iran meant that U.S. influence in the Gulf was consequently diluted, also to Iraq's satisfaction. Despite this, relations between the Ba'thist rulers of Iraq and the Islamic revolutionary leaders of Iran soon deteriorated. It quickly became apparent that Ayatollah Khomeini and his followers were strongly opposed to the Iraqi government for a number of reasons, not least of which undoubtedly was Baghdad's compliance with the Shah's 1978 request that Khomeini be ousted from his Iraqi exile.

The two countries also began to exhibit a deep ideological division. Iraq was a secular nationalist and socialist regime while that of Iran was theocratic and based on the Shari'a, the body of Islamic law, and was also intent on spreading the "Islamic revolution" beyond the borders of Iran. In part, this included appeals to Shi'is in other Gulf states to rise up against their governments. Even though half of Iraq's population is Shi'i (while also Arab), Iraqi governments have also been dominated by Sunnis; Iranian exhortations for revolution could have great potential appeal for Iraqi Shi'is. At the same time, the emerging Iraqi–Iranian split reflected older enmities, most prominently the Arab–Persian dichotomy. While the new Iranian leadership emphasized the Islamic nature of the new state, they also were intent on preserving Tehran's control over the country's many ethnic and religious minorities, among whom were the Arabs of Khuzistan. Another challenge to the neighboring Arab world was the brief revivification of old Iranian claims to the Bahrain islands.

The Iraqi response was to play on traditional Arab fears of Iranian expansionism and hegemony, in order to gain support among its Arab Shi'is and from other Arab states. The continued Iranian occupation of the (Arab) islands of Abu Musa and the Tunbs was heavily stressed. A final point of conflict between Iraq and Iran was lingering Iraqi bitterness over the division of the Shatt al-'Arab, which was considered to have been a concession unfairly forced on Iraq through Pahlavi Iran's superior military capabilities and exploitation of internal Iraqi weaknesses. All these factors contributed to the eruption of hostilities between the two countries in mid-1980; the course of the war is summarized in the following chronology.

1980

March 28	Iranian President Abol-Hasan Bani Sadr states that Iran is not interested in good relations with Arab nations in the Gulf
April 9	Iraqi President Saddam Husayn responds angrily to Iranian leader Khomeini's call for his overthrow
Sept. 2	Initial clashes between Iraqi and Iranian forces take place along the border area near Qasr-i Shirin
Sept. 9–16	Iraqis cross Shatt al-'Arab near Khurramshahr and Abadan in the south and launch attacks in Khuzistan from Dezful to Ahwaz further north; offensive bogs down after early successes
Sept. 17	President Husayn declares the 1975 border agreement with Iran void because of Iran's refusal to abide by it; claims that the Shatt al-'Arab is "totally Iraqi"
Sept. 20	Bani Sadr calls up military reservists as clashes continue; Iraq claims to have captured 90 square miles of disputed territory in Iran

1980	
Sept. 22	Iranian Joint Staff announces that "the Strait of Hormuz and the Northern Coasts of the Persian Gulf are under the control" of the Iranian armed forces but does not say that they have been closed; Iraq's prompt response claims that Iran is "controlling and restricting" use of the strait
Sept. 23	Iraq bombs the Iranian oil refinery at Abadan
Sept. 24	Iraq and Iran halt all their oil shipments through the Gulf, largely due to anticipation of further attacks on oil facilities
Sept. 26	Iraq suspends all oil exports and announces capture of Iranian oil port of Khurramshahr
Sept. 28	A unanimous UN Security Council resolution calls upon Iraq and Iran to end hostilities and to accept mediation; Iraq announces capture of town of Khurramshahr and indicates its willingness to negotiate
Oct. 1	Bani Sadr declares Iran's refusal to negotiate as long as Iraq violates Iranian sovereignty
Oct. 4	Control of Khurramshahr reverts to Iran
Oct. 24	Despite stiff resistance, Iraq regains control of Khurramshahr as siege of Abadan continues
Nov. 20	Iraq resumes pumping oil through pipeline to Mediterranean
Dec. 26	Iraqi troops push across border into Iranian province of Kurdistan
1981	
March 6	Iran rejects Islamic Conference initiatives for ceasefire
March 24	Iran announces goal of the "removal of Saddam's regime"
March 29	Husayn responds by pledging full support to national secessionist movements within Iran
April 21	Kuwait extends $2 billion in interest-free loans to Iraq to help rebuild areas destroyed in the war
April 23	Iran accuses Habib Chatti, spokesman of the Islamic Peace Mission, of being pro-Iraqi
May 17	Iran claims its forces have penetrated Iraqi defenses around Abadan and recaptured strategic road
June 14	Kuwait accuses Iran of aggression after air strikes on Kuwait border outposts
June 16	Iraq and Iran exchange 42 prisoners
June 28	UN Envoy Olaf Palme flies to Baghdad from Tehran with peace plan

1981	
July 20	Iran claims to have disabled power plants in Kurdish region of Iraq
July 27	Tehran denies charges that Israeli arms have been shipped to Iran
Aug. 6	Foreign Ministers of Cuba, India and Zambia and a senior PLO official arrive in Tehran in an attempt to mediate the dispute
Sept. 27	Iranian forces advance, claiming to have killed 600 Iraqis and captured 3000
Sept. 29	Iranian gains are overshadowed by deaths of four military leaders in air crash, including Commander-in-Chief Valliollah Fallahi and Defense Minister Musa Namju
Oct. 1	Kuwait accuses Iran of attacking an oil depot and setting it ablaze
Oct. 8	Iraqi President Husayn indicates he will press for peaceful solution to war with Iran
Nov. 11	Iraq claims to have shot down an Iranian F-14 over the Gulf
Nov. 29	Iran opens a broad offensive which eventually leads to the recapture of Bostan, within a few miles of the Iraqi border near Susangard
Dec. 7–11	Series of Iraqi counterattacks fail to regain Bostan, while Iranian attacks gradually encroach on Iraqi positions, especially around Khorramshahr
Dec. 11	Iran launches fifth major offensive of 1981, which apparently meets with limited success
Dec. 20	Iraq claims its troops have forced Iranian forces to retreat in the mountainous border area around Gilan-i Gharb and Sumar in western Iran
1982	
Jan.	Iranians claim to have entered Iraqi territory near the northern town of Nowsud and to have captured many prisoners
Jan. 28	King Husayn of Jordan announces in Amman that Jordan will send volunteers to Iraq to fight alongside Iraqi forces
Feb. 11	Iranians switch to heavier use of artillery as combat losses reduce air force to approximately 75 F-4s and less than a dozen F-14s
Mar. 2	UN Envoy Olaf Palme fails to achieve cease-fire in fifth attempt
Mar. 24	Iranians begin major offensive in Dezful area

1982

Mar. 29	Significant gains reportedly made by Iran in Dezful area
Mar. 31	Iraqi soldiers, their morale broken by the ferocious Iranian offensive, are reported surrendering to Iranians en masse; Iran claims to have captured 15,000 prisoners since war began
Apr. 30	Iran announces major offensive in southern Iran to regain Khuzistan province and claims recapture of Khurramshahr
May 5	Tehran reports that Iranian helicopter gunships attack targets inside Iraq for first time since war began
May 6	Iraq announces it is repulsing Iranian forces along Karun River
May 10	Heavy fighting takes place in and around Khurramshahr, despite Iranian claims to have taken the city; Iraq announces that 30,000 Iraqi troops have retreated to the Shatt al-'Arab
May 19	Iraq attacks Iranian positions in Khurramshahr but is forced to withdraw
May 24	Iran claims to have retaken Khurramshahr, along with 30,000 Iraqi prisoners
May 30	Iraq bombs main Iranian oil refinery on Kharg Island, in retaliation for Iranian shelling of Iraqi civilian areas
June 5	As its conditions for ending the war, Iran demands $150 billion in reparations and the trial of Iraqi President Saddam Husayn as a war criminal
June 10	Iraq offers to withdraw its forces from Iran if Iran honors a cease-fire and does not cross Iraqi borders
June 22	Iran rejects Iraqi withdrawal as sole condition for ending fighting
June 30	Iraq withdraws last of its troops from Iranian territory
July 14	Iran invades Iraqi territory near Basra
July 15	Iranian forces penetrate 10 miles into Iraq and are reportedly headed for Basra
July 16	Heavy fighting continues in north near Qasr-i Shirin and near Basra; Iraq claims to have killed 6,070 Iranians; U.S. State Department announces that U.S. would be prepared to hold joint maneuvers with Gulf states if requested owing to the "potentially serious consequences" of the Iran-Iraq war
July 20	Iranian advance is brought to a stop; both sides suffer heavy casualties but U.S. analysts claim that Iraq has gained upper hand in conflict
July 21	Iranian jets bomb Baghdad

1982	
July 26	Iran continues to shell Basra
July 29	Iran claims to have advanced 8 miles towards Baghdad; Iraq denies claim
July 31	Baghdad announces that Iraqi forces have killed 27,255 Iranians in past 18 days of fighting around Basra
Aug. 3	Prime Minister Ulusu of Turkey announces that he will fly to Baghdad and Tehran in attempt at mediation
Aug. 7	Iraq announces that it has completely withdrawn from Iranian soil; Iran reports that 2 Iraqi outposts near Qasr-i Shirin have been captured and 200 Iraqi troops killed
Aug. 15	President Husayn warns that Iraq will bomb "vital targets" in Iran, including oil installations unless Iran agrees to halt the fighting
Aug. 18	Iraq reportedly shells Iranian oil terminals on Kharg Island
Aug. 25	Iraq again claims to have shelled Kharg Island, and nearby tankers refuse to approach the island
Sep. 5	Iraq claims to have sunk four ships near Kharg Island oil terminal; Iranian forces retaliate with rocket attack on Iraq's offshore terminal at Khawr al-Umayya
Sep. 13	Iraq claims to have sunk 2 Iranian naval vessels in the Gulf
Sep. 16	Iran claims to have shot down an Iraqi MiG-25 in the first indication of Iraq's possession of such aircraft
Oct. 1	Iran claims to have recaptured 60 square miles of its territory near Sumar; the heaviest fighting since the 13 July Iranian offensive takes place 100 miles northeast of Baghdad
Oct. 4	Sudan announces it will send ground forces to Iraq to help repel the invasion; Iraq asserts that 20,000 volunteers are fighting with their forces, coming from Morocco, Tunisia, Egypt and Jordan; Iran rejects UN Security Council call for immediate end to hostilities and full withdrawal of forces from all occupied territories
Oct. 6	Iraq claims to have repelled an Iranian invasion near Sumar, killing 2,540 Iranians; Iraq also reported to have sunk 2 Iranian naval vessels near Kharg Island; both sides claim important gains in fierce fighting along Shatt al-'Arab
Oct. 12	Iran begins new assault on Basra, using as many as 100,000 troops
Oct. 14	Iraq claims to have repelled major Iranian offensive in Mandali, killing 2,297 Iranian troops

1982	
Oct. 28	Iran makes claim that Iraqi air attack on Dezful resulted in 21 deaths, mostly women and children
Nov. 3	Iran reports that it has recaptured a 95-square-mile tract of land from Iraq, killing 200 Iraqis and destroying 40 enemy tanks
Nov. 7	Iran drives 6 miles into Iran in the general area of Musian in what appears to be Iran's second major offensive of the year
Nov. 9	Iranian President Ali Khamenei vows that Iranian forces will proceed onward to Baghdad whenever it is deemed necessary
Nov. 14	Military analysts gauge that Iraq's air superiority will almost surely be retained as Baghdad continues to receive MiG-23s and French-built Gazelle helicopter gunships; as of 1 September, Iraq thought to possess 200 usable military aircraft to Iran's 50
Nov. 23	Iraq claims to push back Iranians in major counter-offensive in which 945 Iranians are reportedly killed along central front
Nov. 25	In two battles along the Musian sector of the battle front, Iran claims to have killed 900 Iraqis while Iraq counters with reports of 2,262 Iranian deaths
Dec. 4.	An attempt on the life of Iraqi President Saddam Husayn reported in the town of al-Dujayl; rumors of severe reprisals against the town reportedly prompt; Saudi Arabia and Kuwait to withhold $6.5 billion in economic aid to Iraq
Dec. 10	Iraq reports that 3 Iranian naval vessels are destroyed near Iranian port of Bandar Khomeini; Lloyd's of London confirms the destruction of 2 Iranian tankers lying offshore of Kharg Island after Iraqi attack

Bibliographic Sources for this Appendix:
45, 60, 69, 78, 81, 117, 118, 143, 144, 192, 226, 227, 228, 247, 271, 306, 310, 312, 348, 389.

APPENDIX G

The Rapid Deployment Force (RDF)

MISSION

Originated in 1977 directive for a "quick-reaction" force to be deployed on short notice to any crisis area outside NATO and Korea but with emphasis on Southwest Asia. Events in Iran and Afghanistan in 1979 and the enunciation of the Carter Doctrine in 1980 served as an impetus to greatly expand both the size and the role of the force. Measures were taken at this time to increase the speed with which the force could be deployed, to establish a new command headquarters and to secure access to military facilities in the region surrounding the Gulf. The purpose of the force, as stated by its first commander, was

> to plan the employment of designated forces, to jointly train and exercise them, and to ultimately deploy and employ them in response to contingencies threatening U.S. interests anywhere in the world. In essence, to provide the essential command and control that will bring together, in a synergistic way, the capabilities of our four services.[1]

COMPONENTS

The Rapid Deployment Joint Task Force (RDJTF) was established on 1 March 1980 with headquarters at MacDill Air Force Base, Tampa, Florida. Until October 1981, the force was subordinate during peacetime to the U.S. Readiness Command (REDCOM), also headquartered at MacDill, which was responsible for planning and training. Since then, it has been designated as a separate Joint Task Force, with the RDJTF commander reporting directly to the Secretary of Defense through the Joint Chiefs of Staff. Necessary manpower was to come from a reservoir of some 200,000 troops drawn from all four services. Ground forces would be drawn from army and marine units with the navy and air force providing support or air

1. Lt. Gen. P. X. Kelley, quoted in Jeffrey Record, *The Rapid Deployment Force and U.S. Military Intervention in the Persian Gulf* (Cambridge, Mass.: Institute for Foreign Policy Analysis, February 1981), p. 52.

cover functions. Units tentatively assigned to the RDJTF included the following:[2]

Army

82nd Airborne Division
101st Air Assault Division
24th Mechanized Infantry Division
6th Combat Brigade (Air Cavalry)
Various Ranger and Special Forces units

Air Force

1st Tactical Fighter Wing (F-15)
27th Tactical Fighter Wing (F-111)
347th Tactical Fighter Wing (F-4)
354th Tactical Fighter Wing (A-10)
366th Tactical Fighter Wing (F-111)
552nd Airborne Warning and Control Wing (E-3A/AWACS)
150th Tactical Fighter Group, Air National Guard (A-7)
121st Tactical Fighter Wing, Air National Guard (A-7)
Various Reconnaissance, Strategic and Tactical Airlift Squadrons

Navy

3 Aircraft Carrier Battle Groups
1 Surface Action Group
5 Squadrons of Antisubmarine Warfare Patrol Aircraft (P-3)
1 Amphibious Ready Group (ARG)
13 ENTPS Ships (NTPF)

Marine Corps

1 Marine Amphibious Force (Air-Ground Task Force)
7th Marine Amphibious Brigade (ENTPS MAB)

The amount of time necessary to deploy the force has been cut by maritime prepositioning of equipment. In July 1980, a Near-Term Prepositioned Ship (NTPS) force was deployed in the Indian Ocean, consisting of seven ships based on Diego Garcia Island with sufficient equipment and supplies to support one Marine Amphibious Brigade of approximately 12,000 men and several Air Force fighter squadrons for a period of 15 days. Plans called for the upgrading of prepositioning capabilities by 1987 to 12 specially designed Maritime Prepositioning Ships (MPS) providing equipment and supplies for a full Marine Corps division for a period of 30 days.

2. Maxwell Orme Johnson, *The Military as an Instrument of U.S. Policy in Southwest Asia: The Rapid Deployment Joint Task Force, 1979–1982* (Boulder, Colo.: Westview Press, 1983), p. 103.

REGIONAL FACILITIES

Agreement has been reached with three countries in the western Indian Ocean region for limited access of U.S. forces to, and the prepositioning of some equipment at, the following military facilities:

Kenya	Mombasa	port
	Nairobi	airfield
	Nanyuki	airfield
Oman	al-Masira	airfield
	Muscat	port
	Salala	airfield
	al-Sib	airfield
	Thamarit	airfield
Somalia	Berbera	airfield and port
	Mogadishu	airfield and port

Factilities may also be available for use in Egypt and Israel, in addition to the permanent base on the central Indian Ocean island of Diego Garcia.

OPERATION BRIGHT STAR

A series of military manuevers held in Egypt, Sudan, Somalia and Oman in late 1981 and designed to test the RDF's ability to project limited power in the Gulf region. The exercise involved 6000 American troops in conjunction with forces from the host countries and cost $69,000,000. A smaller exercise was carried out by American forces in Oman in late 1982.

U.S. CENTRAL COMMAND

The RDF was redesignated a "unified command" on 1 January 1983. As the U.S. Central Command (USCENTCOM), its specified theater of operation includes Southwest Asia and Northeast Africa, and its commander has equal standing with the commanders of all American forces in Europe or the Pacific. As a theater command, it has been given reponsibility for essentially all U.S. military activity within this region, including military planning, exercises involving U.S. and regional forces, administration of security assistance, and other representational activities. Full deployment could involve as many as 300,000 personnel, drawn from the units listed above. Headquarters for USCENTCOM remain at MacDill A.F.B., Florida.

Bibliographic Sources for this Appendix:
50, 94, 97, 147, 155, 167, 181, 266, 267, 268, 331, 334, 376.

APPENDIX H

The Middle Eastern Oil Producing States: Basic Data

ALGERIA

Legal Name: Democratic and Popular Republic of Algeria
Capital: Algiers
Size: 2,460,500 km^2
Population: 20,030,000 (July 1982)
Ethnic Divisions: 99% Arab-Berbers, less than 1% Europeans
Type of State: Republic
Government Leader: Col. Chadli Benjadid; President
GDP: $41.0 billion (1981 est.); $1720 per capita

BAHRAIN

Legal Name: State of Bahrain
Capital: al-Manama
Size: 596 km^2 plus group of 32 smaller islands
Population: 380,000 (July 1982)
Ethnic Divisions: 63% Bahraini, 10% other Arab, 13% Asian, 8% Iranian, 6% other
Type of State: Amirate (monarchy)
Government Leader: 'Isa ibn Salman Al Khalifa; Amir
GDP: $1.8 billion (1980 est.); $4600 per capita

IRAN

Legal Name: Islamic Republic of Iran
Capital: Tehran
Size: 1,647,240 km^2
Population: 41,203,000 (July 1982)
Ethnic Divisions: 63% ethnic Persians, 3% Kurds, 13% other Iranian, 18% Turkic, 3% Arab and other Semitic, 1% other
Type of State: Republic
Government Leader: Ayatollah Rouhollah Khomeini; *faqih* or leader; 'Ali

Khamene'i; President
GNP: $81.7 billion (1979); $2170 per capita

IRAQ

Legal Name: Republic of Iraq
Capital: Baghdad
Size: 445,480 km^2
Population: 14,034,000 (July 1982)
Ethnic Divisions: 70.9% Arabs, 18.3% Kurds, 0.7% Assyrians, 2.4% Turkomans, 7.7% other
Type of State: Republic
Government Leader: Saddam Husayn; President
GNP: $35.2 billion (1979 est.); $2730 per capita

KUWAIT

Legal Name: State of Kuwait
Capital: Kuwait
Size: 16,058 km^2 (excluding Neutral Zone)
Population: 1,553,000 (July 1982)
Ethnic Divisions: 42% Kuwaitis, 41% other Arabs, 7% South Asians, 4% Iranians, 6% other
Type of State: Amirate (monarchy)
Government Leader: Jabir Al Ahmad Al Sabah; Amir
GDP: $27.2 billion (1980); $18,390 per capita

LIBYA

Legal Name: Socialist People's Libyan Arab Jamahiriya
Capital: Tripoli
Size: 1,758,610 km^2
Population: 3,425,000 (July 1982)
Ethnic Divisions: 97% Berber and Arab with some Negro stock; some Greeks, Maltese, Jews, Italians, Egyptians, Pakistanis, Turks, Indians, and Tunisians
Type of State: Republic
Government Leader: Col. Mu'ammar al-Qadhdhafi; no official title
GDP: roughly $24.5 billion (1981 est.); $6960 per capita

OMAN

Legal Name: Sultanate of Oman
Capital: Muscat
Size: 212,380 km^2
Population: 948,000 (July 1982)

Ethnic Divisions: Almost entirely Arab with small groups of Iranians, Baluchis, and Indians
Type of State: Sultanate (monarchy)
Government Leader: Qabus ibn Sa'id Al Bu Sa'id; Sultan
GNP: $5.2 billion (1980); $5780 per capita

QATAR

Legal Name: State of Qatar
Capital: al-Doha
Size: 10,360 km^2
Population: 258,000 (July 1982)
Ethnic Divisions: 25% Qataris, 20% other Arabs, 34% South Asians, 16% Iranians, 5% others
Type of State: Amirate (monarchy)
Government Leader: Khalifa ibn Hamad Al Thani; Amir
GNP: $5.0 billion (1979); $29,900 plus per capita

SAUDI ARABIA

Legal Name: Kingdom of Saudi Arabia
Capital: al-Riyadh
Size: 2,331,000 km^2
Population: 10,112,000 (January 1981)
Ethnic Divisions: 90% Arab, 10% Afro-Asian
Type of State: monarchy
Government Leader: Fahd ibn 'Abd al-'Aziz Al Sa'ud; King
GDP: $77 billion (1979 est.); $115 billion (FY 1980 est.); $9500 per capita

UNITED ARAB EMIRATES (UAE)

Legal Name: United Arab Emirates
Capital: Abu Dhabi
Size: 82,880 km^2
Population: 1,240,000 (July 1982)
Ethnic Divisions: Emirians 19%, other Arabs 23%, South Asians 50% (fluctuating), other expatriates (includes Westerners and East Asians) 8%
Type of State: Federation of amirates; members are Abu Dhabi, 'Ajman, Dubai, al-Fujayra, Ra's al-Khayma, Sharjah, and Umm al-Qaywayn
Government Leader: Zayid ibn Sultan Al Nuhayyan; President (and Amir of Abu Dhabi)
GNP: $30 billion (1980); $32,000 per capita

Note: Population figures are projected estimates for July 1982.
Source: U.S. Central Intelligence Agency, National Foreign Assessment Center, *The World Factbook—1982* (Washington: USGPO, April 1982).

BIBLIOGRAPHY

1. Abdullah, Muhammad Morsy. *The United Arab Emirates: A Modern History*. London: Croom Helm; New York: Barnes & Noble, 1978.
2. Abir, Mordechai. *Oil, Power, and Politics*. London: Frank Cass, 1975.
3. Abrahamian, E. "Structural Causes of the Iranian Revolution." *MERIP Reports*, Vol. 10, No. 4 (May 1980), pp. 21–26.
4. Abu 'Adhirah, Sa'id Salman. "Sedentarisation and Settlement of the Bedouin." *Arabian Studies*, Vol. 4 (1978), pp. 1–5.
5. Abu al Khail, Mohammed. "The Oil Price in Perspective." *International Affairs* (London), Vol. 55, No. 4 (October 1979), pp. 517–530.
6. Ahrari, Mohammed. "OAPEC and 'Authoritative' Allocation of Oil: An Analysis of the Arab Oil Embargo." *Studies in Comparative International Development*, Vol. 14, No. 1 (1979), pp. 9–21.
7. Ajomo, M. A. "An Appraisal of the Organization of the Petroleum Exporting Countries." *Texas International Law Journal*, Vol. 13, No. 11 (1977), pp. 11–34.
8. Akehurst, John. *We Won a War: The Campaign in Oman, 1965–1975*. London: Michael Russell, 1983.
9. Akins, James. "Politics and Saudi Oil Policy." *Middle East Economic Survey*, Vol. 24, No. 52 (October 12, 1981), pp. i–vii.
10. Alaolmolki, Nozar. "Revolution and Legitimacy: The Islamic Republic of Iran." *Australian Outlook*, Vol. 36 (1982), pp. 17–22.
11. Alexander, Nathan. "The Foreign Policy of Libya: Inflexibility and Change." *Orbis*, Vol. 24, No. 4 (1981), pp. 819–846.
12. ———. "Libya: The Continuous Revolution." *Middle Eastern Studies*, Vol. 17, No. 2 (April 1981), pp. 210–227.
13. Alford, Jonathan. "Les Occidentaux et la securité du Golfe." *Politique Étrangère*. Vol. 46, No. 3 (September 1981), pp. 677–690.
14. Allan, J. A. *Libya: The Experience of Oil*. London: Croom Helm; Boulder, Colo.: Westview Press, 1981.
15. Allen, Loring. "Not So Wild a Dream: OPEC's Amazing Rise to World Economic Power." *Harvard Magazine*, Vol. 80, No. 5 (May–June 1978), pp. 22–28.
16. Amin, S. H. "The Iran-Iraq Conflict: Legal Implications." *International and Comparative Law Quarterly*, Vol. 31 (January 1982), pp. 167–188.
17. Amirsadeghi, Hossein, ed. *The Security of the Persian Gulf*. London: Croom Helm; New York: St. Martin's Press, 1981. Contents include:
   - A. John C. Campbell, "The Gulf Region in the Global Setting," pp. 1–25;
   - B. Geoffrey Kemp, "The Strategic Balance and the Control of the Persian Gulf," pp. 26–42;
   - C. Shahram Chubin, "The Soviet Union and the Persian Gulf," pp. 45–56;
   - D. C. D. Carr, "The United States-Iranian Relationship 1948–1978: A Study in Reverse Influence," pp. 57–84;
   - E. Lewis A. Dunn, "Persian Gulf Nuclearisation: Prospects and Implications," pp. 85–99;

F. Steven L. Canby, "The Iranian Military: Political Symbolism versus Military Usefulness," pp. 100–130;
G. Stephanie G. Neuman, "Arms Transfers, Indigenous Defence Production and Dependency: The Case of Iran," pp. 131–150;
H. Richard Haass, "Saudi Arabia and Iran: The Twin Pillars in Revolutionary Times," pp. 151–169;
I. John Duke Anthony, "The Persian Gulf in Regional and International Politics: The Arab Side of the Gulf," pp. 170–196;
J. Edmund Ghareeb, "Iraq: Emergent Gulf Power," pp. 197–230;
K. L. P. Elwell-Sutton, "The Iranian Revolution: Triumph or Tragedy?" pp. 231–254;
L. Fereidun Fesharaki, "Revolution and Energy Policy in Iran: International and Domestic Implications," pp. 255–280.

18. Amuzegar, Jahangir. "Oil Wealth: A Very Mixed Blessing." *Foreign Affairs,* Vol. 60, No. 4 (1982), pp. 814–835.
19. ———. "A World Without OPEC." *Washington Quarterly,* Vol. 5 (1982), pp. 60–70.
20. Anderson, Lisa. "Libya and American Foreign Policy." *Middle East Journal,* Vol. 36, No. 4 (1982), pp. 516–534.
21. ———. "Religion and Politics in Libya." *Journal of Arab Affairs,* Vol. 1, No. 1 (October 1981), pp. 53–77.
22. Anthony, John Duke. *Arab States of the Lower Gulf: People, Politics, Petroleum.* Washington: Middle East Institute, 1975.
23. ———, ed. *The Middle East: Oil, Politics and Development.* Washington: American Enterprise Institute for Public Policy Research, 1975. Contents include:

A. Edith Penrose, "International Oil Companies and Governments in the Middle East," pp. 3–19;
B. Ronald S. Ritchie, "The Impact of High Oil Prices on the Consuming Nations," pp. 29–35;
C. Yusif A. Sayigh, "Oil in Arab Developmental and Political Strategy: An Arab View," pp. 37–44.
D. George Lenczowski, "The Middle East: Politics and the Energy Crisis," pp. 55–64;
E. Muhammed Loutfi, "Prospects for Development and Investment for Oil-Producing Countries," pp. 67–78;
F. John Duke Anthony, "The Impact of Oil on Political and Socioeconomic Change in the United Arab Emirates," pp. 79–98;
G. Edith Penrose, "Afterword: Oil Revenues—Hopes and Caveats," pp. 99–101.

24. Arjomand, Said Amir. "Shi'ite Islam and the Revolution in Iran." *Government and Opposition,* Vol. 16, No. 3 (1981), pp. 293–316.
25. Askari, Hossein, John T. Cummings, and Howard C. Reed. "The Gulf: Gold Rush or Economic Development?" *Journal of Arab Affairs,* Vol. 1, No. 2 (April 1982), pp. 263–282.
26. Atkin, Muriel. "The Kremlin and Khomeini." *Washington Quarterly,* Vol. 4, No. 2 (1981), pp. 50–67.
27. Atta, Jacob K. "The Oil Policies of Kuwait: An Outsider's Interpretation." *Journal of Energy and Development,* Vol. 6, No. 1 (1980), pp. 153–164.
28. Attiga, Ali. "The Economic Development of the Oil Exporting Countries." *Middle East Economic Survey,* Vol. 24, No. 52 (12 October 1981), supplement pp. 1–5.

29. ———. "The Future for Oil: A View from OAPEC." *Arab Perspectives,* Vol. 2 (September 1981), pp. 31–40.
30. Ayoob, Mohammed. "Blueprint for a Catastrophe: Conducting Oil Diplomacy by 'Other Means' in the Middle East and the Persian Gulf." *Australian Outlook,* Vol. 33, No. 3 (December 1979), pp. 265–273.
31. ———. "The Superpowers and Regional 'Stability': Parallel Responses to the Gulf and the Horn." *The World Today,* Vol. 35, No. 5 (May 1979), pp. 197–205.
32. Ayubi, Nazih. "OPEC Surplus Funds and Third World Development: The Egyptian Case." *Journal of South Asian and Middle Eastern Studies,* Vol. 5, No. 4 (1982), pp. 39–56.
33. Bahry, Louay. "The New Saudi Woman: Modernizing in an Islamic Framework." *Middle East Journal,* Vol. 36, No. 4 (1982), pp. 502–515.
34. Ball, Nicole, and Milton Leitenberg. "The Iranian Domestic Crisis: Foreign Policy Goals of the United States." *Journal of South Asian and Middle Eastern Studies,* Vol. 2, No. 3 (1979), pp. 36–56.
35. Bayat, Mangol. "The Iranian Revolution of 1978–79: Fundamentalist or Modern?" *Middle East Journal,* Vol. 37, No. 1 (1983), pp. 30–42.
36. ———. "Islam in Pahlavi and Post-Pahlavi Iran: A Cultural Revolution?" In John L. Esposito, ed., *Islam and Development: Religion and Sociopolitical Change* (Syracuse: Syracuse University Press, 1980), pp. 87–106.
37. Beling, Willard A., ed. *King Faisal and the Modernisation of Saudi Arabia.* London: Croom Helm; Boulder, Colo.: Westview Press, 1980. Contents include:

Part I: The Creation of the Saudi State

A. George Rentz, "The Saudi Monarchy," pp. 15–34;
B. Ralph Braibanti, "Saudi Arabia in the Context of Political Development Theory," pp. 35–57;
C. Fouad Abdul-Salam Al-Farsy, "King Faisal and the First Five Year Development Plan," pp. 58–71;

Part II: The Modernisation Process

D. Abdulrahman M. Al-Sadhan, "The Modernisation of the Saudi Bureaucracy," pp. 75–89;
E. Federico S. Vidal, "Development of the Eastern Province: A Case Study of al-Hasa Oasis," pp. 90–101;
F. John Duke Anthony, "The US-Saudi Arabian Joint Commission on Economic Co-operation," pp. 102–109;

Part III: Social Change and Traditional Islam

G. George M. Baroody, "The Practice of Law in Saudi Arabia," pp. 113–124;
H. William A. Rugh, "Saudi Mass Media and Society in the Faisal Era," pp. 125–144;
I. Catherine Parssinen, "The Changing Role of Women," pp. 145–170;

Part IV: The Islamic Foundations of Saudi Foreign Policy

J. David E. Long, "King Faisal's World View," pp. 173–183;
K. Abdullah M. Sindi, "King Faisal and Pan-Islamism," pp. 184–201;
L. Abdulaziz H. Al-Sowayyegh, "Saudi Oil Policy During King Faisal's Era," pp. 202–229;
M. Malcolm C. Peck, "The Saudi-American Relationship and King Faisal," pp. 230–247.

38. Bill, James A. "The Politics of Extremism in Iran." *Current History,* Vol. 81, No. 471 (January 1982), pp. 9–13, 36.
39. ———. "Power and Religion in Revolutionary Iran." *Middle East Journal,* Vol. 36, No. 1 (1982), pp. 22–47.

40. Binder, Leonard. "United States Policy in the Middle East: Toward a Pax Saudiana." *Current History,* Vol. 81, No. 471 (January 1982), pp. 1–4, 40–42, 48.

41. Bonine, Michael E., and Nikki Keddie, eds. *Modern Iran: The Dialectics of Continuity and Change.* Albany, N.Y.: State University of New York Press, 1981. Contents include:

I. Overviews in Time

A. Nikki Keddie, "Religion, Society, and Revolution in Modern Iran," pp. 19–34;
B. Mangol Bayat-Phillip, "Tradition and Change in Iranian Socio-Religious Thought," pp. 35–56;
C. Byron J. Good, "The Transformation of Health Care in Modern Iranian History," pp. 57–84;

II. Nomads and Agriculturalists

A. Lois Beck, "Economic Transformations Among the Qashqai Nomads, 1962–1978," pp. 85–108;
B. Paul Barker, "Tent Schools of the Qashqai: A Paradox of Local Initiative and State Control," pp. 109–128;
C. Gene R. Garthwaite, "Khans and Kings: The Dialectics of Power in Bakhtiari History," pp. 129–142;
D. Roger T. Olson, "Persian Gulf Trade and the Agricultural Economy of Southern Iran in the Nineteenth Century," pp. 143–160;
E. Eric J. Hooglund, "Rural Socioeconomic Organization in Transition: The Case of Iran's Bonehs," pp. 161–180;

III. Urban Groups and Classes

A. Ervand Abrahamian, "The Strengths and Weaknesses of the Labor Movement in Iran, 1941–1953," pp. 181–202;
B. Michael E. Bonine, "Shops and Shopkeepers: Dynamics of an Iranian Provincial Bazaar," pp. 203–228;
C. Mary-Jo Delvecchio Good, "The Changing Status and Composition of an Iranian Provincial Elite," pp. 229–248;
D. William R. Royce, "The Shirazi Provincial Elite: Status Maintenance," pp. 249–264;

IV. Aspects of Culture

A. Hamid Naficy, "Cinema as a Political Instrument," pp. 265–284;
B. William O. Beeman, "A Full Arena: The Development and Meaning of Popular Performance Traditions in Iran," pp. 285–306.

42. Bonnenfant, Paul, ed. *La péninsule Arabique d'aujourd'hui.* Paris: Editions du Centre national de la recherche scientifique, 1982. Volumes I and II. For Centre d'études et de recherches sur l'Orient arabe contemporain.

43. Borg, I. Y. "Current Developments Affecting Future Availability of Oil and Gas in the Free World." Livermore, Cal.: University of California, Lawrence Livermore Laboratory, 17 March 1981. Prepared for U.S. Department of Energy, UCRL-53129.

44. Bradley, C. Paul. *Recent United States Policy in the Persian Gulf.* Hamden, Conn.: Shoe String Press, 1982.

45. Brett-Crowther, M. R. "Iran and Iraq at War: The Effect on Development." *Round Table,* No. 281 (January 1981), pp. 61–69.

46. Brodman, John, and Jack Moore. "The Outlook for World Oil Supply and Demand Through 1983." *Journal of Energy and Development,* Vol. 7, No. 1 (1981), pp. 1–12.

47. Brown, Harold. "U.S. Security Policy in Southwest Asia: A Case Study in Complexity." Washington: Johns Hopkins University, School of Advanced International Studies, Johns Hopkins Foreign Policy Institute, 1981. Occasional Paper.
48. Brown, William R. "The Oil Weapon." *Middle East Journal,* Vol. 36, No. 3 (1982), pp. 301–318.
49. Burki, Shahid Javed. "What Migration to the Middle East May Mean for Pakistan." *Journal of South Asian and Middle Eastern Studies,* Vol. 3, No. 3 (1980), pp. 47–66.
50. Burruss, Mildred. *Rapid Deployment Force: A Selected Bibliography.* Washington: National Defense University Library, 1982. Bibliography 82–1.
51. Campbell, W. R., and Djamchid Darvich. "Global Implications of the Islamic Revolution for the Status Quo in the Persian Gulf." *Journal of South Asian and Middle Eastern Studies.* Vol. 5, No. 1 (1981), pp. 31–51.
52. Carnegie Panel on U.S. Security and the Future of Arms Control. *Challenges for U.S. National Security—Assessing the Balance: Defense Spending and Conventional Forces.* A Preliminary Report, Part II. Washington: Carnegie Endowment for International Peace, 1981. Includes section on "The Military Balance in the Persian Gulf," pp. 149–194.
53. Carswell, Robert. "Economic Sanctions and the Iran Experience." *Foreign Affairs,* Vol. 60, No. 2 (1981–1982), pp. 247–265.
54. Cattan, Henry. *The Evolution of Oil Concessions in the Middle East and North Africa.* Dobbs Ferry, NY: Oceana, 1967. Published for the Parker School of Foreign and Comparative Law, Columbia University.
55. Chenery, Hollis B. "Restructuring the World Economy: Round II." *Foreign Affairs,* Vol. 59, No. 5 (1981), pp. 1102–1120.
56. Chibwe, Ephraim C. *Afro-Arab Relations in the New World Order.* London: Croom Helm; New York: St. Martin's Press, 1978.
57. Choucri, Nazli. "OPEC: Calming a Nervous World Oil Market," *OPEC Bulletin,* Vol. 8, No. 2 (March 1982), pp. 13–27.
58. ———. "Power and Politics in World Oil." *Technology Review,* Vol. 85, No. 7 (October 1982), pp. 24–36.
59. Chubin, Shahram. "Gains for Soviet Policy in the Middle East." *International Security,* Vol. 6, No. 4 (1982), pp. 122–152.
60. ———. "La guerre irano-irakienne: Paradoxes et particularités." *Politique Étrangère,* Vol. 47, No. 2 (June 1982), pp. 381–394.
61. ———. "Leftist Forces in Iran." *Problems of Communism,* Vol. 29, No. 4 (July–August 1980), pp. 1–25.
62. ———, ed. *Security in the Persian Gulf, Vol. 1: Domestic Political Factors.* London: International Institute for Strategic Studies; Montclair, New Jersey: Allanheld, Osmun, 1981.
63. ———. "Soviet Policy Towards Iran and the Gulf." *Adelphi Papers,* No. 157. London: International Institute for Strategic Studies, 1980.
64. ———. "U.S. Security Interests in the Persian Gulf in the 1980s." *Daedalus,* Vol. 109, No. 4 (1980), pp. 31–65.
65. Clarke, John I., and Howard Bowen-Jones, eds. *Change and Development in the Middle East: Essays in Honour of W. B. Fisher.* London: Methuen, 1981. Contents include:
    A. J. Stace Birks, "The Impact of Economic Development on Pastoral Nomadism in the Middle East: An Inevitable Eclipse?" pp. 82–94;
    B. Keith McLachlan, "The Oil Industry in the Middle East," pp. 95–112;

C. Gerald H. Blake, "Offshore Politics and Resources in the Middle East," pp. 113–129;
D. William J. Donaldson, "Fisheries of the Arabian Peninsula," pp. 189–198;
E. Roderic W. Dutton, "A Rural Community Development Project in Oman," pp. 199–212;
F. Mohammed el Mehdawi, "The Industrialization of Libya," pp. 252–263;
G. Brian D. Clark, "Urban Planning in Iran," pp. 280–288;
H. Nassir A. Saleh, "Provincial and District Delimitation in the Kingdom of Saudi Arabia," pp. 305–317.

66. Collins, John M., and Clyde R. Mark. "Petroleum Imports from the Persian Gulf: Use of U.S. Armed Force to Ensure Supplies." Washington: Library of Congress, Congressional Research Service, 1979.
67. Conant, Melvin A. *The Oil Factor: U.S. Foreign Policy 1980–1990.* Lexington, Mass.: Lexington Books/D.C. Heath, 1981.
68. Cooley, John K. "The Libyan Menace." *Foreign Policy,* No. 42 (1981), pp. 74–93.
69. Cottam, Richard W. "Revolutionary Iran and the War with Iraq." *Current History,* Vol. 80, No. 462 (January 1981), pp. 5–9, 38–42.
70. ———, David Schoenbaum, Shahram Chubin, Theodore H. Moran, and Richard A. Falk. "The United States and Iran's Revolution." *Foreign Policy,* No. 34 (1979), pp. 3–34.
71. Cottrell, Alvin J., and Frank Bray. *Military Forces in the Persian Gulf.* Beverly Hills and London: Sage Publications, 1978. The Washington Papers, v. VI, 60.
72. ———, gen. ed.; C. Edmund Bosworth, R. Michael Burrell, Keith McLachlan and Roger M. Savory, eds. *The Persian Gulf States: A General Survey.* Baltimore: Johns Hopkins University Press, 1980. Contents include:
A. C. Edmund Bosworth, "The Nomenclature of the Persian Gulf," pp. xvii–xxxiv;

Part I: The History of the Persian Gulf

B. Roger M. Savory, "The Ancient Period," pp. 3–13;
C. *Idem,* "A.D. 600–1800," pp. 14–40;
D. Malcolm Yapp, "The Nineteenth and Twentieth Centuries," pp. 41–69;
E. *Idem,* "British Policy in the Persian Gulf," pp. 70–100;
F. Gerald Blake, Heather Bleaney, David Imrie and Richard Lawless, "Geographers of the Persian Gulf," pp. 101–120;
G. R. Michael Burrell and Keith McLachlan, "The Political Geography of the Persian Gulf," pp. 121–139;
H. Alvin J. Cottrell, Robert J. Hanks and Frank T. Bray, "Military Affairs in the Persian Gulf," pp. 140–171;
I. Ralph H. Magnus, "International Organizations in the Persian Gulf," pp. 172–191;

Part II: Economics and Urban Development in the Persian Gulf

J. Keith McLachlan, "Oil in the Persian Gulf Area," pp. 195–224;
K. Michael E. Bonine, "The Urbanization of the Persian Gulf Nations," pp. 225–278;

Part III: Culture: Religion, Language and Literature

L. James E. Dougherty, "Religion and Law," pp. 281–313;
M. Bruce Ingham, "Languages of the Persian Gulf," pp. 314–333;
N. G. M. Wickens, "Literature in Arabic," pp. 334–353;
O. Robert M. Rehder, "Persian Literature," pp. 354–366;

Part IV: Arts and Society in the Persian Gulf

P. Ralph H. Magnus, "Societies and Social Change in the Persian Gulf," pp. 369–413;

Q. Robert Hillenbrand, "Art in the Persian Gulf," pp. 414–484;
R. Brian D. Clark, "Tribes of the Persian Gulf," pp. 485–509;
S. Michael M. J. Fischer, "Competing Ideologies and Social Structure in the Persian Gulf," pp. 510–538.

73. Cummings, John Thomas, Hossein G. Askari, and Michael Skinner. "Military Expenditures and Manpower Requirements in the Arabian Peninsula." *Arab Studies Quarterly,* Vol. 2, No. 1 (1980), pp. 38–49.
74. Dafter, Ray. "World Oil Production and Security of Supplies." *International Security,* Vol. 4, No. 3 (1979–1980), pp. 154–176.
75. Davis, Eric. "The Political Economy of the Arab Oil-Producing Nations: Convergence with Western Interests." *Studies in Comparative International Development,* Vol. 14, No. 2 (1979), pp. 75–94.
76. Davis, M. Thomas. "The Politics of Begin's Baghdad Raid." *Naval War College Review,* Vol. 35, No. 2 (March–April 1982), pp. 33–40.
77. Dawisha, Adeed I. "Internal Values and External Threats: The Making of Saudi Foreign Policy." *Orbis,* Vol. 23, No. 1 (1979), pp. 129–143.
78. ———. "Iraq and the Arab World: The Gulf War and After." *The World Today,* Vol. 37, No. 5 (May 1981), pp. 188–194.
79. ———. "Iraq: The West's Opportunity." *Foreign Policy,* No. 41 (1980–1981), pp. 134–153.
80. ———. "Saudi Arabia's Search for Security." *Adelphi Papers,* No. 158. London: International Institute for Strategic Studies, 1979–1980.
81. Dawisha, Karen. "Moscow and the Gulf War." *The World Today,* Vol. 37, No. 1 (January 1981), pp. 8–14.
82. ———. "Moscow's Moves in the Direction of the Gulf—So Near Yet So Far." *Journal of International Affairs,* Vol. 34, No. 2 (1980–81), pp. 219–234.
83. ———. "Soviet Decision-Making and the Middle East: The 1973 October War and the 1980 Gulf War." *International Affairs* (London), Vol. 57, No. 1 (1980–1981), pp. 43–59.
84. Deeb, Marius K. "Islam and Arab Nationalism in al-Qaddhafi's Ideology." *Journal of South Asian and Middle Eastern Studies,* Vol. 2, No. 2 (1978), pp. 12–26.
85. Deese, David A., and Joseph S. Nye, eds. *Energy and Security.* Cambridge, Mass.: Ballinger Publishing, 1981. A Report of Harvard's Energy and Security Research Project. Contents include:

A. Joseph S. Nye, "Energy and Security," pp. 3–22;
B. Thomas L. Neff, "The Changing World Oil Market," pp. 23–48;
C. Gary Samore, "The Persian Gulf," pp. 49–110;
D. Marshall I. Goldman, "The Role of Communist Countries," pp. 111–130;
E. Fen Hampson and Kevin J. Middlebrook, "Energy Security in North America," pp. 131–180;
F. David A. Deese and Linda B. Miller, "Western Europe," pp. 181–210;
G. Joseph S. Nye, "Japan," pp. 211–228;
H. David A. Deese, "The Oil-Importing Developing Countries," pp. 229–260;
I. William W. Hogan, "Import Management and Oil Emergencies," pp. 261–302;
J. Alvin A. Alm, E. William Colglazier, and Barbara Kates-Garnick, "Coping with Interruptions," pp. 303–346;
K. Philip K. Verleger, Jr., "Financial Implications of Petroleum Disruptions," pp. 347–364;
L. Geoffrey Kemp, "Military Force and Middle East Oil," pp. 365–390;
M. Joseph S. Nye, David A. Deese, and Alvin L. Alm, "Conclusion: A. U.S. Strategy for Energy Security," pp. 391–426.

86. "Defending the Gulf: A Survey." *The Economist,* 6 June 1981, survey pp. 1–38.
87. DeForth, Peter W. "U.S. Naval Presence in the Persian Gulf: The Mideast Force Since World War II." *Naval War College Review,* Vol. 28, No. 1 (1975), pp. 28–38.
88. Demir, Soliman. *Arab Development Funds in the Middle East.* Elmsford, NY: Pergamon Press for UNITAR, 1979.
89. Dhaher, Ahmad J. "Culture and Politics in the Arab Gulf States." *Journal of South Asian and Middle Eastern Studies,* Vol. 4, No. 4 (1981), pp. 21–36.
90. Doran, Charles F. "OPEC Structure and Cohesion: Exploring the Determinants of Cartel Policy." *Journal of Politics,* Vol. 42, No. 1 (February 1980), pp. 82–101.
91. Eftehari, N. "Crise pétrolière et guerre économique." *Peuples Méditerranéens,* No. 19 (April–June 1982), pp. 69–110.
92. Eilts, Hermann F. "Security Considerations in the Persian Gulf." *International Security,* Vol. 5, No. 2 (1980), pp. 79–113. See the response of J. B. Kelly to this article, and Eilts' rejoinder, in Vol. 5, No. 4 (1981), pp. 186–203.
93. Energy Economics Research, Ltd. and Middle East Economic Survey. *International Crude Oil and Product Prices: A Biannaual Review and Analysis of Oil Price Trends in World Markets, July 1982.* Nicosia: Middle East Petroleum and Economic Publications, 1982.
94. Epstein, Joshua M. "Soviet Vulnerabilities in Iran and the RDF Deterrent." *International Security,* Vol. 6, No. 2 (1981), pp. 126–159.
95. Erb, Richard D., guest ed. "The Arab Oil-Producing States of the Gulf: Political and Economic Developments." *AEI Foreign Policy and Defense Review,* Vol. 2, Nos. 3 and 4 (1980), entire issue. Contents include:
    A. Richard D. Erb, "Introduction," pp. 2–4;
    B. *Idem,* "The Gulf Oil Producers: Overview and Oil Policy Implications," pp. 5–13;
    C. Bayly Winder, "Saudi Arabia: Sociopolitical Developments," pp. 14–20;
    D. Richard D. Erb, "Saudi Arabia: Economic Developments," pp. 21–30;
    E. Phebe Marr, "Iraq: Sociopolitical Developments," pp. 30–38.
    F. Basil al-Bustany, "Iraq: Economic Developments," pp. 38–44;
    G. Naseer H. Aruri, "Kuwait: Sociopolitical Developments," pp. 44–50;
    H. Carl Bazarian, "Kuwait: Economic Developments," pp. 50–55;
    I. John Duke Anthony, "United Arab Emirates: Sociopolitical Developments," pp. 56–60;
    J. Carl Bazarian, "United Arab Emirates: Economic Developments," pp. 61–65;
    K. Emile A. Nakhleh, "Bahrain and Qatar: Sociopolitical Developments," pp. 65–67;
    L. Carl Bazarian, "Bahrain and Qatar: Economic Developments," pp. 68–70;
    M. Richard D. Erb and Nebil N. al-Shawaf, "Economic Integration: Nascent Linkages," p. 70–76;
    N. Emile A. Nakhleh, "Political Relations Among the Arab Gulf States: Developments and Prospects," pp. 77–81;
    O. Les Janka, "Security Risks and Reactions," pp. 82–88.
96. Exxon Corporation. Public Affairs Department. *Middle East Oil.* N.p., September 1980. Exxon Background Series.
97. Fabyanic, Thomas A. "Conceptual Planning and the Rapid Deployment Joint Task Force." *Armed Forces and Society,* Vol. 7, No. 3 (1981), pp. 343–365.
98. Farah, Tawfic. "Inculcating Supportative Attitudes in an Emerging State: The

Case of Kuwait." *Journal of South Asian and Middle Eastern Studies,* Vol. 2, No. 4 (1979), pp. 56–68.

99. Farid, Abdul Majid, ed. *Oil and Security in the Arabian Gulf.* London: Croom Helm; New York: St. Martin's Press, 1981. In association with the Arab Research Centre, London. Contents include:
    A. Abdullah Al-Tariki, "What do the Arab Oil Producers Want from the Consumers," pp. 15–17;
    B. Fred Halliday, "The Iranian Revolution in International Affairs: Programme and Practice," pp. 18–35;
    C. Robert J. Hanks, "Oil and Security in US Policy Towards the Arabian Gulf—Indian Ocean Area," pp. 36–56;
    D. Ruben N. Andreasyan, "Oil and Soviet Policy in the Arabian Gulf—Indian Ocean Area," pp. 57–81;
    E. Robert Pranger, "The American Answer," pp. 82–85;
    F. James E. Akins, "American Domestic and Foreign Policy and the Arab-Israeli Conflict," pp. 86–95;
    G. Ahmed Baha El-Din, "An Arab View of Superpower 'Security' in the Gulf," pp. 96–98;
    H. Douglas Hurd, "Britain and the Gulf," pp. 99–101;
    I. Michael Field, "Britain, the EEC and the Middle East," pp. 102–106;
    J. Michel Jobert, "Oil Prospects in the Gulf: The EEC and the Gulf," pp. 107–120;
    K. Pierre Desprairies, "Is a European Policy Possible?" pp. 121–125;
    L. Indu Prakash Singh, "Oil and Great Power Rivalry in the Gulf—Indian Ocean Area: Attitude of Third World Countries," pp. 126–129;
    M. Tim Niblock, "Can the Indian Ocean States Keep Out the Great Powers?" pp. 130–133;
    N. Mohammad El-Rumaihi, "Oil and Security in the Gulf: An Arab Point of View," pp. 134–137;
    O. "The Gulf War," pp. 141–152;
    P. Arab Research Centre, "Conclusion: Analysis of Policy Implications," pp. 153–158.

100. Al-Farsy, Fouad. *Saudi Arabia: A Case Study in Development.* London: Stacey International, 1978.

101. Fatemi, Khosrow. "Leadership by Distrust: The Shah's Modus Operandi." *Middle East Journal,* Vol. 36, No. 1 (1982), pp. 48–61.

102. Fathaly, Omar I., and Fathi S. Abusadra. "The Impact of Socio-political Change on Economic Development in Libya." *Middle Eastern Studies,* Vol. 16, No. 3 (October 1980), pp. 225–235.

103. ———, and Monte Palmer. "Opposition to Change in Rural Libya." *International Journal of Middle East Studies,* Vol. 11, No. 2 (April 1980), pp. 247–261.

104. ———, Monte Palmer and Richard Chackerian. *Political Development and Bureaucracy in Libya.* Lexington, Mass.: Lexington Books, 1977.

105. ———, and Monte Palmer. *Political Development and Social Change in Libya.* Lexington, Mass.: Lexington Books/D.C. Heath, 1980.

106. Fekrat, M. Ali. "OPEC and the International Economic Balance." *Journal of South Asian and Middle Eastern Studies,* Vol. 3, No. 4 (1980), pp. 3–8.

107. ———. "OPEC: Bargaining for the World Product." *Journal of South Asian and Middle Eastern Studies,* Vol. 2, No. 3 (1979), pp. 26–35.

108. Feldman, Shai. "The Bombing of Osiraq—Revisited." *International Security,* Vol. 7, No. 2 (1982), pp. 114–142.

109. Fukuyama, Francis. *The Soviet Union and Iraq Since 1968.* Santa Monica,

Calif.: Rand Corporation, 1980. Note N-1524-AF.

110. Gazit, Shlomo. "Risk, Glory and the Rescue Operation." *International Security,* Vol. 6, No. 1 (1981), pp. 111–135.

111. Gholamnezhad, Abdolhamid. "Critical Choices for OPEC Members and the United States." *Journal of Conflict Resolution,* Vol. 25 (March 1981), pp. 115–143.

112. Gitelson, Susan Aurelia. "Arab Aid to Africa: How Much and at What Price?" *Jerusalem Quarterly,* No. 19 (1981), pp. 120–127.

113. Graham, Helga. *Arabian Time Machine: Self-Portrait of an Oil State.* London: Heinemann, 1977; New York: Holmes & Meier, 1978.

114. Graham, Robert. *Iran: The Illusion of Power.* London: Croom Helm, 1978. 2nd rev. ed.: 1979.

115. Greig, Ian. "The Security of Persian Gulf Oil." *Atlantic Community Quarterly,* Vol. 18, No. 2 (1980), pp. 193–200.

116. Griffin, James M., David J. Teece, *et al. OPEC Behavior and World Oil Prices.* London: George Allen & Unwin, 1982. Contents include:
   A. James M. Griffin and David J. Teece, "Introduction," pp. 1–36;
   B. M. A. Adelman, "OPEC as a Cartel," pp. 37–63;
   C. David J. Teece, "OPEC Behavior: An Alternative View," pp. 64–93;
   D. Theodore Moran, "Modeling OPEC Behavior: Economic and Political Alternatives," pp. 94–130;
   E. John H. Lichtblau, "The Limitations to OPEC's Pricing Policy," pp. 131–144;
   F. George Daly, James M. Griffin and Henry B. Steele, "Recent Price Escalations: Implications for OPEC Stability," pp. 145–174;
   G. Robert S. Pindyck, "OPEC Oil Pricing, and the Implications for Consumers and Producers," pp. 175–185;
   H. William W. Hogan, "Policies for Oil Importers," pp. 186–206;
   I. James M. Griffin and David J. Teece, "Conclusions," pp. 207–219.

117. Grummon, Stephen R. *The Iran-Iraq War: Islam Embattled.* New York: Praeger, 1982. Published with the Center for Strategic and International Studies, Georgetown University. The Washington Papers, No. 92.

118. "Gulf War: The War That Won't Lie Down and Die." *The Economist,* May 9, 1981, pp. 31–34.

119. Habib, Henry. "Libyan Foreign Policy." *Journal of South Asian and Middle Eastern Studies,* Vol. 4, No. 2 (1980), pp. 71–79.

120. Haendel, Dan. "The Persian Gulf: Geopolitics Revisited." *International Security Review,* Vol. 6, No. 1 (1981), pp. 79–93.

121. Hajjar, Sami G. "The Jamahiriya Experiment in Libya: Qadhafi and Rousseau." *Journal of Modern African Studies,* Vol. 18, No. 2 (1980), pp. 181–200.

122. Halliday, Fred. *Arabia Without Sultans.* New York: Random House, 1975.

123. ———. *Iran: Dictatorship and Development.* Harmondsworth, England: Penguin, 1979. 2nd ed.: 1979.

124. ———. "The Iranian Revolution in International Affairs: Programme and Practice." *Millennium: Journal of International Studies,* Vol. 9, No. 2 (1981), pp. 108–121.

125. Hallwood, Paul, and Stuart Sinclair. *Oil, Debt, and Development: OPEC in the Third World.* London: George Allen & Unwin, 1981.

126. ———, and Stuart Sinclair. "OPEC's Developing Relationships with the Third World." *International Affairs* (London), Vol. 58, No. 2 (1982), pp. 271–286.

127. Hanks, Robert J. "Rapid Deployment in Perspective." *Strategic Review,* Vol. 9, No. 2 (1981), pp. 17–23.

128. Hartshorn, J. E. "From Multinational to National Oil: The Structural Change." *Journal of Energy and Development,* Vol. 5, No. 2 (1980), pp. 207–220.
129. Hensel, Howard M. "Soviet Policy Towards the Rebellion in Dhofar." *Asian Affairs* (London), Vol. 13 (O.S. 69), Part II (June 1982), pp. 183–207.
130. Hetherington, Norriss S. "Industrialization and Revolution in Iran: Forced Progress or Unmet Expectation?" *Middle East Journal,* Vol. 36, No. 3 (1982), pp. 362–373.
131. Holden, David, Richard Johns and James Buchan. *The House of Saud: The Rise and Rule of the Most Powerful Dynasty in the Arab World.* London: Sidgwick & Jackson; New York: Holt, Rinehart, and Winston, 1981.
132. Hooglund, Eric J. *Land and Revolution in Iran, 1960–1980.* Austin: University of Texas Press, 1982. Modern Middle East Series.
133. ———. "Rural Participation in the [Iranian] Revolution." *MERIP Reports,* Vol. 10, No. 4 (May 1980), pp. 3–6.
134. Hooglund, Mary. "One Village in the [Iranian] Revolution." *MERIP Reports,* Vol. 10, No. 4 (May 1980), pp. 7–12.
135. Hottinger, Arnold. "L'expansionnisme Libyen: Machrek, Maghreb et Afrique noire." *Politique Étrangère,* Vol. 46, No. 1 (March 1981), pp. 137–149.
136. Hurewitz, J. C., ed. *Oil, the Arab-Israeli Dispute, and the Industrial World: Horizons of Crisis.* Boulder, Colo.: Westview Press, 1976. Contents include:

A. J. C. Hurewitz, "The Middle East and the International World," pp. 1–6;

Part I: National Responses to the Energy Crisis

B. Lawrence Scheinman, "U.S. International Leadership," pp. 9–20;
C. Ted Greenwood, "Canada's Quest for Energy Autarky," pp. 21–33;
D. Wolfgang Hager, "Western Europe: The Politics of Muddling Through," pp. 34–51;
E. Masao Sakisaka, "Japan's Long-Term Vulnerabilities," pp. 52–64;
F. Melvin A. Conant, "Access to Oil," pp. 65–75;

Part II: Changing National Perspectives on the Arab-Israel Dispute

G. Robert H. Trice, "American Interest Groups after October 1973," pp. 79–95;
H. Janice Gross Stein, "Canada: Evenhanded Ambiguity," pp. 96–109;
I. Hans Maull, "The Strategy of Avoidance: Europe's Middle East Policies after the October War," pp. 110–137;
J. Kazushige Hirasawa, "Japan's Tilting Neutrality," pp. 138–146;
K. Paul Jabber, "Petrodollars, Arms Trade, and the Pattern of Major Conflicts," pp. 149–164;
L. Yair Evron, "The Abiding Threat of War: Perspectives in Israel," pp. 165–176;
M. Ian Smart, "Patterns of Middle East Politics in the Coming Decade," pp. 177–192;
N. Benjamin J. Cohen, "Mixing Oil and Money," pp. 195–211;
O. Guy F. Erb and Helen C. Low, "Resource Transfers to the Developing World," pp. 212–230;
P. T. R. McHale, "Changing Financial Institutions in the Arab Oil States," pp. 231–245;
Q. Wilfrid L. Kohl, "The International Energy Agency: The Political Context," pp. 246–257;
R. Kiichi Saeki, "Japan's Energy-Security Dilemma," pp. 258–265;
S. William D. Nordhaus, "Energy and Economic Growth," pp. 266–282;
T. J. C. Hurewitz, "The Pervasive Crisis," pp. 283–301.

137. ———. *The Persian Gulf: After Iran's Revolution.* New York: Foreign Policy

Association, April 1979. Headline Series, No. 244.

138. Hurt, Harry, III. "The New Energy Crisis: OPEC vs. OPEC." *Texas Monthly,* Vol. 10, No. 8 (August 1982), pp. 103–105, 184–194.

139. Ibrahim, Saad Eddin. *The New Arab Social Order.* Boulder, Colo.: Westview Press, 1982.

140. Ioannides, Christos P. "The Hostages of Iran: A Discussion with the Militants." *Washington Quarterly,* Vol. 3, No. 3 (1980), pp. 12–35.

141. "Iran Two Years After." *MERIP Reports,* No. 98 (July–August 1981), entire issue. Contents include:
   A. Eric Rouleau, "The War and the Struggle for the State," pp. 3–8;
   B. Patrick Clawson, "Iran's Economy: Between Crisis and Collapse," pp. 11–15;
   C. Philip Shehadi, "Economic Sanctions and Iranian Trade," pp. 15–16.

142. Ismael, Jacqueline S. *Kuwait: Social Change in Historical Perspective.* Syracuse: Syracuse University Press, 1982.

143. Ismael, Tareq Y. *Iraq and Iran: Roots of Conflict.* Syracuse: Syracuse University Press, 1982.

144. ———. "The Iraq-Iran Conflict." *Behind the Headlines,* Vol. 39, No. 3 (1981). Toronto: Canadian Institute of International Affairs, 1981.

145. Jabber, Paul. "Conflict and Cooperation in OPEC: Prospects for the Next Decade." *International Organization,* Vol. 32, No. 2 (1978), pp. 377–399.

146. Jaidah, Ali M. "Downstream Operations and the Development of OPEC Member Countries." *Journal of Energy and Development,* Vol. 4, No. 2 (1979), pp. 304–312.

147. Johnson, Maxwell Orme. *The Military as an Instrument of U.S. Policy in Southwest Asia: The Rapid Deployment Joint Task Force, 1979–1982.* Boulder, Colo.: Westview Press, 1983. A Westview Replica Edition.

148. Johnson, Thomas M., and Raymond T. Barrett. "Omani Navy Operating in Troubled Waters." *United States Naval Institute Proceedings,* Vol. 108, No. 947 (March 1982), pp. 99–103.

149. Kanovsky, Eliyahu. "Saudi Arabia in the Red." *Jerusalem Quarterly,* No. 16 (1980), pp. 137–144.

150. Kashkett, Steven B. "Iraq and the Pursuit of Nonalignment." *Orbis,* Vol. 26 (1982), pp. 477–494.

151. Keddie, Nikki R. "Iran: Change in Islam, Islam and Change." *International Journal of Middle East Studies,* Vol. 11, No. 4 (July 1980), pp. 527–542.

152. ———, with a section by Yann Richard. *Roots of Revolution: An Interpretive History of Modern Iran.* New Haven: Yale University Press, 1981.

153. Kelidar, Abbas, ed. *The Integration of Modern Iraq.* London: Croom Helm; New York: St. Martin's Press, 1979. Contents include:
   A. M. M. al-Adhami, "The Elections for the Constituent Assembly in Iraq, 1922–4," pp. 13–31;
   B. A. Shikara, "Faisal's Ambitions of Leadership in the Fertile Crescent: Aspirations and Constraints," pp. 32–45;
   C. H. H. Kopietz, "The Use of German and British Archives in the Study of the Middle East: The Iraqi Coup d'Etat of 1936," pp. 46–62;
   D. David Pool, "From Elite to Class: The Transformation of Iraqi Political Leadership," pp. 63–87;
   E. Paul P. J. Hemphill, "The Formation of the Iraqi Army, 1921–33," pp. 88–110;
   F. Atheel Al-Jomard, "Internal Migration in Iraq," pp. 111–122;
   G. Shakir M. Issa, "The Distribution of Income in Iraq, 1971," pp. 123–134;

H. Keith McLachlan, "Iraq: Problems of Regional Development," pp. 135–149;
I. Edith Penrose, "Industrial Policy and Performance in Iraq," pp. 150–170;
J. Sa'ad N. Jawad, "The Kurdish Problem in Iraq," pp. 171–182;
K. A. R. Kelidar, "Aziz al-Haj: A Communist Radical," pp. 183–192.

154. ———. "The Problems of Succession in Saudi Arabia." *Asian Affairs,* Vol. 65 (N.S. 9), Pt. 1 (February 1978), pp. 23–30.

155. Kelley, P. X. "A Discussion of the Rapid Deployment Force with Lieutenant General P. X. Kelley." *AEI Special Analysis,* No. 80-4. Washington: American Enterprise Institute for Public Policy Research, 1980.

156. Kelly, J. B. *Arabia, the Gulf and the West: A Critical View of the Arabs and Their Oil Policy.* London: Weidenfeld & Nicolson; New York: Basic Books, 1980.

157. Kerr, Malcolm H., and El Sayed Yassin, eds. *Rich and Poor States in the Middle East: Egypt and the New Arab Order.* Boulder, Colo.: Westview Press, 1982. Contents include:
A. Malcolm Kerr, "Introduction: Egypt in the Shadow of the Gulf," pp. 1–16;
Part I: The New Arab Social Order
B. Saad Eddin Ibrahim, "Oil, Migration and the New Arab Social Order," pp. 17–70;
C. Georges Sabagh, "Migration and Social Mobility in Egypt," pp. 71–98;
Part II: The New Arab Economic Order
D. Essam Montasser, "The Arab Economy and Its Developing Strategy: A New Arab Economic Order," pp. 99–128;
E. Jeswald W. Salacuse, "Arab Capital and Trilateral Ventures in the Middle East: Is Three a Crowd?" pp. 129–164;
F. Hazem El-Beblawi, "The Predicament of the Arab Gulf Oil States: Individual Gains and Collective Losses," pp. 165–224;
G. Naiem A. Sherbiny and Ismail Serageldin, "Expatriate Labor and Economic Growth: Saudi Demand for Egyptian Labor," pp. 225–258;
H. Gouda Abdel-Khalek, "The Open Door Economic Policy in Egypt: Its Contribution to Investment and Its Equity Implications," pp. 259–284;
I. Galal Ahmad Amin, "External Factors in the Reorientation of Egypt's Economic Policy," pp. 285–318;
Part III: The New Arab Political Order
J. Ali E. Hillal Dessouki, "The New Arab Political Order: Implications for the 1980's," pp. 319–348;
K. Nazih N. M. Ayubi, "Implementation Capability and Political Feasibility of the Open Door Policy in Egypt," pp. 349–414;
L. Paul Jabber, "Oil, Arms, and Regional Diplomacy: Strategic Dimensions of the Saudi-Egyptian Relationship," pp. 415–448;
M. Malcolm H. Kerr, "Egypt and the Arabs in the Future: Some Scenarios," pp. 449–472.

158. Khadduri, Majid. *Socialist Iraq.* Washington: Middle East Institute, 1978.

159. Khalifa, Ali Mohammed. *The United Arab Emirates: Unity in Fragmentation.* London: Croom Helm; Boulder, Colo.: Westview Press, 1979.

160. Khouja, M. W., and P. G. Sadler. *The Economy of Kuwait.* London: Macmillan, 1979.

161. Khuri, Fuad I. *Tribe and State in Bahrain: The Transformation of Social and Political Authority in an Arab State.* Chicago: University of Chicago Press, 1980.

162. Knauerhase, Ramon. "Saudi Arabia: Fifty Years of Economic Change." *Cur-*

rent History, Vol. 82, No. 480 (January 1983), pp. 19–23, 35–36.

163. ———. "Saudi Arabia's Foreign and Domestic Policy." *Current History,* Vol. 80, No. 462 (January 1981), pp. 18–22, 37.
164. Koury, Enver M., and Emile A. Nakhleh, eds. with Thomas W. Mullen. *The Arabian Peninsula, Red Sea and Gulf: Strategic Considerations.* Hyattsville, Md.: Institute of Middle Eastern and North African Affairs, 1979.
165. ———. *The Saudi Decision-Making Body: The House of Al-Saud.* Hyattsville, Md.: Institute of Middle Eastern and North African Affairs, 1978.
166. Krapels, Edward N. "Oil and Security: Problems and Prospects of Importing Countries." *Adelphi Papers,* No. 136. London: International Institute for Strategic Studies, 1977.
167. Krulak, Victor H. "The Rapid Deployment Force: Criteria and Imperatives." *Strategic Review,* Vol. 8, No. 2 (1980), pp. 39–43.
168. Kuniholm, Bruce R. "What the Saudis Really Want: A Primer for the Reagan Administration." *Orbis,* Vol. 25, No. 1 (1981), pp. 107–121.
169. Al-Kuwari, Ali Khalifa. *Oil Revenues in the Gulf Emirates: Patterns of Allocation and Impact on Economic Development.* Epping, U.K.: Bowker; Boulder, Colo.: Westview Press, 1978.
170. Lacey, Robert. *The Kingdom: Arabia and the House of Saud.* London: Hutchinson; New York: Harcourt Brace Jovanovich, 1982.
171. ———. "Saudi Arabia: A More Visible Role in the Middle East." *The World Today,* Vol. 38, No. 1 (January 1982), pp. 4–12.
172. Lackner, Helen. *A House Built on Sand: A Political Economy of Saudi Arabia.* London: Ithaca Press, 1978.
173. Ledeen, Michael, and W. Lewis. "Carter and the Fall of the Shah: The Inside Story." *Washington Quarterly,* Vol. 3, No. 2 (1980), pp. 3–40.
174. Lenczowski, George. "The Arc of Crisis: Its Central Sector." *Foreign Affairs,* Vol. 57, No. 4 (1979), pp. 796–820.
175. ———. "The Persian Gulf Crisis and Global Oil." *Current History,* Vol. 80, No. 462 (January 1981), pp. 10–13, 43.
176. ———. "The Soviet Union and the Persian Gulf: An Encircling Strategy." *International Journal,* Vol. 37 (1982), pp. 307–327.
177. Le Vine, Victor T., and Timothy W. Luke. *The Arab–African Connection: Political and Economic Realities.* Boulder, Colo.: Westview Press, 1979.
178. Levy, Walter J. "Oil: An Agenda for the 1980s." *Foreign Affairs,* Vol. 59, No. 5 (1981), pp. 1079–1101.
179. ———. "Oil and the Decline of the West." *Foreign Affairs,* Vol. 58, No. 5 (1980), pp. 999–1015.
180. ———. *Oil Strategy and Politics, 1941–1981.* Edited by Melvin A. Conant. Boulder, Colo.: Westview Press, 1982.
181. Liggett, William R. "Long-range Combat Aircraft and Rapid Deployment Forces." *Air University Review,* Vol. 33 (July–August 1982), pp. 72–81.
182. Litwak, Robert. *Security in the Persian Gulf, Vol. 2: Sources of Inter-State Conflict.* London: International Institute for Strategic Studies; Montclair, New Jersey: Allanheld, Osmun, 1981.
183. Long, David. *The Persian Gulf.* 2nd ed. Boulder, Colo.: Westview Press, 1978.
184. Longrigg, Stephen H. *Oil in the Middle East: Its Discovery and Development.* 3rd ed. London: Oxford University Press, 1968.
185. Looney, Robert E. *Economic Origins of the Iranian Revolution.* New York: Pergamon Press, 1982.
186. ———. *Saudi Arabia's Development Potential: Application of an Islamic Growth Model.* Lexington, Mass.: Lexington Books/D.C. Heath, 1982.

187. ———. "Saudi Arabia's Islamic Growth Model." *Journal of Economic Issues,* Vol. 16, No. 2 (June 1982), pp. 453–459.
188. Luttwak, Edward. "Cubans in Arabia? Or, the Meaning of Strategy." *Commentary,* Vol. 68, No. 6 (December 1979), pp. 62–66.
189. Mabro, Robert, ed. *World Energy Issues and Policies.* Oxford: Oxford University Press for the Oxford Energy Seminar, St. Catherine's College, 1980. Proceedings of the First Oxford Energy Seminar, September 1979. Included among the contents are:
    A. Francisco R. Parra, "OPEC Oil: Recent Developments and Problems of Supplies," pp. 3–11;
    B. Jack E. Hartshorn, "The Special Characteristics of OPEC and Importing Countries' National Oil Companies," pp. 157–165;
    C. C. C. Pocock, "World Energy Outlook: Perceptions of an Oilman," pp. 213–215;
    D. Ali Jaidah, "World Energy Outlook: an OPEC Perspective," pp. 217–224;
    E. René Ortiz, "Crude Oil: Issues and Policies for Oil-Exporting Countries," pp. 283–291;
    F. Aziz al-Watari, "Refining and Petrochemicals: Developments in Some Oil-Exporting Countries," pp. 292–322;
    G. George Corm, "The Economic Future of OPEC Countries: Constraints and Opportunities," pp. 323–335;
    H. Ali Attiga, "Oil and Regional Co-operation Among the Arab Countries," pp. 337–347;
    I. Humberto Calderon Berti, "OPEC and the Energy Crisis," pp. 349–353;
    J. Ali Khalifa Al Sabah, "Conceptual Perspective for a Long-Range Oil Production Policy," pp. 355–361.
190. Mabro, Robert, and Elizabeth Monroe. "Arab Wealth from Oil: Problems of Its Investment." *International Affairs,* Vol. 50, No. 1 (January 1974), pp. 15–27.
191. McHale, T. R. "A Prospect of Saudi Arabia." *International Affairs* (London), Vol. 56, No. 4 (1980), pp. 622–647.
192. Maddy-Weitzman, Bruce. "Islam and Arabism: The Iran–Iraq War." *Washington Quarterly,* Vol. 5 (1982), pp. 181–188.
193. Maechling, Eugenie. "Security Risks to Energy Production and Trade: The Problems of the Middle East." *Energy Policy,* Vol. 10 (June 1982), pp. 120–130.
194. Malik, Hafeez. "Struggle for Power and Oil in the Persian Gulf." *Journal of South Asian and Middle Eastern Studies,* Vol. 3, No. 3 (1980), pp. 3–14.
195. El Mallakh, Ragaei. *The Economic Development of the United Arab Emirates.* London: Croom Helm; New York: St. Martin's Press, 1981.
196. ———. *Kuwait: Trade and Investment.* London: Croom Helm; Boulder, Colo.: Westview Press, 1979.
197. ———. *Qatar: Development of an Oil Economy.* London: Croom Helm; New York: St. Martin's Press, 1979.
198. ———, and Jacob K. Atta. *The Absorptive Capacity of Kuwait: Domestic and International Perspectives.* Lexington, Mass.: Lexington Books/D.C. Heath, 1981.
199. Mansfield, Peter. *The New Arabians.* Chicago: J. G. Ferguson, 1981. Distributed by Doubleday & Co.
200. Mason, John P. "Qadhdhafi's 'Revolution' and Change in a Libyan Oasis Community." *Middle East Journal,* Vol. 36, No. 3 (1982), pp. 319–335.
201. Maull, Hanns W. "The Control of Oil." *International Journal,* Vol. 36, No. 2 (1981), pp. 273–293.

202. Melikian, Levon H., and Juhaina S. Al-Easa. "Oil and Social Change in the Gulf." *Journal of Arab Affairs,* Vol. 1, No. 1 (October 1981), pp. 79–98.
203. Menashri, David. "The Shah and Khomeini: Conflicting Nationalisms." *Crossroads,* No. 8 (1982), pp. 53–79.
204. Mikdashi, Zuhayr. "Oil-exporting Countries and Oil-importing Countries: What Kind of Interdependence?" *Millennium: Journal of International Studies,* Vol. 9, No. 1 (1980), pp. 1–20.
205. ———. "Oil Prices and OPEC Surpluses: Some Reflections." *International Affairs* (London), Vol. 57, No. 3 (1981), pp. 407–427.
206. Miller, Aaron David. *Search for Security: Saudi Arabian Oil and American Foreign Policy, 1939–1949.* Chapel Hill: University of North Carolina Press, 1980.
207. Molyneux, Maxine. *State Policies and the Position of Women Workers in the People's Democratic Republic of Yemen, 1967–1977.* Geneva: International Labour Office, 1982.
208. Moran, Theodore H. "Modeling OPEC Behavior: Economic and Political Alternatives." *International Organization,* Vol. 35, No. 2 (1981), pp. 241–272.
209. Mortimer, Robert. "Algeria: Which Way Ahead?" *Africa Report,* Vol. 24, No. 3 (May–June 1979), pp. 9–14.
210. Mossavar-Rahmani, Bijan. "OPEC and NOPEC: Oil in South–South Relations." *Journal of International Affairs,* Vol. 34, No. 1 (1980), pp. 40–58.
211. Mottahedeh, Roy. "Iran's Foreign Devils." *Foreign Policy,* No. 38 (1980), pp. 19–34.
212. Nakhleh, Emile. *Bahrain: Political Development in a Modernizing Society.* Lexington: D. C. Heath, Lexington Books, 1976.
213. ———. *The Persian Gulf and American Policy.* New York: Praeger, 1982. Praeger Special Studies.
214. Nashashibi, Hikmat S. "The Development of Capital Markets in the Gulf." *Journal of Arab Affairs,* Vol. 1, No. 1 (October 1981), pp. 99–112.
215. Nellis, J. R. "Socialist Management in Algeria." *Journal of Modern African Studies,* Vol. 15, No. 4 (1977), pp. 529–554.
216. Nelson, Harold D., *et al. Algeria: A Country Study.* 3rd ed. Washington: USGPO, 1979. American University, Foreign Area Studies.
217. ———, *et al. Libya: A Country Study.* 2nd ed. Washington: USGPO, 1979. American University, Foreign Area Studies.
218. Newsom, David D. "America EnGulfed." *Foreign Policy,* No. 43 (1981), pp. 17–32.
219. Niblock, Tim, ed. *Social and Economic Development in the Arab Gulf.* London: Croom Helm; New York: St. Martin's Press, 1980. Contents include:
    - A. J. J. Malone, "Involvement and Change: The Coming of the Oil Age to Saudi Arabia," pp. 20–48;
    - B. Mohamed G. Rumaihi, "The Mode of Production in the Arab Gulf Before the Discovery of Oil," pp. 49–60;
    - C. Rosemarie Said Zahlan, "Hegemony, Dependence and Development in the Gulf," pp. 61–79;
    - D. Keith McLachlan, "Natural Resources and Development in the Gulf States," pp. 80–94;
    - E. John Townsend, "Problems Confronting the Establishment of a Heavy Industrial Base in the Arab Gulf," pp. 95–105;
    - F. Donald P. Cole, "Pastoral Nomads in a Rapidly Changing Economy: The Case of Saudi Arabia," pp. 106–121;
    - G. J. C. Wilkinson, "Changes in the Structure of Village Life in Oman," pp. 122–134;

H. J. S. Birks and C. A. Sinclair, "Economic and Social Implications of Current Development in the Arab Gulf: The Oriental Connection," pp. 135–160;
I. Emile A. Nakhleh, "Political Participation and the Constitutional Experiments in the Arab Gulf: Bahrain and Qatar," pp. 161–176;
J. Naomi Sakr, "Federalism in the United Arab Emirates: Prospects and Regional Implications," pp. 177–186;
K. Tim Niblock, "The Prospects for Integration in the Arab Gulf," pp. 187–209;
L. Fred Halliday, "The Gulf Between Two Revolutions: 1958–1979," pp. 210–238.

220. ———, ed. *State, Society and Economy in Saudi Arabia.* London: Croom Helm, 1981. Contents include:
A. Tim Niblock, "Introduction," pp. 11–22;
B. Derek Hopwood, "The Ideological Basis: Ibn Abd al-Wahhab's Muslim Revivalism," pp. 23–35;
C. Peter Sluglett and Marion Farouk-Sluglett, "The Precarious Monarchy: Britain, Abd Al-Aziz Ibn Saud and the Establishment of the Kingdom of Hijaz, Najd and Its Dependencies, 1925–1932," pp. 36–57;
D. Rosemarie Said Zahlan, "Abd al-Aziz Ibn Saud's Changing Relationship with the Gulf States During the 1930s," pp. 58–74;
E. Tim Niblock, "Social Structure and the Development of the Saudi Arabian Political System," pp. 75–105;
F. James Buchan, "Secular and Religious Opposition in Saudi Arabia," pp. 106–124;
G. Fred Halliday, "A Curious and Close Liaison: Saudi Arabia's Relations with the United States," pp. 125–147;
H. John Duke Anthony, "Aspects of Saudi Arabia's Relations with Other Gulf States," pp. 148–170;
I. Shirley Kay, "Social Change in Modern Saudi Arabia," pp. 171–185;
J. Ugo Fabietti, "Sedentarization as a Means of Detribalization: Some Policies of the Saudi Arabian Government Towards the Nomads," pp. 186–197;
K. J. S. Birks and C. A. Sinclair, "The Domestic Political Economy of Development in Saudi Arabia," pp. 198–213;
L. Paul Stevens, "Saudi Arabia's Oil Policy in the 1970s—Its Origins, Implementation and Implications," pp. 214–234;
M. H. G. Hambleton, "The Saudi Arabian Petrochemical Industry: Its Rationale and Effectiveness," pp. 235–277;
N. Rodney Wilson, "The Evolution of the Saudi Banking System and Its Relationship with Bahrain," pp. 278–300.

221. Noreng, Øystein. "Friends or Fellow Travelers? The Relationship of Non-OPEC Exporters with OPEC." *Journal of Energy and Development,* Vol. 4, No. 2 (1979), pp. 313–335.

222. ———. *Oil Politics in the 1980s: Patterns of International Cooperation.* New York: McGraw-Hill, 1978. 1980s Project/Council on Foreign Relations.

223. Noyes, James H. *The Clouded Lens: Persian Gulf Security and U.S. Policy.* Stanford, Cal.: Hoover Institution Press, 1979.

224. Nye, Roger P. "Political and Economic Integration in the Arab States of the Gulf." *Journal of South Asian and Middle Eastern Studies,* Vol. 2, No. 1 (1978), pp. 3–21.

225. Nyrop, Richard F., *et al. Area Handbook for Saudi Arabia.* Washington: USGPO, 1977. American University, Foreign Area Studies.

226. ———, *et al. Area Handbook for the Persian Gulf States.* Washington: USGPO, 1977. American University, Foreign Area Studies.
227. O'Ballance, Edgar. "The Iraqi–Iranian War: The First Round." *Parameters,* Vol. 11 (March 1981), pp. 54–59.
228. ———. "The Kurdish Factor in the Gulf War." *Military Review,* Vol. 61, No. 6 (June 1981), pp. 3–20.
229. O'Brien, W., J. Moore and W. Mallison. "Israel's Attack on Osirak." *Freedom at Issue,* No. 63 (November–December 1981), pp. 3–11.
230. Ochsenwald, William. "Saudi Arabia and the Islamic Revival." *International Journal of Middle East Studies,* Vol. 13, No. 3 (August 1981), pp. 271–286.
231. Odell, Peter R. "The Significance of Oil." *Journal of Contemporary History,* Vol. 3, No. 3 (July 1968), pp. 93–110.
232. ———. "Toward a Geographically Reshaped World Oil Industry." *The World Today,* Vol. 37, No. 12 (December 1981), pp. 447–453.
233. ———, and Luis Vallenilla. *The Pressures of Oil: A Strategy for Economic Revival.* London: Harper & Row, 1978.
234. *Oil Diplomacy: The Atlantic Nations in the Oil Crisis of 1978–79.* Philadelphia: Foreign Policy Research Institute, 1980. Originally published as special issue of *Orbis,* Vol. 23, No. 4 (1980). Contents include:
    A. Alexander M. Haig, Jr., "Reflections on Energy and Western Security," pp. ix–xiii;
    B. Paul Kemezis, "The Permanent Crisis: Changes in the World Oil System," pp. 3–26;
    C. Richard B. Mancke, "The American Response: 'On the Job Training'?" pp. 27–43;
    D. Hanns W. Maull, "Western Europe: A Fragmented Response to a Fragmenting Order," pp. 45–66;
    E. Richard E. Bissell, "The West in Concert: A Very Complex Score," pp. 67–85;
    F. Harvey Sicherman, "Politics of Dependence: Western Europe and the Arab–Israeli Conflict," pp. 87–99;
    G. Mason Willrich and Bijan Mossavar-Rahmani, "Oil on Troubled Waters: The Industrial World and the OPEC Middle East," pp. 101–116;
    H. Alberto M. Bagnasco, "Oil and Money: A Note on the Western Financial Community," pp. 117–127;
    I. "Getting Out of the Crisis: Interviews," pp. 129–134;
    J. "An Oil Crisis Lexicon," pp. 135–137.
235. Olayan, Suliman S. "A Saudi View of Oil." *Washington Quarterly,* Vol. 3, No. 4 (1980), pp. 175–179.
236. Ortiz, René G. "The World Energy Outlook in the 1980s and the Role of OPEC." *Journal of Energy and Development,* Vol. 4, No. 2 (1979), pp. 197–211.
237. Paul, Jim. "Insurrection at Mecca." *MERIP Reports,* No. 91 (October 1980), pp. 3–4.
238. Penrose, Edith. "OPEC's Importance in the World Oil Industry." *International Affairs,* Vol. 55, No. 1 (January 1979), pp. 18–32.
239. ———, and E. F. Penrose. *Iraq: International Relations and National Development.* Boulder, Colo.: Westview Press, 1978.
240. Perkins, K. "Oman 1975: The Year of Decision." *Journal of the Royal United Service Institution,* Vol. 124, No. 1 (March 1979), pp. 38–45.
241. Peterson, J. E. "Conflict in the Yemens and Superpower Involvement." *Occasional Paper.* Washington; Georgetown University, Center for Contemporary Arab Studies, 1981.

242. ———. "Guerrilla Warfare and Ideological Confrontation in the Arabian Peninsula: The Rebellion in Dhufar." *World Affairs,* Vol. 139, No. 4 (1977), pp. 278–295.
243. ———. *Oman in the Twentieth Century: Political Foundations of an Emerging State.* London: Croom Helm; New York: Barnes & Noble, 1978.
244. ———. "Tribes and Politics in Eastern Arabia." *Middle East Journal,* Vol. 31, No. 3 (1977), pp. 297–312.
245. Peterzell, Jay. "Reagan's Covert Action Policy (III)." *First Principles,* Vol. 7, No. 5 (March 1982), pp. 1–7.
246. Petrossian, Vahe. "Dilemmas of the Iranian Revolution." *The World Today,* Vol. 36, No. 1 (January 1980), pp. 19–25.
247. ———. "The Gulf War." *The World Today,* Vol. 36, No. 11 (November 1980), pp. 415–417.
248. ———. "Iran's Crisis of Leadership." *The World Today,* Vol. 37, No. 2 (February 1981), pp. 39–44.
249. Pindyck, Robert S. "Some Long-Term Problems in OPEC Oil Pricing." *Journal of Energy and Development,* Vol. 4, No. 2 (1979), pp. 259–272.
250. Piscatori, James P. "The Role of Islam in Saudi Arabia's Political Development." In John L. Esposito, ed., *Islam and Development: Religion and Sociopolitical Change* (Syracuse: Syracuse University Press, 1980), pp. 123–138.
251. Plaut, Steven E. "OPEC Is Not a Cartel." *Challenge,* Vol. 24, No. 5 (November–December 1981), pp. 18–24.
252. Pranger, Robert J., and Dale R. Tahtinen. "American Policy Options in Iran and the Persian Gulf." *AEI Foreign Policy and Defense Review,* Vol. 1, No. 2 (1979), entire issue.
253. Presley, J. R. "Saudi Arabia: A Decade of Economic Progress." *Three Bank Review,* No. 127 (September 1980), pp. 25–40.
254. Price, David Lynn. "Moscow and the Persian Gulf." *Problems of Communism,* Vol. 28, No. 2 (March–April 1979), pp. 1–13.
255. Quandt, William B. "The Middle East Crises." *Foreign Affairs,* Vol. 58, No. 3 (1980), pp. 540–562.
256. ———. "Riyadh Between the Superpowers." *Foreign Policy,* No. 44 (1981), pp. 37–56.
257. ———. *Saudi Arabia in the 1980s: Foreign Policy, Security, and Oil.* Washington: Brookings Institution, 1981.
258. al-Qudsi, Sulayman S. "Pre- and Post-Fiscal Distributional Patterns in Kuwait." *Middle Eastern Studies,* Vol. 17 (July 1981), pp. 393–407.
259. *Race and Class,* Vol. 21, No. 1 (1979). Special issue on Iranian revolution. Contents include:
    A. Eqbal Ahmad, "The Iranian Revolution: A Landmark for the Future," pp. 3–12;
    B. Nikki R. Keddie, "Oil, Economic Policy and Social Conflict in Iran," pp. 13–30;
    C. Mansour Farhang, "Resisting the Pharaohs: Ali Shariati on Oppression," pp. 31–40;
    D. Richard Falk, "Iran and American Geopolitics in the Gulf," pp. 41–56;
    E. William A. Dorman, "Iranian People vs. U.S. News Media: A Case of Libel," pp. 57–66;
    F. Stuart Schaar, "Orientalism at the Service of Imperialism," pp. 67–80;
    G. Fred Halliday, "Theses on the Iranian Revolution," pp. 81–90.
260. Ramazani, Nesta. "Behind the Veil: Status of Women in Revolutionary Iran." *Journal of South Asian and Middle Eastern Studies,* Vol. 4, No. 2 (1980), pp. 27–36.

261. Ramazani, Rouhollah K. "Iran's Revolution: Patterns, Problems and Prospects." *International Affairs* (London), Vol. 56, No. 3 (1980), pp. 443–457.
262. ———. *The Persian Gulf and the Strait of Hormuz.* Alphen aan den Rijn: Sijthoff and Nordhoff, 1979. International Straits of the World, Vol. 3.
263. ———. "Security in the Persian Gulf." *Foreign Affairs,* Vol. 57, No. 4 (1979), pp. 821–835.
264. ———. "The Strait of Hormuz: The Global Chokepoint." In Larry W. Bowman and Ian Clark, eds., *The Indian Ocean in Global Politics* (Boulder, Colo.: Westview Press; Nedlands: University of Western Australia, 1981), pp. 7–20.
265. ———. "Who Lost America? The Case of Iran." *Middle East Journal,* Vol. 36, No. 1 (1982), pp. 5–21.
266. Record, Jeffrey. *The Rapid Deployment Force and U.S. Military Intervention in the Persian Gulf.* Cambridge, Mass., and Washington, D.C.: Institute for Foreign Policy Analysis, February 1981.
267. ———. "The RDF: Is the Pentagon Kidding?" *Washington Quarterly,* Vol. 4, No. 3 (1981), pp. 41–51.
268. ———. "The Rapid Deployment Force: Problems, Constraints, and Needs." In *National Security Policy for the 1980s, Annals of the American Academy of Political and Social Science,* Vol. 457 (September 1981), pp. 109–120.
269. Richardson, John. "Europe in the Middle East: Shaping a Political Role." *SAIS Review,* No. 3 (1981–82), pp. 107–117.
270. Ricks, Thomas M. "The Iranian People's Revolution: Its Nature and Implications for the Gulf States." *CCAS Reports.* Washington: Georgetown University, Center for Contemporary Arab Studies, April 1979.
271. Rondot, Philippe. "La guerre du Golfe." *Défense Nationale,* Vol. 37 (June 1981), pp. 79–95.
272. Ross, Arthur. "OPEC's Challenge to the West." *Washington Quarterly,* Vol. 3, No. 1 (1980), pp. 50–58.
273. Ross, Dennis. *Considering Soviet Threats to the Gulf.* Washington: The Wilson Center, 1981. International Security Studies Program, No. 29.
274. ———. "Considering Soviet Threats to the Persian Gulf." *International Security,* Vol. 6, No. 2 (1981), pp. 159–180.
275. Rouleau, Eric. "Khomeini's Iran." *Foreign Affairs,* Vol. 59, No. 1 (1980), pp. 1–20.
276. Rubin, Barry. "Iran's Year of Turmoil." *Current History,* Vol. 82, No. 480 (January 1983), pp. 28–31, 42.
277. ———. *Paved With Good Intentions: The American Experience and Iran.* London: Oxford University Press, 1980.
278. Rubinstein, Alvin. "The Soviet Union and the Arabian Peninsula." *The World Today,* Vol. 35, No. 11 (November 1979), pp. 442–452.
279. Rustow, Dankwart A. "U.S.–Saudi Relations and the Oil Crises of the 1980s." *Foreign Affairs,* Vol. 55, No. 3 (April 1977), pp. 494–516.
280. al-Sabah, Salem al-Jabir. *Les Emirats du Golfe.* Paris: Fayard, 1980.
281. al-Sabah, Y. S. F. *The Oil Economy of Kuwait.* London: Kegan Paul International, 1980.
282. Said Zahlan, Rosemarie. *The Creation of Qatar.* London: Croom Helm; New York: Barnes & Noble, 1979.
283. ———. *The Origins of the United Arab Emirates: A Political and Social History of the Trucial States.* London: Macmillan, 1978.
284. St. John, Ronald Bruce. "The Soviet Penetration of Libya." *The World Today,* Vol. 38, No. 4 (April 1982), pp. 131–138.
285. Salameh, Ghassane. "Political Power and the Saudi State." *MERIP Reports,* No. 91 (October 1980), pp. 5–22. Tr. from the French by Vivian Steir.

286. ———. "Saudi Arabia: Development and Dependence." *Jerusalem Quarterly,* No. 20 (1981), pp. 109–122. Translation of article published originally in *Dirasat 'Arabiya,* February 1980.

287. Sale, Richard. "Carter and Iran: From Idealism to Disaster." *Washington Quarterly,* Vol. 3, No. 4 (1980), pp. 75–87.

288. Al-Salem, Faisal. "The Issue of Identity in Selected Arab Gulf States." *Journal of South Asian and Middle Eastern Studies,* Vol. 4, No. 4 (1981), pp. 3–20.

289. Samii, Massod V. "Oil Supply and the Absorptive Capacity of OPEC: Reflection into the Future." *Journal of Energy and Development,* Vol. 6, No. 1 (1980), pp. 72–85.

290. Sampson, Anthony. *The Seven Sisters: The Great Oil Companies & The World They Shaped.* New York: Viking Press, 1975; New York: Bantam Books, 1976.

291. Sankari, Farouk A. "Oil, Human Resources, and the Development Syndrome: The Libyan Case." *Studies in Comparative International Development,* Vol. 16, No. 1 (1981), pp. 55–74.

292. Sayigh, Yusif. "The Social Cost of Oil Revenues." *Jerusalem Quarterly,* Vol. 23 (1982), pp. 134–144.

293. El-Serafy, Salah. "Absorptive Capacity, the Demand for Revenues, and the Supply of Petroleum." *Journal of Energy and Development,* Vol. 7, No. 1 (1981), pp. 73–88.

294. ———. "The Oil Price Revolution of 1973–1974." *Journal of Energy and Development,* Vol. 4, No. 2 (1979), pp. 273–290.

295. Shahshahani, Ahmad, and Mihssen Kadhim. "Development Problems of an Energy-Based Economy." *Journal of South Asian and Middle Eastern Studies,* Vol. 2, No. 3 (1979), pp. 57–83.

296. Sherbiny, Naiem A., and Mark A. Tessler, eds. *Arab Oil: Impact on the Arab Countries and Global Implications.* New York: Praeger, 1976. Contents include:

Part I: Introduction

A. Naiem A. Sherbiny and Mark A. Tessler, "The Impact of Arab Oil," pp. 3–18;

Part II: Economic Issues in Arab Oil-Producing Countries

B. L. Donald Fixler and Robert L. Ferrar, "An Application of Economic Theory to Middle Eastern Oil Pricing," pp. 21–34;

C. Naiem A. Sherbiny, "Arab Oil Production Policies in the Context of International Conflicts," pp. 35–57;

D. Ragaei El Mallakh, "Industrialization in the Arab World: Obstacles and Prospects," pp. 58–74;

Part III: Political Development and Cultural Change in Arab Oil-Producing Countries

E. John P. Entelis, "Oil Wealth and Prospects for Democratization in the Arabian Peninsula: The Case of Saudi Arabia," pp. 77–111;

F. Clement Henry Moore, "Petroleum and Political Development in the Maghreb," pp. 112–129;

G. Mark A. Tessler, "Development, Oil, and Cultural Change in the Maghreb," pp. 130–160;

Part IV: Arab Oil and the International Economy

H. Alfred Steinherr, "Optimal Petroleum Production and Trade: Policies for Oil-Exporting and -Importing Countries," pp. 163–176;

I. John H. Makin, "International Financial Implications of Increased Flows of Funds to and from Petroleum-Producing Countries," pp. 177–201;

J. Zuhayr Mikdashi, "Surplus Funds and the Strategy of Oil-Exporting Countries," pp. 202–224;

K. Richard A. Paull, "Present and Future World Energy Needs and the Role of the Arab Countries," pp. 225–254;

Part V: Arab Oil and International Politics

L. Yusif A. Sayigh, "Oil in Arab Developmental and Political Strategy: An Arab View," pp. 257–264;

M. Farouk A. Sankari, "The Character and Impact of Arab Oil Embargoes," pp. 265–278;

N. William B. Quandt, "U.S. Energy Policy and the Arab-Israeli Conflict," pp. 279–294;

O. David Garnham, "The Oil Crisis and U.S. Attitudes Toward Israel," pp. 295–304;

P. Robert O. Freedman, "The Soviet Union and the Politics of Middle Eastern Oil," pp. 305–327.

297. Shihata, Ibrahim F. I. "OPEC Aid, the OPEC Fund, and Cooperation with Commercial Development Finance Sources." *Journal of Energy and Development,* Vol. 4, No. 2 (1979), pp. 291–303.

298. ———. "OPEC Countries as a Donor Group." *An-Nahar Arab Report and MEMO,* March 2 and 9, 1981.

299. Shwadran, Benjamin. *Middle East Oil: Issues and Problems.* Cambridge, Mass.: Schenkman, 1977.

300. Singer, S. Fred, and Stephen Stamas. "An End to OPEC?" *Foreign Policy,* No. 45 (1981–1982), pp. 115–125.

301. Smart, Ian. "Oil, the Super-powers and the Middle East." *International Affairs* (London), Vol. 53, No. 1 (January 1977), pp. 17–35.

302. Smolansky, Oles M. "Soviet Policy in Iran and Afghanistan." *Current History,* Vol. 80, No. 468 (October 1981), pp. 321–324, 339.

303. Sobhan, Rehman. "Institutional Mechanisms for Channelling OPEC Surpluses Within the Third World." *Third World Quarterly,* Vol. 2 (October 1980), pp. 721–745.

304. ———. "OPEC's Political Options: Case for Collective Self-reliance within the Third World." *Alternatives,* Vol. 7, No. 1 (1981), pp. 43–60.

305. Soffan, Linda Usra. *The Women of the United Arab Emirates.* London: Croom Helm; New York: Barnes & Noble, 1980.

306. Staudenmaier, William O. "Military Policy and Strategy in the Gulf War." *Parameters,* Vol. 12, No. 2 (June 1982), pp. 25–35.

307. Stern, Jonathan P. "Gulf Oil Strategy." *Washington Quarterly,* Vol. 3, No. 2 (1980), pp. 67–72.

308. Stewart, Richard A. "Oman: The Next Crisis?" *United States Naval Institute Proceedings,* Vol. 106, No. 4 (April 1980), pp. 97–102.

309. Stone, Russell A., ed. *OPEC and the Middle East: The Impact of Oil on Societal Development.* New York: Praeger, 1977. Praeger Special Studies in International Politics and Government. Contents include:

Part I: The Challenge of Societal Development

A. Russell A. Stone, "Dimensions of Social Change," pp. 3–18;

Part II: Political Dimensions of Change

B. Don Peretz, "Middle East Oil and Changing Political Constellations," pp. 21–35;

C. I. Robert Sinai, "Problems of Transition to Economic Growth and Modernization in Middle Eastern Societies," pp. 36–61;

Part III: Economic Development: Projects and Policies

D. Girgis B. Ghobrial, "OPEC: Its International Economic Significance, 1974–75," pp. 65–101;

E. Bertrand P. Boucher, "Development Projects in the Middle East: Domestic Investments Utilizing Oil Revenues," pp. 102–114;
F. Marwan M. El Nasser, "Energy Interdependence and OPEC Government Use of the Foreign Sector," pp. 115–129;
G. Farouk A. Sankari, "Higher Oil Prices and the Consuming Developing Countries: Consequences, Initiatives, and Potential Energy Policy Actions," pp. 130–150;

Part IV: Country Studies

H. Khalid M. Jaffer, "Kuwait's Economic Development Policies," pp. 153–157;
I. Abdo I. Baaklini, "Patterns of Aid: Kuwait, the Middle East, and the Developing World," pp. 158–196;
J. Mohamed El Behairy, "Egypt and OPEC," pp. 197–210;
K. Abdul H. Raoof, "Ideology and Politics in Iraqi Oil Policy: The Nationalization of 1972," pp. 211–224;
L. James A. Paul, "Algeria's Oil Economy: Liberation or Neo-colonialism?" pp. 225–247;
M. Manuchehr Afzal, "The Impact of Oil Revenue on Education in Iran," pp. 248–257.

310. Stork, Joe. "The War in the Gulf." *MERIP Reports,* No. 97 (June 1981), pp. 3–18.
311. Szaz, Z. Michael, ed. *The Impact of the Iranian Events Upon Persian Gulf and United States Security.* Washington: American Foreign Policy Institute, 1979. Studies on Middle East Problems. Contributions include:
A. Robert J. Hanks, "The Geopolitical Situation in the Persian Gulf," pp. 1–18;
B. Rouhollah K. Ramazani, "Iran's Revolution in Perspective," pp. 19–38;
C. Edmund Ghareeb, "Iraq and Gulf Security," pp. 39–64;
D. J. E. Peterson, "Kuwait, Soviet Activities and Gulf Security," pp. 65–88;
E. James E. Akins, "Saudi Arabia, Soviet Activities and Gulf Security," pp. 89–110;
F. Emile Nakhleh, "Bahrain and Persian Gulf Security," pp. 111–127;
G. Robert G. Neumann, "Afghanistan Under the Red Flag," pp. 128–148;
H. Enver M. Koury, "The United Arab Emirates: National Interests and Security Problems," pp. 149–164;
I. J. E. Peterson, "Oman, Persian Gulf and U.S. Security," pp. 165–182;
J. Jesse A. Helms, "Washington's Bankrupt Middle East Policy," pp. 183–195.

312. Tahir-Kheli, Shirin, and Shaheen Ayubi, eds. *The Iran–Iraq War: Old Conflicts, New Weapons.* New York: Praeger, 1983.
313. ———. "Proxies and Allies: The Case of Iran and Pakistan." *Orbis,* Vol. 24, No. 2 (1980), pp. 339–352.
314. ———, and William O. Staudenmaier. "The Saudi–Pakistani Military Relationship: Implications for U.S. Policy." *Orbis,* Vol. 26, No. 1 (1982), pp. 155–171.
315. Tahtinen, Dale R. *National Security Challenges to Saudi Arabia.* Washington: American Enterprise Institute for Public Policy Research, 1978.
316. Taylor, Alan R. "The Euro–Arab Dialogue." *Middle East Problem Paper,* No. 15. Washington: Middle East Institute, 1978.
317. Thompson, W. Scott. "The Persian Gulf and the Correlation of Forces." *International Security,* Vol. 7, No. 1 (1982), pp. 157–180.
318. Townsend, John. "L'industrie dans les Emirats arabes unis." *Maghreb-*

Machrek, No. 87 (January–March 1980), pp. 23–34.
319. ———. *Oman: The Making of a Modern State.* London: Croom Helm; New York: St. Martin's Press, 1977.
320. Tucker, Robert W. "American Power and the Persian Gulf." *Commentary,* Vol. 70, No. 5 (November 1980), pp. 25–41.
321. ———. "Oil and American Power—Six Years Later." *Commentary,* Vol. 68, No. 3 (September 1979), pp. 35–42.
322. ———. "Oil and American Power—Three Years Later." *Commentary,* Vol. 63, No. 1 (January 1977), pp. 29–36.
323. ———. "Oil: The Issue of American Intervention." *Commentary,* Vol. 59, No. 1 (January 1975), pp. 21–31.
324. ———. "The Purposes of American Power." *Foreign Affairs,* Vol. 59, No. 2 (1980–1981), pp. 241–274.
325. Tucker, William. "The Energy Crisis is Over." *Harper's Magazine,* Vol. 263 (November 1981), pp. 25–29, 32–36.
326. Tuma, Elias H. "Strategic Resources & Viable Interdependence: The Case of Middle Eastern Oil." *Middle East Journal,* Vol. 33, No. 3 (1979), pp. 269–287.
327. Turner, Arthur Campbell. "Iraq: Pragmatic Radicalism in the Fertile Crescent." *Current History,* Vol. 81, No. 471 (January 1982), pp. 14–17.
328. Turner, Louis, and James Bedore. "Saudi Arabia: The Power of the Purse-Strings." *International Affairs* (London), Vol. 54, No. 3 (July 1978), pp. 405–420.
329. U.S. Central Intelligence Agency. *The International Energy Situation: Outlook to 1985.* Washington: USGPO, April 1977.
330. ———. National Foreign Assessment Center. *The World Oil Market in the Years Ahead: A Research Paper.* Washington: USGPO, August 1979.
331. U.S. Congress. Congressional Budget Office. *The Marine Corps in the 1980s: Prestocking Proposals, The Rapid Deployment Force, and Other Issues.* Washington: USGPO, 1980.
332. ———. *The World Oil Market in the 1980s: Implications for the U.S.* Washington: USGPO, May 1980.
333. U.S. Congress. House of Representatives, Committee on Banking, Finance and Urban Affairs. *Iran: The Financial Aspects of the Hostage Settlement Agreement.* Washington: USGPO, 1981.
334. ———. Committee on the Budget. *Military Readiness and the Rapid Deployment Joint Task Force (RDJTF);* Hearings, 30 September and 1 October 1980. Washington: USGPO, 1980.
335. ———. Committee on Foreign Affairs. *The Iran Hostage Crisis: A Chronology of Daily Developments.* Report prepared by the Foreign Affairs and National Defense Division, Congressional Research Service, Library of Congress. Washington: USGPO, 1981.
336. ———. *The Iran Hostage Crisis: A Chronology of Daily Developments January 1–15, 1981.* Report prepared by Foreign Affairs and National Defense Division, Congressional Research Service, Library of Congress. Washington: USGPO, 1981.
337. ———. *Iran's Seizure of the United States Embassy;* Hearings. Washington: USGPO, 1981.
338. ———. *Regional Stability in Northern Africa;* Report of a Study Mission to Northern Africa, Spain, and Malta, 4–14 April 1980. Washington: USGPO, 1980.
339. ———. *U.S. Security Interests in the Persian Gulf;* Report of a Staff Study Mission to the Persian Gulf, Middle East, and Horn of Africa, October 21–November 13, 1980. Washington: USGPO, 1981.

340. ———. Subcommittee on Africa. *Libya-Sudan-Chad Triangle: Dilemma for United States Policy;* Hearings. Washington: USGPO, 1982.
341. ———. Subcommittee on Europe and the Middle East. *Activities of the United States Army Corps of Engineers in Saudi Arabia;* Hearing, 25 June 1979. Washington: USGPO, 1979. Appendix 2: "U.S. Army Corps of Engineer Work for Foreign Governments, Saudi Arabia."
342. ———. *Saudi Arabia and the United States: The New Context in an Evolving "Special Relationship";* Report prepared by the Foreign Affairs and National Defense Division, Congressional Research Service, Library of Congress. Washington: USGPO, 1981.
343. ———. *U.S. Interests in, and Policies Toward, the Persian Gulf, 1980;* Hearings. Washington: USGPO, 1980.
344. ———. Subcommittees on Europe and the Near East, and on International Economic Policy and Trade. *Export of Frigate Engines to Iraq;* Hearings, 14 May 1980. Washington: USGPO, 1981.
345. ———. Subcommittees on International Security and Scientific Affairs, and on Europe and the Middle East. *Israeli Attack on Iraqi Nuclear Facilities;* Hearings, 17 and 25 June 1981. Washington: USGPO, 1981.
346. ———. *Proposed U.S. Arms Sales to Saudi Arabia;* Hearing, 12 December 1979. Washington: USGPO, 1980.
347. ———. Committee on Government Operations. Environment, Energy and Natural Resources Subcommittee. *Alternatives to Dealing with OPEC;* Hearings. Washington: USGPO, 1979.
348. ———. *Effect of Iraqi–Iranian Conflict on U.S. Energy Policy;* Hearing, 30 September 1980. Washington: USGPO, 1981.
349. ———. Committee on International Relations. *Oil Fields as Military Objectives: A Feasibility Study;* Prepared by the Congressional Research Service, Library of Congress. Washington: USGPO, 1975.
350. ———. *United States Arms Policies in the Persian Gulf and Red Sea Areas: Past, Present, and Future;* Report of a Staff Survey Mission to Ethiopia, Iran and the Arabian Peninsula. Washington: USGPO, 1977.
351. ———. Permanent Select Committee on Intelligence. Subcommittee on Oversight. *Intelligence on the World Energy Outlook and Its Policy Implications;* Hearings, 17 and 18 October 1979. Washington: USGPO, 1980.
352. ———. Select Committee on Intelligence. Subcommittee on Collection, Production, and Quality. *U.S. Intelligence Analysis and the Oil Issue, 1973–74;* Staff Report. Washington: USGPO, 1977.
353. U.S. Congress. Joint Economic Committee. *Economic Consequences of the Revolution in Iran;* Compendium of Papers. Washington: USGPO, 1980. Contents include:
    A. Bernard Reich, "The United States and Iran: An Overview," pp. 5–21;
    B. Leonard Binder, "Iran's Unfinished Revolution: Possible Scenarios," pp. 22–46;
    C. Theodore H. Moran, "Budgetary Austerity in Iran: Revenue Needs, Expenditure Options, and Oil Export Policy for the Early 1980s," pp. 47–64;
    D. Rouhollah K. Ramazani, "Iran's Foreign Policy: Perspectives and Projections," pp. 65–97;
    E. W. H. Donnelly, C. A. Phillips, A. Reifman, and J. P. Riva, Jr., "Global Oil and Energy Implications of the Iranian Revolution," pp. 98–125;
    F. Edward A. Kolodziej, "The Iranian Revolution and Inter-allied Relationships: The United States, Western Europe and Japan," pp. 126–159;
    G. William H. Cooper, "The Impact of the Iranian Revolution on the Soviet Union," pp. 160–178;

H. Bernard Reich, David E. Pollock, and Sally Ann Baynard, "The Iranian Revolution and Its Effects on the Middle East," pp. 179–207;
I. Richard P. Cronin, "Impact of the Iranian Revolution on South Asia," pp. 208–222;
J. Clyde R. Mark, "The Revolution in Iran—Chronology of Events," pp. 223–227.

354. U.S. Congress. Joint Economic Committee. *The Political Economy of the Middle East: 1973–78; A Compendium of Papers.* Washington: USGPO, 1980. Contents include:
A. Fern Racine Gold and Charles K. Ebinger, "Economic Change in the Oil Exporting Countries of the Middle East," pp. 91–103;
B. David Long, "Saudi Arabia," pp. 180–190;
C. John Duke Anthony, "Gulf States and Iraq," pp. 210–231;
D. David Curry, "Petrodollars and the International Payments System," pp. 306–312;
E. Werner J. Feld, "Western European Foreign Policies Towards the Middle East: Responses to the Oil Crisis," pp. 400–422.

355. U.S. Congress. Senate. Committee on Armed Services. *Military and Technical Implications of the Proposed Sale of Air Defense Enhancements to Saudi Arabia;* Report of the Hearings, . . . together with Additional Views. Washington: USGPO, 1981.

356. ———. *Military and Technical Implications of the Proposed Sale to Saudi Arabia of Airborne Warning and Control System (AWACS) and F-15 Enhancements;* Hearings, 28 and 30 September 1981. Washington: USGPO, 1981.

357. U.S. Congress. Senate. Committee on Energy and Natural Resources. *The Geopolitics of Oil;* a Staff Report. Washington: USGPO, 1980.

358. ———. *World Petroleum Outlook—1981;* Hearing, 22 January, 1981. Washington: USGPO, 1981.

359. U.S. Congress. Senate. Committee on Foreign Relations. *Arms Sales Package to Saudi Arabia;* Hearings, 1, 5, 6, 14 and 15 October 1981. Parts I and II. Washington: USGPO, 1981.

360. ———. *The Iran Agreements;* Hearings. Washington: USGPO, 1981.

361. ———. *The Israeli Air Strike;* Hearings, 18, 19 and 25 June 1981. Washington: USGPO, 1981.

362. ———. *Persian Gulf Situation;* Hearing. Washington: USGPO, 1981.

363. ———. *The Proposed AWACS/F-15 Enhancement Sale to Saudi Arabia;* Staff Report, September 1981. Washington: USGPO, 1981.

364. ———. *The Situation in Iran*; Top Secret Hearing on 8 May 1980; sanitized and printed on 18 February 1981. Washington: USGPO, 1981.

365. ———. *United States Foreign Policy Objectives and Overseas Military Installations*; Prepared by Congressional Research Service, Library of Congress. Washington: USGPO 1979. Contents include:
A. "Southern Mediterranean and Middle East," pp. 66–83;
B. "The Indian Ocean, Arabian Peninsula, and Persian Gulf," pp. 84–121.

366. ———. Subcommittee on African Affairs and Subcommittee on Near Eastern and South Asian Affairs. *Libyan Activities*; Hearing, 8 July 1981. Washington: USGPO, 1981.

367. ———. Subcommittee on International Economic Policy. *The Future of Saudi Arabian Oil Production;* Staff Report. Washington: USGPO, 1979.

368. ———. Subcommittee on Near Eastern and South Asian Affairs. *U.S. Security Interests and Policies in Southwest Asia*; Hearings. Washington: USGPO, 1980.

369. U.S. Congress. Senate. Committee on the Judiciary. *The Undercover Investigation of Robert L. Vesco's alleged attempts to reverse a State Department ban preventing the export of planes to Libya*; Staff Report. Washington: USGPO, 1982.
370. ———. Subcommittee to Investigate the Activities of Individuals Representing the Interests of Foreign Governments. *Inquiry into the Matter of Billy Carter and Libya*; Hearings. Washington: USGPO, 1981.
371. U.S. General Accounting Office. *Critical Factors Affecting Saudi Arabia's Oil Decisions*; Report by the Comptroller General of the U.S. Washington: General Accounting Office, 12 May 1978.
372. Van Hollen, Christopher. "Don't Engulf the Gulf." *Foreign Affairs,* Vol. 59, No. 5 (1981), pp. 1064–1078.
373. Vernant, Jacques. "L'Occident et la sécurité du Golfe." *Défense Nationale,* Vol. 37 (May 1981), pp. 135–141.
374. Vernon, Raymond, ed. *The Oil Crisis.* New York: W. W. Norton, for the American Academy of Arts and Sciences, 1976. First published as special issue of *Daedalus,* Vol. 104, No. 4 (1975), "The Oil Crisis in Perspective." Contents include:

Part I: Introduction

A. Raymond Vernon, "An Interpretation," pp. 1–14;

Part II: The Crisis

B. Joel Darmstadter and Hans H. Landsberg, "The Economic Background," pp. 15–37;
C. Edith Penrose, "The Development of Crisis," pp. 39–57;

Part III: The States as Actors

D. George Lenczowski, "The Oil-Producing Countries," pp. 59–72;
E. James W. McKie, "The United States," pp. 73–90;
F. Romano Prodi and Alberto Clô, "Europe," pp. 91–112;
G. Yoshi Tsurumi, "Japan," pp. 113–127;
H. Marshall I. Goldman, "The Soviet Union," pp. 129–143;
I. Norman Girvan, "Economic Nationalism," pp. 145–158;

Part IV: The Multinational Oil Companies

J. Mira Wilkins, "The Oil Companies in Perspective," pp. 159–178;
K. Robert B. Stobaugh, "The Oil Companies in the Crisis," pp. 179–202;

Part V: The International Organizations

L. Zuhayr Mikdashi, "The OPEC Process," pp. 203–215;
M. Ulf Lantzke, "The OECD and Its International Energy Agency," pp. 217–227;

Part VI: Synthesis

N. Klaus Knorr, "The Limits of Economic and Military Power," pp. 229–243;
O. Raymond Vernon, "The Distribution of Power," pp. 245–257;
P. Ian Smart, "Uniqueness and Generality," pp. 259–281.

375. Von Schack, Alard. "Black Africa and Its Arab Neighbours to the North." *Aussenpolitik,* Vol. 28, No. 1 (1977), pp. 105–118.
376. Waltz, Kenneth N. "A Strategy for the Rapid Deployment Force." *International Security,* Vol. 5, No. 4 (1981), pp. 49–73.
377. Waterbury, John, and Ragaei El Mallakh. *The Middle East in the Coming Decade: From Wellhead to Well-Being?* New York: McGraw-Hill, 1978. 1980s Project/Council on Foreign Relations.
378. Weinbaum, M. G., and Gautam Sen. "Pakistan Enters the Middle East." *Orbis,* Vol. 22, No. 3 (1978), pp. 595–612.
379. Weiner, Myron. "International Migration and Development: Indians in the

Persian Gulf." *Population and Development Review,* Vol. 8, No. 1 (March 1982), pp. 1–36.

380. Whelan, John, ed. "Japan and the Middle East." *MEED Special Report* (December 1981), pp. 3–88.
381. ———, and Shakib Otaqui, eds. "Kuwait and the Middle East." *Middle East Economic Digest,* Special Report (May 1982), pp. 3–104.
382. Whetten, Laurence. "Security and South-West Asia: Security Implications of Recent Political Change." *Round Table,* No. 273 (January 1979), pp. 31–40.
383. Whittle, Richard. "U.S. Security Interests in the Persian Gulf." *Congressional Quarterly Weekly Report,* Vol. 39 (17 October 1981), pp. 2009–2016.
384. Wilson, Rodney J. S. *The Economics of the Middle East.* New York: Holmes and Meier, 1979.
385. Wohlstetter, Albert. "Half-Wars and Half-Policies in the Persian Gulf." In W. Scott Thompson, ed., *From Weakness to Strength* (San Francisco: Institute for Contemporary Studies, 1980).
386. ———. "Les États-Unis et la sécurité du Golfe." *Politique Étrangère,* Vol. 46, No. 1 (March 1981), pp. 75–88.
387. ———. "Meeting the Threat in the Persian Gulf." *Survey,* Vol. 25, No. 2 (1980), pp. 128–188.
388. Wolfe, Ronald G., ed. *The United States, Arabia, and the Gulf.* Washington: Georgetown University, Center for Contemporary Arab Studies, 1980. CCAS Studies in Arab–American Relations. Includes:

Part I: Social and Political Conditions

A. Ahmad Bahaeddine, "Political Dynamics in the Region," pp. 3–7;
B. Yousef Al-Shirawi, "The Impact of Development on Gulf Society and Politics," pp. 8–13;
C. Asad Abdul-Rahman, "The Palestinians in Kuwait: Their Political Significance," pp. 14–19;

Part II: Political Economy of Arabia and the Gulf

D. Ibrahim M. Oweiss, "Economic Integration in the Gulf," pp. 21–28;
E. Thomas Stauffer, "The Political Uses of Arab Oil," pp. 29–34;

Part III: International Political and Strategic Dimensions

F. Charles W. Yost, "Superpower Rivalry in the Gulf," pp. 37–43;
G. Hussein Sirriyeh, "Security and Stability in the Gulf: Background to United States Policy," pp. 44–48;
H. Marmaduke G. Bayne, "United States Strategic Planning for the Gulf," pp. 49–56;

Part IV: Current United States Policy: Conflicting Views

I. David D. Newsom, "United States Policy Towards the Gulf," pp. 59–64;
J. William J. Porter, "Tensions in the Gulf and the Inadequacy of United States Policy," pp. 65–68.

389. Wright, Claudia. "Implications of the Iraq–Iran War." *Foreign Affairs,* Vol. 59, No. 2 (1980–1981), pp. 275–303.
390. ———. "Iraq—New Power in the Middle East." *Foreign Affairs,* Vol. 58, No. 2 (1979–1980), pp. 257–277.
391. Yamani, Ahmad Z. "Energy Outlook: The Year 2000." *Journal of Energy and Development,* Vol. 5, No. 1 (1979), pp. 1–8.
392. ———. "OPEC's Role in the North–South Dialogue." *MEES Supplement,* Vol. 25, No. 7 (November 1981), pp. 1–4.
393. Yegnes, Tamar. "Saudi Arabia and the Peace Process." *Jerusalem Quarterly,* No. 18 (1981), pp. 101–120.
394. Yodfat, Aryeh Y. "The USSR and the Persian Gulf." *Australian Outlook,* Vol. 33, No. 1 (April 1979), pp. 60–71.

395. ———, and M. Abir. *In the Direction of the Persian Gulf: The Soviet Union and the Persian Gulf.* London: Frank Cass; Totowa, N.J.: Biblio Distribution Center, 1977.
396. Yorke, Valerie. "Bid for Gulf Unity." *The World Today,* Vol. 37, Nos. 7–8 (July–August 1981), pp. 246–249.
397. ———. *The Gulf in the 1980s.* London: The Royal Institute of International Affairs, 1980. Chatham House Papers, No. 6.
398. ———. "Oil, the Middle East and Japan's Search for Security." *International Affairs* (London), Vol. 57, No. 3 (1981), pp. 428–448.
399. ———. "Security in the Gulf: A Strategy of Pre-emption." *The World Today,* Vol. 36, No. 7 (July 1980), pp. 239–250.
400. Younger, Sam. "Ideology and Pragmatism in Algerian Foreign Policy." *The World Today,* Vol. 34, No. 3 (March 1978), pp. 107–114.
401. Zabih, Sepehr. *Iran's Revolutionary Upheaval: An Interpretative Essay.* San Francisco: Alchemy Books, 1979.
402. Al-Zamil, A. "The Petrochemical Industry in the Kingdom of Saudi Arabia." *Journal of Energy and Development,* Vol. 6, No. 2 (1981), pp. 205–212.
403. Zartman, I. William. "The Power of American Purposes." *Middle East Journal,* Vol. 35, No. 2 (1981), pp. 163–177.
404. Anthony, John Duke. "The Gulf Cooperation Council." *Journal of South Asian and Middle Eastern Studies,* Vol. 5, No. 4 (1982), pp. 3–18.
405. Askari, Hossein G. "OPEC and International Aid: An Appraisal." *SAIS Review,* No. 3 (1981–1982), pp. 133–148.
406. Christeller, Christian. *Une nouvelle stratégie arabo-iranienne de gestion pétrolière.* Paris: Economica, 1980.
407. Chubin, Shahram. *Security in the Persian Gulf, Vol. 4: The Role of Outside Powers.* London: International Institute for Strategic Studies; Montclair, New Jersey: Allanheld, Osmun, 1982.
408. "Inheritance: A Survey of Economic Development in the Arabian Peninsula." *The Economist,* 19 February 1983, survey pp. 1–34
409. Mortimer, Robert. "Algeria's New Sultan." *Current History,* Vol. 80, No. 470 (December 1981), pp. 418–421, 433–434.
410. Mossavar-Rahmani, Bijan. *Energy Policy in Iran: Domestic Choices and International Implications.* New York: Pergamon Press, 1981.
411. Owen, Roger. "Income Distribution and Social Welfare in the Arab World." *Journal of the Institute for Socioeconomic Studies,* Vol. 6, No. 4 (1981–82), pp. 46–60.
412. Plascov, Avi. *Security in the Persian Gulf, Vol. 3: Modernization, Political Development and Stability.* London: International Institute for Strategic Studies; Montclair, New Jersey: Allanheld, Osmun, 1982.
413. Singer, S. Fred. "Saudi Arabia's Oil Crisis." *Policy Review,* No. 21 (1982), pp. 87–100.
414. Sowell, Lewis C., Jr. *Base Development and the Rapid Deployment Force: A Window to the Future.* Washington: National Defense University Press, 1982. National Security Affairs Monograph Series, 82–5.
415. Tétreault, Mary Ann. *The Organization of Arab Petroleum Exporting Countries: History, Policies, and Prospects.* Westport Conn.: Greenwood Press, 1981.

USEFUL PERIODICALS

Adelphi Papers (London)
Africa (London)
African Affairs (London)
L'Afrique et l'Asie Modernes (Paris)
American–Arab Affairs (Washington, D.C.)
American Enterprise Institute, *Foreign Policy and Defense Review* (Washington, D.C.)
Annals of the American Academy of Political and Social Science (Philadelphia)
Arab Studies Quarterly (Detroit)
Aramco World Magazine (New York)
Armed Forces and Society (Chicago)
Asian Affairs (London)
Australian Outlook (Victoria)
Commentary (New York)
Comparative Studies in Society and History (New York)
Conflict: An International Journal for Conflict and Policy Studies (New York)
Current History (Philadelphia)
Défense Nationale (Paris)
Economic Development and Cultural Change (Chicago)
The Economist (London)
Economist Intelligence Unit, *Quarterly Economic Review* (London)
Etudes internationales (Quebec)
Foreign Affairs (New York)
Foreign Policy (Washington, D.C.)
Geographical Journal (London)
India Quarterly (New Delhi)
International Affairs (London)
International Journal (Toronto)
International Journal of Middle East Studies (Cambridge)
International Labour Review (Geneva)
International Security (Cambridge, Mass.)
International Studies (New Delhi)
Jerusalem Journal of International Relations (Jerusalem)
Jerusalem Quarterly (Jerusalem)
Jeune Afrique (Paris)
Journal of Arab Affairs (Fresno, Cal.)
Journal of Contemporary History (Beverly Hills, Cal.)
Journal of Developing Areas (Macomb, Ill.)
Journal of Energy and Development (Boulder, Colo.)
Journal of International Affairs (New York)
Journal of Modern African Studies (New York)
Journal of South Asian and Middle Eastern Studies (Villanova, Pa.)
Maghreb-Machrek (Paris)
MERIP Reports (Washington, D.C.)
The Middle East (London)
Middle East Economic Digest (London)
Middle East Economic Survey (Nicosia)
Middle East International (London)
Middle East Policy Survey (Washington, D.C.)
Middle East Review (New York)
Middle Eastern Studies (London)
Mideast Observer in Washington (Washington, D.C.)
Military Review (Fort Leavenworth, Kansas)
New Left Review (New York)
OAPEC News Bulletin (Kuwait)
OPEC Bulletin (Vienna)
OPEC Review (Vienna)
Oil and Arab Cooperation (Kuwait)
Oil and Gas Journal (Tulsa, Okla.)
Orbis (Philadelphia)
Orient (Opladen, Germany)
Parameters: Journal of the U.S. Army War College (Carlisle Barracks, Pa.)
Petroleum Economist (London)
Petroleum Intelligence Weekly (New York)
Platt's Oilgram News (New York)
Politique Étrangère (Paris)
Problems of Communism (Washington, D.C.)
Race and Class (London)
RIPEH/The Review of Iranian Political Economy and History (Washington, D.C.)
Revue française de science politique (Paris)
Revue juridique et politique

indépendance et coopération (Paris)
Round Table (London)
Studies in Comparative International Development (New Brunswick, N.J.)
Third World Quarterly (London)
U.S. Naval Institute Proceedings (Annapolis, Md.)
U.S. Naval War College Review (Newport, R.I.)
Washington Quarterly (Washington, D.C.)
Washington Report on Middle East Affairs (Washington, D.C.)
World Affairs (Washington, D.C.)
World Oil (Houston)
World Politics (Princeton, N.J.)
The World Today (London)

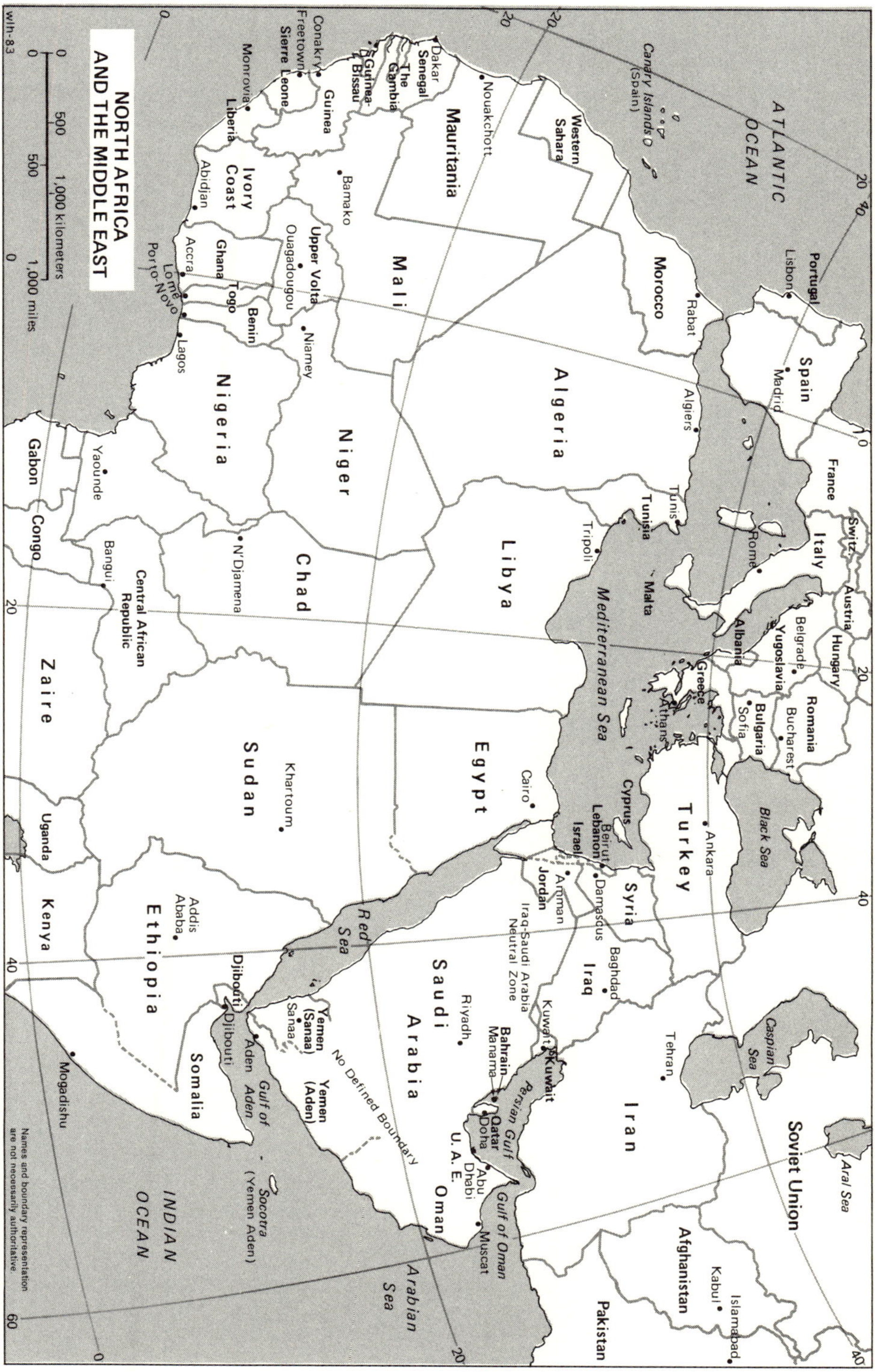
NORTH AFRICA
AND THE MIDDLE EAST
0 500 1,000 kilometers
0 500 1,000 miles
wlh-83
Names and boundary representation
are not necessarily authoritative.
ATLANTIC
OCEAN
Canary Islands
(Spain)
Western
Sahara
Mauritania
Nouakchott
Dakar
Senegal
The
Gambia
Guinea-
Bissau
Conakry
Freetown
Sierre Leone
Guinea
Monrovia
Liberia
Ivory
Coast
Abidjan
Bamako
Mali
Upper Volta
Ouagadougou
Ghana
Accra
Togo
Lomé
Porto-Novo
Benin
Niamey
Lagos
Nigeria
Niger
Morocco
Rabat
Portugal
Lisbon
Spain
Madrid
Algeria
Algiers
France
Tunis
Tunisia
Italy
Switz.
Rome
Tripoli
Libya
Malta
Austria
Hungary
Yugoslavia
Belgrade
Albania
Romania
Bucharest
Bulgaria
Sofia
Greece
Athens
Mediterranean Sea
Black Sea
Turkey
Ankara
Cyprus
Beirut
Lebanon
Israel
Syria
Damascus
Amman
Jordan
Egypt
Cairo
Chad
N'Djamena
Yaoundé
Gabon
Congo
Bangui
Central African
Republic
Zaire
Sudan
Khartoum
Uganda
Kenya
Addis
Ababa
Ethiopia
Red
Sea
Iraq-Saudi Arabia
Neutral Zone
Iraq
Baghdad
Kuwait
Kuwait
Saudi
Arabia
Riyadh
Bahrain
Manama
Qatar
Doha
Persian Gulf
U. A. E.
Abu
Dhabi
Oman
Muscat
Gulf of Oman
Yemen
(Sanaa)
Sanaa
Yemen
(Aden)
Aden
No Defined Boundary
Gulf of
Aden
Djibouti
Djibouti
Somalia
Mogadishu
Socotra
(Yemen Aden)
INDIAN
OCEAN
Arabian
Sea
Iran
Tehran
Caspian
Sea
Soviet Union
Aral Sea
Afghanistan
Kabul
Islamabad
Pakistan

The Middle East (Southwest Asia)

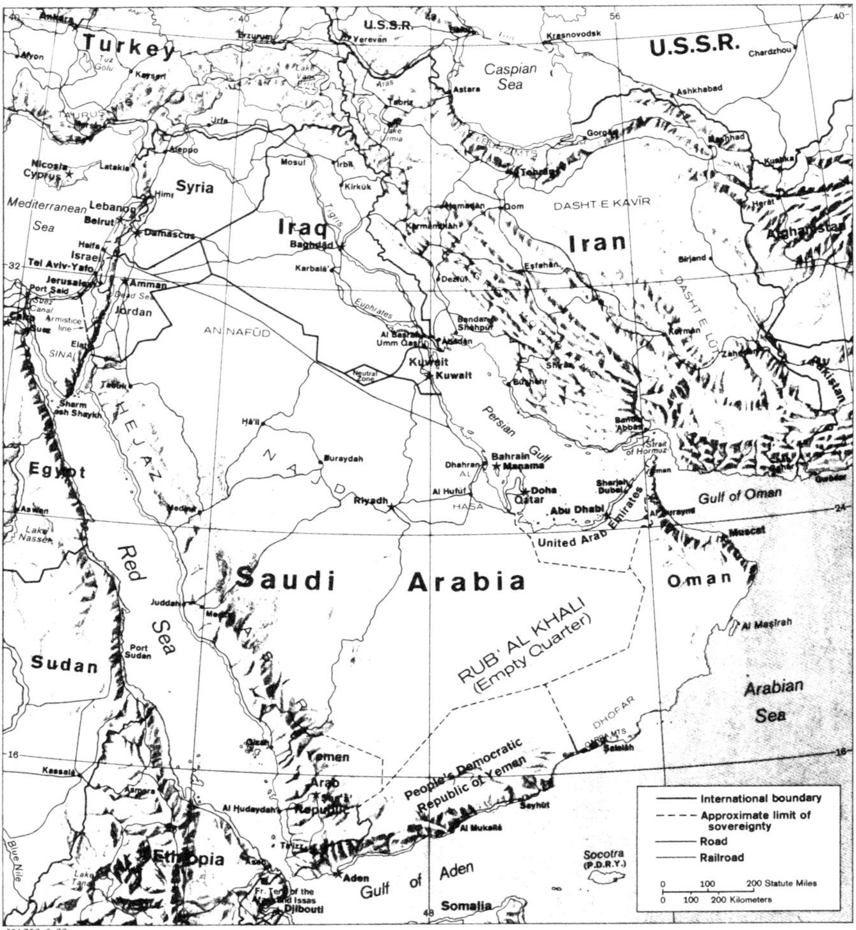

501763 6-73

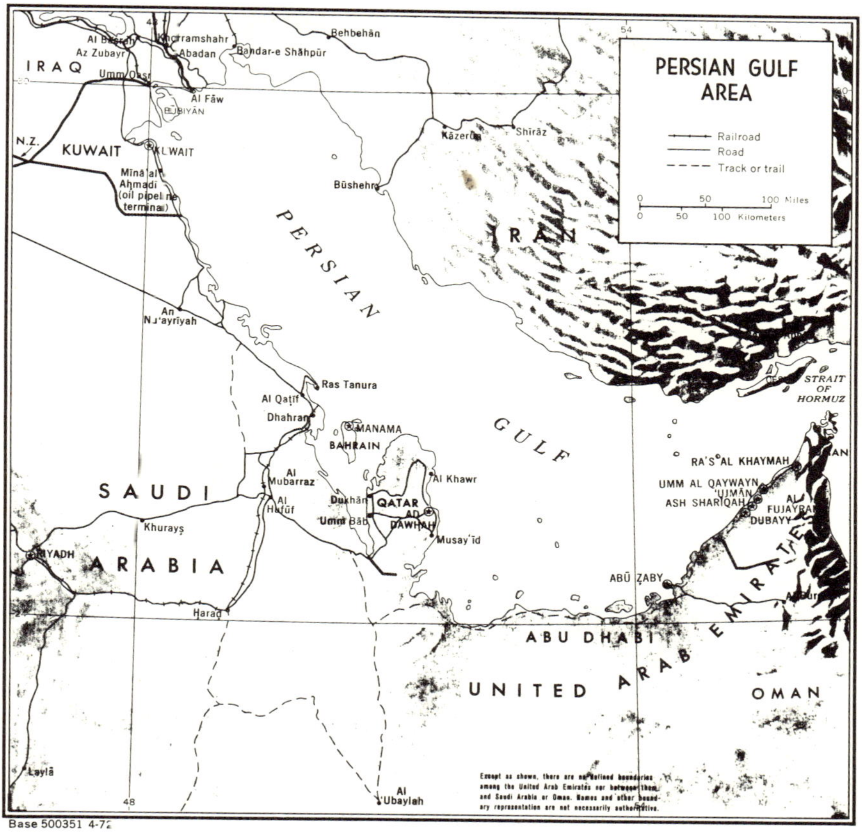

Base 500351 4-72

The Superpowers and the Gulf

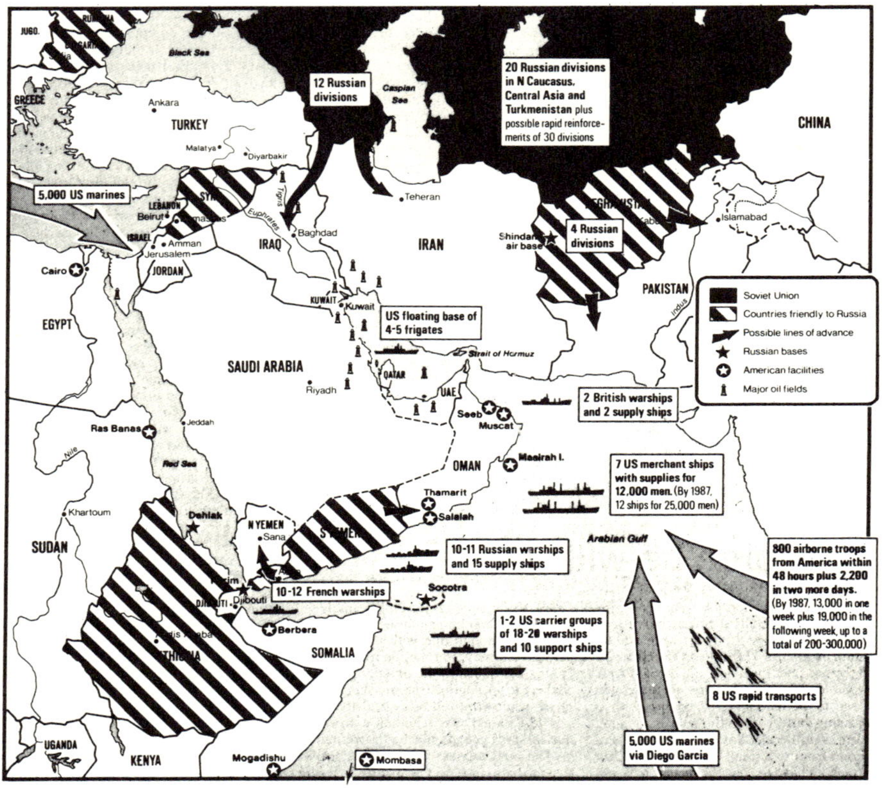

Reprinted with permission from THE ECONOMIST, June 6, 1981.

Principal Oil Movements By Sea

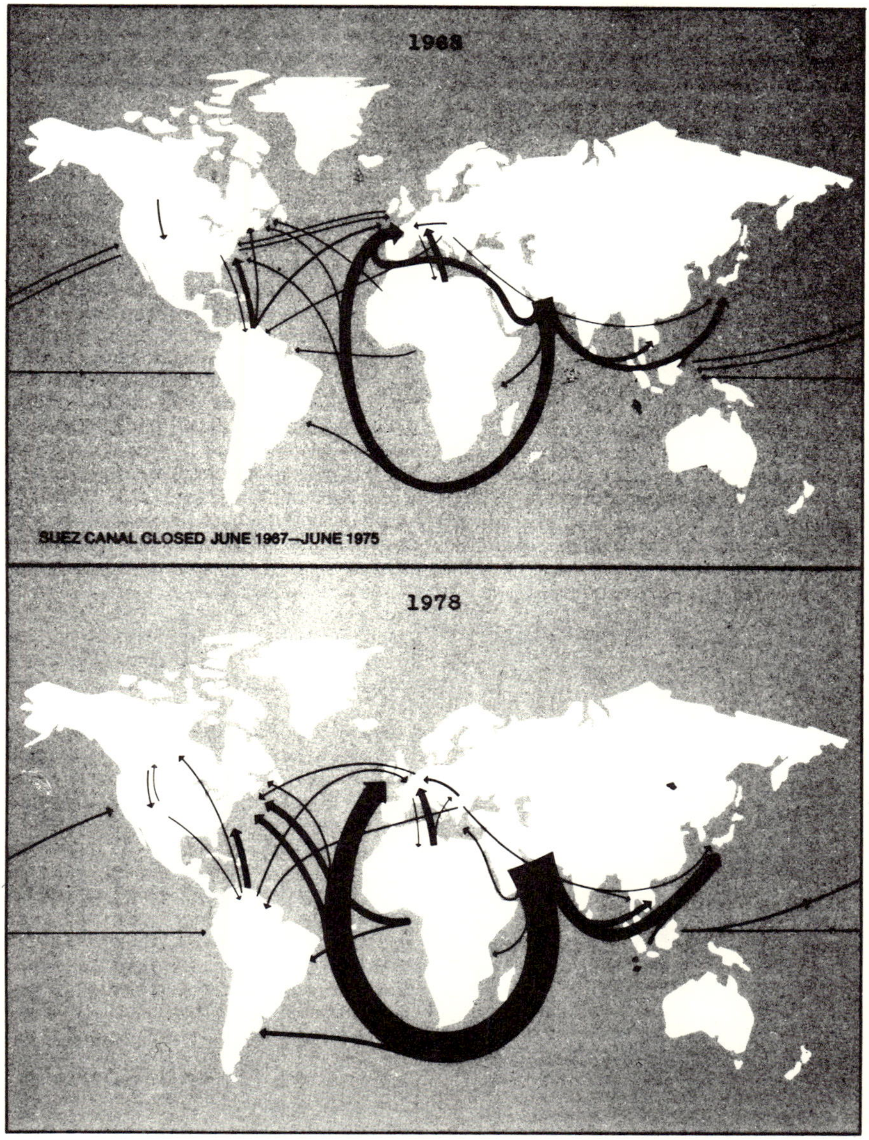

Exports from the Middle East region expanded rapidly as world oil consumption grew. The widths of the arrows in the maps above are proportional to the volumes of crude oil and petroleum products moved in international trade. The patterns changed significantly when the Suez Canal was closed to shipping in 1967.

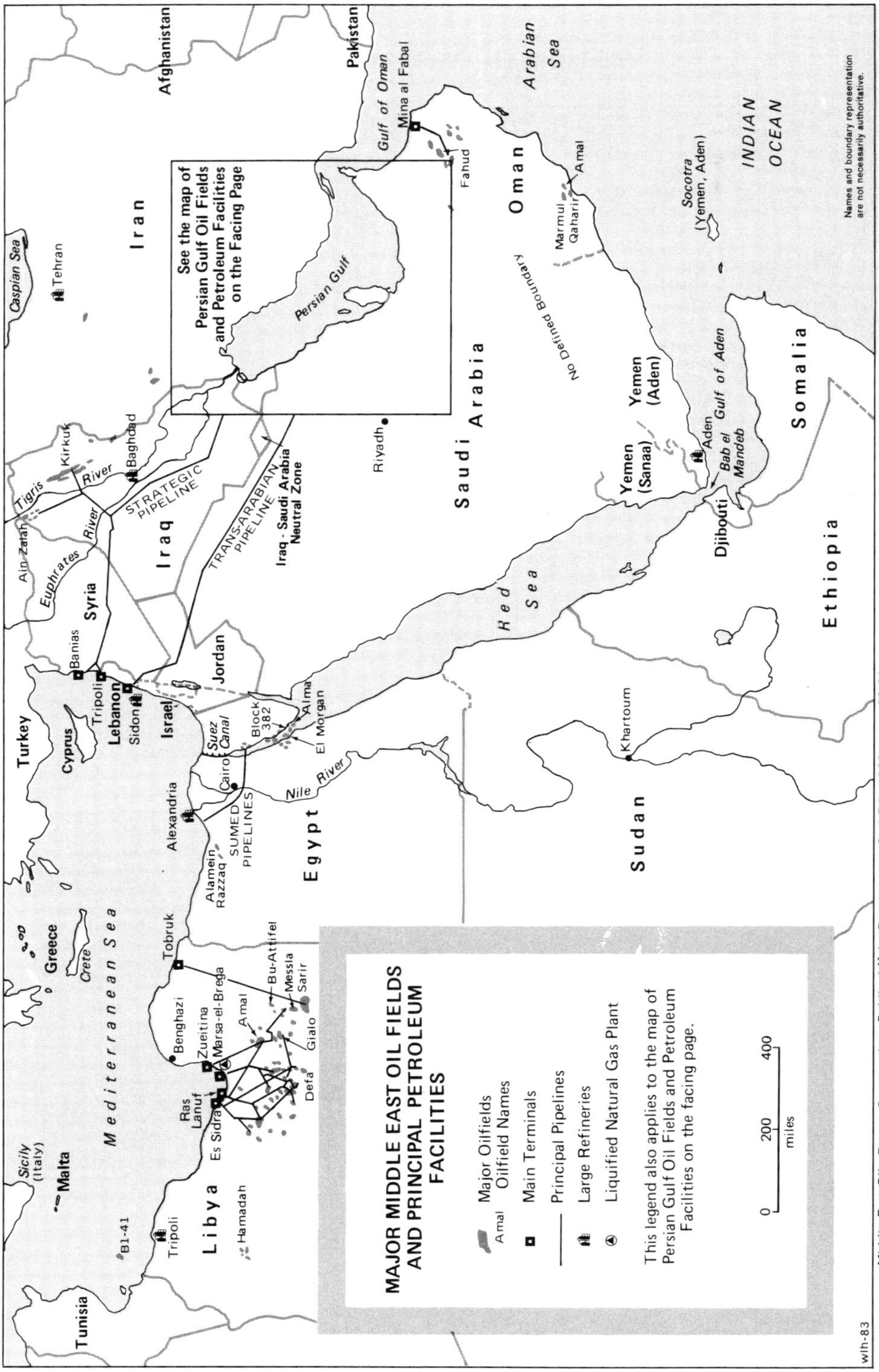

Corporation, Public Affairs Department, 2nd Ed. 1980, pages 20-21.

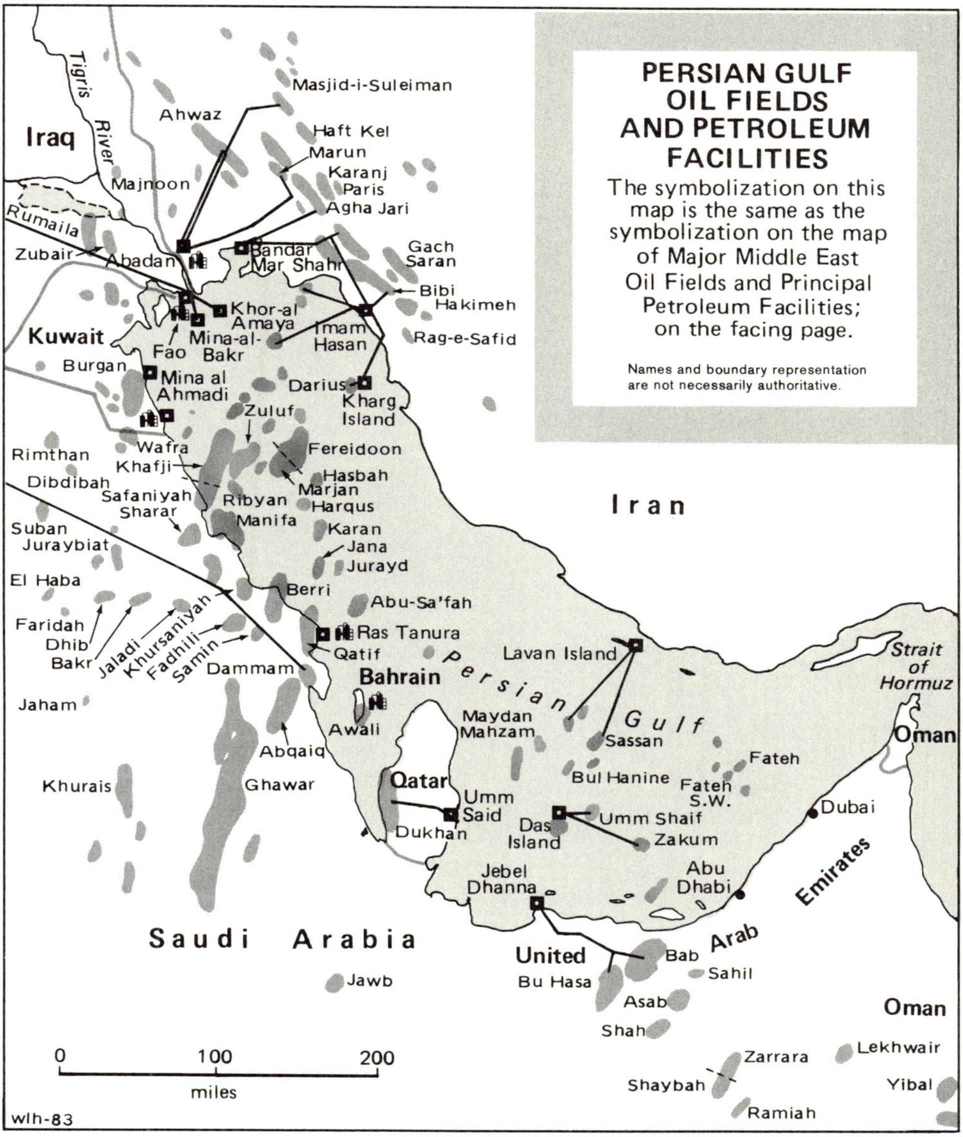

Adapted from: *Middle East Oil*, Exxon Corporation, Public Affairs Department, 2nd Ed., 1980, pages 20-21.